Epi-Informatics

Epi-Informatics

Discovery and Development of Small Molecule Epigenetic Drugs and Probes

Edited by

José L. Medina-Franco, PhD

Faculty of Chemistry, Department of Pharmacy,
National Autonomous University of Mexico, Mexico City, Mexico

ELSEVIER

AMSTERDAM • BOSTON • HEIDELBERG • LONDON
NEW YORK • OXFORD • PARIS • SAN DIEGO
SAN FRANCISCO • SINGAPORE • SYDNEY • TOKYO
Academic Press is an imprint of Elsevier

Academic Press is an imprint of Elsevier
125 London Wall, London EC2Y 5AS, UK
525 B Street, Suite 1800, San Diego, CA 92101-4495, USA
50 Hampshire Street, 5th Floor, Cambridge, MA 02139, USA
The Boulevard, Langford Lane, Kidlington, Oxford OX5 1GB, UK

British Library Cataloguing-in-Publication Data
A catalogue record for this book is available from the British Library

Library of Congress Cataloging-in-Publication Data
A catalog record for this book is available from the Library of Congress

ISBN: 978-0-12-802808-7

Publisher: Mica Haley
Acquisition Editor: Kristine Jones
Editorial Project Manager: Molly McLaughlin
Production Project Manager: Kirsty Halterman and Karen East
Designer: Maria Inés Cruz

Typeset by TNQ Books and Journals
www.tnq.co.in

Dedication

This book is dedicated to my loving wife, Karina Martínez Mayorga.

Contents

List of Contributors

Rodrigo Aguayo-Ortiz Departamento de Farmacia, Universidad Nacional Autónoma de México, México City, México

Paola B. Arimondo Unité de Service et de Recherche CNRS-Pierre Fabre, ETaC, CRDPF, Toulouse, France

J. Correa-Basurto Laboratorio de Modelado Molecular, Bioinformática y Diseño de Fármacos de la Escuela Superior de Medicina, Instituto Politécnico Nacional, Mexico City, Mexico

Alberto Del Rio Institute of Organic Synthesis and Photoreactivity (ISOF), National Research Council (CNR), Bologna, Italy

Alfonso Dueñas-González Unidad de Investigación Biomédica en Cáncer, Instituto de Investigaciones Biomédicas, Universidad Nacional Autónoma de México and Instituto Nacional de Cancerología, Mexico City, México

Alexandre Erdmann Unité de Service et de Recherche CNRS-Pierre Fabre, ETaC, CRDPF, Toulouse, France

Eli Fernández-de Gortari Departamento de Farmacia, Universidad Nacional Autónoma de México, México City, México

Y. George Zheng Department of Pharmaceutical and Biochemical Sciences, College of Pharmacy, University of Georgia, Athens, GA, USA

D.E. González-Juárez Laboratorio de Modelado Molecular, Bioinformática y Diseño de Fármacos de la Escuela Superior de Medicina, Instituto Politécnico Nacional, Mexico City, Mexico

Dominique Guianvarc'h Sorbonne Universités—UPMC Univ Paris, École Normale Supérieure—PSL Research University, Département de Chimie, CNRS UMR 7203 LBM, Paris, France

Manfred Jung Institute of Pharmaceutical Sciences, Albert-Ludwigs-University of Freiburg, Freiburg, Germany

Berin Karaman Institute of Pharmacy, Martin-Luther University of Halle-Wittenberg, Halle/Saale, Germany

Dmitri Kireev Center for Integrative Chemical Biology and Drug Discovery, UNC Eshelman School of Pharmacy, University of North Carolina at Chapel Hill, Chapel Hill, NC, USA

Cheng Luo Drug Discovery and Design Center, State Key Laboratory of Drug Research, Shanghai Institute of Materia Medica, Chinese Academy of Sciences, Shanghai, China

Wenchao Lu Drug Discovery and Design Center, State Key Laboratory of Drug Research, Shanghai Institute of Materia Medica, Chinese Academy of Sciences, Shanghai, China

Wencong Lu Department of Chemistry, College of Sciences, Shanghai University, Shanghai, China

Karina Martinez-Mayorga Instituto de Química, Universidad Nacional Autónoma de México, Mexico City, México

José L. Medina-Franco Facultad de Química, Departamento de Farmacia, Universidad Nacional Autónoma de México, Mexico City, México

Oscar Méndez-Lucio Centre for Molecular Informatics, Department of Chemistry, University of Cambridge, Cambridge, UK

Fanwang Meng Drug Discovery and Design Center, State Key Laboratory of Drug Research, Shanghai Institute of Materia Medica, Chinese Academy of Sciences, Shanghai, China; Department of Chemistry, College of Sciences, Shanghai University, Shanghai, China

Vedran Miletić Department of Informatics and Faculty of Engineering, University of Rijeka, Rijeka, Croatia

Carolina Peña Montes Facultad de Química, Universidad Nacional Autónoma de México, Mexico City, Mexico

J. Jesús Naveja Facultad de Medicina, PECEM, Universidad Nacional Autónoma de México, Mexico City, México; Facultad de Química, Departamento de Farmacia, Universidad Nacional Autónoma de México, Mexico City, México

Patrik Nikolić Department of Biotechnology, University of Rijeka, Rijeka, Croatia

Ivica Odorcić Department of Biotechnology, University of Rijeka, Rijeka, Croatia

Kun Qian Department of Pharmaceutical and Biochemical Sciences, College of Pharmacy, University of Georgia, Athens, GA, USA

Rino Ragno Rome Center for Molecular Design, Sapienza University of Rome, Rome, Italy

Wolfgang Sippl Institute of Pharmacy, Martin-Luther University of Halle-Wittenberg, Halle/Saale, Germany

Željko M. Svedružić Department of Biotechnology and Faculty of Medicine, University of Rijeka, Rijeka, Croatia

Greta Varchi Institute of Organic Synthesis and Photoreactivity (ISOF), National Research Council (CNR), Bologna, Italy

Chen Wang Drug Discovery and Design Center, State Key Laboratory of Drug Research, Shanghai Institute of Materia Medica, Chinese Academy of Sciences, Shanghai, China

Wei Wan Drug Discovery and Design Center, State Key Laboratory of Drug Research, Shanghai Institute of Materia Medica, Chinese Academy of Sciences, Shanghai, China

Jakyung Yoo Life Science Research Institute, Daewoong Pharmaceutical Co., Ltd, Gyeonggi-do, Republic of Korea

O.J. Zacarías-Lara Laboratorio de Modelado Molecular, Bioinformática y Diseño de Fármacos de la Escuela Superior de Medicina, Instituto Politécnico Nacional, Mexico City, Mexico

Preface

Epigenetic drug discovery has emerged as a promising strategy for the treatment of cancer and several other diseases. Epigenetics, however, is enormously complex and requires a multidisciplinary effort to accelerate the identification and development of molecular probes and drug candidates that can enter into the clinic. As part of the multidisciplinary efforts, a broad range of computational methods are being applied in synergy with experimental approaches to further advance the discovery of compounds that modulate epigenetic targets.

The purpose of this book is to put together successful strategies and examples of chemical information and computational approaches that experts in the field use to address the increasing needs of epigenetic drug and probe discovery. The first chapter presents an introduction to epigenetic targets and the current status of epi-drugs and epi-probes. The next two chapters cover an introduction to the major computational approaches used so far in epigenetic research and examples of the major contributions that molecular modeling and chemoinformatics have made to this field. The following seven chapters discuss detailed applications of computational methods to specific epigenetic targets. The next chapter addresses drug repurposing in epigenetic research. This chapter is followed by an in-depth discussion of computational analysis of structure–activity relationships of small molecules directed to epigenetic molecular targets. The next chapter covers an innovative contribution to the emerging foodinformatics field to uncover small molecules from food chemicals with epigenetic activity.

I sincerely thank all contributing authors of this book for their excellent work and commitment to developing the chapters. I also want to thank Catherine Van Der Laan, Kristine Jones, Karen East, Kirsty Halterman, and each person from Elsevier who made this book possible. In particular many thanks to Molly M. McLaughlin for the continued support and dedication to this book. My hope is that the joint effort from all participants can be translated into a source of information for those working on epigenetics, a source of inspiration to scientists, and motivation to students to further advance science and improve human health.

José L. Medina-Franco
Universidad Nacional Autónoma de México
UNAM
Mexico City, Mexico
August, 2015

CHAPTER 1

Introduction of Epigenetic Targets in Drug Discovery and Current Status of Epi-Drugs and Epi-Probes

Alfonso Dueñas-González[1], J. Jesús Naveja[2,3], José L. Medina-Franco[3]

[1]Unidad de Investigación Biomédica en Cáncer, Instituto de Investigaciones Biomédicas, Universidad Nacional Autónoma de México and Instituto Nacional de Cancerología, Mexico City, México; [2]Facultad de Medicina, PECEM, Universidad Nacional Autónoma de México, Mexico City, México; [3]Facultad de Química, Departamento de Farmacia, Universidad Nacional Autónoma de México, Mexico City, México

CONTENTS

1. INTRODUCTION

1.1 Current Status of Epigenetic Drugs

It took 34 years from the initial observation in 1980 that the nucleoside analog 5-azacytidine possessed DNA demethylating activity (Jones and Taylor, 1980) to its approval in 2004 by the US Food and Drug Administration (FDA) against myelodysplastic syndrome (MDS), a clinical entity where abnormalities in DNA methylation seem to play an important role in its molecular pathogenesis (Derissen et al., 2013). At the time of this writing, a total of six epigenetic drugs (two DNA demethylating and four histone deacetylase inhibitors), azacitidine included, have received FDA approval for cancer indications. The other DNA demethylating nucleoside analog, decitabine (Derissen et al., 2013), is also indicated for MDS treatment; the histone deacetylase inhibitors include vorinostat (Mann et al., 2007) and romidepsin (Bertino and Otterson, 2011), approved for the treatment of cutaneous T-cell lymphoma (CTCL), and the newer belinostat, for peripheral T-cell lymphoma (Lee et al., 2015), and panobinostat for multiple myeloma (Anon, 2015).

Table 1 summarizes the agents, dates, and indications of these epigenetic drugs. It is important to remark that the first four of these drugs were discovered based on observation of cell phenotypes and was not until several years later that their biochemical targets were discovered. These discoveries and their subsequent development into therapeutic agents indicated the potential of epigenetic therapies. As a result, epigenetic drug discovery has rapidly intensified across both industry and academia. In fact, belinostat and panobinostat were the first agents developed after their biochemical target was identified. Currently, epigenetic targets are now abundant in drug discovery pipelines of

Epi-Informatics. http://dx.doi.org/10.1016/B978-0-12-802808-7.00001-0

Table 1 FDA-Approved Epigenetic Agents

Agent	Class	Indication	Date
Azacitidine	DNMT inhibitor	First-line for all subtypes of MDS	2004
Decitabine	DNMT inhibitor	First-line for all subtypes of MDS	2006
Vorinostat	HDAC inhibitor	Pretreated cutaneous T-cell lymphoma	2006
Romidepsin	HDAC inhibitor	Pretreated cutaneous T-cell lymphoma	2009
Belinostat	HDAC inhibitor	Relapsed or refractory peripheral T-cell lymphoma	2014
Panobinostat	HDAC inhibitor	Pretreated multiple myeloma in combination with bortezomib and dexamethasone	2015

FDA, Food and Drug Administration; DNMT, DNA methyltransferases; HDAC, Histone Deacetylases.

multiple biotechnology companies. The fact that the six approved drugs are approved against cancer by no means implies that epigenetic alterations are limited to malignant conditions. It merely reflects that malignant diseases were the first to be studied in the epigenetic field. Moreover, the numerous studies of epigenetic traits in cancer lead to the recent consideration of epigenetic alterations as another cancer hallmark. Indeed, the study of epigenetics in many chronic diseases has rapidly evolved and now a huge number of studies have been published on the epigenetic pathogenesis of these diseases, including but not limited to neurodegenerative diseases (Alzheimer's disease, Parkinson's disease, Huntington's disease, multiple sclerosis, epilepsy, schizophrenia, bipolar disorders, and other psychiatric conditions), respiratory conditions (asthma, chronic obstructive pulmonary disease, pulmonary hypertension, and lung fibrosis), cardiovascular conditions (atherosclerosis, coronary artery disease, and heart failure), metabolic diseases (diabetes mellitus type 1 and 2), and autoimmune disorders (systemic lupus erythematosus, rheumatoid arthritis, systemic sclerosis and Sjögren's syndrome) (Best and Carey, 2010; Stein, 2014; Mau and Yung, 2014; Lu, 2013; Doehringer et al., 2011; Gay and Wilson, 2014; Slomko et al., 2012; Lovinsky-Desir and Miller, 2012; Mehler, 2008). Therefore, it may be just a matter of time for the approval of drugs targeting epigenetics against nonmalignant conditions.

1.2 Epigenetics

The term epigenetics was coined in the 1940s by Conrad Waddington, a British embryologist and geneticist (Waddington, 1942) who was attempting to describe "the interactions of genes with their environment, which brings the phenotype into being." Since then, this definition has changed to encompass the study of heritable phenotypic traits that result from modifications to a chromosome that do not alter the underlying genetic code (Berguer et al., 2009). An increasing awareness of the importance of epigenetics can easily be appreciated by just searching the PubMed database for the word "epigenetics" and looking at the number of publications that result. In 1994 there were six publications and in 2013—20 years later—3191 works were published.

The human genome contains approximately three billion base pairs (bps) of DNA, which are organized in 23 chromosomes, hence each diploid cell with 46 chromosomes contains six billion bps of DNA. This amount of DNA, if fully linearized, would mean that each cell would have at least 2 m long of DNA. Thus, a primary function of epigenetics is to package DNA into cells. In this regard, histone proteins are key players responsible for organizing the long fibers of DNA within the nucleus. The complex of DNA with histones is referred to as chromatin and the unit of chromatin is a nucleosome, which is formed by an octamer of histone (histone proteins H1, H2A, H2B, H3, and H4). Approximately 147 bps of superhelical DNA is wrapped around dimers of histones H2A, H2B, H3, and H4 comprising the nucleosome core particle. The nucleosome can be considered the functional unit of the DNA since its status standing alone and within the nuclear context determines whether or not a gene is expressed (You and Jones, 2012). Epigenetics, therefore, can be referred to as the study of all elements that participate in nucleosome–chromatin regulation as determinants of gene expression.

In the 1970s, the concept of DNA methylation emerged. It was demonstrated that the addition of a methyl group at the fifth-position of the cytosine in a CpGdinucleotide could inactivate the expression of genes (Jurkowska et al., 2011). Thus DNA methylation driven by DNA methyltransferase enzymes gained momentum as the most important epigenetic factor (Davie and Hendzel, 1994). In the 1990s, the transcriptional regulatory effect of histone acetylases and deacetylases added another level of complexity by regulating the acetylation status of the lysine tails at the histone core of the nucleosomes (Heyn and Esteller, 2012). In addition, it was uncovered that these enzymes were also able to bind and regulate DNA methyltransferases and other proteins having a DNA methylation binding domain (Esteller, 2007; Villar-Garea and Imhof, 2006). The recognition that histone proteins can undergo modifications such as acetylation/deacetylation, methylation/demethylation, phosphorylation, and other modifications led to the discovery of the enzymes responsible for these modifications, as well as numerous interactions among histones and DNA itself. Protein complexes with stand-alone nucleosome-remodeling activities (Cosgrove, 2007; Iorio et al., 2010) are recognized as equally important epigenetic players. Noncoding RNAs are another set of epigenetic players that are themselves regulated by epigenetic machinery elements. Lately, the discovery of DNA demethylases (Ito et al., 2011)—the ten–eleven translocation (tet1,tet2, and tet3) enzymes—which convert 5-methylcytosine to 5-hydroxymethylcytosine, have resulted in an unanticipated higher complexity of how epigenetics drives the fine-tuned regulation of gene expression (Sesenberger et al., 2013).

1.2.1 DNA Methylation

In terms of simplicity there are three groups of epigenetic modifiers from the therapy standpoint as most of them are targetable: those that methylate and demethylate DNA, a huge number of histone modification enzymes, and noncoding RNAs (ncRNAs). DNA methylation directly affects the genomic DNA. It involves the covalent addition of a methyl group to the carbon atom 5 of the cytosine–pyrimidine ring in a CpG (cytosine–guanine) dinucleotide. Methyl groups in the dinucleotide CpGs of gene promoters regulate interactions between DNA and the transcription machinery of the cell and, in general, DNA methylation correlates with transcriptional silencing and typically occurs in DNA sequences that contain cytosines adjacent to a guanine nucleotide (known as a CpG site). Approximately 30 million CpG sites exist in the human genome (Cocozza et al., 2011). However, certain DNA regions are rich in CpG sites, known as CpG islands. The methylation of CpG islands is associated with the silencing of the genes that are in proximity to the CpG islands. Normal differentiation requires proper DNA methylation/demethylation patterns and abnormal DNA methylation limits the capacity of a cell to differentiate into tissue-/cell-specific lineages and may ultimately induce a state of disease (Fisher and Fisher, 2011).

DNA methylation plays a role in the silencing of genes that are necessary for early stages of development, but no longer needed for subsequent stages of development. Failure to properly methylate a specific region of DNA may lead to abnormal gene expression. Alternatively, aberrant DNA methylation may lead to inappropriate gene silencing, which may promote disease. Interestingly, DNA methylation also plays an opposite role in the activation of specific genes (Rishi et al., 2010). Mammalian cells possess multiple pathways to establish, maintain, and modify DNA methylation throughout the genome. Enzymes in mammalian cells, known as DNA methyltransferases (DNMT1, DNMT3A, DNMT3B), are responsible for establishing and maintaining DNA methylation while DNMT2 and DNMT3B may participate in DNA methylation in a catalytic-independent manner (Schaefer and Lyko, 2010). Since DNA methylation is dynamic, mammalian cells also possess the ability to remove DNA methylation.

Passive DNA demethylation was the first to be described. Because it is passive, it depends on DNA replication and cell division plus the subsequent lack of action of DNA methylation maintenance pathways (Piccolo and Fisher, 2014). In contrast, active DNA demethylation is replication-independent and occurs through the active enzymatic removal of the methylcytosine (Piccolo and Fisher, 2014). Among DNA demethylases stand the enzyme activation-induced cytidine deaminase (AID) via deamination of 5-mC, which yields a thymidine residue that is replaced with an unmethylated cytosine by the base-excision repair (BER) pathway that replaces the thymidine residue with an unmethylated cytosine. Thus, AID may promote aberrant gene expression by decreasing the promoter DNA methylation of specific genes (Isobe et al., 2013; Munoz et al., 2013). The family of tet1, tet2, and tet3 (ten–eleven translocation) proteins are also considered as active DNA demethylases. These enzymes carry out the hydroxylation of 5-mC to 5-hmC (Ito et al., 2010); 5-hmC in turn is replaced with an unmethylated cytosine by the BER pathway (Hackett et al., 2013). Recent data demonstrate that several proteins bind to 5-hmC, revealing the possibility that specific proteins may be able to interpret the 5-hmC epigenetic mark and subsequently influence chromatin structure and gene expression (Mellen et al., 2012; Iurlaro et al., 2013).

Taken together, the DNA methylation/demethylation machinery allows mammalian cells to regulate gene expression throughout development. Although abnormal DNA methylation signals may contribute to disease, the reversible nature of epigenetic alterations makes the DNA methylation machinery an exciting therapeutic target (Wright, 2013), hence, to the former targets, DNMT1, DNMT3A, and DNMT3B proteins, now tet1, tet2, and tet3 proteins as well as activation-induced cytidine deaminase (AID) protein, add to the armamentarium of potential targets for the DNA methylation/demethylation process. In addition, methyl-CpG-binding domain (MBD) proteins, which read and interpret the methylation moieties on DNA and are critical mediators of many epigenetic processes, may also be targetable. These include at least MBD1, MBD2, MBD3, MBD4, and MeCP2 (Sansom et al., 2007).

1.2.2 Histone Modification

Histone proteins are key players responsible for organizing the long fibers of DNA within the nucleus, and the complex of DNA with histones is referred as chromatin. The unit of chromatin is a nucleosome. The level of chromatin compaction defines euchromatin and heterochromatin; the former has relatively loose compaction and is typically transcriptionally permissive, whereas heterochromatin (facultative or constitutive) is more condensed and typically transcriptionally repressive (Li and Reionbergh, 2011; Li et al., 2005). Though several mechanisms are being described to influence the level of chromatin compaction, the chemical modifications of histones are of major relevance; therefore, the targeting of these enzymes is a promising research avenue in epigenetic therapy. As stated earlier, there are now four HDAC inhibitors approved for cancer treatment. A huge number of histone modifications are being described but acetylation/deacetylation and methylation/demethylation in lysine tails are the most studied. The first demonstrations of the effect of histone modification upon gene transcription occurred in 1996 with the discovery of histone acetylase and histone deacetylase, p55/Gcn5 and HDAC1/Rpd3, respectively. A large body of data now supports the notion that histone modifications mainly work through the recruitment of effector proteins harboring specialized reader domains (Rundlett et al., 1996; Ogryzko et al., 1996). While the histone code or pattern of histone

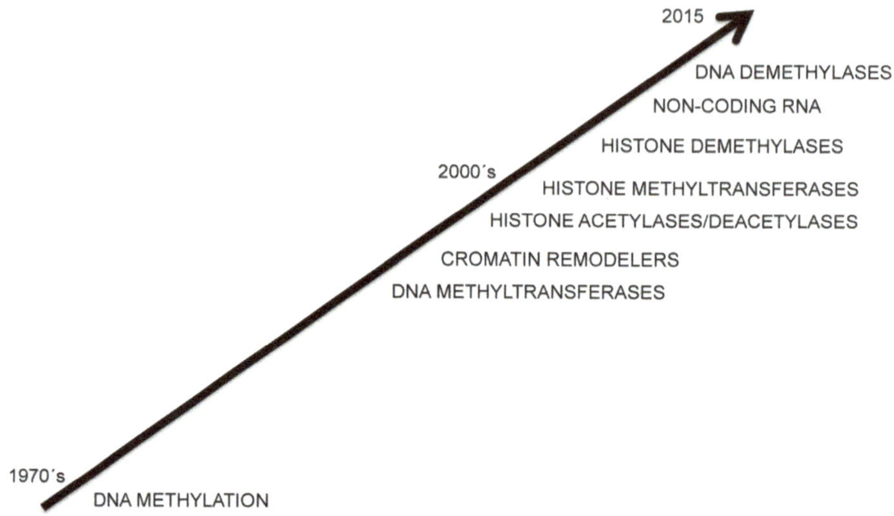

FIGURE 1 Increasing complexity of epigenetics over time.
In the 1970s, DNA methylation was the deemed the most important mechanism regulating gene transcription for several years. Around 20 years later, human DNA methyltransferases were cloned and since then, the pace of discovery of new epigenetic players has dramatically increased. Because of their multiple interactions, the complexity of the regulation of gene transcription has raised almost exponentially.

modifications in health and disease provided a simplistic explanation of how modified histones regulate the transcription, deepest knowledge reveals that the context where histones exert their transcriptional regulation is much more complex than originally anticipated. Not only do histone modifications per se have a regulatory role, but also noncanonical histones and nuclear compartmentalizations also are great contributors of transcriptional regulation (Talbert and Henikoff, 2010; Wu and Yao, 2013). Figure 1 depicts in a schematic manner the increasing complexity of epigenetics over time.

Soon after histone acetylases/deacetylases were shown to regulate transcription by modifying histones, in the year 2000, the first histone methyltransferase (KMT) was discovered and subsequently, a number of these were identified through homology searching of the enzymatic SET (Su(var)3–9, Enhancer of zeste, and Trithorax) domain of this enzyme family (Dillon et al., 2005). The SET domain is the catalytic domain, having 130 amino acids shared by the two families, Su(var)3–9, E(z) (enhancer of zeste), and trithorax (Jenuwein, 2006). Others lack the SET domain, however, like KMT4 (Okada et al., 2005), but both classes use S-adenosyl-L-methionine (SAM) as the methyl donor group (Nguyen and Zhang, 2011). In general KMTs appear to be highly specific for methylating some lysines though, this specificity can be modified according to their interacting partners or degree of activity. To mention some examples, KMT1A/B trimethylates histone 3 lysine 9 (H3K9me3) from monomethylated H3K9me1 (Peters et al., 2001, Ng et al., 2009), while the H3K9 methyltransferase KMT1C (also known as G9a) methylates H3K9me1 to H3K9me2 (Tachibana et al., 2002). In addition, KMT2A (also known as MLL1) catalyzes the methylation of H3K4 to H3K4me2 (Milne et al., 2002; Nakamura et al., 2002).

As for DNA methylation regarding the absence of active demethylation for many years, Shi and Tsukada (2013). first uncovered the existence of histone demethylases, being the first KDM, LSD1/KDM1A (Shi et al., 2004). KDM1A demethylase work through a flavin adenine dinucleotide (FAD)-dependent amine oxidase domain that demethylates H3K4me2 and H3K4me1 and modulates gene expression (Shi et al., 2004). Afterward, a second class of histone demethylases was discovered (KDM2A/B proteins, JHDM1A and JHDM1B histone demethylases) that utilize the JmjC domain, use as cofactors salfa-ketoglutarate, molecular oxygen, and Fe(II) for demethylation through the oxidation of methyl groups (Shi and Whetstine, 2007). The action of these demethylases releases formaldehyde from methylated histones (Tsukada et al., 2006; Yamane et al., 2006). Though the KDM2 and KDM3 families were unable to demethylate trimethylated lysines, discovery of the first tridemethylase family, KDM4AKDM4D, showed that indeed they perform these reactions but do not remove H3K9me1 or H3K36me1 (Trojer et al., 2009; Whetstine et al., 2006).

Beyond histone modification histone methylases and demethylases directly interact with a number of factors including RNA polymerase and RNA polymerase cofactors, other histone methyltransferases and demethylases, as well as CG-rich DNA and CpG islands (Eissenberg and Shilatifard, 2010; Müller and Kassis, 2006; Ley et al., 2010). For example, the zinc finger CxxC domain of KDM2A has increased affinity for nonmethylated CpG islands (Blackledge et al., 2010). Demethylases and methyltransferases can also be recruited to target regions through interactions with noncoding RNAs such as the archetypal long noncoding (lncRNA) Xist (Payer and Lee, 2008). Xist derives at least part of its function from the ability to recruit PRC2 through interactions with KMT6 (EZH2) and Suz12, which results in H3K27me3 and transcriptional silencing (Zhao et al., 2008). Noncoding RNA can also be involved in transcriptional activation; the HOTTIP lncRNA is expressed from the HoxA locus and recruits MLL via interaction with WDR5, which results in the transcriptional active mark H3K4me3 at the promoter and activation of the HoxA cluster (Wang et al., 2011).

In summary, the complex distribution and dynamics of histone methylation marks in order to regulate transcription require that both methylases and demethylases be at certain times and space (Smith and Shilatifard, 2010). Despite this complexity, some patterns of modification are relatively known. While enrichment in acetylation of histone tails is typically associated with transcriptional activation of genes, the functional consequences of methylation depend on the number of methyl groups added and the residue location within the histone tail. For example, histone marks associated with open chromatin and active gene expression are histone 3 lysine 4 di- and trimethylation (H3K4me2 and H3K4me3) and histone 3 lysine 9 monomethylation (H3K9me1); on the contrary, histone 3 lysine 27 di- and trimethylation (H3K27me2 and H3K27me3) and histone 3 lysine 9 di- and trimethylation (H3K9me2 and H3K9me3) are associated with inactive chromatin and repression of gene expression. In addition, some marks, such as histone 3 lysine 4 monomethylation (H3K4me1) and histone 3 lysine 27 acetylation (H3K27ac) are found in the enhancer elements of genes and can influence gene expression even at large distances from the gene. Active enhancers are enriched with H3K27ac, while those that only bear H3K4me1 are poised for activation in response to a stimuli (Chervona and Costa, 2012).

Other less studied histone modifications but perhaps equally important for health and disease include citrullination, ubiquitination, sumoylation, ADP-ribosylation, proline isomerization, and serine/threonine/tyrosine phosphorylation. As mentioned above, histone modifications are not restricted to lysine tails but also occur in the histone fold or globular domains that regulate histone–histone and histone–DNA interactions (Cosgrove et al., 2004; Cosgrove and Wolberger, 2005). In general, as compared to acetylation/deacetylation and methylation/demethylation, most of these modifications occur in amino acids other than lysine. Examples of these are methylation of arginines (mono-, symmetric, or asymmetric dimethylation), and phosphorylation of serine and threonine residues (Turner, 2005; Wang et al., 2007). Kinases that phosphorylate H3S10 are either part of kinase signaling pathways, such as MAPK, or involve components of cell-cycle control systems, such as Aurora B (Cheung et al., 2000; Hirota et al., 2005). Histone methyltransferases on arginine (PRMT and CARM1) as well (Sims et al., 2003). Arginine methylation is reversed through the peptidylarginine desiminase PADI4 that converts monomethylated arginines to citrulline (Cuthbert et al., 2004). At least PRMT5 member of this protein family mediates methylation of histone H4R3 recruits DNMT3A, the de novo DNA methyltransferase, coupling histone arginine, and DNA methylation in gene silencing. Histone tails are also subject to additional modifications such as sumoylation, ubiquitination, ADP ribosylation, and proline isomerization. Polyubiquitination is generally a signal for proteasome-mediated degradation, but monoubiquitination of histones, exemplified by monoubiquitination of H2BK120, is associated with transcription (Fischle, 2014; Cha and Jho, 2012; Wolf, 2009).

1.2.3 Noncoding RNAs

Noncoding RNAs (ncRNAs) are a third player in epigenetic regulation. Interestingly, they are epigenetically regulated themselves (Keller and Bühler, 2013). So far, known types of ncRNAs are two: microRNAs (miRNAS), short-interfering RNAs (siRNAs), and PIWI-interacting RNAs (piRNAs). The other type is the long noncoding RNAs (lncRNAs), which commonly are 200 nucleotides or more in length. ncRNAs participate in more than one way in the regulation of gene expression (Fu, 2014). It has been shown, for instance, that the expression of HOTAIR (HOX antisense intergenic RNA), a lncRNA, is found overexpressed in breast cancer and induces transcriptional modifications triggered by the polycomb repressive complex via trimethylation of H3K27 (Wu et al., 2015). ncRNAs are also able to recruit DNA methyltransferases such as DNMT3B, promoting de novo methylation of gene promoter and gene silencing associated with heterochromatin formation and a pattern of histone modification that impedes transcription (Sandhu et al., 2012). Noncoding RNAs are also importantly implicated in the control of inappropriate expression of noncoding repetitive DNA sequences throughout the genome, such as LINE elements. In this case this is performed by ncRNAs named piRNA (for Piwi-containing proteins) (Huang et al., 2013). Another set of players are implicated in genome host-defense mechanisms, RNA- and DNA-guided endonucleases of the argonaute family (Ago). These proteins are critical for transcriptional regulation via complexing with RISC (RNA-induced silencing complex) as well as via recruiting chromatin modifiers that methylate either H3K27 or H3K4 (Aravin et al., 2008).

Small noncoding RNAs (sncRNAs) also regulate gene expression of miRNAs. The first miRNA to be identified with this action was lin-4 in *Caenorhabditis elegans* (Moazed, 2009). Years later, the latest release of the Sanger miRNA Registry (http://microrna.sanger.ac.uk) shows more than 700 human miRNA annotations (release 11.0), though it has been predicted that the human genome may contain more than 1000 different miRNAs (Berezikov et al., 2005), estimating that at least 30% of all human genes are regulated by miRNAs (Bartel, 2004).

As it occurs with protein-coding genes, an aberrant pattern of methylation of CpG islands near or within miRNA genes could result in a misregulated expression of key miRNAs and ultimately in pathogenic alterations, including tumorigenesis (El-Osta, 2004). Supporting these observations, a study analyzing miRNAs that are aberrantly expressed in ovarian cancer identified a number of hypomethylated miRNA (including miR-21, miR-203, and miR-205) with the encoded miRNAs displaying up-modulated expression (Iorio et al., 2005). With this short overview of how epigenetic players are themselves regulated epigenetically, it is clear that the picture is more and more complex, as miRNAs genes are subject to hypermethylation and hypomethylation in a tumor- and tissue-specific manner. Moreover, if we take into account the functional importance of microRNAs in maintaining cell fate, and the fact that the repressive complex PRC2 component Suz12 binds to several miRNA gene clusters (Lee et al., 2006), it confirms the existence of additional layers of control by miRNA function linked to the epigenetic regulators.

As expected because of their regulation by epigenetic mechanisms, chromatin-modifying drugs can alter the expression patterns of some miRNAs (Saito et al., 2006), a study that combined inhibitors of DNA methylation and histone deacetylation to induce the expression of particular miRNAs. Simultaneous treatment of T24 human bladder cancer cells with decitabine, a DNA-demethylating agent and 4-phenylbutyric acid (PBA, a HDACi) resulted in more than threefold up-regulation in approximately 5% of 313 miRNAs analyzed (Saito et al., 2006).

1.3 The Genetics of Epigenetics

Early estimates based on the exponential increase of cancer incidence as a function of age suggested that there are around 10 rate-limiting events in cancer (Renan, 1993). These estimations have been largely confirmed by studies taking advantage of the emerging platforms for sequencing. A number of studies sequencing complete cancer genomes across a wide range of human tumors have shown that common human cancers possess between 1×10^3 and 1×10^3 somatic base substitutions in their genomes (Wood et al., 2007; Greenman et al., 2007), which raises several fold further if tumors have mismatch repair and consequent microsatellite instability (Timmermann et al., 2010; Greenman et al., 2007). An average colorectal or breast carcinoma contains approximately 80 nonsynonymous mutations per tumor, of which 15 at most are expected to represent driver mutations (functionally important alterations that are selected for), whereas the rest are passenger mutations arising during repeated cycles of cell division (Wood et al., 2007). Though historically the attractiveness of targeting epigenetic alterations in cancer at least, stem from their reversibility as these alterations were touted as functional and not structural, recent studies have demonstrated that in fact, coding genes of epigenetic enzymes are frequent targets of somatic mutations as well (Peltomäki, 2012). Among these are included enzymes responsible for methylating or demethylating DNA and histones, as well as other posttransductional modifiers of histones and coding genes for subunits of chromatin remodeling complexes.

So far, more than a dozen epigenetic genes have been reported having mutations, mostly somatic (Table 2), but a few reported mutated in the germline are responsible for congenital developmental syndromes such as the ICF syndrome (immunodeficiency, centromere instability, and facial anomalies) for mutations in DNMT3b (Xu et al., 1999); the Rubinstein-Taybi syndrome (mental and growth retardation plus dysmorphic features and cancer predisposition) for the histone acetyl transferases CBP and EP300 (Roelfsema et al., 2005) and mutations at MLL2, a histone methyltransferase of the trithorax group (Ng et al., 2010) as a cause of Kabuki syndrome (facial, cardiac, and skeletal abnormalities plus immunological and mental defects); and DNMT1 mutations, which cause hereditary sensory neuropathy with dementia and hearing loss (Klein et al., 2011). Table 2 summarizes representative epigenetic players that are found mutated in malignancies.

Table 2 Selected Epigenetic Players Found Mutated in Malignancies

Enzyme-coding Gene	Function	Disease
DNMT1	Maintenance-DNA methylation	Colorectal carcinoma
DNMT3a	de novo DNA methylation	Acute myeloid leukemia, thymoma
EP300/CREBBP	Histone acetylation	Acute lymphoblastic leukemia, lymphomas, epithelial cancers
HDAC2	Histone deacetylation	Colon and endometrial carcinomas
EZH2	Histone methyltransferase (H3K27)	Lymphomas
KMT2 MLL family	Histone methyltransferase (H3K4)	A number of hematological and solid neoplasias
RIZ1	Histone methyltransferase (H3K9)	Colorectal carcinoma, lymphoma
SETD2	Histone methyltransferase	Renal cell carcinoma
UTX	Histone demethylase (H3K27)	Multiple myeloma, other solid tumors
JARID1C	Histone demethylase (H3K4)	Renal cell carcinoma
SNF5	SWI/SNF chromatin remodeler	Malignant rhabdoid tumor

Among somatic mutations, most are truncating with evidence of biallelic loss but also missense mutations can occur. Likewise, most lead to a protein with loss-of-function, though some with gain-of-function mutations have been reported. So far, mutations have been reported in almost every type of epigenetic player. In the DNA methylation machinery both DNMT1 and DNMT3A (DNA methyltransferase) gene mutations have been seen in colorectal cancer (Kanai et al., 2003) and leukemias (Ley et al., 2010), respectively. Mutations at histone methyltransferases (HMT) or their cofactors include *MLL2* (Parsons et al., 2011; Morin et al., 2011); HMT that methylates at H3K4, *MLL3* (Parsons et al., 2011) at H3K4, *RIZ* (Chadwick et al., 2000) at H3K4; *EZH2*(Morin et al., 2010) at H3K27, *SETD2* (Dalgliesh et al., 2010) at H3K36, and *MEN1* (Guru et al., 1998) component of MLL1 and MLL2 at H3K4. Mutations at histone demethylases are seen in *JARID1C*, which demethylates H3K4, and *UTX* at H3K27 (Dalgliesh et al., 2010; van Haaften et al., 2009). Among histone acetylation modifiers, both histone actyltransferases and histone deacetylases coding genes are found mutated, *EP300/CREBBP* and *HDAC2*, respectively (Gayther et al., 2000; Ropero et al., 2006). Finally, the four chromatin remodelers on which mutations are reported include *SNF5* (INI1,SMARCB1), *PBRM1* (FAF180), *ARID1A* (BAF250A), and *DAXX/ATRX* (Jackson et al., 2009; Varela et al., 2011; Wiegand et al., 2010; Jiao et al., 2011). The first three belong to the SWI/SNF family, whereas the last one belongs to the SNF2 ATPAse family.

These mutations are seen in both hematological and solid malignancies in frequencies varying from 7% in colorectal cancer for DNMT1 to almost 100% (98%) in malignant rhabdoid tumors for *SNF5*. As the finding of genetic alterations in epigenetic players is relatively novel as compared to classical alterations in oncogenes and tumor suppressor genes, it is still debatable at least for some cases whether these mutations are drivers or passengers. Nevertheless, the fact that in some cases mutations are detectable even in preneoplastic stages as is the case for ARID1A mutations in endometriosis tissue adjacent to clear cells ovarian cancer (Wiegand et al., 2010) may suggest that they participate in the carcinogenesis process.

1.4 Epigenomics

As for genetics, the Human Genome Project, whose first draft sequence of the entire DNA genome was published in 2001 and declared complete in 2003, parallel efforts were undertaken by the International Human Epigenome Project (IHEP), which was launched in 2010 (American Association for Cancer Research Human Epigenome Task Force; European Union, Network of Excellence, Scientific Advisory Board). Likewise, the Encyclopedia of DNA Elements (ENCODE) project (ENCODE Project Consortium, 2004) was aimed to catalog the regulatory elements in human cells by studying the epigenomic signatures of cells in culture. Thus, ENCODE has advanced our understanding of the principles of genome, epigenome, and chromatin organization, identifying hundreds of thousands of potential regulatory regions and transcription factor binding sites. Part of the ENCODE consortium, GENCODE, has annotated the human genome with novel transcripts including new noncoding RNAs and pseudogenes, highlighting transcriptional complexity (Harrow et al., 2012). Many disease variants identified in genomewide association studies are located within putative enhancer regions defined by the ENCODE project. The ultimate goal of the IHEP is to understand the patterns of DNA methylation and posttranslational histone modifications that regulate the access of specific genes to transcriptional machinery within a cell. Recently, the Roadmap Epigenomics Project as part of the IHEP reached a milestone in a study describing the integrative analysis of 111 reference human epigenomes in which the profiling of histone modification patterns, DNA accessibility, DNA methylation, and RNA expression was achieved. The results show that disease and trait-associated genetic variants are enriched in predicted tissue-specific epigenomic marks, revealing biologically relevant cell types for diverse human traits and providing a resource for interpreting the molecular basis of human disease (Roadmap Epigenomics Consortium et al., 2015; Romanoski et al., 2015). Understanding the principles of chromatin and epigenome organization will help to identify new disease mechanisms, biomarkers, and drug targets, particularly as ongoing epigenome mapping projects generate data for primary human cell types that play important roles in disease.

1.5 Conclusions

From this short introductory chapter there are some issues to be reviewed.

1. The past simple view of DNA promoter methylation as the main molecular pathogenetic mechanism to inactivate the expression of tumor suppressor genes in cancer (regardless of the known—and possibly deleterious—global DNA hypomethylation in tumors) and, as a consequence, a natural pharmacological target aimed to reactivate the expression of these genes belongs to the past. Currently, the increasing complexity of epigenetics, with a high number of players and interactions participating in normal and disease processes such as cancer, has made it almost impossible to arrive at educated guesses on how to modulate by pharmacological means this cascade of events: that is, how to best make use of pharmacological interventions to modify in a beneficial way the processes of cancer or any other disease without at the same time producing undesirable epigenetic changes that could favor disease development. This simplistic view of DNA gene promoter methylation as a strategy to treat complex diseases like cancer honors the saying, "simple is beauty." We could think that if someone, now in light of so many complexities regulating gene expression, would rethink to use current unspecific transcriptome-modifying agents such as DNA methyltransferase inhibitors.
2. Even though the pace of knowledge in basic mechanisms of epigenetics and gene regulation not only in the cell but in a whole organism context is advancing rapidly, we are still far from an in-depth understanding that could allow us to modify epigenetic and epigenomic landscapes with therapeutic goals using either broad spectrum or selective agents.
3. As it happens in biomedical research, the pace of discovery of novel mechanisms, molecules, pathways, and novel technologies for prediction, diagnosis, and therapeutic development of any kind or drugs is much more rapid and overcomes the capacity of clinicians for testing them in clinical trials due to a number of scientific, but mostly economically driven reasons that are not discussed here. Just to mention some gross data on epigenetic therapies, Wigle compiled from literature and published (Wigle, 2011) at least 232 confirmed or putative chromatin-modifying enzymes being potentially targetable. These include in descending order, lysine methyltransferases (52), arginine

methyltransferases (44), other (42), histone acetyltransferases (28), lysine demethylases (26), histone deacetylases (18), kinases (17), and arginine deaminases (5). It is likely that in the last 4 years this list has been greatly expanded. In addition, at least two more epigenetic classes of agents are in development. The ones that do not target enzymes per se, but are readers of the chromatin modifications such as BET and not BET bromodomains proteins, as well as lysine reader proteins having chromo, tudor, and PHD domains (Owen and Trzupek, 2014). Moreover, the growing importance of epigenetic ncRNAs for normal regulation of transcription and disease development has opened a new field in therapy, epigenomic engineering, which potentially would reduce the target-off effects of current epigenetic drugs by the bioengineering of artificial DNA binding domains to manipulate and putatively normalize the altered landscape of the cancer epigenome (Falahi et al., 2015).

1.6 Perspectives

The road to find effective drugs and other therapeutic tools to successfully treat not only malignant diseases but other multifactorial chronic degenerative diseases with epigenetic therapy is long. We are just at the beginning of fully (if ever) understanding the role of epigenetics/epigenomics in health and disease. Nevertheless, now having first-generation epigenetic drugs approved—for cancer at least—is a vivid demonstration that we are going ahead. As stated earlier, numerous technological advances are rendering a new wave of potentially more selective and efficacious epigenetic drugs. To crystallize these advances for human treatment there are at least two immediate actions that can simultaneously be pursued. One is to greatly increase the number of small clinical exploratory trials with existing drugs not expecting to find that they work in all patients, but to accompany these studies with pharmacogenetic, pharmacogenomic, and other "-omic" technologies to identify the predictive factors for response and/or toxicity, which means personalized medicine. The other approach, which also can be cataloged as personalized medicine, is to continue the study of diseases to find out epigenetic drivers unique to certain diseases and then proceed to clinical trials in highly selected populations of patients. In either case, new paradigms in drug development are needed. Currently, this process seems to exclusively belong to "big pharma." Advocacy groups of the civil society, patients, and philanthropic organizations may change the balance toward the democratization of drug development.

ACKNOWLEDGMENTS

A. D-G acknowledges the support for this manuscript to Fonsec CONACyT grant 161915, and UNAM PAPIIT grant IT206611. J. L. M-F. thanks UNAM for PAIP grant 5000–9163.

REFERENCES

Anon, 2015. Panobinostat approved for multiple myeloma. Cancer Discov. 5 OF4.

Aravin, A.A., Sachidanandam, R., Bourc'his, D., Schaefer, C., Pezic, D., Toth, K.F., et al., 2008. A piRNA pathway primed by individual transposons is linked to de novo DNA methylation in mice. Mol. Cell 31, 785–799.

Bartel, D.P., 2004. MicroRNAs: genomics, biogenesis, mechanism, and function. Cell 116, 281–297.

Berezikov, E., Guryev, V., van de Belt, J., Wienholds, E., Plasterk, R.H.A., Cuppen, E., 2005. Phylogenetic hadowing and computational identification of human microRNA genes. Cell 120, 21–24.

Berger, S.L., Kouzarides, T., Shiekhattar, R., Shilatifard, A., 2009. An operational definition of epigenetics. Genes Dev. 23, 781–783.

Bertino, E.M., Otterson, G.A., 2011. Romidepsin: a novel histone deacetylase inhibitor for cancer. Expert Opin. Investig. Drugs 20, 1151–1158.

Best, J.D., Carey, N., 2010. Epigenetic therapies for non-oncology indications. Drug Discov. Today 15, 1008–1014.

Blackledge, N.P., Zhou, J.C., Tolstorukov, M.Y., Farcas, A.M., Park, P.J., Klose, R.J., 2010. CpG islands recruit a histone H3 lysine 36 demethylase. Mol. Cell 38, 179–190.

Cha, B., Jho, E.H., 2012. Protein arginine methyltransferases (PRMTs) as therapeutic targets. Expert Opin. Ther. Targets 16, 651–664.

Chadwick, R.B., Jiang, G.L., Bennington, G.A., Yuan, B., Johnson, C.K., Stevens, M.W., et al., 2000. Candidate tumor suppressor RIZ is frequently involved in colorectal carcinogenesis. Proc. Natl. Acad. Sci. U.S.A. 97, 2662–2667.

Chervona, Y., Costa, M., 2012. Histone modifications and cancer: biomarkers of prognosis? Am. J. Cancer Res. 2, 589–597.

Cheung, P., Tanner, K.G., Cheung, W.L., Sassone-Corsi, P., Denu, J.M., Allis, C.D., 2000. Synergistic coupling of histone H3 phosphorylation and acetylation in response to epidermal growth factor stimulation. Mol. Cell 5, 905–915.

Cocozza, S., AkhtarMM, Miele, G., Monticelli, A., 2011. CpG islands undermethylation in human genomic regions under selective pressure. PLoS One 6, e23156.

Cosgrove, M.S., 2007. Histone proteomics and the epigenetic regulation of nucleosome mobility. Expert Rev. Proteomics 4, 465–478.

Cosgrove, M.S., Boeke, J.D., Wolberger, C., 2004. Regulated nucleosome mobility and the histone code. Nat. Struct. Mol. Biol. 11, 1037–1043.

Cosgrove, M.S., Wolberger, C., 2005. How does the histone code work? Biochem. Cell Biol. 83, 468–476.

Cuthbert, G.L., Daujat, S., Snowden, A.W., Erdjument-Bromage, H., Hagiwara, T., Yamada, M., et al., 2004. Histone deimination antagonizes arginine methylation. Cell 118, 545–553.

Dalgliesh, G.L., Furge, K., Greenman, C., Chen, L., Bignell, G., Butler, A., et al., 2010. Systematic sequencing of renal carcinoma reveals inactivation of histone modifying genes. Nature 463, 360–363.

Davie, J.R., Hendzel, M.J., 1994. Multiple functions of dynamic histoneacetylation. J. Cell Biochem. 55, 98–105.

Derissen, E.J., Beijnen, J.H., Schellens, J.H., 2013. Concise drug review: azacitidine and decitabine. Oncologist 18, 619–624.

Dillon, S.C., Zhang, X., Trievel, R.C., Cheng, X., 2005. The SET-domain protein superfamily: protein lysine methyltransferases. Genome Biol. 6, 227.

Doehring, A., Geisslinger, G., Lötsch, J., 2011. Epigenetics in pain and analgesia: an imminent research field. Eur. J. Pain 15, 11–16.

ENCODE Project Consortium, 2004. The ENCODE (ENCyclopedia Of DNA Elements) Project. Science 306, 636–640.

Eissenberg, J.C., Shilatifard, A., 2010. Histone H3 lysine 4 (H3K4) methylation in development and differentiation. Dev. Biol. 339, 240–249.

El-Osta, A., 2004. Understanding the consequences of epigenetic mechanisms and its effects on transcription in health and disease. Cancer Biol. Ther. 3, 816–818.

Esteller, M., 2007. Cancer epigenomics: DNA methylomes and histone-modification maps. Nat. Rev. Genet. 8, 286–298.

Falahi, F., Sgro, A., Blancafort, P., 2015. Epigenome engineering in cancer: fairytale or a realistic path to the clinic? Front. Oncol. 5, 22.

Fisher, C.L., Fisher, A.G., 2011. Chromatin states in pluripotent, differentiated, and reprogrammed cells. Curr. Opin. Genet. Dev. 21, 140–146.

Fischle, W., 2014. Molecular mechanisms of histone modification function. Biochim. Biophys. Acta 1839, 621–622.

Fu, X.D., 2014. Non-coding RNA: a new frontier in regulatory biology. Natl. Sci. Rev. 1, 190–204.

Gay, S., Wilson, A.G., 2014. The emerging role of epigenetics in rheumatic diseases. Rheumatology (Oxford) 53, 406–414.

Gayther, S.A., Batley, S.J., Linger, L., Bannister, A., Thorpe, K., Chin, S.F., et al., 2000. Mutations truncating the EP300 acetylase in human cancers. Nat. Genet. 24, 300–303.

Greenman, C., Stephens, P., Smith, R., Dalgliesh, G.L., Hunter, C., Bignell, G., et al., 2007. Patterns of somatic mutation in human cancer genomes. Nature 446, 153–158.

Guru, S.C., Goldsmith, P.K., Burns, A.L., Marx, S.J., Spiegel, A.M., Collins, F.S., Chandrasekharappa, S.C., 1998. Menin, the product of the MEN1 gene, is a nuclear protein. Proc. Natl. Acad. Sci. U.S.A. 95, 1630–1634.

Hackett, J.A., Sengupta, R., Zylicz, J.J., Murakami, K., Lee, C., Down, T.A., Surani, M.A., 2013. Germline DNA demethylation dynamics and imprint erasure through 5-hydroxymethylcytosine. Science 339, 448–452.

Harrow, J., Frankish, A., Gonzalez, J.M., Tapanari, E., Diekhans, M., Kokocinski, F., et al., 2012. GENCODE: the reference human genome annotation for The ENCODE Project. Genome Res. 22, 1760–1774.

Heyn, H., Esteller, M., 2012. DNA methylation profiling in the clinic: applications and challenges. Nat. Rev. Genet. 13, 679–692.

Hirota, T., Lipp, J.J., Toh, B.H., Peters, J.M., 2005. Histone H3 serine 10 phosphorylation by Aurora B causes HP1 dissociation from heterochromate. Nature 438, 1176–1180.

Huang, X.A., Yin, H., Sweeney, S., Raha, D., Snyder, M., Lin, H., 2013. A major epigenetic programming mechanism guided by piRNAs. Dev. Cell 24, 502–516.

Van Haaften, G., Dalgliesh, G.L., Davies, H., Chen, L., Bignell, G., Greenman, C., et al., 2009. Somatic mutations of the histone H3K27 demethylase gene UTX in human cancer. Nat. Genet. 41, 521–523.

Iorio, M.V., Ferracin, M., Liu, C.G., Veronese, A., Spizzo, R., Sabbioni, S., et al., 2005. MicroRNA gene expression deregulation in human breast cancer. Cancer Res. 65, 7065–7070.

Iorio, M.V., Piovan, C., Croce, C.M., 2010. Interplay between microRNAs and the epigenetic machinery: an intricate network. Biochim. Biophys. Acta 1799, 694–701.

Isobe, T., Song, S.N., Tiwari, P., Ito, H., Yamaguchi, Y., Yoshizaki, K., 2013. Activation-inducedcytidine deaminase auto-activates and triggers aberrant gene expression. FEBS Lett. 587, 2487–2492.

Ito, S., D'Alessio, A.C., Taranova, O.V., Hong, K., Sowers, L.C., Zhang, Y., 2010. Role of Tet proteins in 5mC to 5hmC conversion, ES-cell self-renewal and inner cell mass specification. Nature 466, 1129–1133.

Ito, S., Shen, L., Dai, Q., et al., 2011. Tet proteins can convert 5-methylcytosine to 5-formylcytosine and 5-carboxylcytosine. Science 333, 1300–1303.

Iurlaro, M., Ficz, G., Oxley, D., Raiber, E.A., Bachman, M., Booth, M.J., et al., 2013. A screen for hydroxymethylcytosine and formylcytosine binding proteins suggests functions in transcription and chromatin regulation. Genome Biol. 14, R119.

Jackson, E.M., SievertAJ, G.X., Hakonarson, H., Judkins, A.R., Tooke, L., et al., 2009. Genomic analysis using high-density single nucleotide polymorphism-based oligonucleotide arrays and multiplex ligation-dependent probe amplification provides a comprehensive analysis of INI1/SMARCB1 in malignant rhabdoid tumors. Clin. Cancer Res. 15, 1923–1930.

Jenuwein, T., 2006. The epigenetic magic of histone lysine methylation. FEBS J. 273, 3121–3135.

Jiao, Y., Shi, C., Edil, B.H., De Wilde, R.F., Klimstra, D.S., Maitra, A., et al., 2011. DAXX/ATRX, MEN1, and mTOR pathway genes are frequently altered in pancreatic neuroendocrine tumors. Science 331, 1199–1203.

Jones, P.A., Taylor, S.M., 1980. Cellular differentiation, cytidine analogs and DNA methylation. Cell 20, 85–93.

Jurkowska, R.Z., Jurkowski, T.P., Jeltsch, A., 2011. Structure and function of mammalian DNA methyltransferases. ChemBioChem 12, 206–222.

Kanai, Y., Ushijima, S., Nakanishi, Y., Sakamoto, M., Hirohashi, S., 2003. Mutation of the DNA methyltransferase (DNMT) 1 gene in human colorectal cancers. Cancer Lett. 192, 75–82.

Keller, C., Bühler, M., 2013. Chromatin-associated ncRNA activities. Chromosome Res. 21, 627–641.

Klein, C.J., Botuyan, M.V., Wu, Y., Ward, C.J., Nicholson, G.A., Hammans, S., et al., 2011. Mutations in DNMT1 cause hereditary sensory neuropathy with dementia and hearing loss. Nat. Genet. 43, 595–600.

Lee, T.I., Jenner, R.G., Boyer, L.A., Guenther, M.G., Levine, S.S., Kumar, R.M., et al., 2006. Control of developmental regulators by polycomb in human embryonic stem cells. Cell 125, 301–313.

Lee, H.Z., Kwitkowski, V.E., Del Valle, P.L., Ricci, M.S., Saber, H., Habtemariam, B.A., et al., 2015. FDA approval: belinostat for the treatment of patients with relapsed or refractory peripheral T-cell lymphoma. Clin. Cancer Res. 21, 2270–2666. http://dx.doi.org/10.1158/1078-0432.CCR-14-3119.

Ley, T.J., Ding, L., Walter, M.J., McLellan, M.D., Lamprecht, T., Larson, D.E., et al., 2010. DNMT3A mutations in acute myeloid leukemia. N. Engl. J. Med. 363, 2424–2433.

Li, B., Pattenden, S.G., Lee, D., Gutiérrez, J., Chen, J., Seidel, C., et al., 2005. Preferential occupancy of histone variant H2AZ at inactive promoters influences local histone modifications and chromatin remodeling. Proc. Natl. Acad. Sci. U.S.A. 102, 18385–18390.

Li, G., Reionberg, D., 2011. Chromatin higher-order structures and gene regulation. Curr. Opin. Genet. Dev. 21, 175–186.

Lovinsky-Desir, S., Miller, R.L., 2012. Epigenetics, asthma, and allergic diseases: a review of the latest advancements. Curr. Allergy Asthma Rep. 12, 211–220.

Lu, Q., 2013. The critical importance of epigenetics in autoimmunity. J. Autoimmun. 41, 1–5.

Mann, B.S., Johnson, J.R., Cohen, M.H., Justice, R., Pazdur, R., 2007. FDA approval summary: vorinostat for treatment of advanced primary cutaneous T-cell lymphoma. Oncologist 12, 1247–1252.

Mau, T., Yung, R., 2014. Potential of epigenetic therapies in non-cancerous conditions. Front. Genet. 5, 438.

Mehler, M.F., 2008. Epigenetics and the nervous system. Ann. Neurol. 64, 602–617.

Mellen, M., Ayata, P., Dewell, S., Kriaucionis, S., Heintz, N., 2012. MeCP2 binds to 5hmC enriched within active genes and accessible chromatin in the nervous system. Cell 151, 1417–1430.

Milne, T.A., Briggs, S.D., Brock, H.W., Martin, M.E., Gibbs, D., Allis, C.D., et al., 2002. MLL targets SET domain methyltransferase activity to Hox gene promoters. Mol. Cell 10, 1107–1117.

Moazed, D., 2009. Small RNAs in transcriptional gene silencing and genome defence. Nature 457, 413–420.

Morin, R.D., Johnson, N.A., Severson, T.M., Mungall, A.J., An, J., Goya, R., et al., 2010. Somatic mutations altering EZH2 (Tyr641) in follicular and diffuse large B-cell lymphomas of germinal-center origin. Nat. Genet. 42, 181–185.

Morin, R.D., Mendez-Lago, M., Mungall, A.J., Goya, R., Mungall, K.L., Corbett, R.D., et al., 2011. Frequent mutation of histone-modifying genes in non-Hodgkin lymphoma. Nature 476, 298–303.

Müller, J., Kassis, J.A., 2006. Polycomb response elements and targeting of Polycomb group proteins in Drosophila. Curr. Opin. Genet. Dev. 16, 476–484.

Munoz, D.P., Lee, E.L., Takayama, S., Coppe, J.P., Heo, S.J., Boffelli, D., Di Noia, J.M., Martin, D.I., 2013. Activation-induced cytidine deaminase (AID) is necessary for the epithelial– mesenchymal transition in mammary epithelial cells. Proc. Natl. Acad. Sci. U.S.A. 110, E2977–E2986.

Nakamura, T., Mori, T., Tada, S., Krajewski, W., Rozovskaia, T., Wassell, R., et al., 2002. ALL-1 is a histone methyltransferase that assembles a supercomplex of proteins involved in transcriptional regulation. Mol. Cell 10, 1119–1128.

Ng, S.S., Yue, W.W., Oppermann, U., Klose, R.J., 2009. Dynamic protein methylation in chromatin biology. Cell Mol. Life Sci. 66, 407–422.

Ng, S.B., Bigham, A.W., Buckingham, K.J., Hannibal, M.C., McMillin, M.J., Gildersleeve, H.I., et al., 2010. Exome sequencing identifies MLL2 mutations as a cause of Kabuki syndrome. Nat. Genet. 42, 790–793.

Nguyen, A.T., Zhang, Y., 2011. The diverse functions of Dot1 and H3K79 methylation. Genes Dev. 25, 1345–1358.

Ogryzko, V.V., Schiltz, R.L., Russanova, V., Howard, B.H., Nakatani, Y., 1996. The transcriptional coactivators p300 and CBP are histone acetyltransferases. Cell 87, 953–959.

Okada, Y., Feng, Q., Lin, Y., Jiang, Q., Li, Y., Coffield, V.M., et al., 2005. hDOT1L links histone methylation to leukemogenesis. Cell 121, 167–178.

Owen, D.R., Trzupek, J.D., 2014. Epigenetic drugs that do not target enzyme activity. Drug Discov. Today Technol. 12, e29–34.

Parsons, D.W., Li, M., Zhang, X., Jones, S., Leary, R.J., Lin, J.C., et al., 2011. The genetic landscape of the childhood cancer medulloblastoma. Science 331, 435–439.

Payer, B., Lee, J.T., 2008. X chromosome dosage compensation: how mammals keep the balance. Annu. Rev. Genet. 42, 733–772.

Peltomäki, P., 2012. Mutations and epimutations in the origin of cancer. Exp. Cell. Res. 318, 299–310.

Peters, A.H., O'Carroll, D., Scherthan, H., Mechtler, K., Sauer, S., Schöfer, C., et al., 2001. Loss of the Suv39h histone methyltransferases impairs mammalian heterochromatin and genome stability. Cell 107, 323–337.

Piccolo, F.M., Fisher, A.G., 2014. Getting rid of DNA methylation. Trends Cell Biol. 24, 136–143.

Renan, M.J., 1993. How many mutations are required for tumorigenesis? Implications from human cancer data. Mol. Carcinog. 7, 139–146.

Rishi, V., Bhattacharya, P., Chatterjee, R., Rozenberg, J., Zhao, J., Glass, K., et al., 2010. CpG methylation of half-CRE sequences creates C/EBP alpha binding sites that activate some tissue-specific genes. Proc. Natl. Acad. Sci. U.S.A. 107, 20311–20316.

Roadmap Epigenomics Consortium, Kundaje, A., Meuleman, W., Ernst, J., Bilenky, M., Yen, A., Heravi-Moussavi, A., et al., 2015. Integrative analysis of 111 reference human epigenomes. Nature 518, 317–330.

Roelfsema, J.H., White, S.J., Ariyürek, Y., Bartholdi, D., Niedrist, D., Papadia, F., et al., 2005. Genetic heterogeneity in Rubinstein-Taybi syndrome: mutations in both the CBP and EP300 genes cause disease. Am. J. Hum. Genet. 76, 572–580.

Romanoski, C.E., Glass, C.K., Stunnenberg, H.G., Wilson, L., Almouzni, G., 2015. Epigenomics: Roadmap for regulation. Nature 518, 314–316.

Ropero, S., Fraga, M.F., Ballestar, E., Hamelin, R., Yamamoto, H., Boix-Chornet, M., et al., 2006. A truncating mutation of HDAC2 in human cancers confers resistance to histone deacetylase inhibition. Nat. Genet. 38, 566–569.

Rundlett, S.E., Carmen, A.A., Kobayashi, R., Bavykin, S., Turner, B.M., Grunstein, M., 1996. HDA1 and RPD3 are members of distinct yeast histone deacetylase complexes that regulate silencing and transcription. Proc. Natl. Acad. Sci. U.S.A. 93, 14503–14508.

Saito, Y., Liang, G., Egger, G., Friedman, J.M., Chuang, J.C., Coetzee, G.A., 2006. Specific activation of microRNA-127 with downregulation of the proto-oncogene BCL6 by chromatin-modifying drugs in human cancer cells. Cancer Cell 9, 435–443.

Sandhu, R., Rivenbark, A.G., Coleman, W.B., 2012. Loss of post-transcriptional regulation of DNMT3b by microRNAs: a possible molecular mechanism for the hypermethylation defect observed in a subset of breast cancer cell lines. Int. J. Oncol. 41, 721–732.

Sansom, O.J., Maddison, K., Clarke, A.R., 2007. Mechanisms of disease: methyl-binding domain proteins as potential therapeutic targets in cancer. Nat. Clin. Pract. Oncol. 4, 305–315.

Schaefer, M., Lyko, F., 2010. Solving the Dnmt2 enigma. Chromosoma 119, 35–40.

Seisenberger, S., Peat, J.R., Hore, T.A., et al., 2013. Reprogramming DNA methylation in the mammalian life cycle: building and breaking epigenetic barriers. Philos. Trans. R. Soc. Lond. B Biol. Sci. 368, 20110330.

Shi, Y.G., Lan, F., Matson, C., Mulligan, P., Whetstine, J.R., Cole, P.A., Casero, R.A., et al., 2004. Histone demethylation mediated by the nuclear amine oxidase homolog LSD1. Cell 119, 941–953.

Shi, Y.G., Tsukada, Y., 2013. The discovery of histone demethylases. Cold Spring Harb. Perspect. Biol. 5, pii:a017947.

Shi, Y., Whetstine, J.R., 2007. Dynamic regulation of histone lysine methylation by demethylases. Mol. Cell 25, 1–14.

Sims 3rd, R.H., Nishioka, K., Reinberg, D., 2003. Histone lysine methylation: a signature for chromatin function. Trends Genet. 19, 629–639.

Slomko, H., Heo, H.J., Einstein, F.H., 2012. Minireview: Epigenetics of obesity and diabetes in humans. Endocrinology 153, 1025–1030.

Smith, E., Shilatifard, A., 2010. The chromatin signaling pathway: diverse mechanisms of recruitment of histone-modifying enzymes and varied biological outcomes. Mol. Cell 40, 689–701.

Stein, R.A., 2014. Epigenetic therapies - a new direction in clinical medicine. Int. J. Clin. Pract. 68, 802–811.

Tachibana, M., Sugimoto, K., Nozaki, M., Ueda, J., Ohta, T., Ohki, M., et al., 2002. G9a histone methyltransferase plays a dominant role in euchromatic histone H3 lysine 9 methylation and is essential for early embryogenesis. Genes Dev. 16, 1779–1791.

Talbert, P.B., Henikoff, S., 2010. Histone variants: ancient wrap artists of the epigenome. Nat. Rev. Mol. Cell Biol. 11, 264–275.

Timmermann, B., Kerick, M., Roehr, C., Fischer, A., Isau, M., Boerno, S.T., et al., 2010. Somatic mutation profiles of MSI and MSS colorectal cancer identified by whole exome next generation sequencing and bioinformatics analysis. PLoS One 5, e15661.

Trojer, P., Zhang, J., Yonezawa, M., Schmidt, A., Zheng, H., Jenuwein, T., et al., 2009. Dynamic histone H1 isotype 4 methylation and demethylation by histone lysine methyltransferase G9a/KMT1C and the Jumonji domain-containing JMJD2/KDM4 proteins. J. Biol. Chem. 284, 8395–8405.

Tsukada, Y., Fang, J., Erdjument-Bromage, H., Warren, M.E., Borchers, C.H., Tempst, P., et al., 2006. Histone demethylation by a family of JmjC domain-containing proteins. Nature 439, 811–816.

Turner, B.M., 2005. Reading signals on the nucleosome with a new nomenclature for modified histones. Nat. Struct. Mol. Biol. 12, 110–112.

Varela, I., Tarpey, P., Raine, K., Huang, D., Ong, C.K., Stephens, P., et al., 2011. Exome sequencing identifies frequent mutation of the SWI/SNF complex gene PBRM1 in renal carcinoma. Nature 469, 539–542.

Villar-Garea, A., Imhof, A., 2006. The analysis of histone modifications. Biochim. Biophys. Acta 1764, 1932–1939.

Waddington, C.H., 1942. The epigenotype. Endeavour 18–20.

Wang, G.G., Allis, C.D., Chi, P., 2007. Chromatin remodeling and cancer. Part I. Covalent histone modifications. Trends Mol. Med. 13, 363–372.

Wang, K.C., Yang, Y.W., Liu, B., Sanyal, A., Corces-Zimmerman, R., Chen, Y., et al., 2011. A long noncoding RNA maintains active chromatin to coordinate homeotic gene expression. Nature 472, 120–124.

Whetstine, J.R., Nottke, A., Lan, F., Huarte, M., Smolikov, S., Chen, Z., et al., 2006. Reversal of histone lysine trimethylation by the JMJD2 family of histone demethylases. Cell 125, 467–481.

Wiegand, K.C., Shah, S.P., Al-Agha, O.M., Zhao, Y., Tse, K., Zeng, T., et al., 2010. ARID1A mutations in endometriosis-associated ovarian carcinomas. N. Engl. J. Med. 363, 1532–1543.

Wigle, T.J., 2011. Promoting illiteracy in epigenetics: an emerging therapeutic strategy. Curr. Chem. Genomics 5 (Suppl. 1), 48–50.

Wolf, S.S., 2009. The protein arginine methyltransferase family: an update about function, new perspectives and the physiological role in humans. Cell Mol. Life Sci. 66, 2109–2121.

Wood, L.D., Parsons, D.W., Jones, S., Lin, J., Sjöblom, T., Leary, R.J., et al., 2007. The genomic landscapes of human breast and colorectal cancers. Science 318, 1108–1113.

Wright, J., 2013. Epigenetics: reversible tags. Nature 498, S10–S11.

Wu, F., Yao, J., 2013. Spatial compartmentalization at the nuclear periphery characterized by genome-wide mapping. BMC Genomics 14, 591.

Wu, Y., Zhang, L., Wang, Y., Li, H., et al., 2015. Long non-coding RNA HOTAIR promotes tumor cell invasion and metastasis by recruiting EZH2 and repressing E-cadherin in oral squamous cell carcinoma. Int. J. Oncol. 46, 2586–2594.

Xu, G.L., Bestor, T.H., Bourchis, D., Hsieh, C.L., Tommerup, N., Bugge, M., et al., 1999. Chromosome instability and immunodeficiency syndrome caused by mutations in a DNA methyltransferase gene. Nature 402, 187–191.

Yamane, K., Toumazou, C., Tsukada, Y., Erdjument-Bromage, H., Tempst, P., Wong, J., et al., 2006. JHDM2A, a JmjC-containing H3K9 demethylase, facilitates transcription activation by androgen receptor. Cell 125, 483–495.

You, J.S., Jones, P.A., 2012. Cancer genetics and epigenetics: two sides of the same coin? Cancer Cell 22, 2–9.

Zhao, J., Sun, B.K., Erwin, J.A., Song, J.J., Lee, J.T., 2008. Polycomb proteins targeted by a short repeat RNA to the mouse X chromosome. Science 322, 750–756.

CHAPTER 2

Overview of Computer-Aided Drug Design for Epigenetic Targets

Rodrigo Aguayo-Ortiz, Eli Fernández-de Gortari

Departamento de Farmacia, Universidad Nacional Autónoma de México, México City, México

1. INTRODUCTION

Computational approaches have become indispensable tools to accelerate the development of epigenetic inhibitors helping in the selection, design, and lead identification of novel compounds. Computational methods are also useful for optimization phases and to decrease the costs of industrial drug development.

During the past few years, several in silico approaches such as similarity searching, molecular docking, virtual screening, pharmacophore modeling, and molecular dynamics (MD) have been applied to help understand the activity of known compounds and to design novel epigenetic inhibitors. Chemoinformatics tools are used to select compounds that are already characterized to identify compounds that have an increased opportunity of having activity against a target during the experimental tests. Following these processes, and depending on the information available, more detailed (and usually costly) techniques are further employed to identify relationships between structure and activity, the binding modes and affinities with docking and MD or to perform searches based in pharmacophore modeling.

This chapter provides an overview of the main techniques used in computational drug design to date. Special emphasis is made on the application of computational approaches within the efforts for the design of bioactive molecules against targets related to epigenetic regulation, which are known to be involved in diseases such as cancer, heart disease, psychiatric, among others. In order to facilitate the understanding of computer-aided drug design (CADD) techniques, the discussion in this chapter has been divided into two principal approaches: ligand-based drug design (LBDD) and structure-based drug design (SBDD) methods. Table 1 summarizes computational methods employed for the design of epigenetic inhibitors.

CONTENTS

Epi-Informatics. http://dx.doi.org/10.1016/B978-0-12-802808-7.00002-2

Table 1 Brief Description of the Computational Methods Employed for the Design of Epigenetic Inhibitors

Method	Approach	Brief Description	References
LBDD	Chemoinformatics	Large scientific discipline, standing on the interface between chemistry, biology, and computer science, that deals with the storage, organization, management, retrievals, analysis, dissemination, visualization, and use of chemical information.	Agrafiotis et al. (2007)
	Structure similarity search	Group of methods employed to identify drug-like small molecules based on the assumption that molecules that are globally similar in structure should exhibit similar biological activity.	Vogt and Bajorath (2012)
	Pharmacophore modeling	This tool extracts common chemical features from 3D structures of a set of known ligands representative of essential interactions between the ligands and a specific macromolecular target.	Yang (2010)
	QSAR	Method focused on finding a statistical model that gives a mathematical equation for correlating the biological property to structure within a family of compounds.	Dudek et al. (2006)
SBDD	Homology modeling	Methodology to predict protein structure from its sequence based on the general observation that proteins with similar sequences have similar structures.	Cavasotto and Phatak (2009)
	Docking	Molecular docking is a computational chemistry tool that can find the structure and binding energy of a protein-ligand complex when the structure of both is known.	Novikov and Chilov (2009)
	Molecular dynamics	In the context of drug design, molecular dynamics simulation is a computational tool that helps in the understanding of the physical basis of the structure and function of biological macromolecules.	Karplus and McCammon (2002)

2. LIGAND-BASED DRUG DESIGN

The chemical structures of experimentally established epigenetic inhibitors represent a key source to conduct structure–activity relationships (SAR) and ligand-based virtual screening studies for the identification of novel inhibitors. Chemical databases annotated with bioactivity data are an increasingly important information source available, among which PubChem (Wang et al., 2012b), ChEMBL (Bento et al., 2014), and Binding Database (Liu et al., 2007) contain the largest chemical collection and activity data of epigenetic inhibitors. Moreover, the epigenetic enzyme and modulator database (HEMD)

(Huang et al., 2012) contains 4377 modulators (activators, inhibitors, and regulators) annotated with detailed descriptions of binding sites, pharmacological effects, and therapeutic uses. It is noteworthy that some of the epigenetic modulators in HEMD are not covered by the aforementioned databases, revealing it as one of the most important epigenetic databases to date. The storage of these and other databases has enhanced the knowledge of the SAR by performing the computational techniques described below.

2.1 Chemoinformatics Analysis and Structure Similarity

Chemoinformatics, also referred to in the literature as cheminformatics or chemical information science, is one of the most recent theoretical chemistry disciplines used in drug discovery. This discipline has been defined as "the application of informatics methods to solve chemical problems" (Gasteiger, 2006). Informatics tools for chemical research are centered in the creation and application of a compound library with large chemical diversity for structural and biological analysis, based on the structural descriptors (Xu and Hagler, 2002). These descriptors are employed to transform the chemical information into a numerical code, in order to represent molecules as objects in a chemical space for the manipulation of the two-dimensional (2D) or three-dimensional (3D) stored chemical structures (Todeschini and Consonni, 2009; Willett, 2011).

The characterization of the chemical space and molecular diversity of DNA methyltransferase (DNMT)-focused libraries has been one of the first applications of chemoinformatics methods to identify DNMT inhibitors (ChemDiv, 2011; Medina-Franco and Yoo, 2013b). Yoo and Medina-Franco reported a chemoinformatics analysis of four screening collections (i.e., DNMT-focused library, NCI diversity set, and two natural product databases) using physicochemical properties, molecular fingerprints, and molecular scaffolds (Yoo and Medina-Franco, 2011a). The DNMT-focused library and the two natural product collections have molecules with properties similar to approved drugs, also appearing to be attractive sources to identify DNMT inhibitors with high structure similarity with novel chemical scaffolds. It was concluded that the focused library is also particularly attractive for experimental screening to identify novel inhibitors of DNMTs. Subsequently, Yoo and Medina-Franco reported a comprehensive similarity profile of these four screening collections with the chemical structure of **SGI-1027** (Figure 1), a promising lead compound with activity against DNMT (Yoo and Medina-Franco, 2012a). The rationale of the ligand-based virtual screening is based on the similarity property principle (SPP), which establishes that similar molecules should have similar biological properties (Vogt and Bajorath, 2012; Willett, 2006). In this sense, the DNMT-focused library showed the highest similarity profile and represents the most suitable candidate collection to identify additional DNMT inhibitors. Similarity searching led to the identification of 34 hits as promising candidates for experimental evaluation. In a separate study, Medina-Franco and Yoo reported the similarity profile of five chemical

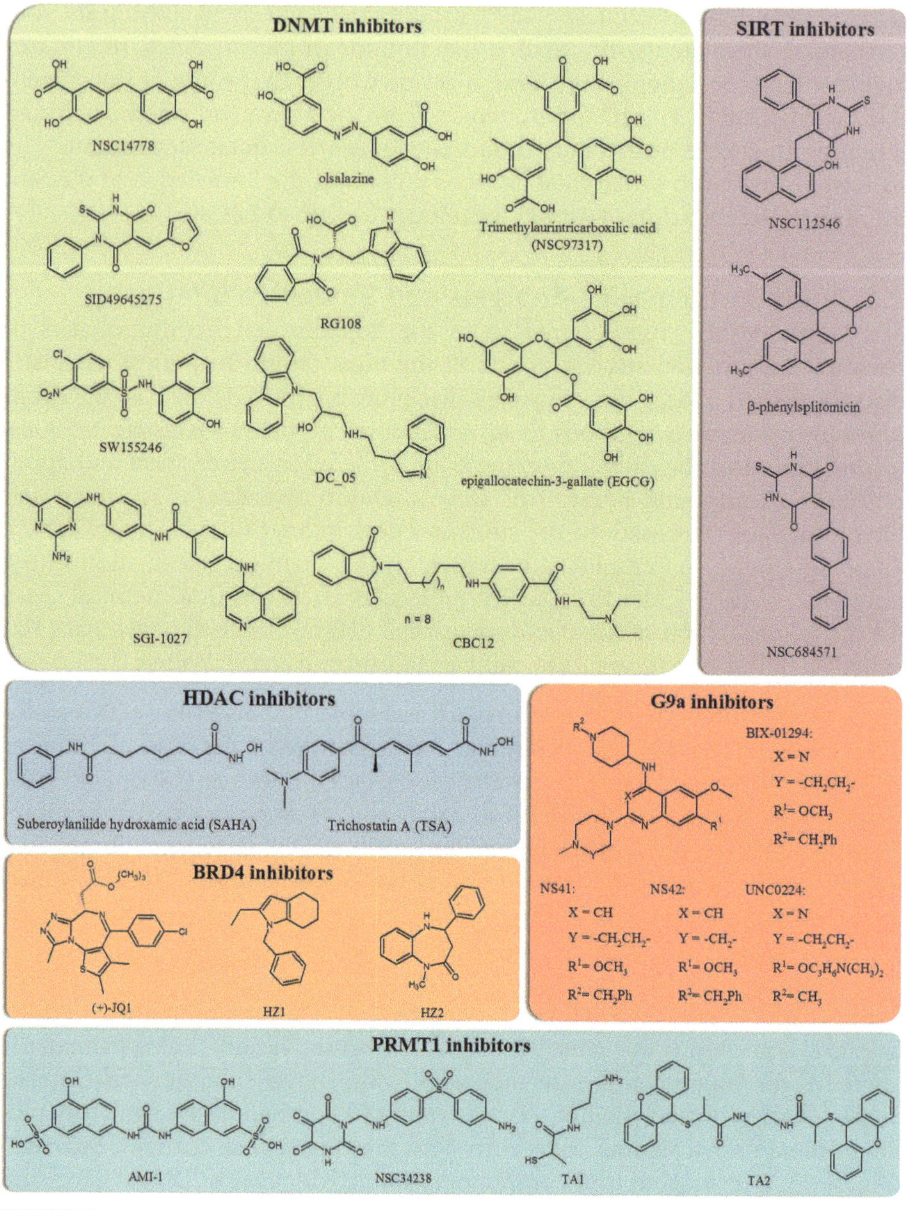

FIGURE 1

Chemical structures of epigenetic inhibitors mentioned in the text.

collections (DNMT-focused, DrugBank, and three natural product databases) to the reference inhibitor **SID49645275** (Figure 1), using MACCS keys and two similarity metrics (*Tanimoto* and *Tanimoto Sub-Structure*) (Medina-Franco and Yoo, 2013a). A total of 562 selected similarity searching hits were subject to further computational screening using docking (described in the Structure-Based Drug Design section). Recently, Méndez-Lucio et al. reported a 3D structure similarity search of a database of approved drugs to a previously identified inhibitor of DNMTs, compound **NSC14778** (Figure.1) (Méndez-Lucio et al., 2014). As a result, olsalazine, an approved anti-inflammatory drug, was identified due to its high structural similarity to that of **NSC14778**. Although the ability of olsalazine to inhibit DNMTs in biochemical assays remains to be determined, it was concluded that it is likely that the drug inhibits more than one DNMT isoform.

Chemoinformatics and similarity techniques have also been used to identify new NAD^+-dependent deacetylase sirtuin (SIRT) inhibitors by searching in commercial databases and subsequent docking of the hits into the binding site. These studies yielded two novel classes of inhibitors based on the high structure similarity with the compounds **NSC112546** (thiobarbiturate scaffold, Figure 1) and β-phenylsplitomicin (splitomicin scaffold, Figure 1). The hits obtained have been employed as leads for the optimization and design of new SIRT inhibitors (Heinke et al., 2011; Lawson et al., 2010).

Hafenbradl and coworkers employed various similarity searching methods for finding inhibitors for the protein lysine methyl transferase (PKMT) G9a with the entire BioFocus compound collection (approximately 900,000 compounds) using as templates **BIX-01294** and **UNC0224** (Figure 1), which bind to the substrate binding pocket (Ahrens et al., 2011). The integrated chemoinformatics-based toolkits used in that work, also called the Chemoinformatics toolbox, included the following: fingerprint similarity searches/Tanimoto metrics (Vogt and Bajorath, 2012), TurboSim (Hert et al., 2005), ring generalization, ring modification, bioisostere enumeration (Wagener and Lommerse, 2006), and Murcko Scaffolds (Bemis and Murcko, 1996). A total of 736 compounds were selected using the chemoinformatics toolbox, of which 50 compounds were identified as hits after verification and only seven were validated experimentally. These final compounds showed IC_{50} values lower than 55 μM.

It is noteworthy that similarity-based searching can also be performed using the Modulator search tool of HEMD, wherein the query compound is matched to the known epigenetic modulators in HEMD (Huang et al., 2012).

2.2 Pharmacophore Modeling

One of the most common strategies employed to screen millions of chemical compounds is the pharmacophore modeling approach, in which the molecule structures are matched to a given pharmacophore (Moura Barbosa and Del Rio, 2012). According to the official IUPAC definition, a pharmacophore is an ensemble of steric and electronic features that is necessary to ensure the optimal supramolecular interactions with a specific biological target and to trigger (or block) its biological response (Wermuth et al., 1998).

Furthermore, pharmacophore screenings on compound databases require all the most probable conformational states of each molecule to match with the different tridimensional pharmacophore features such as hydrogen-bond acceptor (HBA), hydrogen-bond donor (HBD), hydrophobic interactions (HYP), ring aromatic (RA), positive ionizable area (P), negative ionizable area (N), and so on. Additionally, different pharmacophore screening algorithms have been developed with root-mean-square deviation (RMSD)-based and overlay-based scoring functions in order to increase the success of hit compound identification (Sanders et al., 2012). Pharmacophore models have been used in high-throughput virtual screening (HTVS) studies in order to search novel epigenetic inhibitors, some of which are described below.

Vadivelan et al. used 20 well-defined histone deacetylase (HDAC) inhibitors to generate pharmacophore models with the HypoGen module of Catalyst. The best model had four pharmacophore features (HBA, HYP, and two RA centers), which further validation using a large set of 378 HDAC inhibitors gives an *r* value of 0.897 as well as an enrichment factor of 2.68 against a maximal value of 3. The model was further used to retrieve molecules from NCI databases, in which a total 297 molecules were identified as highly active of the 4638 molecules selected from a pool of 238,819 compounds. Nevertheless, the authors suggested carrying out consensus studies to minimize the errors in the false positive and false negative results of the HTVS study and generate a highly predictive tool for estimating biological activity (Vadivelan et al., 2008). Later, Kandakatla and Ramakrishnan used a total of 48 HDAC2 inhibitors to generate pharmacophore models using the HypoGen algorithm module of Discovery Studio. The best model had four pharmacophore features (HBA, HBD, HYP, and RA) validated against 20 test set compounds and employed as a 3D query for HTVS against NCI and Maybridge databases. The selected hits were further employed for molecular docking studies by which eight hit compounds were considered as novel potential HDAC2 inhibitors. From these, compounds **NSC108392** and **MFCD01935795** (Figure 2) showed the best docking scores and estimated activity values (Kandakatla and Ramakrishnan, 2014).

In order to identify novel small molecules as protein arginine methyltransferases (PRMTs) inhibitors, Wang et al. combined pharmacophore-based virtual screening (VS) methods with radioactive methylation assays. Nine hypotheses were generated with a correlation coefficient greater than 0.91 using 17 reported compounds as a training set. Four models with shared features were chosen as queries against the SPECS database, which contained 90,000 compounds, among which top-ranked compounds were selected from 40 groups (120 candidates) and only 102 were tested for inhibitory activity against PRMT1. Finally, six hits were identified as PRMT1 inhibitors within a high inhibitory percentage, of which compounds **AE-848/32000014** and **AE-848/34542007** (Figure 2) showed the lowest in vitro IC_{50} values (41.7 and 12.0 μM, respectively) (Wang et al., 2012a).

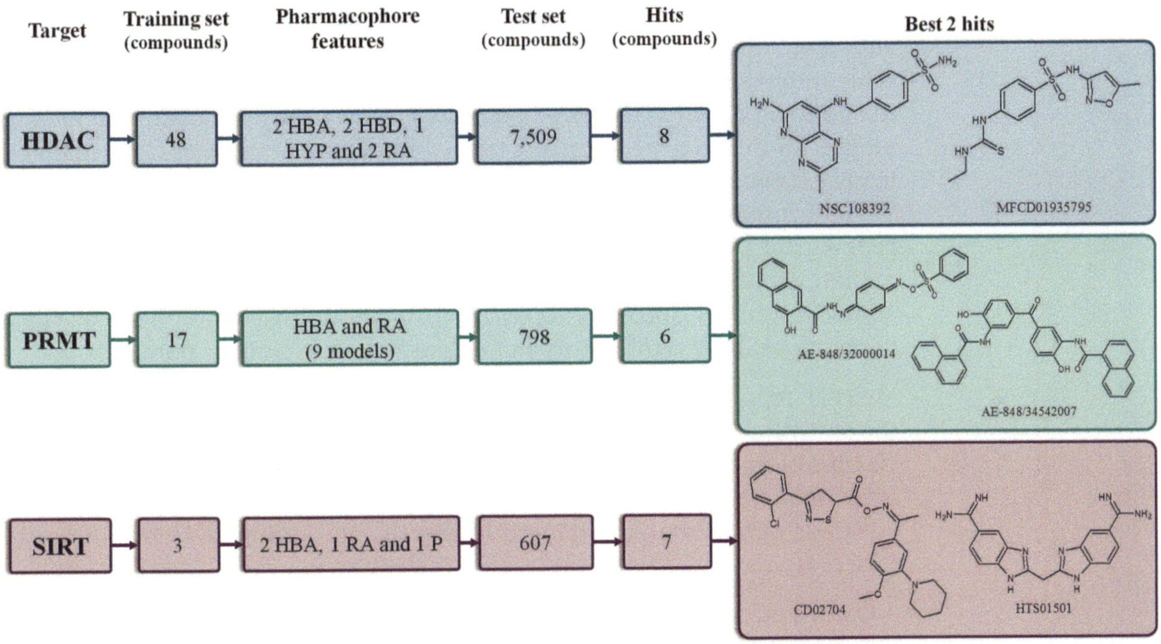

FIGURE 2
Schematic workflows and chemical structures of the hits obtained by pharmacophore modeling approach for three different epigenetic targets.

Sakkiah et al. carried out a pharmacophore modeling study using SIRT1 activators in order to treat age-related disorders like obesity, metabolic syndrome, type II diabetes mellitus, and Parkinson's disease. In that study, the authors employed the HipHop module of CATALYST to identify a series of 3D ligand-based pharmacophore models using three molecules from Sitris Pharmaceuticals as a training set. Five different hypotheses were developed, from which the best model presented four pharmacophore features (two HBA, one RA, and one P) and screened a high number of active molecules from a test set of 607 sirtuin activator molecules. Furthermore, HTVS against the Maybridge database revealed seven molecules that contain all pharmacophore features, among which two new molecules (**CD02704** and **HTS01501**, Figure 2) show similarity with training molecules and could be used as new scaffolds to identify novel SIRT1 activators (Sakkiah et al., 2009).

Furthermore, the pharmacophore modeling approach has been used in conjunction with 3D-QSAR and molecular docking studies (vide infra) in order to improve the identification of novel epigenetic inhibitors.

2.3 Quantitative Structure–Activity Relationships

Classical quantity structure-activity relationships (QSAR) studies are focused on finding a statistical model that gives a mathematical equation for correlating the biological property to structure within a family of compounds. The equation is later used to predict properties such as bioactivity, LogP, toxicity, and so on. Molecular descriptors are used to represent the structure of the molecule into a set of numerical values, some of which are useful for explaining the desired property. 2D-QSAR and 3D-QSAR are the most common methodologies for model generation; the type of employed descriptors is the main difference between them. The first method is independent of the 3D orientation of the compound and describes the constituent entities by topological and geometrical properties, while the second depends on the molecular conformation and alignment for the calculation of descriptors (Dudek et al., 2006; Todeschini and Consonni, 2009). Both methodologies have been useful to design novel HDAC inhibitors and enhance their inhibitory potencies.

Xie et al. reported a comprehensive QSAR study of 124 compounds to identify the structural determinants for HDAC inhibition by using various QSAR and classification methods. A highly predictive QSAR model with r^2 of 0.76 and leave-one-out (LOO) cross-validated r^2 of 0.73 with an overall rate around 92% was obtained and employed to identify and optimize HDAC inhibitors (Xie et al., 2004).

A QSAR study reported by Wang et al. for a series of new trichostatin A (TSA)-like hydroxamic acids for the inhibition cell proliferation of the PC-3 cell line was developed using descriptors from Qikprop and electronic structure calculations. From a "backward-stepping" approach, a QSAR equation with an r^2 of 0.92 and F of 59.25 was developed using the PM3 atomic charge on the carbon atom of the CONHOH moiety (QcO), globularity (Glob), and hydrophilic component of the solvent-accessible surface area (FISA) as independent variables to describe the IC_{50} values of 19 inhibitors. That QSAR study was useful to determine that the PC-3 response was driven by a net contribution of a mixture of different HDAC enzymes as well as a contribution from the transport properties of the evaluated compounds (Wang et al., 2004). Similarly, Dessalew et al. employed the structure and HDAC inhibitory data of a series of aminophenylbenzamides and acrylamide derivatives to carry out a 2D-QSAR study using the tool for structure–activity relationship (TSAR) package. The best model showed a good predictive ability with a cross-validated correlation coefficient of 0.594. The model also explained that the antiproliferative activity is predominantly explained by the steric factors of the substituent (Dessalew, 2007).

3D-QSAR has been one of the most used methods for the study of SARs. CoMFA and CoMSIA methods have been applied in various HDAC inhibitor series (Chen et al., 2003; Guo et al., 2005; Juvale et al., 2006; Liu et al., 2005). The study developed by Juvale et al. was applied on a set of 57 hydroxamate analogs. In that study, the derived models exhibited good conventional and cross-validated values of 0.910 and 0.502 for CoMFA, and 0.987 and 0.534 for CoMSIA, while the CoMFA models exhibited good predictivity with 17 external molecules as compared with that of CoMSIA models. Recently, Bajpai et al. performed a simple multiple regression analysis (MRA) with the same hydroxamate analogues used by Juvale. In this study, the best model showed good conventional, cross-validated (from LOO), and predictive correlation coefficients of 0.925, 0.769, and 0.765, respectively. Similar to the 3D-QSAR study, the resulting equation suggested that the steric properties show great efficiency inhibiting HDAC1 (Agarwal et al., 2013).

Ragno et al. reported the first GRID/GOLPE 3D-QSAR study on four congeneric series of HDAC inhibitors: TSA analogs, suberoylanilide hydroxamic acid (SAHA) analogs, and two series of aroyl-pyrrolyl-hydroxy-amides derivatives. The compound alignments were generated based on the TSA and SAHA (Figure 1) conformations extracted from its crystallographic structures. Twelve 3D-QSAR models were developed and displayed high statistical coefficient values, of which one showed to be a valuable tool for future design of new HDAC inhibitors. Additionally, the molecular docking study showed that the 3D-QSAR maps cannot be employed to extract information about ligand/receptor interactions; however, the model interpretation using the HDAC1 structure allowed comparison between the 3D-QSAR and enzyme residues to have potent HDAC inhibitors (Ragno et al., 2006).

3D chemical-feature-based QSAR pharmacophore models have been developed with the structure–functional requirements for the binding of the HDAC inhibitors. Chen et al. built quantitative pharmacophore models from a training set of 30 hydroxamic acid compounds, in which the best pharmacophore model showed a correlation coefficient of 0.94 using five pharmacophore features (HBD, HBA, and three hydrophobic interactions) and validated by an external test of 25 compounds. That model reflected the structural requirements for interaction with the target, a powerful tool to undertake 3D searches on large databases (Chen et al., 2008). Subsequently, in another study using pharmacophore features, five-point pharmacophore models were generated to derive a predictive atom-based 3D-QSAR analysis of 48 aminophenyl benzamide derivatives as HDAC inhibitors. In that work, Mahipal et al. obtained a model with a correlation coefficient value of 0.99 along with good statistical significance according to a Fisher ratio of 631.80, and a high value of cross-validated correlation coefficient ($q^2 = 0.85$) exhibiting a good predictive power. This model also demonstrated the positive contribution in HDAC inhibitory potency of electron withdrawing, hydrophobic, and hydrogen bond donor moieties in the molecular structure to enhance the activity of this group of molecules (Mahipal et al., 2010).

Katritzky et al. carried out 2D-QSAR and 3D-QSAR studies with the same protein in order to compare both QSAR approaches and correlate the inhibitory activity of a series of indole amide hydroamic acids. The models generated in each of the methods presented good statistical parameters and external predictive power, also showing that the steric effects make a higher contribution to the biological activity (Katritzky et al., 2007).

Clark et al. employed an electronic structure-based algorithm to predict the nature of the expected inhibition of novel oligodeoxynucleotide inhibitors of DNMT1. The electronic structure–activity relationship (ESAR) model was useful to determine the effects of the proposed inhibitors into the methyltransferase action and for deciding whether nucleophilic attack or methyltransfer will be affected in inhibition (Clark et al., 2003).

3. STRUCTURE-BASED DRUG DESIGN

This approach is used when the 3D structure of the current epigenetic target is known. The structure can be obtained using X-ray crystallography, Nuclear Magnetic Resonance (NMR), or Homology Modeling studies. The protein structure is further subjected to molecular docking studies in order to predict the binding mode and binding energy of a compound into its binding site. Additionally, MD simulations have been employed to evaluate the protein structure conformational space and to analyze the affinity of protein-ligand complexes. In the next sections we describe some of the most recent works in which these computational tools were used for the optimization and design of novel epigenetic inhibitors.

3.1 Structural Data of Epigenetic Targets

The HEMD platform contains information for 269 enzymes annotated with a detailed description of epigenetic mechanisms, catalytic processes, and related diseases, also allowing the prediction of conserved features for proteins, and classifies them as ideal targets for experimental validation (Huang et al., 2012). This integrated repository covers four classes of epigenetic enzymes, of which 19 are from DNA methylation, 219 from histone modification, 28 from chromatin remodeling, and three from noncoding RNA. A total of 142 epigenetic enzymes (929 redundant proteins) have been experimentally resolved and extracted from the Protein Data Bank (PDB) (Berman et al., 2000), while the remaining 127 enzymes were constructed using the homology modeling technique. Some homology models contained in this library had also been employed in other studies (Lee et al., 2005; Medina-Franco and Yoo, 2013b; Ragno et al., 2007; Yoo and Medina-Franco, 2011b).

The structural data, classes, and families of some of the most important epigenetic enzymes and proteins associated with the modulation of the epigenetic code have been extensively reviewed elsewhere (Andreoli et al., 2013).

3.2 Docking

Molecular docking is a computational chemistry technique widely used in the protein–ligand complex simulation that is applied to rational drug design. The ultimate goal is to predict the ligand's conformations inside the target, in calculating the potential energy of the protein–ligand complex. However, given the complexity of such systems and the current limitations of our calculation capacity, it is impossible to achieve these goals accurately following, for example, an ab-initio methodology. Still, the low cost of molecular docking allows filtering large chemical compounds databases to identify active and inactive molecules; whereas more accurate, and therefore more expensive, techniques prove to be unmanageable. Thus, docking can be understood as a series of approximations that can generally be divided into two fundamental parts: the scoring function and the conformational search algorithm. Broadly speaking, the scoring function is defined by the surface of potential energy where optimization is carried out. The conformational search consists of a series of different algorithms (i.e., Monte Carlo, Genetic algorithms, Incremental construction, Graph search) that in principle can sample a set of conformations (populations) covering the potential energy search space of the complex. It is important to note that if the scoring function is not sufficiently accurate, it will lead to a deformation of the potential energy surface and therefore to an incorrect estimate of energy and structure. Furthermore, a poor conformational search will result in a loss of the best candidates (Novikov and Chilov, 2009).

Docking has proven useful in a wide variety of applications as in the prediction of whether a ligand will be an enzymatic inhibitor through a competitive mechanism. To achieve this, docking can determine ligand molecular poses, the most important protein–ligand interactions, or act as a constitutive part of VS.

To do docking it is necessary to have the target in a 3D structure, requiring crystallography or nuclear magnetic resonance (NMR). If the information is not available it can also be generated by homology modeling (Erwing and Kuntz, 1997). However, the crystallographic structure itself is not necessarily comparable with the structure present in the organism. One technique that solves this problem is the study and minimization of the target's structure through MD or using any of the minimization algorithms available today. Once the structure of the target and the ligand has been minimized, we must consider that docking only partially takes into account system conformational flexibility, generally allowing the movement of the ligand but not of the structure of the target. Besides, docking does not consider the solvent, which can result in an uncertain characterization of the system, as in the cases where the solvent molecules are part of the protein structure or are essential in the coupling process between the ligand and its binding site (Novikov and Chilov, 2009). If all of these aspects of docking are taken into account, a forecast can be made for an approximate pose within the binding site of interest. Among the most commonly used programs to perform docking are AutoDock (Morris et al., 2009), AutoDock Vina (Trott and Olson, 2009), and DOCK (Erwing et al., 2001).

Several docking studies related with epigenetic targets have been published (Ragno et al., 2004; Ran et al., 2015; Subha and Kumar, 2008; Vidler et al., 2013; Wang et al., 2005; Zhao et al., 2014). For example, Aldawasari et al. reported a series of resveratrol derivatives that present selectivity inhibition against 3A and 3B isoforms of DNMT. This study showed the favorable interactions between the catalytic and cofactor domains of DNMT3 and such derivatives, also compared with the DNMT1 domains. From these results, two selective and potent DNMT3A inhibitors, called **FSA9** and **FSA10** (Figure 3) proved to be more potent than resveratrol (Aldawsari et al., in press).

It is a common practice that experimental studies use docking to support or explain some of their results (a posteriori modeling). An example of this is the work developed by Fang et al., which used this technique to corroborate experimental results. Specifically, the authors reported that (−)-epigallocatechin-3-gallate (**EGCG**, Figure 1), one of the major components of green tea, is an inhibitor of DNMT and can reactivate silenced genes in cancer cells. Through the poses and the energy obtained from MD calculation, they also suggested that this effect is achieved by the formation of hydrogen bonds between different residues in the active site of DNMT (Fang et al., 2003).

Another instance that reflects the usefulness of docking can be seen in the work of Medina-Franco et al., which corroborates that one of most powerful DNMT inhibitors found through high throughput screening (**SID49645275**) actually binds to the site of this enzyme cofactor (Medina-Franco and Yoo, 2013a).

Docking has also been used to support the determination of ligand crystallographic structures. For example, Chang et al. reported the crystal structure of the catalytic SET domain of G9a-like protein (GLP) in complex with **BIX-01294** and S-adenosyl-L-homocysteine (Chang et al., 2009). Therein, they evaluated the relative energies of the possible boat–boat, chair–boat, and chair–chair conformations with respect to the piperidine and the diazepane rings of compound **BIX-01294**. The lowest energy chair–chair conformer was submitted to extra-precision flexible Glide docking (Friesner et al., 2004) of the protonated forms with the binding site of GLP. The binding mode identified with this conformer showed a potential hydrogen bond between the polar hydrogen of the protonated piperidine and Asp1131, which is consistent with the experimental electron density (PDB: 3FPD).

Based on the GLP–**BIX-01294** and G9a–**UNC0224** crystallographic structures, Srimongkolpithak designed and synthesized a variety of novel G9a inhibitors in order to better define the pharmacophoric features associated with the core heterocycle (Srimongkolpithak et al., 2014). Those efforts resulted in the identification of two potent and selective G9a/GLP inhibitors based on a quinoline scaffold, herein named **NS41** and **NS42** (Figure 1). The proposed binding

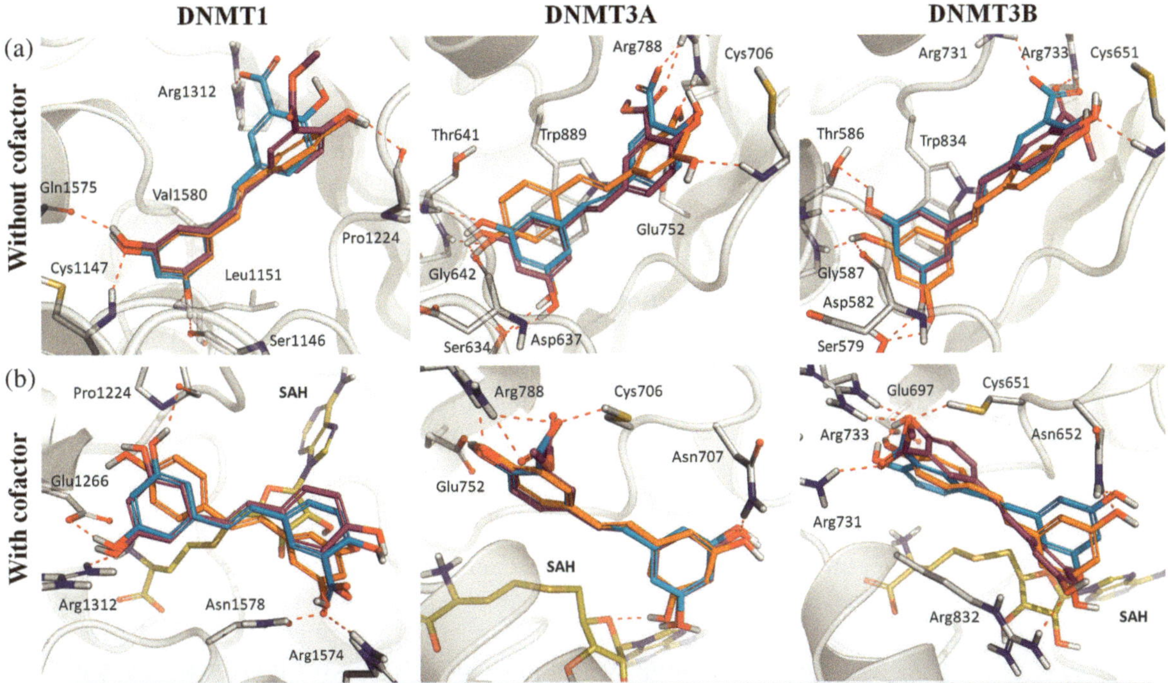

Compounds	DNMT1			DNMT3A			DNMT3B		
	IC_{50} (μM)	Docking score		IC_{50} (μM)	Docking score		IC_{50} (μM)	Docking score	
		A	B		A	B		A	B
FSA9	> 300	-8.13	-7.67	40	-9.66	-8.17	52	-8.85	-7.54
FSA10	> 300	-7.43	-6.42	25	-9.27	-7.01	62	-8.99	-8.82
Resveratrol	> 300	-7.74	-5.83	105	-9.03	-6.57	65	-7.95	-6.22

FIGURE 3

Comparison of the binding modes calculated for compounds FSA9 (blue), FSA10 (purple), and resveratrol (orange) as predicted by AutoDock 4.2, in the absence (a) and in the presence (b) of the cofactor SAH (yellow) in the active site of DNMT1, DNMT3A, and DNMT3B enzymes. The table beneath the figure summarizes the docking scores and the IC_{50} values using a cell-free biochemical assay *(Adapted from Aldawsari et al., in press).*

modes, validated through the crystallographic binding pose of **UNC0224** (PDB: 3K5K) (Liu et al., 2009), showed the importance of maintaining the dimethoxy groups on the benzenoid ring and the protonated basic nitrogen at Position 1 for the hydrogen bond interaction with G9a.

In the initial phases of the design of HDAC competitive inhibitors, Butler K. V. (Butler et al., 2010), has proved that docking is a helpful tool in modeling the sirtuin cofactor NAD (Trapp et al., 2006). Other epigenetic targets to be studied using this technique included the Histone Acetyltransferases (HAT) and the Histone Methyltransferases (HMT) (Angeles et al., 2005).

Using an efficient shape-based search algorithm and the AScore scoring function, Subha and Kumar carried out a molecular docking study over four class I HDACs with 12 known inhibitors currently under the phase I/II trials (Subha and Kumar, 2008). From these compounds, only TSA satisfied almost all the properties (i.e., binding affinity score, drug likeness value, and drug score), also showing the best binding mode in which the hydroxyl group fits well into the active site. Based on this, the authors suggest the design of better hydroxamate derivatives as potential inhibitors of class-I HDACs.

The output of this technique is useful as preliminary information to determine the reaction mechanics of a known active compound against a specific target. Such a study was reported by Singh et al., showing the key DNMT amino acid residues that interact with hydralazine as well as the responsible functional groups in the energetically favorable interaction (Singh et al., 2009).

3.2.1 Filtering Methods

Filtering methods include a series of computational tools that allow the treatment of larger databases of chemical compounds. These algorithms are capable of selecting those chemical compounds that meet certain criteria such as biological activity, toxicity, structure, physicochemical properties, and others, according to the particular research objectives. There are two general types of filtering strategies. One strategy contemplates exhaustive computational methodologies of the entire database in question, but they are expensive and so their use will depend on the resources, the size of the database, and the research objectives. The other strategy refers to the low-cost computational methods initially applied to the database in order to reduce the size of the molecular set, but this may cause losing information due to the low demands of the methods and thus requires the application of accurate methods to the resulting subset. The most common techniques used in filtering methodologies are consensus docking and docking-based virtual screening (DBVS).

3.2.1.1 Consensus Docking

VS is a robust strategy for discovering macromolecule inhibitors. This method can be divided into a series of steps that includes one of the aforementioned docking techniques. The accuracy in the calculation of affinity in these methods is closely linked to the predicted poses. A problem that might arise from this is that different poses can be associated with similar affinities, which makes it difficult to have a criterion of pose selection. Therefore, a wrong prediction in VS poses will lead to false positives and consequently a decrease in the number of hits found (Tuccinardi et al., 2014).

Pose prediction using docking programs has a reported success rate of approximately 59%, whereas with a combination of different methodologies, pose and affinity prediction, can reach an 80% success rate. The simplest way to combine results called consensus docking is as follows:

1. Choose two docking methodologies that use different search algorithms and scoring functions.
2. Calculate poses and scoring with both methodologies.
3. Find clustering based on RMSD (approx. 2 Å).
4. For the best candidates, calculate RMSD with regards to the conformation present in some cocrystallized ligands.
5. Calculate an average scoring value for the representative structures of the best clusters.

To debug the results, the affinity of the best candidates can be calculated by using a more expensive methodology like MD, semi-empirical methods, among others (Houston and Walkinshaw, 2013).

Epigenetic target studies using consensus docking so far have been used primarily to search for inhibitors against DNMTs. For instance, we can refer to the studies made by Kuck et al. in which the initial step was to perform a VS search for about 260,000 compounds from the National Cancer Institute (NCI) database. The filtering method output a subset of 65,000 compounds, of which a selection was made of the lead-like molecules. A calculation of poses and scoring for each of the lead-like molecules was performed with three different docking methodologies. At the end of the process they reported 13 compounds, of which seven showed activity against DNMT1 and the remaining six presented selective inhibition since they do not inhibit DNMT3B (Kuck et al., 2010).

Consensus docking as a part of the VS methodology has proven to be an effective tool in the systematic search of active natural products against epigenetic targets. This finding was reported by Medina-Franco and Yoo (Medina-Franco and Yoo, 2013b), who identified a group of 53 consensus hits based on three different docking methodologies. Notably the methodology identified a methylenedisalicylic acid derivative that had already been found using VS on the NCI database (Kuck et al., 2010).

Increasing the VS pose prediction success based on linking different docking methodologies has proved to be a valuable tool in the search for computer hits. Also with the combination of more exact experimental or computational methods this methodology significantly reduces costs for pharmaceutical research, particularly by laboratories with limited resources.

3.2.1.2 Docking-Based Virtual Screening

Several techniques currently facilitate the identification of hits. high-throughput screening (HTS), for example, is frequently used as a means of testing activity on thousands of compounds. Nevertheless, it has the limitation of being highly expensive, prompting a search for lower-cost alternatives. This has led to a computational methodology called VS, which is faster and less expensive than HTS. VS is normally used in a stage before HTS in order to decrease the size of the molecular set that will be tested experimentally (Cosconati et al., 2010).

VS can be classified as either ligand-based or receptor-based. In the case of receptor-based VS, the structure of the target in question must be known in advance from homology models or experimental methods (X-ray crystallography and NMR) (Erwing and Kuntz, 1997).

The most commonly used programs to carry out the molecular coupling for VS for epigenetic targets are DOCK (Erwing et al., 2001), FlexX (Claussen et al., 2001), GOLD (Verdonk et al., 2003), AD (Morris et al., 2009), ICM (Abagyan et al., 1994), and LigandFit (Venkatachalam et al., 2003), among others (Tuccinardi, 2009).

With the 3D experimental structure of the receptor, the use of compound databases, and any of the programs mentioned above, it is possible to perform VS-based docking and to select the best candidates in terms of scoring, RMSD, or both of these criteria. Examples of compounds databases include ZINC (Irwin et al., 2012), CHEMBL (Gaulton et al., 2011), Binding Database (Liu et al., 2007), and HEMD (Huang et al., 2012).

Spannhoff et al. carried out the first target-based approach to PRMT inhibitors using homology models of human and *Aspergillus nidulans* PRMT1 (Spannhoff et al., 2007a). A total of 1630 compounds, retrieved from the NCI diversity set, were docked into the binding pocket of the fungal PRMT model with the program GOLD. Using a pharmacophore model and visual interaction inspection, 36 compounds were selected and tested afterward in an in vitro assay for their ability to inhibit human PRMT1. From these, compounds **AMI-1** and **NSC34238** showed the lowest in vitro IC_{50} values (1.2 and 1.7 μM, respectively), but higher concentrations of these are required for cellular effects. Additionally, the docking results showed that **AMI-1** is likely to interact with the cosubstrate binding site rather than with the substrate pocket, as compound **NSC34238** does.

The same research group identified a thioglycolic amide (compound **TA1**, Figure 1) as a PRMT inhibitor via the fragment-based virtual screening (FBVS) approach (Spannhoff et al., 2007b). To do this, they filtered the Hans-Knöll-Database (HKI, including about 9000 compounds) and docked the resulting compounds (900 structures) into a homology model of a fungal PRMT, using the GOLD program. As it is chemically not very stable and is not a druglike molecule, the authors searched for structurally related compounds that also contain an α-methylthioglycolic amide substructure. Based on this, an in-house molecule (compound **TA2**, Figure 1) was identified as an inhibitor of both fungal and human PRMTs, also suggesting a mode of bisubstrate inhibition, which opens up interesting possibilities for future structural variations toward new potent and selective PRMT inhibitors.

Vilder et al. employed similarity searching, literature KAc mimetics, substructure searching, and docking to identify novel potential bromodomain (BRD) type 4 inhibitors based on the structure and binding mode of **(+)-JQ1** (K_d = 50 nM, Figure 1) (Vidler et al., 2013). With these results, the authors selected and tested 143 lead compounds, discovering six hits with moderate potency. Furthermore, they solved the crystal structure of four hits (PDB IDs: 4MEP, 4MEN, 4MEQ, and 4MEO), determined the binding mode, and improved their potency through the synthesis of analogous compounds.

Using fragment-based VS, Zhao et al. identified compounds **HZ1** and **HZ2** (Figure 1) as two novel small molecules that bind to the first bromodomain of BRD4 with low micromolar affinity (IC_{50} = 7.0 and 7.5, respectively) and favorable binding free energies, also showing to be selective over other families of bromodomains (Zhao et al., 2014). This information is valuable to further develop novel compounds for the treatment of hematological malignancies and solid cancers, for which GRD4 is a therapeutic target.

Chen et al. performed a VS-based docking study for DNMT1 protein (PDB ID: 4DA4) with Glide and a database of 111,121 compounds (SPECS) previously selected for its physicochemical properties. Activity testing was carried out on the resulting 51 compounds, finding that the compound (**DC_05**) was active against DNMT1 with an IC_{50} of 10.3 μM. Using compound **DC_05** they carried out selectivity tests on other isoforms and a structural similarity search study (Chen et al., 2014).

On DNMT, Siedlecki et al. conducted VS on a subset of chemical diversity (NCI Diversity Set) with 1990 compounds. The compounds were selected on the basis of criteria such as hydrophobicity, size, and the presence of rare atoms. Of seven hits two were tested in vitro, showing an activity between 10 and 100 μm, of which compound **RG108** revealed novel characteristics and substantial developmental potential as an antitumor drug (Siedlecki et al., 2006).

Kuck et al. presented a VS methodology based on consensus docking by screening the NCI database in a DNMT1 homology model. Of 260,071 compounds, those that were non–lead-like were removed, leaving a subset of 65,375. HTVS docking and Glide standard precision was subsequently applied, selecting those compounds with a score ≤−5.9 and ≤−7.5, respectively. These steps resulted in a subset of 227 compounds on which consensus docking was applied using

Glide XP, Gold, and AutoDock. At the end 24 consensus hits were obtained, 13 of which were experimentally tested and compared to the scores obtained for **RG108** (a reference inhibitor of DNMT1, Figure 1) (Kuck et al., 2010). Experimental tests performed with DNMT1 and DNMT3B show that six of the 13 compounds obtained by the VS methodology are selective against DNMT1 but with a low potency. This methodology also found new scaffolds that had not been reported in literature as inhibitors of DNMT. The new scaffolds can be used for experimental optimization or as a starting point for chemical space exploration.

3.2.2 Docking Considering Protein Flexibility

Despite the usefulness of the rigid docking approach, it is well known that protein flexibility plays an important role in the protein-ligand recognition due to the different conformation that the receptor adopts upon ligand binding. These conformational changes can range from minor arrangements (for example, side chain rotations) to the complete rearrangement of the tertiary structure of the protein. To address this issue, different computational approaches have been employed for the design and evaluation of epigenetic inhibitors and are described elsewhere (Cavasotto and Abagyan, 2004; Sherman et al., 2006; Sinko et al., 2013).

Saha et al. carried out flexible docking studies with human histone deacetylase (HDAC8) using a series of triazole-linked hydroxamic acid derivatives (Saha et al., 2010). The lowest energy conformers of each molecule were compared with the experimental bound SAHA coordinates. From these, only two analogs adopt an acceptable conformation within the active site; however, neither had a lower energy than the reference compound.

Synthetic analogs of the VS hit **NSC319745** gave rise to the identification of four compounds with inhibitory activity against DNMT3A (Kabro et al., 2013; Kuck et al., 2010). Among them, the compound **AK1** (Figure 4(a)) showed an EC_{50} of 36 μM against DNMT3A. A flexible docking study suggests that this compound has a high affinity to SAM pocket and can be used as a lead in the development and optimizations of new selective inhibitors against DNMT3A (Figure 4(b)).

Yoo and Medina-Franco performed a flexible docking study of the inhibitor trimethylaurintricarboxylic acid (**NSC97317**, Figure 1) with the catalytic domain of DNMT1 (Yoo and Medina-Franco, 2012b). The docking study showed that **NSC97317** can bind into the catalytic site by interaction with the amino acid residues that participate in the mechanism of DNA methylation. Furthermore, the best docking poses were selected and compared to the structure-based pharmacophore described above, in which only one carboxylic acid group of the compound matches with the negatively charged feature of the model (considering a distance matching tolerance of 2.0 Å) (Yoo and Medina-Franco, 2011b).

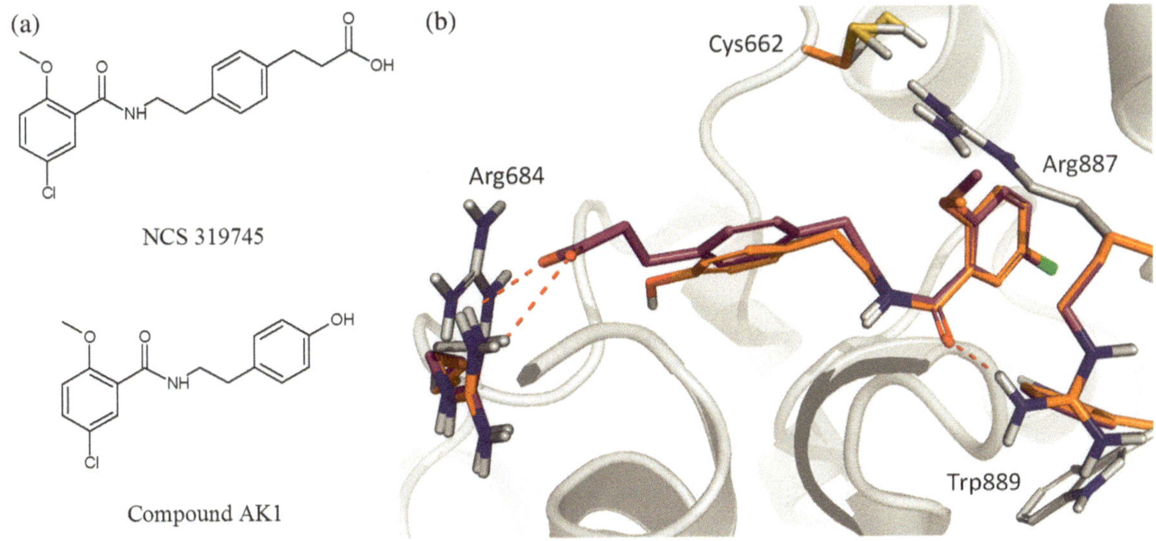

FIGURE 4
(a) Chemical structures and (b) binding mode of NCS319745 (purple) and compound AK1 (orange) as predicted by AutoDock Vina in the active site of DNMT3A.

Induced fit-docking (IFD) has been one of the most employed flexible docking methods for the modeling of novel DNMT inhibitors. In this technique, the receptor is modeled with flexibility to accommodate the induced fit associated with ligand binding. IFD is useful when the DNMT conformation must change to accommodate the correct binding conformation of the ligand (Sherman et al., 2006).

Yoo et al. conducted an IFD of **SGI-1027** and **CBC12**, compounds with long scaffolds, with the cofactor and substrate binding site of the MTase domain of human DNMT1 and DNMT3A in the presence and absence of other domains (Yoo et al., 2013). Docking with DNMT1 revealed that both compounds are able to occupy the cofactor- and substrate-binding sites depending on the presence of other domains. Moreover, **SGI-1027** occupies the cofactor binding site in the binding model with DNMT3A, whereas **CBC12** is docked in the substrate binding site as well as the cofactor site. The docking study also shows that the binding score of **SGI-1027** is more favorable than the corresponding score of SAH and that this is a nonselective inhibitor. These results were in excellent agreement with the experimental results.

Recently, Medina-Franco and coworkers reported the binding model of **SW155246** with human DNMT1 that explains the inhibitory activity using an IFD protocol previously reported (Medina-Franco et al., 2014). IFD had a significant impact on the docking scores and the predicted binding modes as compared to rigid docking; showing that compound **SW155246** preferably interacts with the cofactor binding site as well as the catalytic site. Asgatay et al. also carried out an IFD study of two constrained analogs of **RG108** using a crystallographic structure of murine catalytic DNMT1, which also showed an interaction with both binding sites (Asgatay et al., 2014).

Medina-Franco et al. suggested that other approaches could be used in future work for modeling protein flexibility using conformations of human DNMT1 obtained through crystallographic information or MD. Currently, MD has been used to analyze larger conformational changes of the protein due to the binding of a ligand through a determined period of time. This tool has been used to evaluate the stability and affinity of a ligand in the protein-ligand complex in order to observe the most important interactions and changes in the binding mode. The basis of this technique and some examples with epigenetic targets are described in what follows.

3.2.2.1 Ensemble Docking

Proteins are dynamic systems and the changes produced by their movement are related to their activity and molecular recognition. In most of the traditional docking simulations the flexibility of the receptor is not taken into consideration due to the high computational cost involved and the lack of knowledge about the best way to include this behavior within simulations. Unfortunately when docking is performed on a single conformation, prediction of the coupling poses are incorrect between 50% and 70% (Nichols et al., 2011; Totrov and Abagyan, 2008).

In order to take into account protein flexibility during molecular docking studies without the computational cost that this entails, a methodology called "ensemble docking" has been proposed (Osguthorpe et al., 2012). This methodology consists of performing MD simulation on the protein structure and exploring the conformational space of a biomolecule. On the obtained trajectories, representative conformations are selected by using clustering methods.

Molecular docking is performed for each of the representative structures, with the best poses selected through RMSD or energies criteria obtained by a scoring function (Campbell et al., 2003; Nichols et al., 2012). The conformations used for docking do not necessarily have to be obtained from MD. The study also can be based on reported crystallographic structures of the same protein and the corresponding principal component analysis (PCA) to map the major conformational differences between these structures and to find a group of numerous conformations of the target (Craig et al., 2010).

The use of ensemble docking is exemplified in the work done by Yoo et al. where the combination of classic docking and VS and ensemble docking was conducted with the ligand **SGI-1027** on DNMT3A and DNMT1 isoforms in the presence of an SAH cofactor, identifying the best poses from residues that interact through hydrogen bonding with the functional groups of the ligand (Yoo et al., 2013).

3.3 Molecular Dynamics

MD simulation provides the necessary tools to analyze a system's atomic movement on a microscopic scale under the approximations of classical mechanics. This is a powerful and widely used tool in medicinal chemistry for biomolecular simulation because it allows the evolution of biological systems to be monitored over time and produces approximations of static and dynamic properties (Karplus and McCammon, 2002).

The observed phenomena on simulation depend directly on time, and this may pose a limiting factor since the computational cost of the simulation is directly proportional to the time of simulation. This is because each of the phenomena that occur at a particular time scale depends on those that occur in minor scales (see Table 2).

Thanks to the accessibility of MD programs such as AMBER MD (Wang et al., 2006) and GROMACS (Pronk et al., 2013), the power of computational calculation and different protein structural information, nanosecond (ns) MD simulations have become common and simulations can even be carried out in microseconds (μs) (Pronk et al., 2013).

Given the huge population of possible conformations derived from the high number of degrees of freedom in a macromolecule, crystallographic information represents only a static picture. MD simulation has been used as a tool to take into account these changes that are known to be closely associated with the formation of complexes and biological activity (Becker et al., 2001; Del Sol et al., 2009).

MD has been used to simulate the conformational changes that occur in different enzymes related to epigenetic regulation. As mentioned above, these changes are linked directly with the biological activity and gives clues that help to define the fundamental differences between two or more isoforms that can be exploited for selective molecule design that potentially reduce side effects. Deschamps et al. reported an example where changes

Table 2 Examples of Characteristic Movements Occurring in a Protein

Type of Movement	Time (s)
Bond vibration	10^{-15}
Bond rotation	10^{-12}
Side chains movement at the protein surface	10^{-10}
Ion channel transport	10^{-9}
Allosteric transitions	10^{-11}–10^{-7}
Protein folding	10^{-5}–1
Membrane proteins folding	10^{3}

in flexibility and conformation of Histone deacetylase isoforms invisible to X-ray crystallography were obtained by MD simulation in order to be considered for future strategies to design new selective ligands. Some examples of the characteristics that can be determined from MD simulation to improve the HDAC drug design include structural water molecules, plasticity of HDAC catalytic channels, dynamics of open and close subchannels, substrate recognition, ligand accommodation and release, and interconversion between open, subopen, or close catalytic sites (Deschamps et al., 2015).

MD can also be used to determine the behavior of an inhibitor in a biological system. For example, Caulfield et al. used an MD simulation of a DNMT3B isoform homology model to report the behavior of the selective inhibitor nanaomycin A with and without cofactor S-adenosyl methionine (SAM). They draw the conclusion that the presence of SAM is fundamental to the inhibitor stabilization through hydrogen bonds in the coupling site (Caulfield and Medina-Franco, 2011). This information also lays the foundations for further analysis focused on the mechanism of action.

Medina-Franco and Yoo found that MD was valuable to refine the ligand's poses obtained through docking in DNMT inhibitor design. They also claimed that starting from MD and crystallographic information of DNMT, active conformations could be obtained, which served as a starting point in molecular coupling analysis (Medina-Franco and Yoo, 2013b).

Based on the best poses obtained for selective HDACS isoforms inhibitors with experimental reported activities, Thangapandian et al. performed MD simulations to determine the specific interactions between the inhibitors and the protein (Thangapandian et al., 2012). The authors found that the differences in experimental activities between the inhibitors could be explained by the presence of one hydrogen bond interaction between the metal binding part of inhibitors and the aromatic interactions observed in the tunnel and surface of the active site. Moreover, changes of one amino acid in different isoform active sites explained the selectivity between inhibitors, which is a very useful information for the design of new HDACS inhibitors.

MD has been used in the preparation and assistance of experimental optimization strategies. Among the optimization studies conducted with MD to develop more selective and less toxic inhibitors, we find the research made by Tambunan et al. (Tambunan et al., 2013). Through known HDAC Class II *Homo sapiens* inhibitor structure suberoylanilide hydroxamic acid (SAHA) used in cancer therapy, Tambunan et al. proposed structural changes in SAHA based on triazole substituents. From MD binding free energies the researchers were able to discriminate between modifications that improve SAHA affinity and those who were less likely to be effective.

Through MD simulations Zhao and Caflisch were able to identify alternative binding modes that had not been observed in crystallographic structures of human bromodomains reported before (Zhao and Caflisch, 2015). In the conserved acetyl-lysine binding site of bromodomains six structural water molecules were present. The authors performed explicit solvent MD simulations in the presence of high concentrations of dimethylsulfoxide, methanol, and ethanol, showing that several water molecules can be displaced. Subsequent analyses lead the researchers to two main conclusions: the existence of a metastable state corresponding to multiple poses and the importance of solvent molecules in molecular recognition.

Table 3 summarizes the compounds identified by the computer-aided drug design approaches for the different epigenetic targets mentioned in this chapter.

Table 3 Compounds Identified to Date for Epigenetic Targets Using Computational Approaches

Target	Compound	Approach	References
DNMT	Olsalazine	Similarity search	Méndez-Lucio et al. (2014)
	RG108	Docking-based virtual screening	Siedlecki et al. (2006)
	NSC14778		Kuck et al. (2010)
	NSC319745		
	DC_05		Chen et al. (2014)
	NSC97317	Docking	Yoo and Medina-Franco (2012b)
	SW155246		Medina-Franco et al. (2014)
	SGI-1027	Induced fit-docking	Yoo et al. (2013)
	CBC12		
SIRT	NSC684571	Similarity search/docking	Heinke et al. (2011)
	CD02704	Pharmacophore modeling	Sakkiah et al. (2009)
	HTS01501		
HDAC	NSC108392	Pharmacophore modeling/docking	Kandakatla and Ramakrishnan (2014)
	MFCD01935795		
PKMT G9a	NS41	Docking	Srimongkolpithak et al. (2014)
	NS42		
BRD	HZ1	Docking-based virtual screening	Zhao et al. (2014)
	HZ2		
PRMT	AE-848/32000014	Pharmacophore modeling/docking	Wang et al. (2012a)
	AE-848/34542007		
	AMI-1	Docking-based virtual screening	Spannhoff et al. (2007a)
	NSC34238		
	TA1		Spannhoff et al. (2007b)
	TA2		

4. COMBINING METHODS

The combined use of computational techniques improves the analysis and selection of compounds in different studies. Some of the most common combinations involve docking techniques with other methods, as shown in the following subsections.

4.1 Docking: 3D-QSAR

Guo et al. conducted docking simulations on a series of 29 substituted hydroxamic acid-based HDAC inhibitors into the active site of human HDAC1 (Guo et al., 2005). Based on the binding mode, predictive 3D-QSAR models were generated, which had conventional r^2 up to 0.982 for CoMFA and 0.954 for COMSIA. Both models have similar predictive capabilities to identify novel molecules with improved activities.

Ragno et al. conducted docking simulations and 3D-QSARs analysis on four series of HDAC inhibitors (Ragno et al., 2006). Twelve 3D-QSAR models were obtained with high statistical coefficient values. The predictive capability of the models was evaluated through two external test sets composed of seven and 12 newly compounds, selecting the most predictive model with SDEP values of 0.96 and 0.99, respectively. The model interpretation was employed in order to correlate the ligand-based and structure-based approaches for the design of more potent HDAC inhibitors. Furthermore, the same group employed three different binding modes of a focused library of small molecule compounds into the catalytic domain of the human PRMT1 in order to represent the structure-based aligned training sets to build 3D-QSAR models (Ragno et al., 2007). The 3D-QSAR models were derived using the GRID/GOLPE procedure. The cross-validation data of the final three models were statistically robust with q^2 and SDEP values ranging from 0.88 to 0.93 and from 0.12 to 0.40, respectively. Therein, the docking and the 3D-QSAR studies allowed a deep binding mode analysis of tested molecules.

Docking simulations into HDAC1and HDAC8 binding sites of about 40 ligands that had been tested for their inhibitory activity against the two isoenzymes were conducted by Ortore et al. (Ortore et al., 2009). The best binding poses were employed as an alignment tool for the development of a 3D-QSAR model. The 3D-QSAR models using C_3, N_2, and O probes showed good correlation coefficient (r^2 ranging from 0.921 to 0.946) and a moderate predictive correlation coefficient (q^2 values from 0.690 to 0.793). This information was useful to extract quantitative information of the ligand–receptor interactions.

Yang et al. reported docking simulation and 3D-QSAR analysis on a series of novel lactam-based HDAC inhibitors (Yang et al., 2012). The conformations of the inhibitors were obtained from the docking model in the active site of hisonte deacetylase-like protein (HDLP) and compared with **TSA** structure. The QSAR equation (training set: $r^2=0.77$, test set: $r^2=0.76$) present six principal descriptors that can be used to improve the HDAC inhibitory activity of lactam-based analogs. This QSAR equation also suggests that small HDAC inhibitors with big surface area of stabilized aromatic cap group would show better HDAC inhibitory activity.

4.2 Docking: Pharmacophore Modeling

Tervo et al. generated a simple pharmacophore query to select potential SIRT2 hits. A receptor-based virtual screening (RBVS) returned 66 candidate compounds (Tervo et al., 2004). The hits were subsequently docked in the SIRT2 binding pocket in order to visually analyze the binding poses and the ability to match the requirements of the search queries. The docking study and pharmacophore model were useful to identify important regions for the inhibitor binding.

Thangapandian et al. generated two different pharmacophore models using ligand- and receptor-based approaches. These models were validated for their ability to distinguish active compounds from inactive compounds (Thangapandian et al., 2010). Binding conformations of docked compounds were analyzed for the essential interactions with the HDAC8 catalytic site. From this study, three compounds were identified as novel leads to be employed as potent HDAC8 inhibitors.

Yoo and Medina-Franco explored the binding mode of 14 DNMT1 inhibitors with a homology model of the catalytic domain of hDNMT1 (Yoo and Medina-Franco, 2011b). The optimized binding modes of the inhibitors were used to construct a structure-based pharmacophore model with five distinct features that is in agreement with the binding profile of the inhibitors obtained in docking. This pharmacophore model was used subsequently to characterize the binding model of trimethylaurintricaboxylic acid (Yoo and Medina-Franco, 2012b). Furthermore, this group also developed a structure-based pharmacophore hypothesis for 16 inhibitors using a similar approach (Yoo et al., 2012). In that study, the crystallographic structure of the catalytic domain of human DNMT1 was used to develop the binding poses and generate the pharmacophore model. The new model contains five features similar to the features of the previous developed pharmacophore model. These results supported the use of the structure-based pharmacophore models to conduct VS.

5. CONCLUDING REMARKS

CADD has become an indispensable tool to accelerate the development of epigenetic inhibitors helping in the selection, design, and lead identification of novel compounds. LBDD approaches employ the chemical structures of experimentally established epigenetic inhibitors to conduct SAR studies. From this, chemoinformatics and similarity tools have been used to identify new epigenetic inhibitors by searching in commercial and focused databases. Pharmacophore and QSAR models have been employed to find statistical models to correlate the biological property to the chemical structure and screen millions of chemical compounds to match these models. On the other hand, SBDD approaches use 3D structures of the current epigenetic target to carry out docking and MD studies. Docking has been used to propose the possible binding mode of the compounds that were found with LBDD techniques, to explain the activity of compounds evaluated experimentally, and to screen a large number of compounds obtained from chemical databases. Additionally, filtering methods and protein flexibility considerations have been employed to select more potential epigenetic inhibitors. Finally, MD has been used to simulate the conformational changes that take place within different epigenetic targets with and without the inhibitor, which allows evaluating the stability of it in the binding site of the target.

These developments are closely related to calculation power, macromolecules structural information, and the biochemical knowledge of concern to epigenetic regulation mechanisms. It is expected that these three components grow up continuously over time, promising that CADD will take an increasingly important role in medicinal chemistry to produce high-selective, low-toxic, and more potent bioactive molecules for the treatment of related epigenetic mechanisms diseases like cancer, psychiatric disorders, and cardiac illness.

ACKNOWLEDGMENTS

R. Aguayo-Ortiz and E. Fernández-de Gortari are grateful to CONACyT for the fellowships granted No. 510728/288862 and No. 348291/240072, respectively. We thank Yannick Bautista González for the English proof of the manuscript and for helpful suggestions. We also want to thank Frida S. Herrera Vázquez and Rodrigo M. Fernández de Gortari for their support.

REFERENCES

Abagyan, R., Totrov, M., Kuznetsov, D., 1994. ICM—a new method for protein modeling and design: applications to docking and structure prediction from the distorted native conformation. J. Comput. Chem. 15, 488–506.

Agarwal, N., Bajpai, A., Srivastava, V., Gupta, S.P., 2013. A quantitative structure-activity relationship and molecular modeling study on a series of biaryl imidazole derivatives acting as H^+/K^+-ATPase inhibitors. Biochem. Res. Int. 1–15.

Ahrens, T., Bergner, A., Sheppard, D., Hafenbradl, D., 2011. Efficient hit-finding approaches for histone methyltransferases: the key parameters. J. Biomol. Screen. 17, 85–98.

Aldawsari, F.S., Aguayo-Ortiz, R., Kapilashrami, K., Yoo, J., Minkui, L., Medina-Franco, J.L., et al. Resveratrol-salicylate derivatives as selective DNMT3 inhibitors and anticancer agents. J. Enzym. Inhib. Med. Chem. (in press).

Andreoli, F., Moura Barbosa, A.J., Parenti, M.D., Del Rio, A., 2013. Modulation of epigenetic targets for anticancer therapy: clinicopathological relevance, structural data and drug discovery perspectives. Curr. Pharm. Des. 19, 578–613.

Angeles, E., Vázquez-Valadéz, V.H., Vázquez-Valadéz, O., Velázquez-Sánchez, A.M., Ramírez, A., Martínez, L., et al., 2005. Computational studies of 1-hydrazinophthalazine (hydralazine) as antineoplasic agent. Docking studies on methyltransferase. Lett. Drug Des. Discov. 2, 282–286.

Asgatay, S., Champion, C., Marloie, G., Drujon, T., Senamaud-Beaufort, C., Ceccaldi, A., et al., 2014. Synthesis and evaluation of analogues of *N*-phthaloyl-l-tryptophan (RG108) as inhibitors of DNA methyltransferase 1. J. Med. Chem. 57, 421–434.

Agrafiotis, D.K., Bandyopadhyay, D., Wegner, J.K., van Vlijmen, H., 2007. Recent advances in chemoinformatics. J. Chem. Inf. Model 47, 1279–1293.

Becker, O.M., MacKerell Jr., A.D., Roux, B., Watanabe, M., 2001. Computational Biochemistry and Biophysics, first ed. Taylor and Francis e-Library, New York.

Bemis, G.W., Murcko, M.A., 1996. The properties of known drugs. 1. Molecular frameworks. J. Med. Chem. 39, 2887–2893.

Bento, A.P., Gaulton, A., Hersey, A., Bellis, L.J., Chambers, J., Davies, M., et al., 2014. The ChEMBL bioactivity database: an update. Nucleic Acids Res. 42, D1083–D1090.

Berman, H.M., Westbrook, J., Feng, Z., Gilliland, G., Bhat, T.N., Weissig, H., et al., 2000. The protein data bank. Nucleic Acids Res. 28, 235–242.

Butler, K.V., Kalin, J., Brochier, C., Vistoli, G., Langley, B., Kozikowski, A.P., 2010. Rational design and simple chemistry yield a superior, neuroprotective HDAC6 inhibitor, tubastatin A. J. Am. Chem. Soc. 132, 10842–10846.

Campbell, S.J., Gold, N.D., Jackson, R.M., Westhead, D.R., 2003. Ligand binding: functional site location, similarity and docking. Curr. Opin. Struct. Biol. 13, 389–395.

Caulfield, T., Medina-Franco, J.L., 2011. Molecular dynamics simulations of human DNA methyltransferase 3B with selective inhibitor nanaomycin A. J. Struct. Biol. 176, 185–191.

Cavasotto, C.N., Abagyan, R.A., 2004. Protein flexibility in ligand docking and virtual screening to protein kinases. J. Mol. Biol. 337, 209–225.

Clark, J., Shevchuk, T., Kho, M.R., Smith, S.S., 2003. Methods for the design and analysis of oligodeoxynucleotide-based DNA (cytosine-5) methyltransferase inhibitors. Anal. Biochem. 321, 50–64.

Claussen, H., Buning, C., Rarey, M., Lengauer, T., 2001. FlexE: efficient molecular docking considering protein structure variations. J. Mol. Biol. 308, 377–395.

Cosconati, S., Froli, S., Perryman, A.L., Harris, R., Goodsell, D.S., Olson, A.J., 2010. Virtual screening with AutoDock: theory and practice. Expert Opin. Drug Discov. 5, 597–607.

Craig, I.R., Essex, J.W., Spiegel, K., 2010. Ensemble docking into multiple crystallographically derived protein structures: an evaluation based on the statistical analysis of enrichments. J. Chem. Inf. Model 50, 511–524.

Chang, Y., Zhang, X., Horton, J.R., Upadhyay, A.K., Spannhoff, A., Liu, J., et al., 2009. Structural basis for G9a-like protein lysine methyltransferase inhibition by BIX-01294. Nat. Struct. Mol. Biol. 16, 312–317.

ChemDiv, 2011. DNMT-focused Library. http://eu.chemdiv.com/.

Chen, H.-F., Kang, J.-H., Li, Q., Zeng, B.-S., Yao, X.-J., Fan, B.-T., et al., 2003. 3D-QSAR study on apicidin inhibit histone deacetylase. Chin. J. Chem. 21, 1596–1607.

Chen, S., Wang, Y., Zhou, W., Li, S., Peng, J., Shi, Z., et al., 2014. Identifying novel selective non-nucleoside DNA methyltransferase 1 inhibitors through docking-based virtual screening. J. Med. Chem. 57, 9028–9041.

Chen, Y.-D., Jiang, Y.-J., Zhou, J.-W., Yu, Q.-S., You, Q.-D., 2008. Identification of ligand features essential for HDACs inhibitors by pharmacophore modeling. J. Mol. Graph Model 26, 1160–1168.

Cavasotto, C.N., Phatak, S.S., 2009. Homology modeling in drug discovery: current trends and applications. Drug Discov. Today 14, 676–683.

Deschamps, N., Simões Pires, C.A., Carrupt, P-A., Nurisso, A., 2015. How the flexibility of human histone deacetylases influences ligand binding: an overview. Drug Discov. Today 20, 736–742.

Dessalew, N., 2007. QSAR study on aminophenylbenzamides and acrylamides as histone deacetylase inhibitors: an insight into the structural basis of antiproliferative activity. Med. Chem. Res. 16, 449–460.

Dudek, A.Z., Arodz, T., Gálvez, J., 2006. Computational methods in developing quantitative structure-activity relationships (QSAR): a review. Comb. Chem. High. Throughput Screen. 9, 213–228.

Erwing, T.J.A., Kuntz, I.D., 1997. Critical evaluation of search algorithms for automated molecular docking and database screening. J. Comput. Chem. 18, 1175–1189.

Erwing, T.J.A., Makino, S., Skillman, A.G., Kuntz, I.D., 2001. DOCK 4.0: search strategies for automated molecular docking of flexible molecule databases. J. Comput. Aided Mol. Des. 15, 411–428.

Fang, M.Z., Wang, Y., Ai, N., Hou, Z., Sun, Y., Lu, H., et al., 2003. Tea polyphenol (-)-epigallocatechin-3-gallate inhibits DNA methyltransferase and reactivates methylation-silenced genes in cancer cell lines. Cancer Res. 63, 7563–7570.

Friesner, R.A., Banks, J.L., Murphy, R.B., Halgren, T.A., Klicic, J.J., Mainz, D.T., et al., 2004. Glide: a new approach for rapid, accurate docking and scoring. 1. Method and assessment of docking accuracy. J. Med. Chem. 47, 1739–1749.

Gasteiger, J., 2006. Chemoinformatics: a new field with a long tradition. Anal. Bioanal. Chem. 384, 57–64.

Gaulton, A., Bellis, L.J., Bento, A.P., Chambers, J., Davies, M., Hersey, A., et al., 2011. ChEMBL: a large-scale bioactivity database for drug discovery. Nucleic Acids Res. 40, D1100–D1107.

Guo, Y., Xiao, J., Guo, Z., Chu, F., Cheng, Y., Wu, S., 2005. Exploration of a binding mode of indole amide analogues as potent histone deacetylase inhibitors and 3D-QSAR analyses. Bioorg. Med. Chem. 13, 5424–5434.

Heinke, R., Carlino, L., Kannan, S., Jung, M., Sippl, W., 2011. Computer- and structure-based lead design for epigenetic targets. Bioorg. Med. Chem. 19, 3605–3615.

Hert, J., Willett, P., Wilton, D.J., Acklin, P., Azzaoui, K., Jacoby, E., et al., 2005. Enhancing the effectiveness of similarity-based virtual screening using nearest-neighbor information. J. Med. Chem. 48, 7049–7054.

Houston, D.R., Walkinshaw, M.D., 2013. Consensus docking: improving the reliability of docking in a virtual screening context. J. Chem. Inf. Model 53, 384–390.

Huang, Z., Jiang, H., Liu, X., Chen, Y., Wong, J., Wang, Q., et al., 2012. HEMD: an integrated tool of human epigenetic enzymes and chemical modulators for therapeutics. PLoS One 7, e39917.

Irwin, J.J., Sterling, T., Mysinger, M.M., Bolstad, E.S., Coleman, R.G., 2012. ZINC: a free tool to discover chemistry for biology. J. Chem. Inf. Model 52, 1757–1768.

Juvale, D.C., Kulkarni, V.V., Deokar, H.S., Wagh, N.K., Padhye, S.B., Kulkarni, V.M., 2006. 3D-QSAR of histone deacetylase inhibitors: hydroxamate analogues. Org. Biomol. Chem. 4, 2858.

Kabro, A., Lachance, H., Marcoux-Archambault, I., Perrier, V., Doré, V., Gros, C., et al., 2013. Preparation of phenylethylbenzamide derivatives as modulators of DNMT3 activity. Med. Chem. Comm. 4, 1562.

Kandakatla, N., Ramakrishnan, G., 2014. Ligand based pharmacophore modeling and virtual screening studies to design novel HDAC2 inhibitors. Adv. Bioinform. 2014, 1–11.

Karplus, M., McCammon, J.A., 2002. Molecular dynamics simulations of biomolecules. Nat. Struct. Biol. 9, 646–652.

Katritzky, A.R., Slavov, S.H., Dobchev, D.A., Karelson, M., 2007. Comparison between 2D and 3D-QSAR approaches to correlate inhibitor activity for a aeries of indole amide hydroxamic acids. QSAR Comb. Sci. 26, 333–345.

Kuck, D., Singh, N., Lyko, F., Medina-Franco, J.L., 2010. Novel and selective DNA methyltransferase inhibitors: docking-based virtual screening and experimental evaluation. Bioorg. Med. Chem. 18, 822–829.

Lawson, M., Uciechowska, U., Schemies, J., Rumpf, T., Jung, M., Sippl, W., 2010. Inhibitors to understand molecular mechanisms of NAD^+-dependent deacetylases (sirtuins). BBA Gene Regul. Mech. 1799, 726–739.

Lee, W.J., Shim, J.-Y., Zhu, B.T., 2005. Mechanisms for the inhibition of DNA methyltransferases by tea catechins and bioflavonoids. Mol. Pharmacol. 68, 1018–1030.

Liu, B., Lu, A.-J., Liao, C.-Z., Liu, H.-B., Zhou, J.-J., 2005. 3D-QSAR of sulfonamide hydroxamic acid HDAC inhibitors. Acta Phys. Chim. Sin. 21, 333–337.

Liu, F., Chen, X., Allali-Hassani, A., Quinn, A.M., Wasney, G.A., Dong, A., et al., 2009. Discovery of a 2,4-diamino-7-aminoalkoxyquinazoline as a potent and selective inhibitor of histone lysine methyltransferase G9a. J. Med. Chem. 52, 7950–7953.

Liu, T., Lin, Y., Wen, X., Jorissen, R.N., Gilson, M.K., 2007. BindingDB: a web-accessible database of experimentally determined protein-ligand binding affinities. Nucleic Acids Res. 35, D198–D201.

Mahipal, Tanwar, O.P., Karthikeyan, C., Moorthy, N.S.H.N., Trivedi, P., 2010. 3D QSAR of aminophenyl benzamide derivatives as histone deacetylase inhibitors. Med. Chem. 6, 277–285.

Medina-Franco, J., Méndez-Lucio, O., Yoo, J., 2014. Rationalization of activity cliffs of a sulfonamide inhibitor of DNA methyltransferases with induced-fit docking. Int. J. Mol. Sci. 15, 3253–3261.

Medina-Franco, J.L., Yoo, J., 2013a. Docking of a novel DNA methyltransferase inhibitor identified from high-throughput screening: insights to unveil inhibitors in chemical databases. Mol. Divers. 17, 337–344.

Medina-Franco, J.L., Yoo, J., 2013b. Molecular modeling and virtual screening of DNA methyltransferase inhibitors. Curr. Pharm. Des. 19, 2138–2147.

Méndez-Lucio, O., Tran, J., Medina-Franco, J.L., Meurice, N., Muller, M., 2014. Toward drug repurposing in epigenetics: olsalazine as a hypomethylating compound active in a cellular context. Chem. Med. Chem. 9, 560–565.

Morris, G.M., Huey, R., Lindstrom, W., Sanner, M.F., Belew, R.K., Goodsell, D.S., et al., 2009. AutoDock 4 and AutoDockTools 4: automated docking with selective receptor flexibility. J. Comput. Chem. 30, 2785–2791.

Moura Barbosa, A.J., Del Rio, A., 2012. Freely accessible database of commercial compounds for high-throughput virtual screenings. Curr. Top. Med. Chem. 12, 866–877.

Nichols, S.E., Baron, R., Ivetac, A., McCammon, J.A., 2011. Predictive power of molecular dynamics receptor structures in virtual screening. J. Chem. Inf. Model 51, 1439–1446.

Nichols, S.E., Baron, R., McCammon, J.A., 2012. On the use of molecular dynamics receptor conformations for virtual screening. Comput. Drug Discov. Des. 819, 93–103.

Novikov, F.N., Chilov, G.G., 2009. Molecular docking: theoretical background, practical applications and perspectives. Mendeleev Commun. 19, 237–242.

Ortore, G., Di Colo, F., Martinelli, A., 2009. Docking of hydroxamic acids into HDAC1 and HDAC8: a rationalization of activity trends and selectivities. J. Chem. Inf. Model 49, 2774–2785.

Osguthorpe, D.J., Sherman, W., Hagler, A.T., 2012. Generation of receptor structural ensembles for virtual screening using binding site shape analysis and clustering. Chem. Biol. Drug Des. 80, 182–193.

Pronk, S., Pall, S., Schulz, R., Larsson, P., Bjelkmar, P., Apostolov, R., et al., 2013. GROMACS 4.5: a high-throughput and highly parallel open source molecular simulation toolkit. Bioinformatics 29, 845–854.

Ragno, R., Mai, A., Massa, S., Cerbara, I., Valente, S., Bottoni, P., et al., 2004. 3-(4-Aroyl-1-methyl-1*H*-pyrrol-2-yl)-*N*-hydroxy-2-propenamides as a new class of synthetic histone deacetylase inhibitors. 3. Discovery of novel lead compounds through structure-based drug design and docking studies. J. Med. Chem. 47, 1351–1359.

Ragno, R., Simeoni, S., Castellano, S., Vicidomini, C., Mai, A., Caroli, A., et al., 2007. Small molecule inhibitors of histone arginine methyltransferases: homology modeling, molecular docking, binding mode analysis, and biological evaluations. J. Med. Chem. 50, 1241–1253.

Ragno, R., Simeoni, S., Valente, S., Massa, S., Mai, A., 2006. 3-D QSAR studies on histone deacetylase inhibitors. A GOLPE/GRID approach on different series of compounds. J. Chem. Inf. Model 46, 1420–1430.

Ran, T., Zhang, Z., Liu, K., Lu, Y., Li, H., Xu, J., et al., 2015. Insight into the key interactions of bromodomain inhibitors based on molecular docking, interaction fingerprinting, molecular dynamics and binding free energy calculation. Mol. Biosyst. 11, 1295–1304.

Del Sol, A., Tsai, C.-J., Ma, B., Nussinov, R., 2009. The origin of allosteric functional modulation: multiple pre-existing pathways. Structure 17, 1042–1050.

Saha, S., Banerjee, S., Ganguly, S., 2010. Molecular docking studies of some novel hydroxamic acid derivatives. Int. J. Chem. Tech. Res. 2, 932–936.

Sakkiah, S., Krishnamoorthy, N., Gajendrarao, P., Thangapandian, S., Lee, Y., Kim, S., et al., 2009. Pharmacophore mapping and virtual screening for SIRT1 activators. Bull. Korean Chem. Soc. 30, 1152–1156.

Sanders, M.P.A., Barbosa, A.J.M., Zarzycka, B., Nicolaes, G.A.F., Klomp, J.P.G., de Vlieg, J., et al., 2012. Comparative analysis of pharmacophore screening tools. J. Chem. Inf. Model 52, 1607–1620.

Sherman, W., Day, T., Jacobson, M.P., Friesner, R.A., Farid, R., 2006. Novel procedure for modeling ligand/receptor induced fit effects. J. Med. Chem. 49, 534–553.

Siedlecki, P., Boy, R.G., Musch, T., Brueckner, B., Suhai, S., Lyko, F., et al., 2006. Discovery of two novel, small-molecule inhibitors of DNA methylation. J. Med. Chem. 49, 678–683.

Singh, N., Dueñas-González, A., Lyko, F., Medina-Franco, J.L., 2009. Molecular modeling and molecular dynamics studies of hydralazine with human DNA methyltransferase 1. Chem. Med. Chem. 4, 792–799.

Sinko, W., Lindert, S., McCammon, J.A., 2013. Accounting for receptor flexibility and enhanced sampling methods in computer-aided drug design. Chem. Biol. Drug Des. 81, 41–49.

Spannhoff, A., Heinke, R., Bauer, I., Trojer, P., Metzger, E., Gust, R., et al., 2007a. Target-based approach to inhibitors of histone arginine methyltransferases. J. Med. Chem. 50, 2319–2325.

Spannhoff, A., Machmur, R., Heinke, R., Trojer, P., Bauer, I., Brosch, G., et al., 2007b. A novel arginine methyltransferase inhibitor with cellular activity. Bioorg. Med. Chem. Lett. 17, 4150–4153.

Srimongkolpithak, N., Sundriyal, S., Li, F., Vedadi, M., Fuchter, M.J., 2014. Identification of 2,4-diamino-6,7-dimethoxyquinoline derivatives as G9a inhibitors. Med. Chem. Commun. 5, 1821–1828.

Subha, K., Kumar, G.R., 2008. Assessment for the identification of better HDAC inhibitor class through binding energy calculations and descriptor analysis. Bioinformation 3, 218–222.

Tambunan, U.S.F., Bakri, R., Prasetia, T., Parikesit, A.A., Kerami, D., 2013. Molecular dynamics simulation of complex histones deacetylase (HDAC) class II *Homo sapiens* with suberoylanilide hydroxamic acid (SAHA) and its derivatives as inhibitors of cervical cancer. Bioinformation 9, 696–700.

Tervo, A.J., Kyrylenko, S., Niskanen, P., Salminen, A., Leppänen, J., Nyrönen, T.H., et al., 2004. An in silico approach to discovering novel inhibitors of human sirtuin type 2. J. Med. Chem. 47, 6292–6298.

Thangapandian, S., John, S., Lee, K.W., 2012. Molecular dynamics simulation study explaining inhibitor selectivity in different class of histone deacetylases. J. Biomol. Struct. Dyn. 29, 677–698.

Thangapandian, S., John, S., Sakkiah, S., Lee, K.W., 2010. Ligand and structure based pharmacophore modeling to facilitate novel histone deacetylase 8 inhibitor design. Eur. J. Med. Chem. 45, 4409–4417.

Todeschini, R., Consonni, V., 2009. Molecular Descriptors for Chemoinformatics. WILEY-VCH, Milano, Italy.

Totrov, M., Abagyan, R., 2008. Flexible ligand docking to multiple receptor conformations: a practical alternative. Curr. Opin. Struct. Biol. 18, 17–184.

Trapp, J., Jochum, A., Meier, R., Saunders, L., Marshall, B., Kunick, C., et al., 2006. Adenosine mimetics as inhibitors of NAD^+-dependent histone deacetylases, from kinase to sirtuin inhibition. J. Med. Chem. 49, 7307–7316.

Trott, O., Olson, A.J., 2009. AutoDock Vina: improving the speed and accuracy of docking with a new scoring function, efficient optimization, and multithreading. J. Comput. Chem. 31, 455–461.

Tuccinardi, T., 2009. Docking-based virtual screening: recent developments. Comb. Chem. High. Throughput Screen 12, 303–314.

Tuccinardi, T., Poli, G., Romboli, V., Giordano, A., Martinelli, A., 2014. Extensive consensus docking evaluation for ligand pose prediction and virtual screening studies. J. Chem. Inf. Model 54, 2980–2986.

Vadivelan, S., Sinha, B.N., Rambabu, G., Boppana, K., Jagarlapudi, S.A.R.P., 2008. Pharmacophore modeling and virtual screening studies to design some potential histone deacetylase inhibitors as new leads. J. Mol. Graph Model 26, 935–946.

Venkatachalam, C.M., Jiang, X., Oldfield, T., Waldman, M., 2003. LigandFit: a novel method for the shape-directed rapid docking of ligands to protein active sites. J. Mol. Graph Model 21, 289–307.

Verdonk, M.L., Cole, J.C., Hartshorn, M.J., Murray, C.W., Taylor, R.D., 2003. Improved protein–ligand docking using GOLD. Proteins Struct. Funct. Bioinf. 52, 609–623.

Vidler, L.R., Filippakopoulos, P., Fedorov, O., Picaud, S., Martin, S., Tomsett, M., et al., 2013. Discovery of novel small-molecule inhibitors of BRD4 using structure-based virtual screening. J. Med. Chem. 56, 8073–8088.

Vogt, M., Bajorath, J., 2012. Chemoinformatics: a view of the field and current trends in method development. Bioorg. Med. Chem. 20, 5317–5323.

Wagener, M., Lommerse, J.P.M., 2006. The quest for bioisosteric replacements. J. Chem. Inf. Model 46, 677–685.

Wang, D.-F., Helquist, P., Wiech, N.L., Wiest, O., 2005. Toward selective histone deacetylase inhibitor design: homology modeling, docking studies, and molecular dynamics simulations of human class I histone deacetylases. J. Med. Chem. 48, 6936–6947.

Wang, D.-F., Wiest, O., Helquist, P., Lan-Hargest, H.-Y., Wiech, N.L., 2004. QSAR studies of PC-3 cell line inhibition activity of TSA and SAHA-like hydroxamic acids. Bioorg. Med. Chem. Lett. 14, 707–711.

Wang, J., Chen, L., Sinha, S.H., Liang, Z., Chai, H., Muniyan, S., et al., 2012a. Pharmacophore-based virtual screening and biological evaluation of small molecule inhibitors for protein arginine methylation. J. Med. Chem. 55, 7978–7987.

Wang, J., Wang, W., Kollman, P.A., Case, D.A., 2006. Automatic atom type and bond type perception in molecular mechanical calculations. J. Mol. Graph Model 25, 247–260.

Wang, Y., Xiao, J., Suzek, T.O., Zhang, J., Wang, J., Zhou, Z., et al., 2012b. PubChem's bioassay database. Nucleic Acids Res. 40, D400–D412.

Wermuth, C.G., Ganellin, C.R., Lindberg, P., Mitscher, L.A., 1998. Glossary of terms used in medicinal chemistry (IUPAC recommendations 1998). Pure Appl. Chem. 52, 1607–1620.

Willett, P., 2011. Chemoinformatics: a history. WIREs Comput. Mol. Sci. 1, 46–56.

Willett, P., 2006. Similarity-based virtual screening using 2D fingerprints. Drug Discov. Today 11, 1046–1053.

Xie, A., Liao, C., Li, Z., Ning, Z., Hu, W., Lu, X., et al., 2004. Quantitative structure-activity relationship study of histone deacetylase inhibitors. Curr. Med. Chem. Anticancer Agents 4, 273–299.

Xu, J., Hagler, A., 2002. Chemoinformatics and drug discovery. Molecules 7, 566–600.

Yang, J.-S., Chun, T.-G., Nam, K.-Y., Kim, H.-M., Han, G.-H., 2012. Structure-activity relationship of novel lactam based histone deacetylase inhibitors as potential anticancer drugs. Bull. Korean Chem. Soc. 33, 2063–2066.

Yoo, J., Choi, S., Medina-Franco, J.L., 2013. Molecular modeling studies of the novel inhibitors of DNA methyltransferases SGI-1027 and CBC12: implications for the mechanism of inhibition of DNMTs. PLoS One 8, e62152.

Yoo, J., Kim, J.H., Robertson, K.D., Medina-Franco, J.L., 2012. Molecular modeling of inhibitors of human DNA methyltransferase with a crystal structure: discovery of a novel DNMT1 inhibitor. Adv. Protein Chem. Struct. Biol. 87, 219–247.

Yoo, J., Medina-Franco, J.L., 2011a. Chemoinformatic approaches for inhibitors of DNA methyltransferases: comprehensive characterization of screening libraries. Comput. Mol. Biosci. 01, 7–16.

Yoo, J., Medina-Franco, J.L., 2011b. Homology modeling, docking and structure-based pharmacophore of inhibitors of DNA methyltransferase. J. Comput. Aided Mol. Des. 25, 555–567.

Yoo, J., Medina-Franco, J.L., 2012a. Towards the chemoinformatic-based identification of DNA methyltransferase inhibitors: 2D- and 3D-similarity profile of screening libraries. Curr. Comput. Aided Drug Des. 8, 317–329.

Yoo, J., Medina-Franco, J.L., 2012b. Trimethylaurintricarboxylic acid inhibits human DNA methyltransferase 1: insights from enzymatic and molecular modeling studies. J. Mol. Model 18, 1583–1589.

Yang, S.-Y., 2010. Pharmacophore modeling and applications in drug discovery: challenges and recent advances. Drug Discov. Today 15, 444–450.

Zhao, H., Caflisch, A., 2015. Molecular dynamics in drug design. Eur. J. Med. Chem. 91, 4–14.

Zhao, H., Gartenmann, L., Dong, J., Spiliotopoulos, D., Caflisch, A., 2014. Discovery of BRD4 bromodomain inhibitors by fragment-based high-throughput docking. Bioorg. Med. Chem. Lett. 24, 2493–2496.

CHAPTER 3

Structure-Guided Optimization of DNA Methyltransferase Inhibitors

Alexandre Erdmann[1], Paola B. Arimondo[1], Dominique Guianvarc'h[2]

[1]*Unité de Service et de Recherche CNRS-Pierre Fabre, ETaC, CRDPF, Toulouse, France;*
[2]*Sorbonne Universités—UPMC Univ Paris, École Normale Supérieure—PSL Research University, Département de Chimie, CNRS UMR 7203 LBM, Paris, France*

1. INTRODUCTION TO DNA METHYLTRANSFERASES

1.1 Role in Cells

Information contained in a mammal's genome has two components: the genetic information provides manufacturing drawings for all proteins necessary for cell life, while the epigenetic information gives the instructions for implementation of this manufacturing, thereby turning genes on or off deciding when and where. Thus, epigenetics refers to heritable changes in gene expression that do not involve changes in the DNA sequence (Berger et al., 2009; Bird, 2007). At least three processes are involved in epigenetic regulation including DNA methylation, histone modifications, and noncoding RNAs. In mammals, DNA methylation is a well-characterized carrier of the epigenetic information and is related to transcriptional silencing when occurring on gene promoters (Feinberg and Tycko, 2004). DNA methylation occurs at position 5 of the cytosine ring within mainly CpG-sequences. CpG dinucleotides are not distributed randomly across the human genome but instead are concentrated in so-called CpG islands located in the promoter of genes and nontranscribed regions of large repetitive sequences such as centromeric repeat and retrotransposons (Bird, 2002; Illingworth and Bird, 2009). Most of the CpG sites in these repetitive sequences are methylated, thus preventing chromosomal instability by silencing transposable DNA elements (Suzuki and Bird, 2008). Furthermore, some CpG island promoters become methylated in a sustainable manner during development, which results in turning gene expression off, a phenomenon observed in X-chromosome inactivation, imprinted genes, and developmental genes. Methylation status of the promoters regulates dynamically gene expression through modification of the recognition by transcription factors and the proteins involved in chromatin remodeling.

CONTENTS

Epi-Informatics. http://dx.doi.org/10.1016/B978-0-12-802808-7.00003-4

When deregulated, DNA methylation is implicated in several pathologies and particularly in cancers (Esteller, 2007; Feinberg, 2007; Jones and Baylin, 2007). Indeed, both hypo- and hypermethylation are commonly observed in malignant cells since a global decrease of CpG methylation occurs, leading to genomic instability. In parallel, some normally unmethylated CpG islands in the promoter regions of specific genes become methylated in various cancers, thus leading to the abnormal silencing of these gene involved in a variety of important cellular pathways. Thanks to the recent technological advances in DNA methylation mapping, an increasing battery of genes has been reported to be inactivated by a DNA methylation mechanism during tumorigenesis, including tumor suppressor genes, DNA damage repair genes, genes involved in cell adhesion, transcription factors, or genes involved in receptor-mediated signaling pathways. Furthermore, it was shown that this hypermethylation also affects the expression of noncoding RNAs such as microRNAs (miRNAs) involved in malignant transformation. Therefore, controlling the DNA methylome represents a challenging new therapeutic perspective.

Conventional chemotherapeutic drugs are cytotoxic agents that induce cell death by various mechanisms acting mostly on rapidly dividing cells; that is, mainly cancerous cells but also normal rapidly dividing cells. Drugs that target the epigenetic pathways rely on a fundamentally different approach. Indeed, by targeting the epigenetic modification these drugs change the gene expression pattern of the cells, reprogram them, and can thus reverse the status of the cancer cell toward a less aggressive, chemosensitive, immunosensitive, and more normal state (Ahuja et al., 2014; Jones and Baylin, 2007; Wee et al., 2014). As proof of concept of the potential of the epigenetic therapies, two DNA demethylating agents and four histone acetylase inhibitors have been approved for clinical use.

The enzymes catalyzing DNA methylation are known as C5-DNA methyltransferases (DNMTs). They catalyze the transfer of a methyl group during a mechanism involving (1) the extruding of the target cytosine from the DNA double helix into the active site of the enzyme (Klimasauskas et al., 1994), (2) the covalent binding of a conserved cysteine to the carbon 6 of the nucleotide, (3) the transfer of the sulfonium methyl moiety of the cofactor *S*-adenosyl-L-methionine (SAM) to the carbon 5 of the cytosine ring, and finally (4) a beta-elimination step allowing reformation of the 5–6 double bond and release of the enzyme (Jurkowska et al., 2011) (Figure 1(a)).

In mammals, four members of DNMTs have been identified: DNMT1, DNMT3a, DNMT3b, and DNMT3L. DNMT1 is the most abundant and is involved in the maintenance of the methylation pattern during DNA replication, predominantly targeting hemimethylated CpGs. DNMT3a and DNMT3b show no preference for hemimethylated DNA and appear to function as de novo methyltransferases catalyzing the methylation of unmodified DNA especially in stem cells (Okano et al., 1999). DNMT3L lacks the catalytic residues

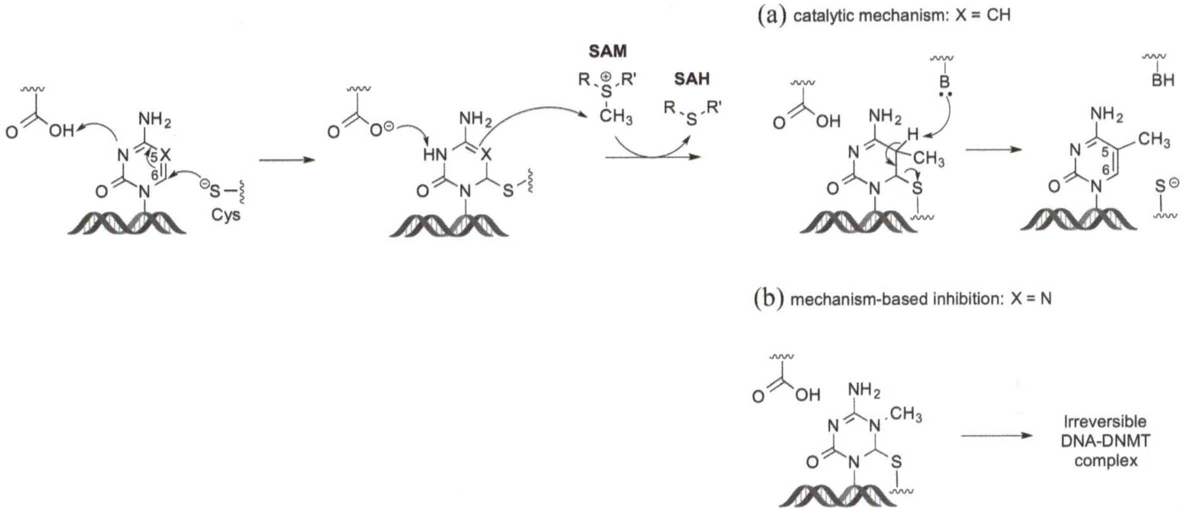

FIGURE 1 Mechanism of DNA methylation by the DNMTs.
(a) on cytosine and (b) on 5-azacytosine. Cys, cysteine; SAM, *S*-adenosyl-L-methionine; SAH, *S*-adenosyl-L-homocysteine.

but is important for cooperating with DNMT3a and 3b and other chromatin actors (Jia et al., 2007). The expression of DNMT1, 3a, and 3b is increased in various tumors including the colon (el-Deiry et al., 1991), prostate (Patra et al., 2002), breast (Girault et al., 2003), liver (Oh et al., 2007), and in leukemia (Vispe et al., 2015). Interestingly, inhibition of DNMTs correlates with reduction in tumorigenicity and increased expression of tumor suppressor genes (Baylin, 2005). Hence, DNMTs are promising targets for the design of novel-specific anticancer strategies.

1.2 Inhibition Approach for Therapy

DNMT inhibitors include two distinct categories of compounds. The first one, known as nucleosides analogs (Figure 2), consists of pro-drugs cytidine analogs that once phosphorylated and incorporated in DNA during replication act as mechanism-based inhibitors, irreversibly trapping DNMT. Indeed, the DNA containing these analogs are first processed by the normal catalytic mechanism mentioned above but it leads to the formation of an irreversible covalent complex between DNMT and DNA via the catalytic cysteine and the C6 of the cytidine analog since the beta-elimination step is prevented due to the chemistry of the analog (Figure 1(b)) (Gabbara and Bhagwat, 1995; Santi et al., 1983, 1984). Among them are 5-aza-cytidine (5-azaC, Vidaza) and 5-aza-deoxycytidine (5-azadC, Dacogen, Decitabine), two analogs of the nucleoside 2′-deoxycytidine, in which the carbon C5 is substituted by a nitrogen preventing the release of the DNMT (Stresemann and Lyko, 2008). The demethylating capacity of these compounds has been known for many years but its approval for the treatment of cancer is related to the more recent understanding of the

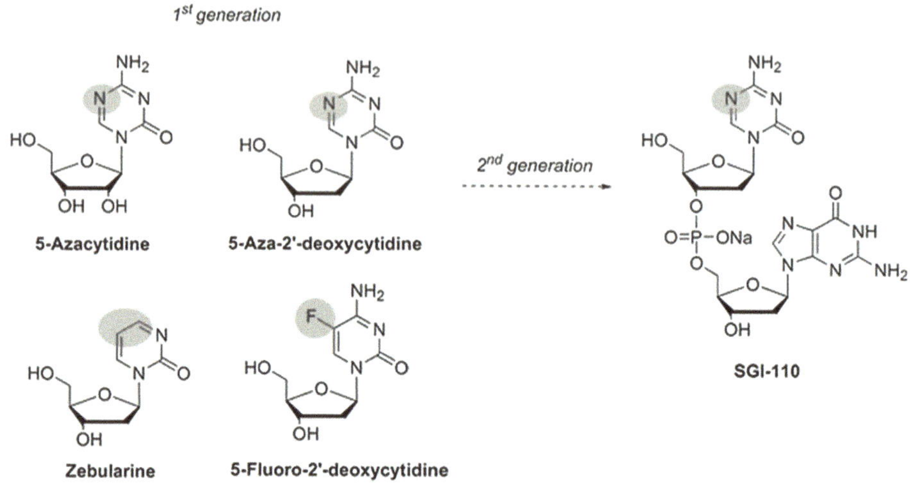

FIGURE 2 Chemical structure of the nucleoside analogs DNMT inhibitors.

fundamental role of epigenetics in tumorigenesis. Between 2004 and 2006, 5-azaC and 5-azadC have been approved by the FDA (US Food and Drug Administration) for the treatment of hematological malignancies including myelodysplastic syndromes, acute myeloid leukemia, and chronic myelomonocytic leukemia (Quintas-Cardama et al., 2010). Their clinical activity against solid tumors is also under investigation (Nie et al., 2014). These compounds are highly effective and are currently the only ones to elicit a significant effect on the global demethylation of genomic DNA (Boumber and Issa, 2011). They do, however, have a number of drawbacks including a lack of selectivity for various DNMTs, a toxicity due to their ability to integrate everywhere in the genome, and a short half-life when exposed to physiological conditions (Stresemann and Lyko, 2008). Some of these drawbacks are being overcome. For example, some efforts were made recently to increase their bioavailability and to increase their pharmacokinetics thanks to the development of a dinucleotide of 5-azadC and deoxyguanosine, named SGI-110 (Chuang et al., 2010; Yoo et al., 2007), that increases the stability of 5-azadC in vivo by protecting it from deamination. This compound is in clinical trials for the treatment of myelodysplastic syndromes, acute myeloid leukemia, and solid tumors. Furthermore, these compounds are currently used in clinical trials in combination with chemotherapies (www.clinicaltrials.gov), since synergistic effects could be obtained and lower doses were used thus minimizing undesired side effects.

Zebularine is also a nucleoside analog that stabilizes the DNMT-DNA complex and shows a lower cytotoxicity (Cheng et al., 2003). However, in comparison to the irreversible analogs, zebularine has to be used at higher concentrations to obtain the same demethylation effect in cells and thus never entered clinical trials (Flotho et al., 2009). Recently, 5-fluoro-2′-deoxycytidine (FdCyd) was also investigated in clinical trial. Unlike the triazine ring of 5-azacytosine, which is likely to be hydrolyzed, FdCyd is chemically stable in physiologic media. However, FdCyd is also a substrate of deaminases and therefore gives several metabolites that inhibit thymidylate synthase and induce undesired cytotoxicity (Beumer et al., 2006). Therefore, FdCyd was studied in combination with a cytidine deaminase inhibitor, tetrahydrouridine, to minimize the formation of these potentially toxic metabolites (Newman et al., 2015). This coadministration approach is also being explored with other cytidine analogs.

Due to these drawbacks, there is a real attraction in the characterization of new families of DNA demethylating agents that would not incorporate into the DNA but interact with the enzyme to prevent it from carrying out the methylation reaction. These compounds constitute the second class of DNMTs inhibitors, which consist of broad heterogeneous chemical families.

Many molecules with poorly elucidated even controversial mechanism of action have been described in the last decade (Erdmann et al., 2015) including natural compounds (tea polyphenol (Rajavelu et al., 2011) (−)-epigallocatechin 3-gallate (EGCG) (Fang et al., 2003), genistein (Fang et al., 2005), nanaomycin A (Kuck et al., 2010a), laccaic acid A (Fagan et al., 2013a)) versus synthetic compounds already used as drug for other therapies (hydralazine (Singh et al., 2009), procainamide (Lee et al., 2005), and procaine (Villar-Garea et al., 2003)). Some compounds were identified in high-throughput screening (HTS) campaigns (flavones and flavanones derivatives (Ceccaldi et al., 2011), dichlone (Ceccaldi et al., 2013), laccaic acid (Fagan et al., 2013a, b), SW155246 (Kilgore et al., 2013)), virtual screening (RG108 (Brueckner et al., 2005; Stresemann et al., 2006), NSC14778 and NSC319745 (Kuck et al., 2010b)), or were designed on rational basis (*S*-adenosyl-L-homocysteine (SAH) analogs (Isakovic et al., 2009; Saavedra et al., 2009)). Today, the first generation of compounds identified in the years 2000–2010 is giving rise to the design of optimized analogs of the lead compounds by SAR studies and molecular modeling (Yoo and Medina-Franco, 2011) thanks to the fairly recent crystal structures determination of DNMT1 (Song et al., 2011, 2012; Takeshita et al., 2011) and Dnmt3a/3L (Jia et al., 2007) (Figure 3). Several examples of hit discovery and optimization will be illustrated.

2. DRUG DISCOVERY

The use of chemoinformatics in drug discovery is increasing, in particular thanks to the resolution of the crystal structure of the target DNMT proteins (Kuck et al., 2010b; Yoo et al., 2012). It constitutes a useful tool in structure-based approaches for the identification of hits (virtual screening), the optimization of compounds, and it is also frequently used to illustrate the mechanism of action of the inhibitors (substrate competition) (Heinke et al., 2011).

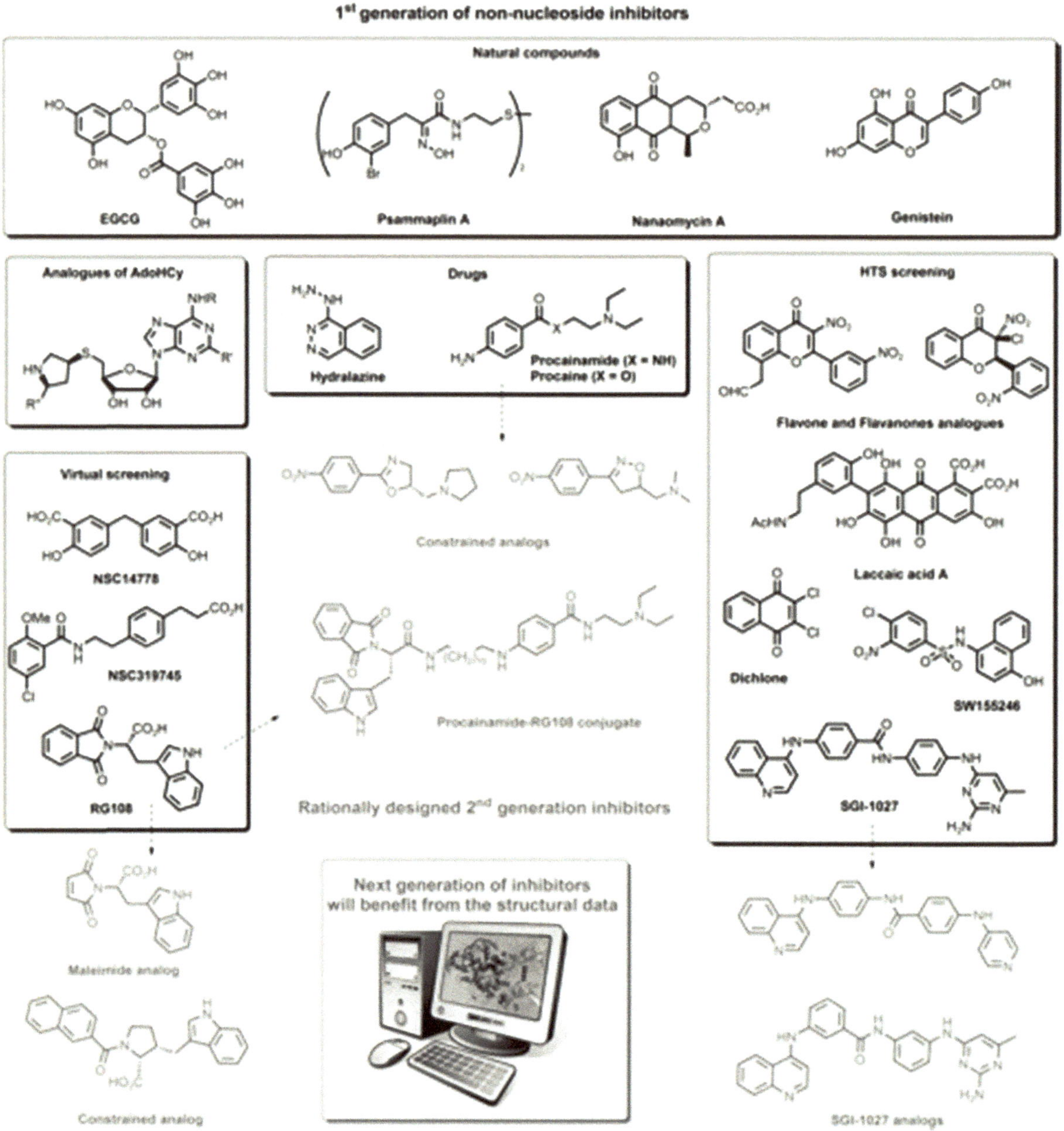

FIGURE 3 Chemical structure of the different families and generations of nonnucleoside DNMT inhibitors.

Herein, we describe the successful use of chemoinformatics through three different examples of nonnucleoside analog inhibitors of DNMTs, covering hit identification, the improvement of the activity, and the mechanism of action (Figure 4).

The use of chemoinformatics in the research field of DNMT inhibitors has long been hampered by the absence of structural data of the most relevant mammalian DNMTs. Indeed, until 2007, X-ray diffraction data were available only for the bacterial C5 DNA methyltransferases M.HhaI from *Haemophilus haemolyticus* (6MHT) (Klimasauskas et al., 1994), and M.HaeIII from *Haemophilus aegyptius* (1DCT) (Reinisch et al., 1995). Based on the fact that DNA methyltransferases show an extensive conservation of their catalytic domain, Siedlecki et al. proposed in 2003 a structural homology model of the

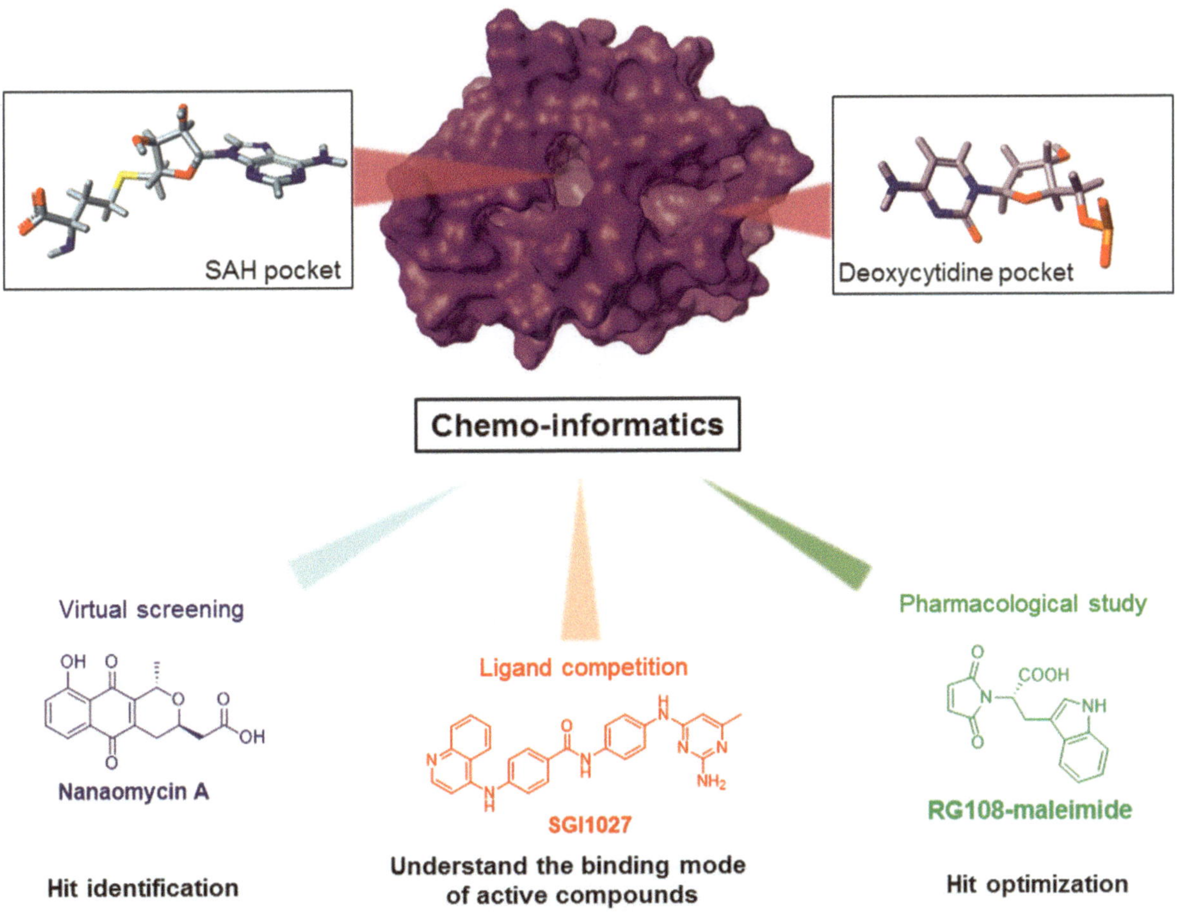

FIGURE 4 Application of chemo-informatics approaches to DNMT inhibitor discovery and optimization.

three-dimensional (3D) DNMT1 catalytic domain based on sequence similarity to proteins with the known structures cited above (Siedlecki et al., 2003). This model was the first to take into account the specificity of the catalytic site of DNMT1 and was used in several chemo-informatics studies as described below. In 2007, Jia et al. solved the structure of the catalytic part of murine Dnmt3a bound to the C-terminal domain of Dnmt3L obtaining a tetramer complex that seems to methylate concomitantly two CpG separated by 10 base pairs (Jia et al., 2007). Interestingly, the C-terminal domain of human DNMT1 was solved in 2011 to reveal an autoinhibition linker allowing the methylation of hemimethylated DNA only (Song et al., 2011). In 2012, the group was able to solve the crystal structure of murin Dnmt1 complexed to fluorodeoxycytidine (Song et al., 2012). This constitutes the first and only crystal structure of the DNMT complex with a DNMT inhibitor. However, no structure is available today of the DNMT/DNA/SAM complex in the presence of a nonnucleoside inhibitor.

2.1 Structure-Based Virtual Screening for Hit Identification

The resolution of the crystal structures of DNMT1 and Dnmt3A in the presence of the DNA and SAH substrates was used to model the active site of the DNMTs and build a pharmacophore model for the identification of new scaffold inhibitors. Indeed, virtual screening represents an effective tool in the search for novel inhibitor lead since it allows the computational screening of very large compound databases at relatively low cost.

The first DNMT inhibitor to be identified by a docking-based virtual screening approach of the National Cancer Institute (NCI) database was RG108 (Figure 3), using a homology model (Siedlecki et al., 2003) of the DNMT1 catalytic domain, since the crystal structure of DNMT1 was not resolved at that time (Brueckner et al., 2005). RG108 was one of the best scoring compounds identified in this screening and was evaluated experimentally, confirming it as DNMT inhibitor, and it was shown to reactivate several tumor suppressor genes in human colon cancer cell line HCT116. This compound was the starting point for pharmacomodulation studies to improve its biological activity as detailed below.

More recently, the natural product nanaomycin A was identified from a multistep structure-based virtual screening of more than 65,000 compounds from the NCI Open Database against the same DNMT1 homology model (Kuck et al., 2010b). In the case of nanaomycin A, the biological assays did not confirm the inhibitory activity against DNMT1, but rather revealed nanaomycin A as the first selective inhibitor of DNMT3B. Molecular docking of nanaomycin A in the active site of DNMT3B suggested the insertion of the inhibitor in the DNA pocket by forming strong interactions with specific amino acids, thus explaining the selectivity observed (Kuck et al., 2010a). In the same screening, compound

NSC319745 (Figure 3) was identified as a weak inhibitor of DNMT. The synthesis of 29 analogs of compound NSC319745 brought further interest for this scaffold with seven compounds, showing some DNMT3A inhibition activity with an EC_{50} of 36 μM for the most potent (Kabro et al., 2013). The inhibitory potency of these compounds needs to be improved but constitute an interesting starting point for selective inhibitors. Indeed, selective inhibitors of one isoform, DNMT1 versus DNMT3, will be of great use as chemical probes to understand the role of each enzyme in tumor formation and maintenance, which still needs to be clarified for each cancer type (Erdmann et al., 2015; Vispe et al., 2015).

Hence, virtual screening has therefore proven to be useful to identify new chemical starting points for more effective compounds (Maldonado-Rojas et al., 2015; Medina-Franco and Yoo, 2013).

2.2 Hit Optimization

These structure-based approaches are also used to guide the optimization of the compounds. These approaches can be rationally-based, as shown by the improvement of the inhibitory potency of RG108 and procainamide by covalent linkage of the two compounds (Figure 3) (Halby et al., 2012) or guided by molecular docking as exemplified below.

DNMTs have two ligands, the SAM cofactor that gives the methyl group and the deoxycytidine that protrudes in the enzyme catalytic pocket and is methylated on position 5 of the pyrimidine ring (Figure 1). It is the product SAH, which is found in the crystal structures of the DNMTs, while the deoxycytidine is much less resolved. The endogenous ligand structures and pockets can be delimited within the 3D structure of DNMTs and used as a model to position the inhibitors into the catalytic site. Once the lead structure of the inhibitor is identified, molecular docking suggests a position hypothesis for the inhibitor into the active site of the enzyme. This is a starting point for further optimization of the compound.

The inhibitor SGI-1027 is a well-illustrated example of this strategy. SGI-1027 is a quinoline-based DNA hypomethylating agent published as a pan-DNMT inhibitor. It was first designed from a DNA ligand containing a quinolinium bisquaternary salt, and was rapidly considered a SAM competitor for access to the cofactor binding site of the enzyme (Datta et al., 2009). However, Yoo et al. (2013) carried out a molecular modeling study of SGI-1027 into the DNMT1 and Dnmt3a catalytic sites suggesting the occupancy of both the SAM and DNA pockets. In agreement with this hypothesis, Rilova et al. (Rilova et al., 2014) docked SGI-1027 in the crystal structure of the bacterial C5-DNA methyltransferase M.Hha I and found it is positioned in both the DNA and SAM pocket (Figure 5).

Interestingly, the pyrimidine motif of SGI1027 was found to fit the SAM pocket, and the quinoline moiety, the deoxycytidine pocket. Guided by this hypothesis, new potent derivatives of SGI-1027 were synthesized (Figure 3), bringing new

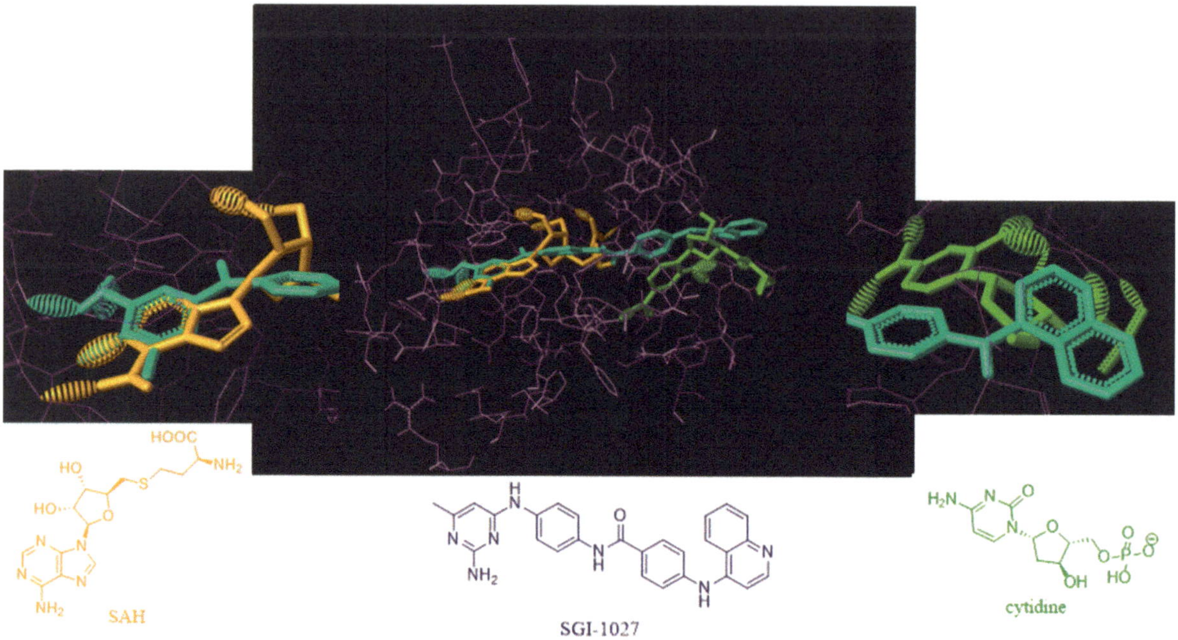

FIGURE 5 Docking of SGI-1027 in the 3D structure of the catalytic pocket of M.Hha I (PDB: 2HR1).
The cocrystalized *S*-adenosyl-L-homocysteine (SAH) product of the methylation reaction is depicted in orange and the deoxycytidine substrate in green. The docked SGI-1027 is in cyan.

insights into the substituents that contribute to the DNMT inhibition activity of this compound. Recently, new insights on the mechanism of action of the SGI-1027 were found by enzymatic competition studies revealing a typical DNA competitor behavior (Gros et al., 2015) and comforting the modeling studies.

If we now consider in particular the SAM pocket for ligand occupancy, it is necessary to analyze its interactions within the enzyme. Indeed the SAM is a universal cofactor, extensively used in cells by a large panel of methyltransferases and more. Therefore, it is critical to identify the specific interactions within the DNMT. Computational approaches allow the visualization of the interactions of the SAH ligand cocrystalized in the catalytic pocket of the DNMT1 (Song et al., 2012) and Dnmt3a (Jia et al., 2007). Interestingly, the comparison of these interactions revealed differences for the two subtypes of enzymes, which could be used to design selective inhibitors.

Figure 6 shows the interaction of the SAH in DNMT1 (PDB: 4DA4) and Dnmt3a (PDB: 2QRV). It clearly highlights that the SAH cofactor interacts in a different manner with DNMT1 than with Dnmt3a. As an example, the Met1172 of DNMT1 interacts by π-stacking with the adenyl group of the SAH. This interaction is not observed between Dnmt3a and the SAH. These differences in the cofactor pocket, identified by chemoinformatics, can be used to optimize the interaction of the ligand in the SAM pocket and, furthermore, for the design

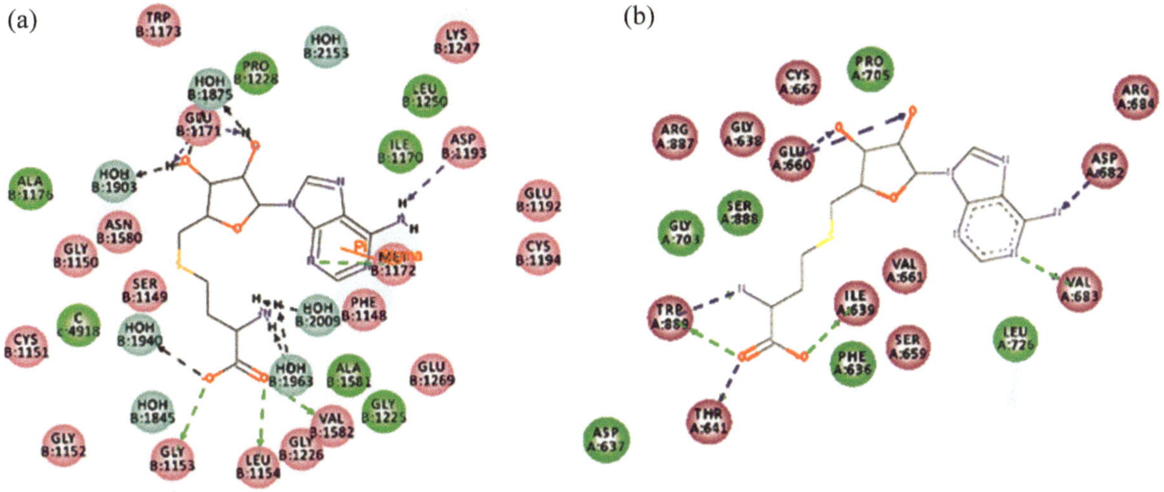

FIGURE 6 **2D diagram of interactions between SAH and (a) DNMT1 (PDB: 4DA4) or (b) Dnmt3a (PDB: 2QRV).** Electrostatic residue interactions, other interactions, and water are represented by pink, green, and blue spheres, respectively. Hydrogen bonds between the ligand and the backbone or side chains are shown in dashed blue, green, and black lines; solid yellow line shows π-stacking interaction.

of selective DNMT inhibitors inspired from the structure of SAH. Indeed, Isakovic et al. reported that chemical modifications of constrained analogs of SAH showed selective profiles of DNMT inhibition (Isakovic et al., 2009). Even if no docking studies were engaged to identify the interaction of Isakovic's compounds within the DNMTs, it seems likely that N^6 alkylation of the adenine ring of the SAH derivative possibly adds π-stacking interactions with the tryptophan residue 889 of Dnmt3 to favor Dnmt3 inhibition activity. In comparison, chloro substitution probably introduces hydrophobic interaction with the valine residue 1582 of DNMT1 retaining a specific DNMT1 inhibition activity.

Other examples of computational studies used to optimize the interactions of the compound within the enzyme are those reporting the modulation of procaine and procainamide. Molecular modeling studies (Singh et al., 2009) and experimental data (Villar-Garea et al., 2003) suggested that procaine and procainamide interact with DNA and target specifically the CpG-rich regions to disrupt the interaction between DNMT and its substrate. In order to increase their potency, oxazoline constrained derivatives of procaine were designed to limit the high flexibility of these drugs and identified as potent DNMT1 inhibitors with an effect on the genome methylation levels in HL60 human myeloid leukemia cells (Castellano et al., 2008). Extensive structure–activity relationships (SARs) led to the synthesis of isoxazoline derivatives as more stable and versatile compounds with a strong and dose-dependent antiproliferative effect against HCT116 human colon carcinoma cells. Surprisingly, the docking studies in the catalytic site of DNMT1 showed its ability to occupy the cofactor pocket.

These results confirm that it is necessary to consider both pockets for chemoinformatics studies to obtain the best computational hypothesis for hit optimization (Castellano et al., 2011).

Furthermore, in the same way, the covalent linkage of analogs RG108 and procainamide was also an interesting strategy to increase the inhibition potency of these DNMT inhibitors (Halby et al., 2012). The aim of this approach was to use procainamide as a DNA binder of the CpG region to guide RG108 to specific DNA sequences, thus increasing the local concentration of RG108 at the targeted sequences of the DNMTs. Procainamide was conjugated to RG108 and, according to the study of several analogs, it was found that the covalent linkage through a dodecan alkyl linker of only the phthalimide moiety of RG108 resulted in a potent DNMT inhibition activity correlated with the cytotoxicity against DU145 and HCT116 tumor cells. Interestingly, saturation transfer difference (STD) NMR spectroscopy experiments confirmed the inability of procainamide to interact with DNMTs, whereas phthalimide and RG108 interacted with the enzymes (Halby et al., 2012). Induced-fit docking was engaged by the Medina-Franco group and showed the ability of this compound to occupy both substrate pockets as expected for this procainamide-RG108 combination strategy (Yoo et al., 2013). Altogether, these data reveal that chemoinformatics studies on the endogenous ligand pockets are important tools to optimize the compounds by improving the positioning and interactions of the compounds into the catalytic complex of the enzyme. As illustrated by the above examples, in particular in the case of DNMT, both DNA and SAM pockets occupancy can lead to an improvement of the potency.

2.3 Mechanism of Action

In the discovery of new DNMT inhibitors, chemoinformatics studies were also used to explain the mechanism underlying the enzymatic or cellular activity. This is particularly useful when the crystal structure of the inhibitors with the enzyme is not available, as it is the case for nonnucleoside inhibitors, or that the mechanism of action cannot be revealed by enzymatic completion studies (Gros et al., 2013), mass spectrometry experiments (Bilgraer et al., 2014; Shi et al., 2014), and other biophysical techniques such as protein thermal shift (Gros et al., 2015).

Suzuki et al. conceived a maleimide derivative of RG108 more potent DNMT1 inhibitor than the RG108 itself (Suzuki et al., 2010). By a docking study they proposed that the maleimide part is placed next to the catalytic cysteine where a Michael addition can occur to form a covalent bond between the cysteine and the maleimide derivative explaining the observed improvement of biological activity for this analog of RG108.

In order to increase the inhibition potency of RG108 and to understand its important feature, Asgatay et al. performed a systematic SAR study (Asgatay et al., 2014). Among the different analogs synthesized, several constrained analogs were designed to partially reduce the flexibility of the compound by introducing a proline ring in the place of the amino acid moiety of RG108, with the aim to restrict it in a bioactive conformer. The most potent inhibitor containing a naphthoyl moiety linked to the amine of the prolinotryptophan showed eightfold better activity than RG108 in the same assay conditions. A docking study on the catalytic site of DNMT1 was engaged to understand the improvement in the biological activity of this new derivative. This study showed that, while RG108 occupies only the SAM pocket (this model was supported by experimental data since STD NMR experiments showed that RG108 and SAH compete in binding to DNMT), the constrained derivative occupied both SAM and DNA pockets. The indole part is in the SAM pocket, the naphthoyl fit toward the cytidine pocket, and the proline acts as a slightly rigid scaffold binding these two moieties (Figure 7).

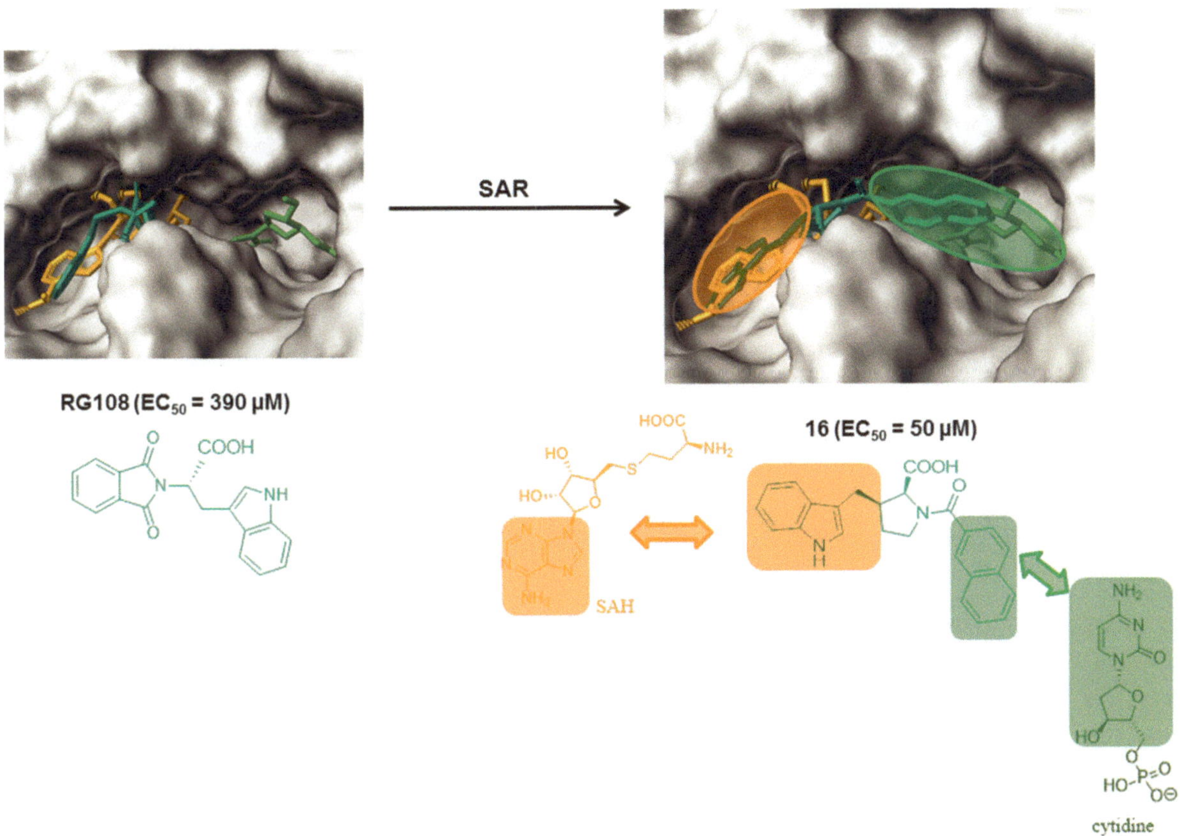

FIGURE 7 Potency improvement of constrained analog 16 in comparison with RG108.
A docking study was realized into the catalytic pocket of DNMT1 (PDB: 4DA4), the SAH is depicted in orange, the deoxycytidine in green, and RG106 and its constraint analog 16 in dark blue. The similarities are highlighted.

These examples clearly showed how chemoinformatic studies can bring valuable information concerning the mechanism of action of DNMT inhibitors, corroborating the potency improvement observed in the biological assays.

3. FURTHER PROMISING PERSPECTIVES

We have recently published our perspectives in the development of new inhibitors of DNMTs (Erdmann et al., 2015; Guianvarc'h and Arimondo, 2014). Here we will focus on those that have been suggested by the molecular modeling and crystallographic studies.

3.1 Bisubstrate Approach

As described above, the DNMT are bisubstrate enzymes and an interesting strategy is to do analogs of each substrate and combine them covalently to increase the affinity for the enzyme and obtain potent inhibitors (Figure 8). This has been done with success for bacterial adenine methyltransferase (Hobley et al., 2012;

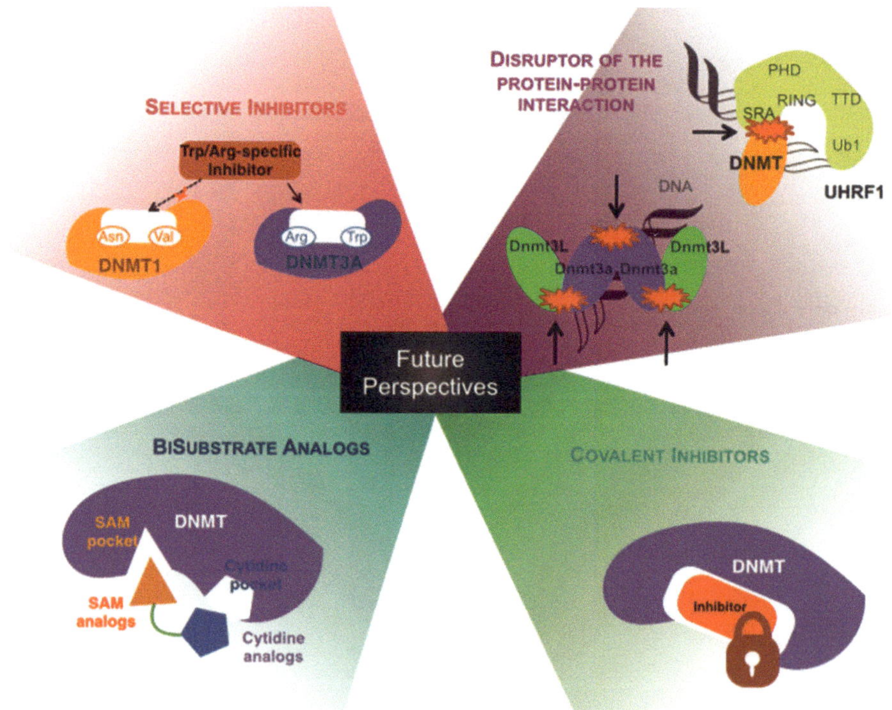

FIGURE 8 Future perspective for DNMT inhibitors. Four strategies are represented. SRA, SET and RING finger associated domain; PHD, plant homeodomain; a Cys_4-His-Cys_3 zinc-finger like motif; TTD, Tudor Tandem Domain; Ub1, ubiquitin related domain; Trp, tryptophan; Arg, arginine; Asn, asparagine; Val, valine.

Mashhoon et al., 2004), catechol-O-methyltransferase (Ellermann et al., 2011), protein lysine, arginine methyltransferase (Daigle et al., 2011; Mori et al., 2010; Osborne et al., 2008), and also DNMT (Wahhab, 2006). Indeed, by this approach, we have obtained new potent inhibitors of DNMT3A (Halby and Arimondo, 2015), able to demethylate and reactivate genes in cancer cells. Chemoinformatics and crystallographic studies will help improve these compounds by optimizing the interactions in each ligand pocket and the size, shape, nature, and flexibility of the linker between the two moieties.

3.2 Selectivity by Interaction with Specific Amino Acids

As discussed above selective inhibitors can be used both to probe the role of each DNMT in cancer cells and to validate the interaction of the inhibitor with the enzyme. The docking of the inhibitor in the 3D structure of the enzyme can suggest which interactions between the compound and the amino acids of the enzyme are important for the selectivity and thus validate the pharmacophore model, as illustrated in Figures 6 and 8. In particular, Figure 8 schematizes that selective inhibitors of DNMT3A can be designed to specifically interact with the arginine 887 and tryptophan 889 residues of Dnmt3a or with asparagine 1580 and valine 1582 residues for DNMT1. Even if these four amino acids do not make interactions with SAH or cytidine, all of them are in the DNMT catalytic pockets and, therefore, can be used to increase the affinity of the inhibitor. Furthermore, clear differences in the physical properties of the pairs Arg/Trp of Dnmt3a and Asp/Val of DNMT1 can explain why these specific amino acids are chosen to induce a subtype selective inhibition.

3.3 Disrupting Protein–Protein Interactions

On one hand, great progress has been made in the targeting of protein–protein interactions both with small molecules (Bier et al., 2015; Laraia et al., 2015) and stapled peptides (Cheray et al., 2014; Higueruelo et al., 2013). On the other hand, clearly the DNMTs are part of complex interaction networks between proteins that regulate gene expression and chromatin compaction. A whole picture of these interactions is still lacking but some partners are already validated and it would be interesting to apply these new technologies targeting protein–protein interactions to disrupt these complexes. For example, the resolution of the crystal structure of the catalytic complex between Dnmt3a and Dnmt3L (Jia et al., 2007) has identified two interfaces, between Dnmt3a and 3L and between Dnmt3a and 3a that could be targeted (Figure 8). In particular, two pairs of phenylalanines seem to be important for the Dnmt3a and 3L interaction. The crystal structure of UHRF1 (Arita et al., 2012) and the characterization of its biological activity (Bashtrykov et al., 2014; Liu et al., 2013; Rothbart et al., 2012) suggest an interaction with the DNMTs, in particular DNMT1 through its SRA domain, guiding DNA methylation (Figure 8). Therefore, compounds disrupting this interaction could inhibit DNA methylation and thus modulate gene expression.

3.4 Covalent Inhibitors

Finally, once compounds highly affine for the DNMTs, selective or not, are found, the insertion of a potential covalent binder group to the thiol of the catalytic cysteine would give promising covalent nonnucleoside inhibitors. This strategy has been applied with success to tyrosine kinase inhibitors (Afatinib and Neratinib) (Kumar and Agrawal, 2014; Lopez-Tarruella et al., 2012), which are FDA-approved. These covalent nonnucleoside inhibitors, once in the DNMT, could form a covalent bond as the azanucleosides without the drawbacks of the incorporation into DNA. The formation of this covalent bond would probably induce an active DNA demethylation process (Erdmann et al., 2015) similar to what happens in the case of 5azaC and 5azadC, which remain today the most potent inhibitors. For this approach and in the absence of the determination of the crystal structure of the inhibitor in the DNMT complex, molecular modeling is of great use to suggest how to modify the compound to add a Michael acceptor or other covalent bond forming groups. For this application, the maleimide derivatives of RG108 described by Suzuki et al. are encouraging since they increase the potency of RG108. Together with new propiophenone derivatives containing a highly reactive Michael acceptor identified in an HTS campaign (Erdmann et al., 2015), they constitute preliminary starting points for further optimization.

In conclusion, structure-based approaches are important guides for medicinal chemists to design and optimize new nonnucleoside inhibitors of DNMTs.

REFERENCES

Ahuja, N., Easwaran, H., Baylin, S.B., 2014. Harnessing the potential of epigenetic therapy to target solid tumors. J. Clin. Invest. 124, 56–63.

Arita, K., Isogai, S., Oda, T., Unoki, M., Sugita, K., Sekiyama, N., Kuwata, K., Hamamoto, R., Tochio, H., Sato, M., Ariyoshi, M., Shirakawa, M., 2012. Recognition of modification status on a histone H3 tail by linked histone reader modules of the epigenetic regulator UHRF1. Proc. Natl. Acad. Sci. U.S.A. 109, 12950–12955.

Asgatay, S., Champion, C., Marloie, G., Drujon, T., Senamaud-Beaufort, C., Ceccaldi, A., Erdmann, A., Rajavelu, A., Schambel, P., Jeltsch, A., Lequin, O., Karoyan, P., Arimondo, P.B., Guianvarc'h, D., 2014. Synthesis and evaluation of analogues of N-phthaloyl-l-tryptophan (RG108) as inhibitors of DNA methyltransferase 1. J. Med. Chem. 57, 421–434.

Bashtrykov, P., Jankevicius, G., Jurkowska, R.Z., Ragozin, S., Jeltsch, A., 2014. The UHRF1 protein stimulates the activity and specificity of the maintenance DNA methyltransferase DNMT1 by an allosteric mechanism. J. Biol. Chem. 289, 4106–4115.

Baylin, S.B., 2005. DNA methylation and gene silencing in cancer. Nat. Clin. Pract. Oncol. 2 (Suppl. 1), S4–S11.

Berger, S.L., Kouzarides, T., Shiekhattar, R., Shilatifard, A., 2009. An operational definition of epigenetics. Genes Dev. 23, 781–783.

Beumer, J.H., Eiseman, J.L., Parise, R.A., Joseph, E., Holleran, J.L., Covey, J.M., Egorin, M.J., 2006. Pharmacokinetics, metabolism, and oral bioavailability of the DNA methyltransferase inhibitor 5-fluoro-2′-deoxycytidine in mice. Clin. Cancer Res. 12, 7483–7491.

Bier, D., Thiel, P., Briels, J., Ottmann, C., 2015. Stabilization of protein-protein interactions in chemical biology and drug discovery. Prog. Biophys. Mol. Biol.

Bilgraer, R., Gillet, S., Gil, S., Evain-Brion, D., Laprevote, O., 2014. A new approach combining LC-MS and multivariate statistical analysis for revealing changes in histone modification levels. Mol. Biosyst. 10, 2974–2983.

Bird, A., 2002. DNA methylation patterns and epigenetic memory. Genes Dev. 16, 6–21.

Bird, A., 2007. Perceptions of epigenetics. Nature 447, 396–398.

Boumber, Y., Issa, J.P., 2011. Epigenetics in cancer: what's the future? Oncology (Williston Park) 25, 220–226, 228.

Brueckner, B., Garcia Boy, R., Siedlecki, P., Musch, T., Kliem, H.C., Zielenkiewicz, P., Suhai, S., Wiessler, M., Lyko, F., 2005. Epigenetic reactivation of tumor suppressor genes by a novel small-molecule inhibitor of human DNA methyltransferases. Cancer Res. 65, 6305–6311.

Castellano, S., Kuck, D., Sala, M., Novellino, E., Lyko, F., Sbardella, G., 2008. Constrained analogues of procaine as novel small molecule inhibitors of DNA methyltransferase-1. J. Med. Chem. 51, 2321–2325.

Castellano, S., Kuck, D., Viviano, M., Yoo, J., Lopez-Vallejo, F., Conti, P., Tamborini, L., Pinto, A., Medina-Franco, J.L., Sbardella, G., 2011. Synthesis and biochemical evaluation of delta(2)-isoxazoline derivatives as DNA methyltransferase 1 inhibitors. J. Med. Chem. 54, 7663–7677.

Ceccaldi, A., Rajavelu, A., Champion, C., Rampon, C., Jurkowska, R., Jankevicius, G., Senamaud-Beaufort, C., Ponger, L., Gagey, N., Ali, H.D., Tost, J., Vriz, S., Ros, S., Dauzonne, D., Jeltsch, A., Guianvarc'h, D., Arimondo, P.B., 2011. C5-DNA methyltransferase inhibitors: from screening to effects on zebrafish embryo development. Chem. Bio. Chem. 12, 1337–1345.

Ceccaldi, A., Rajavelu, A., Ragozin, S., Senamaud-Beaufort, C., Bashtrykov, P., Testa, N., Dali-Ali, H., Maulay-Bailly, C., Amand, S., Guianvarc'h, D., Jeltsch, A., Arimondo, P.B., 2013. Identification of novel inhibitors of DNA methylation by screening of a chemical library. ACS Chem. Biol. 8, 543–548.

Cheng, J.C., Matsen, C.B., Gonzales, F.A., Ye, W., Greer, S., Marquez, V.E., Jones, P.A., Selker, E.U., 2003. Inhibition of DNA methylation and reactivation of silenced genes by zebularine. J. Natl. Cancer Inst. 95, 399–409.

Cheray, M., Nadaradjane, A., Bonnet, P., Routier, S., Vallette, F.M., Cartron, P.F., 2014. Specific inhibition of DNMT1/CFP1 reduces cancer phenotypes and enhances chemotherapy effectiveness. Epigenomics 6, 267–275.

Chuang, J.C., Warner, S.L., Vollmer, D., Vankayalapati, H., Redkar, S., Bearss, D.J., Qiu, X., Yoo, C.B., Jones, P.A., 2010. S110, a 5-Aza-2′-deoxycytidine-containing dinucleotide, is an effective DNA methylation inhibitor in vivo and can reduce tumor growth. Mol. Cancer Ther. 9, 1443–1450.

Daigle, S.R., Olhava, E.J., Therkelsen, C.A., Majer, C.R., Sneeringer, C.J., Song, J., Johnston, L.D., Scott, M.P., Smith, J.J., Xiao, Y., Jin, L., Kuntz, K.W., Chesworth, R., Moyer, M.P., Bernt, K.M., Tseng, J.C., Kung, A.L., Armstrong, S.A., Copeland, R.A., Richon, V.M., Pollock, R.M., 2011. Selective killing of mixed lineage leukemia cells by a potent small-molecule DOT1L inhibitor. Cancer Cell 20, 53–65.

Datta, J., Ghoshal, K., Denny, W.A., Gamage, S.A., Brooke, D.G., Phiasivongsa, P., Redkar, S., Jacob, S.T., 2009. A new class of quinoline-based DNA hypomethylating agents reactivates tumor suppressor genes by blocking DNA methyltransferase 1 activity and inducing its degradation. Cancer Res. 69, 4277–4285.

el-Deiry, W.S., Nelkin, B.D., Celano, P., Yen, R.W., Falco, J.P., Hamilton, S.R., Baylin, S.B., 1991. High expression of the DNA methyltransferase gene characterizes human neoplastic cells and progression stages of colon cancer. Proc. Natl. Acad. Sci. U.S.A. 88, 3470–3474.

Ellermann, M., Paulini, R., Jakob-Roetne, R., Lerner, C., Borroni, E., Roth, D., Ehler, A., Schweizer, W.B., Schlatter, D., Rudolph, M.G., Diederich, F., 2011. Molecular recognition at the active site of catechol-O-methyltransferase (COMT): adenine replacements in bisubstrate inhibitors. Chemistry 17, 6369–6381.

Erdmann, A., Halby, L., Fahy, J., Arimondo, P.B., 2015. Targeting DNA methylation with small molecules: what's next? J. Med. Chem. 58, 2569–2583.

Erdmann, A., Menon, Y., Gros, C., Molinier, N., Novosad, N., Samson, A., Gregoire, J.M., Long, C., Ausseil, F., Halby, L., Arimondo, P.B., 2015. Design and synthesis of new non nucleoside inhibitors of DNMT3A. Bioorg. Med. Chem. http://dx.doi.org/10.1016/j.bmc.2015.06.066.

Esteller, M., 2007. Cancer epigenomics: DNA methylomes and histone-modification maps. Nat. Rev. Genet. 8, 286–298.

Fagan, R.L., Cryderman, D.E., Kopelovich, L., Wallrath, L.L., Brenner, C., 2013a. Laccaic acid A is a direct, DNA-competitive inhibitor of DNA methyltransferase 1. J. Biol. Chem. 288, 23858–23867.

Fagan, R.L., Wu, M., Chedin, F., Brenner, C., 2013b. An ultrasensitive high throughput screen for DNA methyltransferase 1-targeted molecular probes. PLoS One 8, e78752.

Fang, M.Z., Chen, D., Sun, Y., Jin, Z., Christman, J.K., Yang, C.S., 2005. Reversal of hypermethylation and reactivation of p16INK4a, RARbeta, and MGMT genes by genistein and other isoflavones from soy. Clin. Cancer Res. 11, 7033–7041.

Fang, M.Z., Wang, Y., Ai, N., Hou, Z., Sun, Y., Lu, H., Welsh, W., Yang, C.S., 2003. Tea polyphenol (-)-epigallocatechin-3-gallate inhibits DNA methyltransferase and reactivates methylation-silenced genes in cancer cell lines. Cancer Res. 63, 7563–7570.

Feinberg, A.P., 2007. Phenotypic plasticity and the epigenetics of human disease. Nature 447, 433–440.

Feinberg, A.P., Tycko, B., 2004. The history of cancer epigenetics. Nat. Rev. Cancer 4, 143–153.

Flotho, C., Claus, R., Batz, C., Schneider, M., Sandrock, I., Ihde, S., Plass, C., Niemeyer, C.M., Lubbert, M., 2009. The DNA methyltransferase inhibitors azacitidine, decitabine and zebularine exert differential effects on cancer gene expression in acute myeloid leukemia cells. Leukemia 23, 1019–1028.

Gabbara, S., Bhagwat, A.S., 1995. The mechanism of inhibition of DNA (cytosine-5-)-methyltransferases by 5-azacytosine is likely to involve methyl transfer to the inhibitor. Biochem. J. 307 (Pt 1), 87–92.

Girault, I., Tozlu, S., Lidereau, R., Bieche, I., 2003. Expression analysis of DNA methyltransferases 1, 3A, and 3B in sporadic breast carcinomas. Clin. Cancer Res. 9, 4415–4422.

Gros, C., Chauvigne, L., Poulet, A., Menon, Y., Ausseil, F., Dufau, I., Arimondo, P.B., 2013. Development of a universal radioactive DNA methyltransferase inhibition test for high-throughput screening and mechanistic studies. Nucleic Acids Res. 41, e185.

Gros, C., Fleury, L., Nahoum, V., Faux, C., Valente, S., Labella, D., Cantagrel, F., Rilova, E., Bouhlel, M.A., David-Cordonnier, M.H., Dufau, I., Ausseil, F., Mai, A., Mourey, L., Lacroix, L., Arimondo, P.B., 2015. New insights on the mechanism of quinoline-based DNA methyltransferase inhibitors. J. Biol. Chem. 290, 6293–6302.

Guianvarc'h, D., Arimondo, P.B., 2014. Challenges in developing novel DNA methyltransferases inhibitors for cancer therapy. Future Med. Chem. 6, 1237–1240.

Halby, L., Arimondo, P.B., 2015. Quinazoline Derivatives and Their Use as DNA Methyltransferase Inhibitors. Google Patents.

Halby, L., Champion, C., Senamaud-Beaufort, C., Ajjan, S., Drujon, T., Rajavelu, A., Ceccaldi, A., Jurkowska, R., Lequin, O., Nelson, W.G., Guy, A., Jeltsch, A., Guianvarc'h, D., Ferroud, C., Arimondo, P.B., 2012. Rapid synthesis of new DNMT inhibitors derivatives of procainamide. Chem. Bio. Chem. 13, 157–165.

Heinke, R., Carlino, L., Kannan, S., Jung, M., Sippl, W., 2011. Computer- and structure-based lead design for epigenetic targets. Bioorg. Med. Chem. 19, 3605–3615.

Higueruelo, A.P., Jubb, H., Blundell, T.L., 2013. Protein–protein interactions as druggable targets: recent technological advances. Curr. Opin. Pharmacol. 13, 791–796.

Hobley, G., McKelvie, J.C., Harmer, J.E., Howe, J., Oyston, P.C., Roach, P.L., 2012. Development of rationally designed DNA N6 adenine methyltransferase inhibitors. Bioorg. Med. Chem. Lett. 22, 3079–3082.

Illingworth, R.S., Bird, A.P., 2009. CpG islands—'a rough guide'. FEBS Lett. 583, 1713–1720.

Isakovic, L., Saavedra, O.M., Llewellyn, D.B., Claridge, S., Zhan, L., Bernstein, N., Vaisburg, A., Elowe, N., Petschner, A.J., Rahil, J., Beaulieu, N., Gauthier, F., MacLeod, A.R., Delorme, D., Besterman, J.M., Wahhab, A., 2009. Constrained (l-)-S-adenosyl-l-homocysteine (SAH) analogues as DNA methyltransferase inhibitors. Bioorg. Med. Chem. Lett. 19, 2742–2746.

Jia, D., Jurkowska, R.Z., Zhang, X., Jeltsch, A., Cheng, X., 2007. Structure of Dnmt3a bound to Dnmt3L suggests a model for de novo DNA methylation. Nature 449, 248–251.

Jones, P.A., Baylin, S.B., 2007. The epigenomics of cancer. Cell 128, 683–692.

Jurkowska, R.Z., Jurkowski, T.P., Jeltsch, A., 2011. Structure and function of mammalian DNA methyltransferases. Chem. Bio. Chem. 12, 206–222.

Kabro, A., Lachance, H., Marcoux-Archambault, I., Perrier, V., Dore, V., Gros, C., Masson, V., Gregoire, J.-M., Ausseil, F., Cheishvili, D., Laulan, N.B., St-Pierre, Y., Szyf, M., Arimondo, P.B., Gagnon, A., 2013. Preparation of phenylethylbenzamide derivatives as modulators of DNMT3 activity. Med. Chem. Comm. 4, 1562–1570.

Kilgore, J.A., Du, X., Melito, L., Wei, S., Wang, C., Chin, H.G., Posner, B., Pradhan, S., Ready, J.M., Williams, N.S., 2013. Identification of DNMT1 selective antagonists using a novel scintillation proximity assay. J. Biol. Chem. 288, 19673–19684.

Klimasauskas, S., Kumar, S., Roberts, R.J., Cheng, X., 1994. HhaI methyltransferase flips its target base out of the DNA helix. Cell 76, 357–369.

Kuck, D., Caulfield, T., Lyko, F., Medina-Franco, J.L., 2010a. Nanaomycin A selectively inhibits DNMT3B and reactivates silenced tumor suppressor genes in human cancer cells. Mol. Cancer Ther. 9, 3015–3023.

Kuck, D., Singh, N., Lyko, F., Medina-Franco, J.L., 2010b. Novel and selective DNA methyltransferase inhibitors: docking-based virtual screening and experimental evaluation. Bioorg. Med. Chem. 18, 822–829.

Kumar, S., Agrawal, R., 2014. Next generation tyrosine kinase inhibitor (TKI): afatinib. Recent Pat. Anticancer Drug Discov. 9, 382–393.

Laraia, L., McKenzie, G., Spring, D.R., Venkitaraman, A.R., Huggins, D.J., 2015. Overcoming chemical, biological, and computational challenges in the development of inhibitors targeting protein-protein interactions. Chem. Biol. 22, 689–703.

Lee, B.H., Yegnasubramanian, S., Lin, X., Nelson, W.G., 2005. Procainamide is a specific inhibitor of DNA methyltransferase 1. J. Biol. Chem. 280, 40749–40756.

Liu, X., Gao, Q., Li, P., Zhao, Q., Zhang, J., Li, J., Koseki, H., Wong, J., 2013. UHRF1 targets DNMT1 for DNA methylation through cooperative binding of hemi-methylated DNA and methylated H3K9. Nat. Commun. 4, 1563.

Lopez-Tarruella, S., Jerez, Y., Marquez-Rodas, I., Martin, M., 2012. Neratinib (HKI-272) in the treatment of breast cancer. Future Oncol. 8, 671–681.

Maldonado-Rojas, W., Olivero-Verbel, J., Marrero-Ponce, Y., 2015. Computational fishing of new DNA methyltransferase inhibitors from natural products. J. Mol. Graph Model 60, 43–54.

Mashhoon, N., Carroll, M., Pruss, C., Eberhard, J., Ishikawa, S., Estabrook, R.A., Reich, N., 2004. Functional characterization of *Escherichia coli* DNA adenine methyltransferase, a novel target for antibiotics. J. Biol. Chem. 279, 52075–52081.

Medina-Franco, J.L., Yoo, J., 2013. Molecular modeling and virtual screening of DNA methyltransferase inhibitors. Curr. Pharm. Des. 19, 2138–2147.

Mori, S., Iwase, K., Iwanami, N., Tanaka, Y., Kagechika, H., Hirano, T., 2010. Development of novel bisubstrate-type inhibitors of histone methyltransferase SET7/9. Bioorg. Med. Chem. 18, 8158–8166.

Newman, E.M., Morgan, R.J., Kummar, S., Beumer, J.H., Blanchard, M.S., Ruel, C., El-Khoueiry, A.B., Carroll, M.I., Hou, J.M., Li, C., Lenz, H.J., Eiseman, J.L., Doroshow, J.H., 2015. A phase I, pharmacokinetic, and pharmacodynamic evaluation of the DNA methyltransferase inhibitor 5-fluoro-2′-deoxycytidine, administered with tetrahydrouridine. Cancer Chemother. Pharmacol. 75, 537–546.

Nie, J., Liu, L., Li, X., Han, W., 2014. Decitabine, a new star in epigenetic therapy: the clinical application and biological mechanism in solid tumors. Cancer Lett. 354, 12–20.

Oh, B.K., Kim, H., Park, H.J., Shim, Y.H., Choi, J., Park, C., Park, Y.N., 2007. DNA methyltransferase expression and DNA methylation in human hepatocellular carcinoma and their clinicopathological correlation. Int. J. Mol. Med. 20, 65–73.

Okano, M., Bell, D.W., Haber, D.A., Li, E., 1999. DNA methyltransferases Dnmt3a and Dnmt3b are essential for de novo methylation and mammalian development. Cell 99, 247–257.

Osborne, T., Roska, R.L., Rajski, S.R., Thompson, P.R., 2008. In situ generation of a bisubstrate analogue for protein arginine methyltransferase 1. J. Am. Chem. Soc. 130, 4574–4575.

Patra, S.K., Patra, A., Zhao, H., Dahiya, R., 2002. DNA methyltransferase and demethylase in human prostate cancer. Mol. Carcinog. 33, 163–171.

Quintas-Cardama, A., Santos, F.P., Garcia-Manero, G., 2010. Therapy with azanucleosides for myelodysplastic syndromes. Nat. Rev. Clin. Oncol. 7, 433–444.

Rajavelu, A., Tulyasheva, Z., Jaiswal, R., Jeltsch, A., Kuhnert, N., 2011. The inhibition of the mammalian DNA methyltransferase 3a (Dnmt3a) by dietary black tea and coffee polyphenols. BMC Biochem. 12, 16.

Reinisch, K.M., Chen, L., Verdine, G.L., Lipscomb, W.N., 1995. The crystal structure of HaeIII methyltransferase convalently complexed to DNA: an extrahelical cytosine and rearranged base pairing. Cell 82, 143–153.

Rilova, E., Erdmann, A., Gros, C., Masson, V., Aussagues, Y., Poughon-Cassabois, V., Rajavelu, A., Jeltsch, A., Menon, Y., Novosad, N., Gregoire, J.M., Vispe, S., Schambel, P., Ausseil, F., Sautel, F., Arimondo, P.B., Cantagrel, F., 2014. Design, synthesis and biological evaluation of 4-amino-N-(4-aminophenyl)benzamide analogues of quinoline-based SGI-1027 as inhibitors of DNA methylation. Chem. Med. Chem. 9, 590–601.

Rothbart, S.B., Krajewski, K., Nady, N., Tempel, W., Xue, S., Badeaux, A.I., Barsyte-Lovejoy, D., Martinez, J.Y., Bedford, M.T., Fuchs, S.M., Arrowsmith, C.H., Strahl, B.D., 2012. Association of UHRF1 with methylated H3K9 directs the maintenance of DNA methylation. Nat. Struct. Mol. Biol. 19, 1155–1160.

Saavedra, O.M., Isakovic, L., Llewellyn, D.B., Zhan, L., Bernstein, N., Claridge, S., Raeppel, F., Vaisburg, A., Elowe, N., Petschner, A.J., Rahil, J., Beaulieu, N., MacLeod, A.R., Delorme, D., Besterman, J.M., Wahhab, A., 2009. SAR around (l)-S-adenosyl-l-homocysteine, an inhibitor of human DNA methyltransferase (DNMT) enzymes. Bioorg. Med. Chem. Lett. 19, 2747–2751.

Santi, D.V., Garrett, C.E., Barr, P.J., 1983. On the mechanism of inhibition of DNA-cytosine methyltransferases by cytosine analogs. Cell 33, 9–10.

Santi, D.V., Norment, A., Garrett, C.E., 1984. Covalent bond formation between a DNA-cytosine methyltransferase and DNA containing 5-azacytosine. Proc. Natl. Acad. Sci. U.S.A. 81, 6993–6997.

Shi, J., Wang, Y., Zeng, L., Wu, Y., Deng, J., Zhang, Q., Lin, Y., Li, J., Kang, T., Tao, M., Rusinova, E., Zhang, G., Wang, C., Zhu, H., Yao, J., Zeng, Y.-X., Evers, B.M., Zhou, M.-M., Zhou, B.P., 2014. Disrupting the interaction of BRD4 with di-acetylated twist suppresses tumorigenesis in basal-like breast cancer. Cancer Cell 25, 210–225.

Siedlecki, P., Garcia Boy, R., Comagic, S., Schirrmacher, R., Wiessler, M., Zielenkiewicz, P., Suhai, S., Lyko, F., 2003. Establishment and functional validation of a structural homology model for human DNA methyltransferase 1. Biochem. Biophys. Res. Commun. 306, 558–563.

Singh, N., Duenas-Gonzalez, A., Lyko, F., Medina-Franco, J.L., 2009. Molecular modeling and molecular dynamics studies of hydralazine with human DNA methyltransferase 1. Chem. Med. Chem. 4, 792–799.

Song, J., Rechkoblit, O., Bestor, T.H., Patel, D.J., 2011. Structure of DNMT1-DNA complex reveals a role for autoinhibition in maintenance DNA methylation. Science 331, 1036–1040.

Song, J., Teplova, M., Ishibe-Murakami, S., Patel, D.J., 2012. Structure-based mechanistic insights into DNMT1-mediated maintenance DNA methylation. Science 335, 709–712.

Stresemann, C., Brueckner, B., Musch, T., Stopper, H., Lyko, F., 2006. Functional diversity of DNA methyltransferase inhibitors in human cancer cell lines. Cancer Res. 66, 2794–2800.

Stresemann, C., Lyko, F., 2008. Modes of action of the DNA methyltransferase inhibitors azacytidine and decitabine. Int. J. Cancer 123, 8–13.

Suzuki, M.M., Bird, A., 2008. DNA methylation landscapes: provocative insights from epigenomics. Nat. Rev. Genet. 9, 465–476.

Suzuki, T., Tanaka, R., Hamada, S., Nakagawa, H., Miyata, N., 2010. Design, synthesis, inhibitory activity, and binding mode study of novel DNA methyltransferase 1 inhibitors. Bioorg. Med. Chem. Lett. 20, 1124–1127.

Takeshita, K., Suetake, I., Yamashita, E., Suga, M., Narita, H., Nakagawa, A., Tajima, S., 2011. Structural insight into maintenance methylation by mouse DNA methyltransferase 1 (Dnmt1). Proc. Natl. Acad. Sci. U.S.A. 108, 9055–9059.

Villar-Garea, A., Fraga, M.F., Espada, J., Esteller, M., 2003. Procaine is a DNA-demethylating agent with growth-inhibitory effects in human cancer cells. Cancer Res. 63, 4984–4989.

Vispe, S., Deroide, A., Davoine, E., Desjobert, C., Lestienne, F., Fournier, L., Novosad, N., Breand, S., Besse, J., Busato, F., Tost, J., De Vries, L., Cussac, D., Riond, J., Arimondo, P.B., 2015. Consequences of combining siRNA-mediated DNA methyltransferase 1 depletion with 5-aza-2′-deoxycytidine in human leukemic KG1 cells. Oncotarget 6, 15265–15282.

Wahhab, A., 2006. Inhibitors of DNA Methyltransferase. WO2006078752.

Wee, S., Dhanak, D., Li, H., Armstrong, S.A., Copeland, R.A., Sims, R., Baylin, S.B., Liu, X.S., Schweizer, L., 2014. Targeting epigenetic regulators for cancer therapy. Ann. N. Y. Acad. Sci. 1309, 30–36.

Yoo, C.B., Jeong, S., Egger, G., Liang, G., Phiasivongsa, P., Tang, C., Redkar, S., Jones, P.A., 2007. Delivery of 5-aza-2′-deoxycytidine to cells using oligodeoxynucleotides. Cancer Res. 67, 6400–6408.

Yoo, J., Choi, S., Medina-Franco, J.L., 2013. Molecular modeling studies of the novel inhibitors of DNA methyltransferases SGI-1027 and CBC12: implications for the mechanism of inhibition of DNMTs. PLoS One 8, e62152.

Yoo, J., Kim, J.H., Robertson, K.D., Medina-Franco, J.L., 2012. Molecular modeling of inhibitors of human DNA methyltransferase with a crystal structure: discovery of a novel DNMT1 inhibitor. Adv. Protein Chem. Struct. Biol. 87, 219–247.

Yoo, J., Medina-Franco, J.L., 2011. Homology modeling, docking and structure-based pharmacophore of inhibitors of DNA methyltransferase. J. Comput. Aided Mol. Des. 25, 555–567.

CHAPTER 4

Discovery and Development of Small Molecules Targeting Epigenetic Enzymes with Computational Methods

Fanwang Meng[1,2], Chen Wang[1], Wei Wan[1], Wenchao Lu[1], Wencong Lu[2], Cheng Luo[1]

[1]Drug Discovery and Design Center, State Key Laboratory of Drug Research, Shanghai Institute of Materia Medica, Chinese Academy of Sciences, Shanghai, China; [2]Department of Chemistry, College of Sciences, Shanghai University, Shanghai, China

CONTENTS

1. INTRODUCTION

Epigenetics is defined as the study of stably heritable phenotype resulting from changes in gene expression without alterations in the DNA sequence (Berger et al., 2009). Covalent modifications on nucleotides and histones, and RNA-associated silencing are involved in the epigenetic regulation of gene expression profiles, which are often referred to as epigenetic machinery. Specific genes can be activated or repressed at the proper time and placed through dedicate epigenetic machinery that is vital for several biological processes, such as stem cell self-renewal, and differentiation embryo development (Li, 2002). As such, epigenetics regulation has a major role in normal changes in gene expression. Emerging evidence has shown that disrupted epigenetic regulatory can cause gene overexpression or gene silencing, which may finally result in epigenetic diseases such as cancer, neural disorders, metabolic disorders, autoimmune diseases, inflammation, and others (Arrowsmith et al., 2012; Egger et al., 2004; Falkenberg and Johnstone, 2014; Feinberg and Vogelsten, 1983; Jakovcevski and Akbarian, 2012).

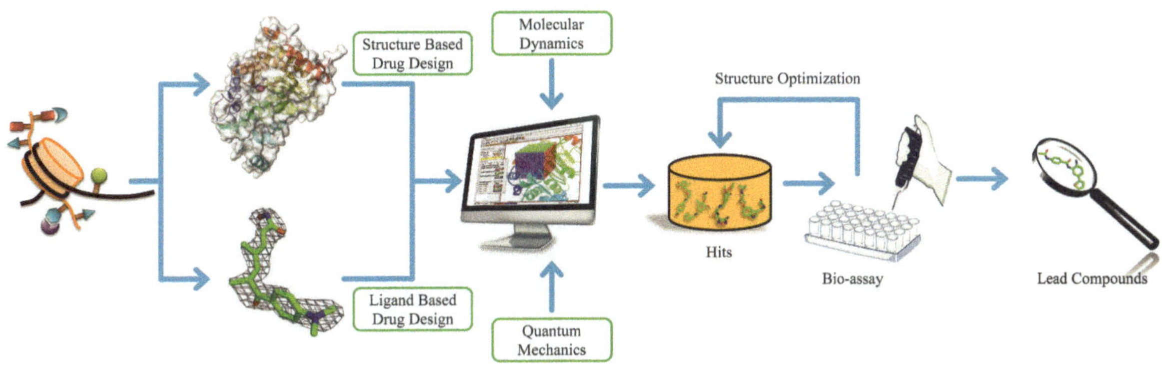

Epi-Informatics. http://dx.doi.org/10.1016/B978-0-12-802808-7.00004-6

Epigenetic regulators can be divided into three categories: epigenetic writers, epigenetic readers, and epigenetic erasers (Arrowsmith et al., 2012; Falkenberg and Johnstone, 2014) (Figure 1). The writer enzymes are responsible for adding an acetyl group or methyl group in histone tails including DNA methyltransferases (DNMT), histone acetyltransferases (HKMT), and protein arginine methyltransferases (PRMT). The writers make chromatin more open and usually result in gene suppression. In contrast, epigenetic eraser proteins such as histone deacetylases and demethylases can reverse covalent modifications within histone tails. Epigenetic readers can recognize modified histone marks and recruit other proteins, which contain PHD figures, bromodomains, Turo domains, and DNA methyl-binding domains.

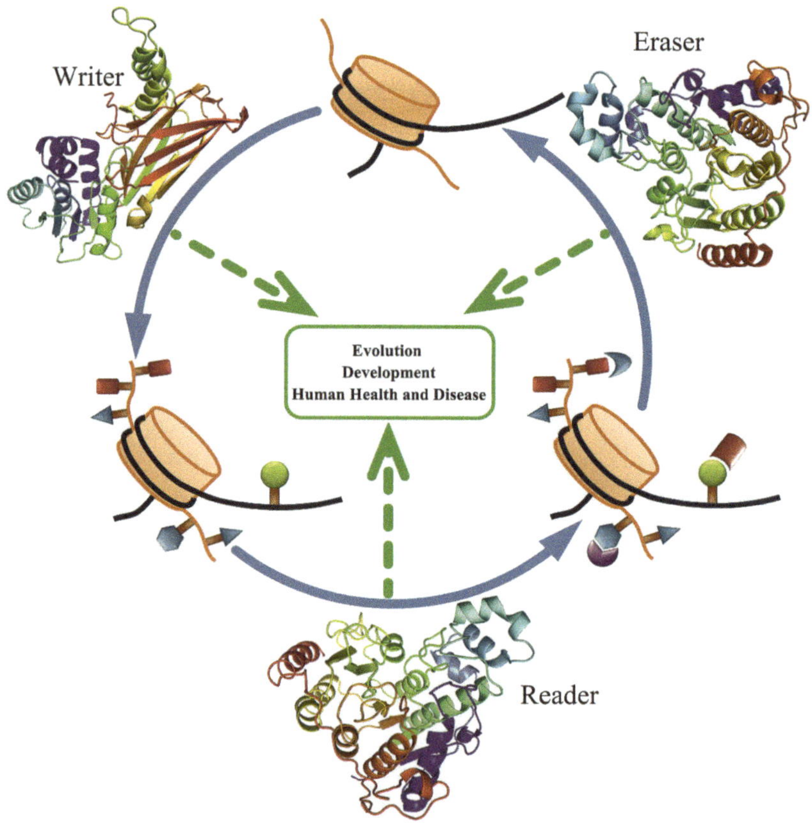

FIGURE 1 Histone writers, readers, and erasers.
Representative structures of histone writers, readers, and erasers are protein arginine PRMT1 (PDB code: 2V74), BRD4 (PDB code: 4OGI), and HDAC2 (PDB code: 4LXZ), which are shown in drawings.  These symbols represent histone modification codes, such as methylation, acetylation, phosphorylation, and so on;  are legends for epigenetic readers. Epigenetic modifying enzymes play significant roles in tissue development, evolution, human health and disease, and so forth.

Notably, epigenetic regulations in cancer are reversible whereas genetic mutations are irreversible. This characteristic feature has inspired both academic and industrial organizations to discover and develop new drugs to reverse the abnormal epigenetic landscape, especially in the field of cancer therapy. Vorinostat (compound 19 in Figure 7) is the first epigenetic drug that was approved by the US FDA in 2006 for the treatment of cutaneous T cell lymphoma (CTCL) targeting histone deacetylase (HDAC) (Arrowsmith et al., 2012), and many other epigenetic drugs are in clinical trials. The largely increased knowledge of substrate selectivity, regulatory signaling pathway, and three-dimensional (3D) structures of epigenetic proteins has paved the way for epigenetic drug discovery and development.

Computer-aided drug design (CADD) has been used extensively in accelerated modern drug design and development. The rapid development of both high-performance hardware and software makes computational methods more convenient and easy to use. Modern computational methods such as docking, pharmacophore, 3D quantitative structure activity relationship (3D-QSAR), molecular dynamics (MD), and quantum mechanics (QM) can also facilitate epigenetic drug design. The principles and methodologies of these computational methods are beyond the scope of this chapter and several excellent reviews and books are recommended (Bello et al., 2013; Honarparvar et al., 2014; Lavecchia and Di Giovanni, 2013; Lionta et al., 2014). In this chapter, we will focus on the applications of computational methods in epigenetic inhibitors and probes design and development; however other aspects of computer-aided methods such as ADME-Tox prediction, bioinformatics, and system biology are not covered here.

2. WRITER

2.1 DNA Methlytransferase

DNA methylation is a major form of epigenetic modification that regulates gene expression by depositing a methyl group on the 5-position of the cytosine ring (Robertson, 2001). DNMTs catalyze DNA methylation and so far five members have been found in mammals: DNMT1, DNMT2, and de novo methyltransferases DNMT3A, DNMT3B, and DNMT3L (Deplus et al., 2002; Goll and Bestor, 2005). Among these, DNMT1 is the most abundant and functions to maintain methylation predominantly at CpG dinucleotides through recognizing the methyl group on the template strand while introducing a new one into the daughter strand during DNA replication (Bestor, 2000). Otherwise, DNMT3A and DNMT3B, working together with DNMT3L, act as de novo methylation enzymes to deposit methyl groups on the unmethylated DNA (Auclair and Weber, 2012; Suetake et al., 2004). Despite the fact that DNMT3L shares a high sequence similarity with DNMT3A, it lacks the catalytic residues for DNMT activity (Cheng and Blumenthal, 2008). Also, although DNMT2 has a conserved structure like other regulators, little knowledge has been obtained in terms of its function (Heil, 2014; Schaefer and Lyko, 2010).

Since previous studies showed aberrant DNA hypermethylation patterns have a high correlation with tumorigenesis due to the repression of some tumor suppressor genes (TSG), DNMTs have become one of the most promising targets of cancers (Feinberg, 2006; Medina-Franco et al., 2015). To date, azacitidine (compound one in Figure 2; approved by the FDA in 2004 for the treatment of Myeloid Dysplasic Syndrome (MDS), Acute Myeloid Leukemia (AML), and Chronic Myelomonocytic Leukemia (CMML)) and decitabine (compound two in Figure 2; approved by the US FDA in 2006 against MDS and AML) are the only two drugs used in clinic (Issa and Kantarjian, 2009). These two inhibitors are nucleotide analogues and can be incorporated into genomic DNA to covalently trap with DNMTs, exhibiting a profound improvement in patient survival. However, both inhibitors show low selectivity and have strong toxicity effects (Hu et al., 2015). Thus, the demand for novel and more potent and specific inhibitors becomes urgent.

2.1.1 Structural and Functional Features

Briefly, DNMT proteins can be divided into two parts: the variable N-terminal regulatory domain and the conserved C-terminal catalytic domain (Bestor, 1988; Yen et al., 1992). The N-terminal part is required for homodimerization and contains a CXXC (Cys-X-X-Cys) domain, which specifically binds to the unmethylated CpG DNA (Fellinger et al., 2009; Pradhan et al., 2008). By solving structures of DNMT1 with unmethylated DNA fraction, Song et al. proposed an autoinhibitory mechanism in DNA methylation maintenance that the CXXC containing linker could position between DNA and the active site of the DNMT1, protected from de novo methylation (Song et al., 2011). Additionally, another mechanism was proposed where the target recognition domain (TRD) within the catalytic part of DNMT recognized the hemimethylated CpG site and maintained DNA methylation (Song et al., 2012). Taken together, these two mechanisms support each other and work in tandem. On the other hand, the ATRX-DNMT3-DNMT3L (ADD) domain of DNMT3A recognized the unmethylated state of lysine 4 in histone H3, indicating that de novo DNMTs were first recruited onto chromatin and then methylated CpG dinucleotides (Otani et al., 2009). S-Adenosylmethionine (SAM) binding site of DNMT1 is shown in Figure 3(a).

2.1.2 Combination of Homology Modeling and Docking-Based Virtual Screening

Siedlecki et al. identified the first DNMT1 inhibitor RG-108 by virtual screening using a homology model (Siedlecki et al., 2006). Since there was no determination of crystal structures of DNMT1 until 2008, a structural

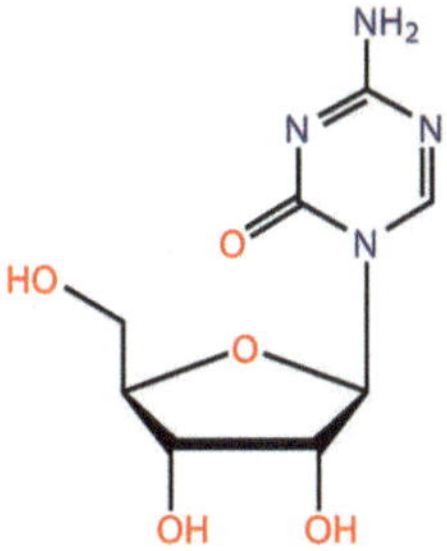

1
Azacitidine

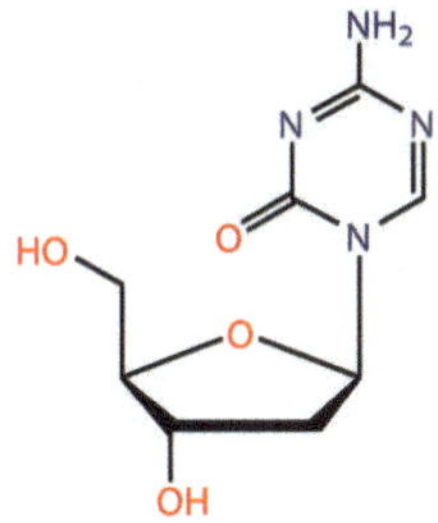

2
Decitabine

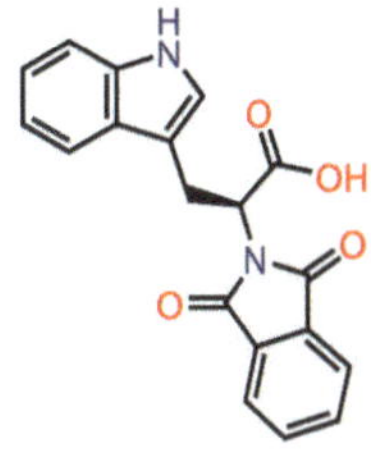

3
RG-108

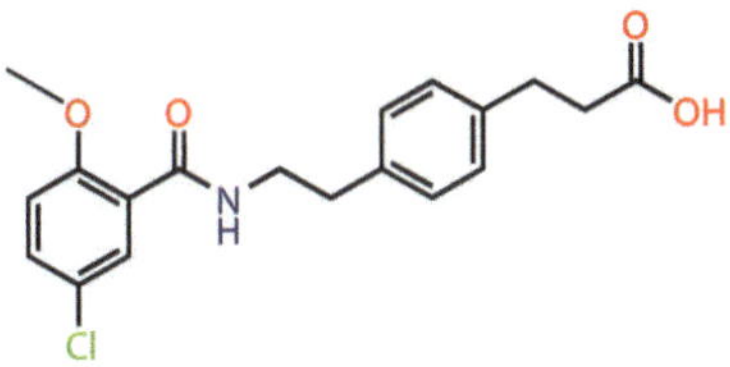

4
NSC 319745

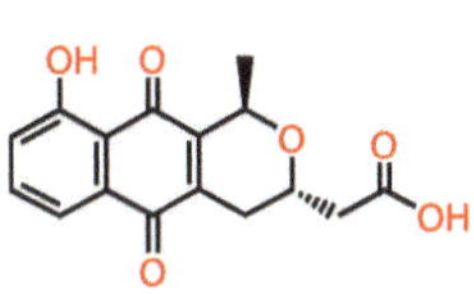

5
Nanaomycin A

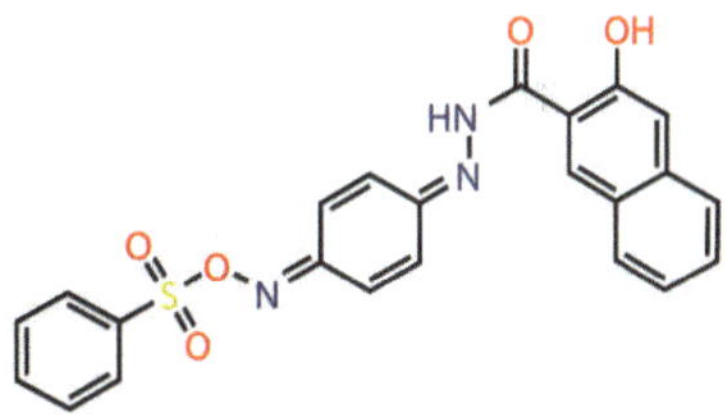

6
A9

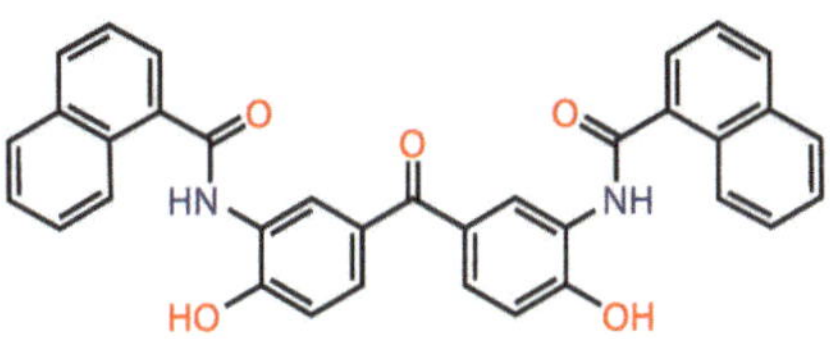

7
A36

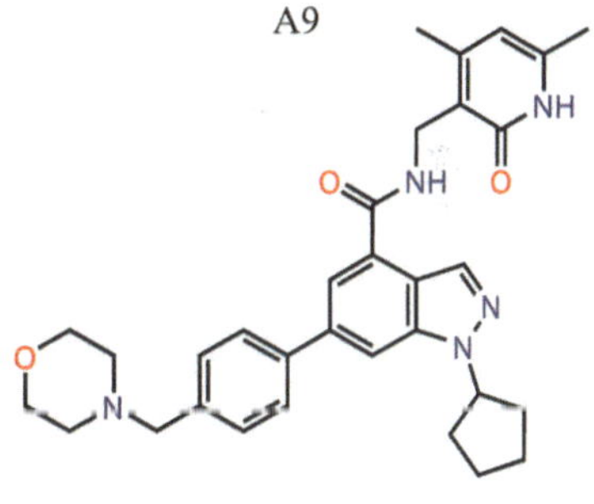

8
EPZ005687

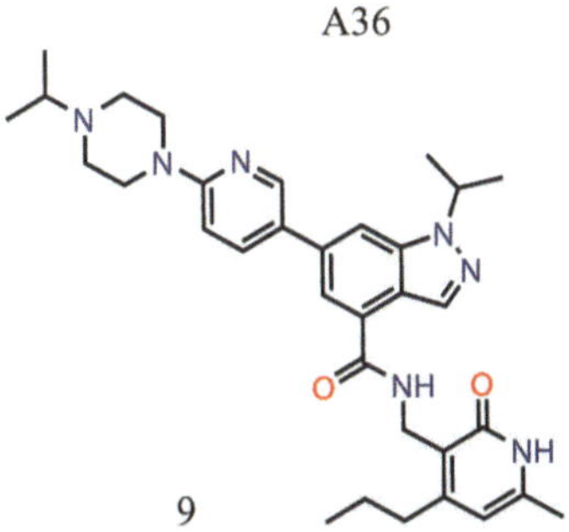

9
UNC1999

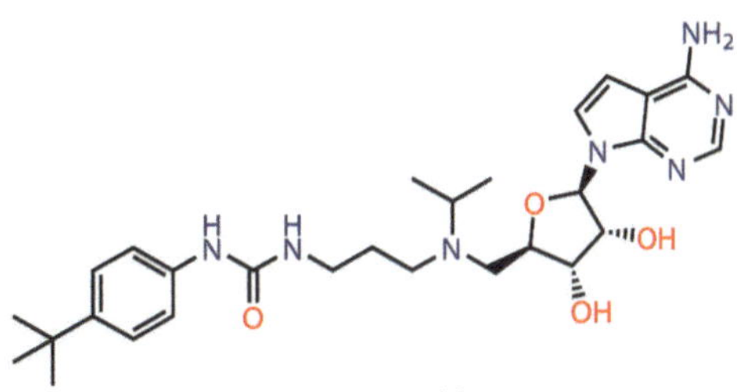

10
EPZ004777

FIGURE 2 **Chemical structures of epigenetic writer inhibitors.**

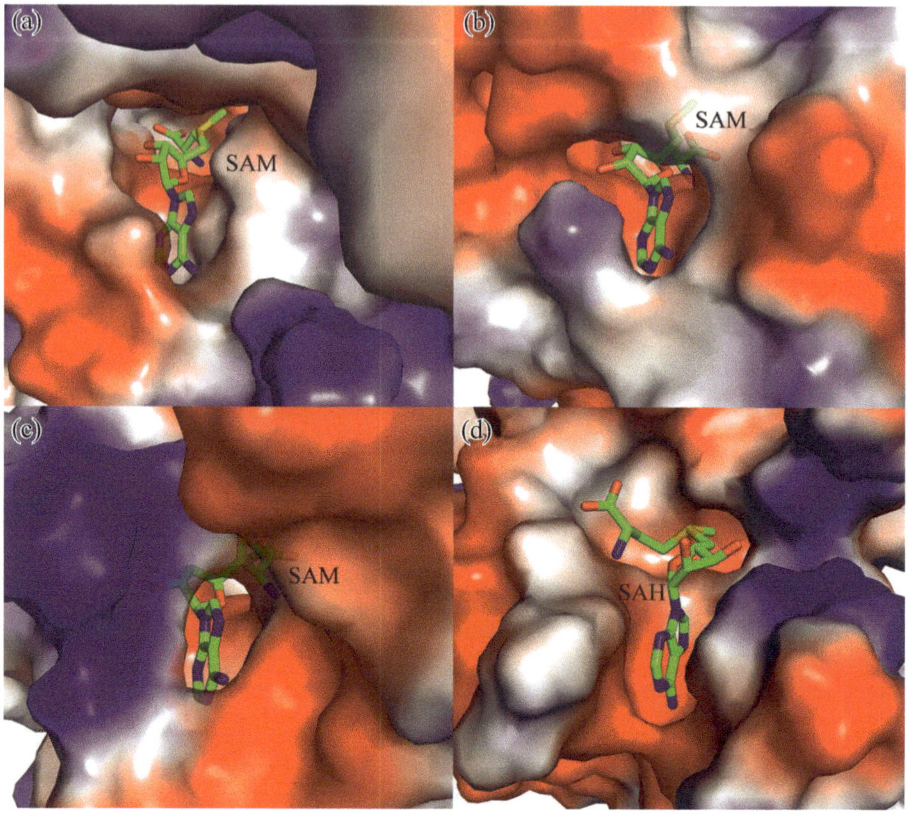

FIGURE 3 S-Adenosyl methionine (SAM) binding site of epigenetic writers.
Histone writers are shown in electrostatic potential surface, including DNMT1 (PDB code: 3AV6) (a), DOT1L (PDB code: 1NW3) (b), PRMT1 (PDB code: 2V74) (c), and SETD7 (PDB code: 1N6A) (d). SAM are shown as sticks as in (a), (b), and (c) and SAH is shown as sticks in (d). Although SAM is the shared cofactor, it can be concluded that the binding sites of different subfamily histone writers differ from each other. Also, the binding conformations of SAM are different. The figure is generated with PyMOL version 1.3r1 (Schrodinger, 2010).

homology modeling of the human DNMT1 catalytic domain was the basis of subsequent virtual screening. Through model establishment and further docking of 1990 compounds into the modeled pocket of the enzyme, RG-108 (compound 3 in Figure 2) was finally obtained. Both in vitro and in vivo methylation assays revealed RG-108's efficacy and the docking data demonstrated its deep binding pattern inside the catalytic site of DNMT1 (Brueckner et al., 2005; Siedlecki et al., 2006). However, because RG-108 shows only low micromolar activity and barely has any selectivity, the compound is not developed further (Suzuki et al., 2010). Following this work, Kuck et al. identified NSC319745 (compound 4 in Figure 2) through a multistep docking strategy using the same homology model (Kuck et al., 2010b). NSC319745 turned out to be not only potent but also selective for DNMT1 in the biochemical assay. Finding these lead candidates like RG-108 and NSC319745 has laid a foundation of further optimization work on more potent and selective inhibitors.

2.1.3 Molecular Dynamics

Caulfield et al. reported the first MD study of DNMT3B bound to nanaomycin A (Caulfield and Medina-Franco, 2011). Nanaomycin A was previously identified by in silico screening searching for DNMT1 inhibitors (Kuck et al., 2010b). However, upon further experimental testing, DNMT3B turned out to be the cellular target of nanaomycin A (compound 5 in Figure 2), making nanaomycin A the first selective inhibitor against DNMT3B (Kuck et al., 2010a). To provide evidence for this selectivity and deeply explore the inhibitory mechanism, Caulfield et al. conducted long time scale (>100 ns) MD simulations of human DNMT3B bound to nanaomycin A with or without SAM. Upon analyzing the binding mode and key interactions of nanaomycin A with DNMT3B, results revealed that cofactor SAM favors the binding of nanaomycin A, and many residues interacting with nanaomycin A are involved in the DNA methylation event. Results further support the hypothesis proposed by previous docking data that the catalytic cysteine perform a nuclephilic Michael 1,4 addition to the α,β-unsaturated carbonyl system of nanaomycin A.

In a 2014 study, Carvalho et al. conducted a series of MD simulations on native or modified (methylated or hetroxymethylated) DNA dodecamers with different sequences (Carvalho et al., 2014). Results showed that modified cytosines exhibit pronounced effects on DNA structural parameters (include twist, roll, and tilt angles) near the methylation site, especially on symmetrically methylated CpG dinucleotides. Finally, the authors proposed that the sequence-dependent conformational changes at certain regions may increase the flexibility and accessibility of DNA at specific sites for proteins like transcription factors to bind.

2.2 Histone Methyltransferase

Histone methyltransferases (HMTs) are histone-modifying enzymes that catalyze the addition of methyl groups to specific residues (lysine or arginine) predominantly on histone H3. Therefore HMTs can be divided into two major types: (1) protein lysine methyltransferases (PKMTs), which can be further divided into SET domains containing enzymes like EZH2 and non-SET domains containing enzymes like DOT1L; and (2) protein arginine methyltransferase (PRMTs) (Allis et al., 2007; Q. Feng et al., 2002; Trievel, 2004). A detailed classification of PKMTs and PRMTs is reviewed elsewhere (Li et al., 2012). HMTs take SAM as their cofactor and by transferring the methyl groups from SAM to the substrate, they biologically function as a regulator of gene expression, genomic stability, lineage development, and so on (Chen et al., 1999; Martin and Zhang, 2005). Besides, a close link between histone methylation patterns and pathogenesis of cancer has gradually been realized (Spannhoff et al., 2009). For example, over-expression of EZH2 has been implicated in breast cancer and prostate cancer while hDOT1L is thought to be related to certain leukemias (Okada et al., 2005; Tao et al., 2014; Vallot et al., 2015). Also, down-regulation of the expression of PRMT1 was demonstrated to suppress Mixed Lineage Leukemia (MLL)-mediated transformation (Cheung et al., 2007). Thus, inhibition of HMTs becomes a potential therapeutic strategy for cancer.

2.2.1 Structural and Functional Features

To date, many crystal structures derived for HMT, HMT-SAM (or S-adenosylhomocysteine (SAH)), HMT-substrate, and HMT-inhibitor are available. Crystal studies showed that SAM or SAH exhibited two different binding modes toward the cofactor-binding pockets of HMTs (Cheng, 2005). For example, SAM adopted a crooked configuration when bound to most PKMTs, which were appropriated for the interaction of its methylsulphonium cation with the basic group of the substrate lysine residues (Cheng et al., 2005). However, in the case of SAH bound to PRMTs, SAH preferred an extended configuration (Yue et al., 2007). Exceptionally, the non-SET domain containing PKMT, DOT1L adopted a rather extended structure compared to that seen in SET-domain containing PKMTs (Richon et al., 2011). Besides the compact SAM-binding pocket, there is also a hydrophobic substrate-binding channel located in the opposite face of the enzyme surface, leaving a cone-shaped active site in between (Zhang et al., 2000). Notably, the structure of the dimerization arm only appeared in PRMTs but not in PKMTs, attributed to the dimer formation, which was crucial for SAM binding and catalytic activities of PRMTs (Zhang and Cheng, 2003).

2.2.2 Pharmacophore-Based Virtual Screening

To identify potent PRMT1 inhibitors with a new scaffold, Wang et al. employed a pharmacophore-based virtual screening from a theoretical soluble ($\log S > -4$) library that contained 90,000 small molecules (Wang et al., 2012). Pharmacophore-based methods are always carried out when a sufficient group of bioactive analogs have already been achieved. After a cluster analysis of the 798 candidate compounds selected from virtual screening, top-ranked compounds of each cluster were selected and 102 small molecules were at last available for further experimental validation. By performing recombinant enzyme inhibition assay, compound A9 and A36 (compounds 6 and 7 in Figure 2) were finally achieved as new chemical leads for further optimizing programs of inhibitors targeting PRMT1. Virtual screening appearing as an enrichment approach of a large size of candidate pool has shown its strength in novel compound identification.

2.2.3 Combination of Medicinal Chemistry Design and Computational Approaches

To search bioavailable chemical probes targeting EZH2, Konze et al. first established an EZH2 homology model based on the 3D structural of G9a-like protein (GLP), an H3K9 mono- and dimethylase (Konze et al., 2013). Through docking of the previously discovered inhibitor EPZ005687 (compound 8 in Figure 2) into this model, the morpholinomethyl moiety of EPZ005687 was found to show no interaction with EZH2 while other moieties presented important interactions including hydrogen bonds and hydrophobic forces. Upon detailed analysis of the binding modes, several compounds were designed by changing the useless morpholinomethyl moiety and keeping the pyridone indazole core for maintaining crucial structural features. Finally, UNC1999 (compound 9 in Figure 2) was obtained by undergoing several experimental validations for its efficacy (for EZH2, $IC_{50} < 10$ nM, in vitro) and selectivity. Notably, by evaluating the pharmacokinetic (PK) properties of UNC1999 in mice, it was found that UNC1999 showed better PK properties than EPZ005687, making itself a suitable oral-taken chemical probe in the study of the function of EZH2 in mouse models.

Daigle et al. identified the first inhibitor of DOT1L (EPZ004777, compound 10 in Figure 2), which showed a high potency ($IC_{50} \approx 400$ pM) and selectivity against MLL-rearranged leukemia (Daigle et al., 2011). This work was initiated upon deep understanding of the reaction mechanism of DOT1L catalysis and followed by a series of optimizing work to seek better SAM-like compounds based on the structural details of the DOT1L active site (Min et al., 2003). EPZ004777, as a lead compound, benefited many following works. For example, Anglin et al. designed and synthesized a total of 55 EPZ004777 analogous for searching DOT1L inhibitors with better selectivity and PK properties. Upon detailed studies on structure-activity relationships (SARs) of these compounds and a long-term optimizing program, several compounds were finally achieved to represent high selectivity (>4500-fold) over other HMTs (Anglin et al., 2012). In another work, by solving the structure of DOT1L in complex with EPZ004777 and performing MD simulations, together with biochemical evaluation, results showed that the binding of EPZ00477 resulted in dramatic conformational rearrangement, which enabled a cavity to open up to hold hydrophobic moiety of EPZ004777 (Basavapathruni et al., 2012). Through combination of medicinal chemistry design and CADD, therapeutic drugs that target HMTs are expected to be developed in the coming years.

3. READERS

Posttranslational modifications (PTMs) of histones are important for the control of chromatin dynamics (Carey and Workman, 2007) and can be inherited by daughter cells as functional epigenetic passengers (Ahmad and Henikoff, 2002). It is assumed that PTMs function as a medium to recruit effectors or readers, hence to be deciphered to accomplish their mission. But it does not mean that the language of PTMs can be decoded by simply matching the histone marks. The functional outcome of certain PTMs are usually correlated with the local chromatin and their cellular context (Berger, 2007). Thus, the structural determination and functional elucidation of the histone modification readers has already been a hotspot (Yun et al., 2011).

To date, there are several kinds of PTMs (except ubiquitination) with corresponding readers reported: lysine methylation, arginine methylation, phosphorylation, and acetylation. The lysine residues of histone may be monomethylated, dimethylated, or trimethylated, each of which forms a specific signal. Thus, lysine recognition modules include diverse reader domains, which are divided into two groups: Plant Homeodomain (PHD) finger and the "Royal Family," composed of Chromo, MBT, Tudor, PWWP, WD40, and so on (Arrowsmith et al., 2012; Maurer-Stroh et al., 2003; Sanchez and Zhou, 2011). Lysine methylation of histone is associated with transcription activation or repression, and is involved in various cancer, intellectual disability and aging (Greer and Shi, 2012). Thus, histone readers have been potential targets in drug discovery and development with some small molecule methyl-lysine (MeK) readership antagonist reported. We will review the inhibitor design of bromodomains.

3.1 Bromodomains

3.1.1 Structural and Functional Features

Bromodomains are small (approximately 110 amino acid) protein domains with conserved sequence and 3D structure, recognizing acetylated lysine of the *N*-terminal tails of histones. The domain adopts a bundle of four alpha helices (αZ-αA-αB-αC), and the differences of ZA and BC loops determine the specificity and affinity of endogenous histone peptides or exogenous small molecule inhibitors (Filippakopoulos et al., 2012) (Figure 4). In 1999, Dhalluin et al. reported the first 3D structure of the bromodomain, which is complexed with H4K8-acetylated peptide (Dhalluin et al., 1999). Since then, other histone modification readers like the chromo domain of HP1 (Bannister et al., 2001), MeK readers (Tudor and MBT) (Kim et al., 2006), and so on were determined with the development of high-throughput approaches. The human whole genome encodes 61 bromodomains in 46 different proteins and over half of their 3D structures have been deposited in the Protein Data Bank (Chung and Tough, 2012). These structures laid foundations of structure-based drug design targeting histone readers.

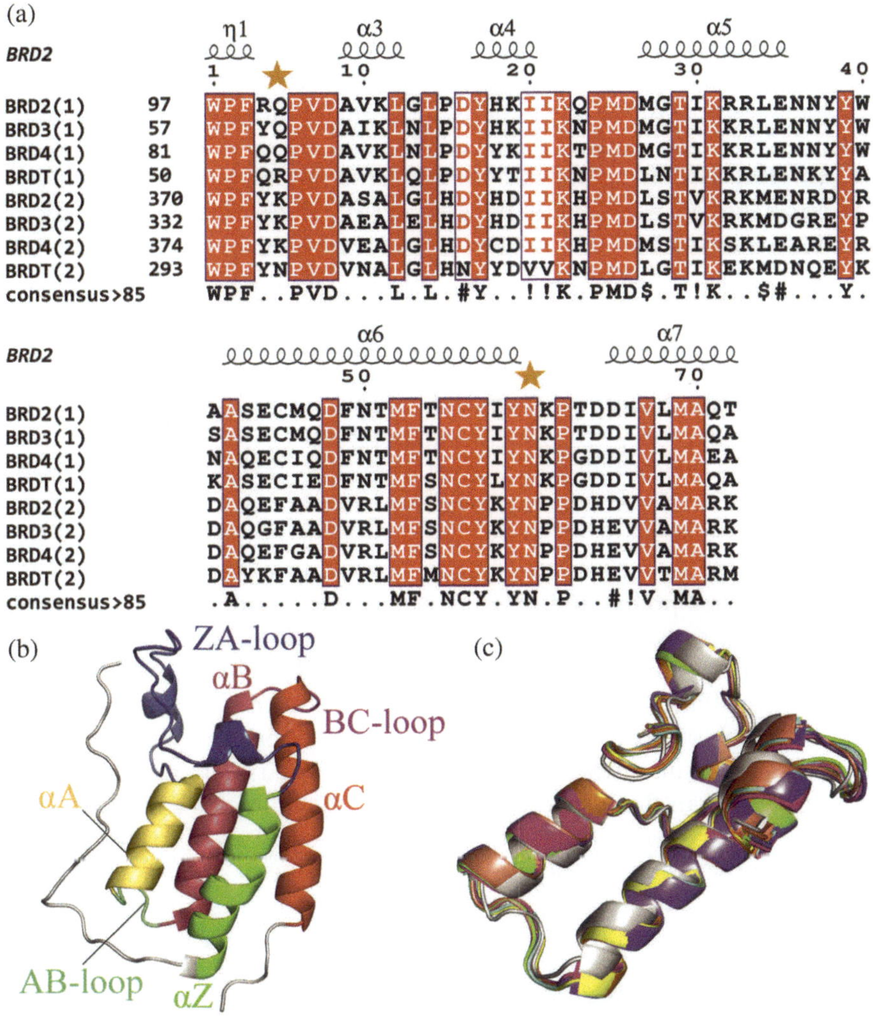

FIGURE 4 Sequence and structure comparisons of BET subfamily.

(a) Protein sequence alignment of two bromodomains among BET subfamily. Conserved and similar residues are highlighted and the conserved glutamine and asparagine at the Kac binding site are annotated with a brown star. BRD2(1) means the first bromodomain of BRD2 and so forth. The alignment is generated with ClustalW2 (Larkin et al., 2007) and ESPript 3.0 (Robert and Gouet, 2014). (b) Overall structure of BRD4(1) (PDB code: 4OGI) and the secondary structure are labeled in the figure. (c) Structure alignment of two bromodomains among BET subfamily, including BRD1(2) (PDB code: 2YEK), BRD2(2) (PDB code: 2DVV), BRD3(1) (PDB code: 3S91), BRD3(2) (PDB code: 3S92), BRD4(1) (PDB code: 4XY9), BRD4(2) (PDB code: 2MJV), BRDT(1) (PDB code: 4FLP), BRDT(2) (PDB code: 2WP1).

Bromodomain-containing proteins (BCPs) play curial roles in regulating histone-directed gene transcription and chromatin remodeling (Mujtaba et al., 2007). Single nucleotide polymorphism (SNP) frequencies analysis and other defined gene mutations or chromosome translocations have provided a large amount of evidence for the association of abnormal BCPs and diseases including cancer, neurological disorders, metabolic disease, and autoimmune/inflammatory diseases (Chung and Tough, 2012). With the discovery of numerous inhibitors of bromodomains, especially BET (bromodomain and extraterminal domain family), the time is ripe for the development of therapeutics of these diseases (Brand et al., 2015; Filippakopoulos et al., 2010; Gallenkamp et al., 2014; Garnier et al., 2014), and several bromodomain inhibitors are in clinical development.

3.1.2 High-Throughput Virtual Screening

Adopting the approach of high-throughput virtual screening (HTvS), Lucas and colleagues selected 22 compounds from a library containing more than 7 million small molecules and discovered seven with significant binding affinity toward BRD4 (Lucas et al., 2013). In the following validation, there were six new scaffolds of potent binders of BRDs. Importantly, XD14 (compound 11 in Figure 5) showed potent and selective activities in the cell models of ovarian and colon cancer (structure shown in Figure 5). Moreover, among the identified hits validated with drug-like physiochemical properties, XD14 showed high binding affinity to BRDs in CREBBP and EP300 and was the first reported inhibitor of EP300 (Lucas et al., 2013). After the generation of Literature Substructures and Similarity Searching, Vidler et al. mined commercially available small molecules in the eMolecules database and conducted a structure-based virtual screening toward BRD4, leading to the discovery of six novel hits (Vidler et al., 2013) including compound 12 with IC_{50} value of 4.7 μM. These results stress the importance of HTvS in the discovery of novel scaffold hits against BRDs, and can be subjected to fragment-based drug design.

3.1.3 Fragment-Based Drug Discovery

Different from HTS, fragment-based drug discovery seeks to find smaller molecules (150–250 MW), weaker binding affinity (mM–30 μM), and smaller libraries (about 1000 molecules) with biophysical screening methods, including NMR (Pellecchia et al., 2002) and X-ray crystallography (Blundell et al., 2002), which provides significant structure information for the optimization of the fragments into high affinity leads (Murray and Rees, 2009; Rees et al., 2004). During the process of the fragment set selection or the enrichment of promising hits, the fast and low-cost method of docking is usually applied.

By a careful visual assessment, application of the Rule-of-Three, clustering and docking, Chung et al. conducted a fragment set selection (Chung et al., 2012). And after screening the fragment set and the following analysis of

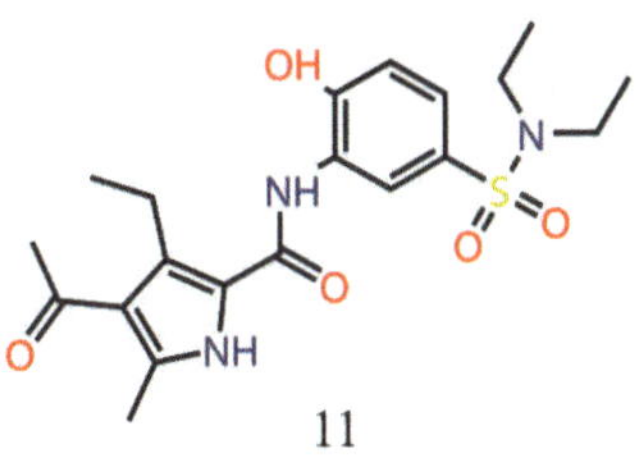

11
XD14

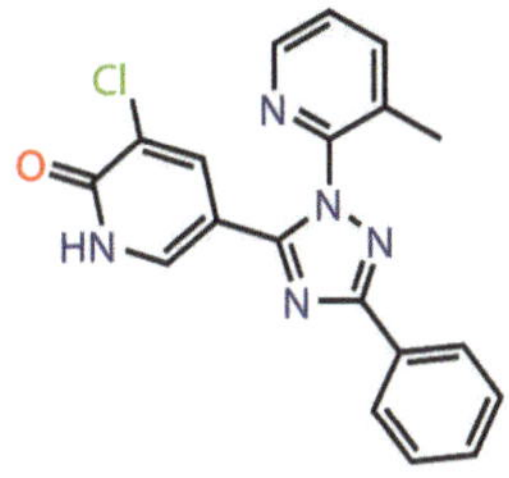

12

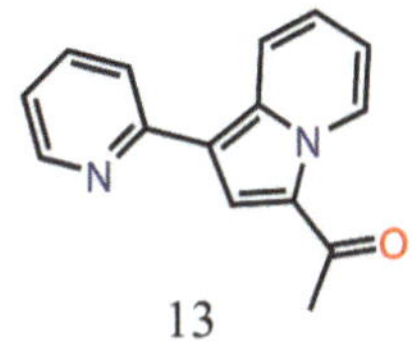

13
Fragment 4

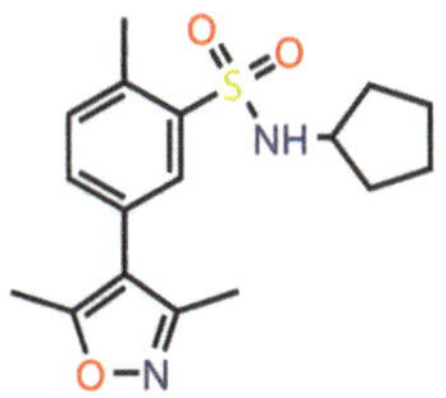

14
Sulfonamide 6a

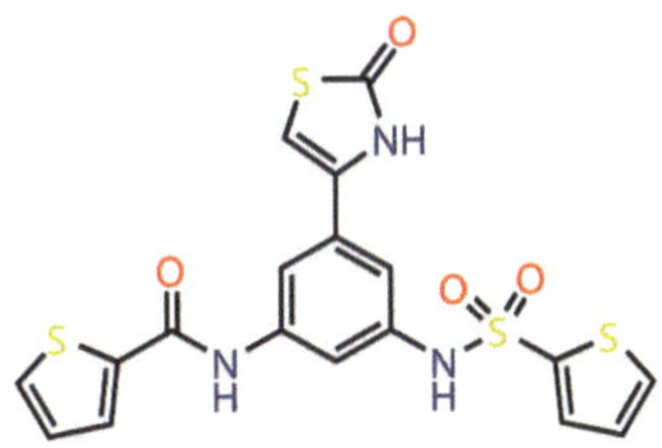

15

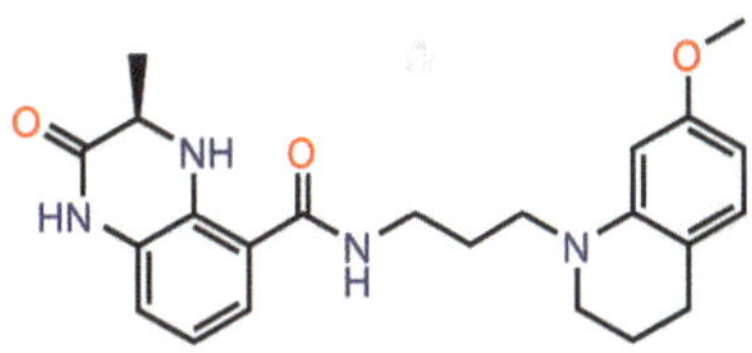

16
(*R*)-2

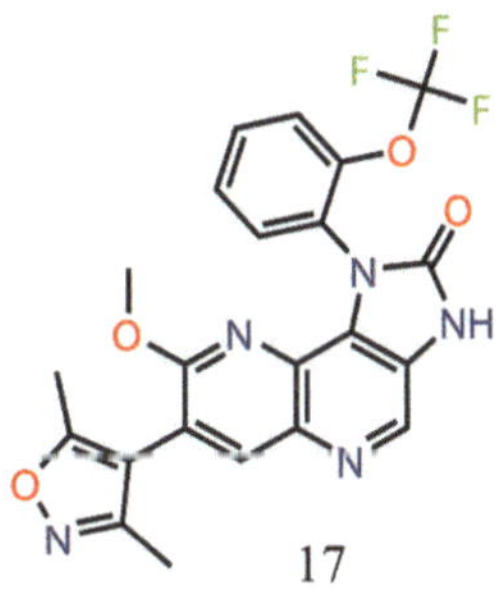

17

FIGURE 5 Inhibitors of epigenetic readers mentioned in this section.

40 BRD2/fragment X-ray complexes, they reported six representative ones in detail, among which two of the fragments are original bromodomain chemotypes. Remarkably, fragment **4** (compound 13 in Figure 5), 1-(1-(Pyridin-2-yl) indolizin-3-yl) ethanone, had good ligand efficiency ($IC_{50} \approx 2\ \mu M$) and reproducible inhibition ($IC_{50} \approx 14\ \mu M$) in cellular assays (Chung et al., 2012). In the subsequent investigation, they selected 4-phenyl-3,5-dimethyl isoxazole 3 as a starting point, and optimized and synthesized a sulfonamide series with good anti-inflammatory activity in cellular assays (compound 14 in Figure 5) (Bamborough et al., 2012). For another case, Zhao et al. designed a fragment library by the application of filtering with empirical rules, clustering, cherry-picking, and docking for enrichment (Zhao et al., 2013). They selected fragment 8 and carried out integrated lead optimization after successive

X-ray crystallography, with two potent compounds of 2-thiazolidinones discovered and confirmed in the cellular assay (compound 15 in Figure 5).

3.1.4 Molecular Dynamics

MD simulations are considerably useful tools for structure-based lead discovery and can give clues to drug design. Huang et al. performed explicit solvent and additional 50 molecules of cosolvent MD simulations on two bromodomains (BAZ2B and CREBBP) to analyze the conserved water around the acetyl-lysine binding site. It showed that the structural water molecules can be briefly replaced by ethanol or methanol, which could help decide which ones to leave for molecular docking and prompted that the introduction of hydroxy substituents in the corresponding position to the ligand might increase its binding affinity (Huang et al., 2014). In another study, Magno et al. simulated the spontaneous and reversible binding process of Kac to the TAF1 bromodomain and analyzed its mechanism and kinetics (Magno et al., 2013). They found two major binding modes interconverting faster than the association/dissociation process: one showed the acetyl-lysine was stabilized by an asparagine side chain through a hydrogen bond that is in correspondence with the available crystallized structure; the other buried deeper and was held by two hydrogen bonds in the loop between helices Z and A. This finding is useful for the design of novel small molecule inhibitors (Magno et al., 2013).

In a study of rational design of cyclic peptide, Gerona-Navarro and coworkers performed 40 ns MD simulations on a complex of CBP BRD and p53K382ac peptide and elucidated the detailed molecular basis of their interactions (Gerona-Navarro et al., 2011). Based on this, they developed a series of cyclic peptides to evaluate its biological behavior complemented with molecular modeling analysis of two ones in complex with CBP BRD. The cyclopeptide 4 exhibited a K_d of 8.0 μM, 24-fold efficacy to the p53K382ac peptide and showed high selectivity for CBP BRD over other BRDs. Importantly, the cyclopeptide demonstrated good efficiency to inhibit p53 transcriptional activity in colorectal carcinoma HCT116 cells in a different mechanism compared to p53 inhibitors reported before.

3.1.5 Quantum Chemistry and Other Approaches

Unlike docking or MD simulations, QM calculations provide accurate descriptions of nonbonding interactions, such as hydrogen-bonding and cation–π interactions. Density functional theory (DFT) calculations indicated an internal hydrogen bond stabilizing the conformation of the dihydroquinoxalinone series to bound to CREBBP bromodomian. And (*R*)-**2** (compound 16 in Figure 5) bind to CREBBP in an induced-fit manner by forming spontaneous cation–π interactions with the side chain R1173. Furthermore compound (R)-2 showed good inhibition in U2OS cells, demonstrating it as an invaluable tool and a promising lead for the discovery of potent and selective inhibitors of CREBBP bromodomain (Rooney et al., 2014). Through the assessment of the nitrogen positions of the naphthyrine compounds by DFT, Mirguet and coworkers found 1,5-naphthyrine series with higher competence than 1,8-naphthyrine or 1,6-naphthyrine ones. This resulted in the synthesis of 1,5-naphthyrine series, and they showed good anti-inflammatory pharmacology in BALB/c mice (compound 17 in Figure 5) (Mirguet et al., 2014).

Although docking and MD are very useful computational tools for the quick discovery of novel bromodomain inhibitors, it remains challenging to develop highly potent and selective ones. And the biggest difficulty is the high similarities of the sequence and 3D structure among acetyl-lysine binding sites across the bromodomains, especially the BET family. To address this, Vidler et al. evaluated druggabilities of the diverse bromodomain family members with available structures by SiteMap and presented their differences (Vidler et al., 2012). And they traced the differences in the unique amino acid signatures of the Kac binding sites, based on which they gave a new classification tree of the bromodomain family. The classification helps medicinal chemists working on a specific bromodomain to identify other ones highly likely to share the similar inhibitors and optimize them for better selectivity.

4. ERASERS

Erasers are in charge of the removal of epigenetic codes that will alter gene expression. Epigenetic erasers consist of histone deacetylases (HDACs), histone serine/threonine/tyrosine phosphatases, histone deubiquitinases, and histone lysine/arginine demethylases. HDACs have been studied extensively and several HDAC inhibitors have been approved by the US FDA for the treatment of cancers. HDAC will be discussed in detail followed by a brief introduction of RNA demethylases.

4.1 Histone Deacetylases

HDACs are enzymes involved in chromatin remodeling and have been studied extensively; they are considered as the promising drug targets for cancer therapy. HDACs are responsible for the removal of acetyl groups in the amino-terminal tail of histone, and play a crucial role in the control of the levels of histone acetylation. The level of acetylation is determined by the activities of histone acetylases (HATs) and HDACs. Histone deacetylation reinforces ionic interactions between positively charged histones and negatively charged DNA, which leads to a more condensed chromatin structure and gene transcription silencing (Yang and Seto, 2007). In addition, the acetylation of histones have a significant role in chromatin assembly, DNA repair, and recombination (Polo and Almouzni, 2005; Vidanes et al., 2005). The substrates of HDACs are not only limited to histone and also to nonhistone proteins, including p53 (Bode and Dong, 2004), NF-κB (Chen and Greene, 2004), AML1 (Yamaguchi et al., 2004), alpha-tubulin (Palazzo et al., 2003), and SHP90 (Kovacs et al., 2005). Accordingly, HDACs have significant effects on protein stability, protein–protein interactions, DNA location,

DNA binding, cell-cycle, progression, differentiation, and apoptosis through the reversible acetylation mechanism (Minucci and Pelicci, 2006).

4.1.1 Structural and Functional Features

HDACs in mammals are classified into four distinct classes according to their sequence homology and catalytic mechanisms (Gregoretti et al., 2004), which differ in subcellular location, structure, and enzymatic function. Class I, II, and IV HDACs are Zn^{2+}-dependent enzymes and they share a similar mechanism, though class IV evolved separately. Class III (also called SIRTs) is a NAD-dependent subfamily, which is structurally unrelated with others. Representative HDAC catalytic domains are illustrated in Figure 6. Class I HDACs (HDAC1, 2, 3, and 8), mainly expressed in the nucleus, are related to yeast PRD3, funding member of

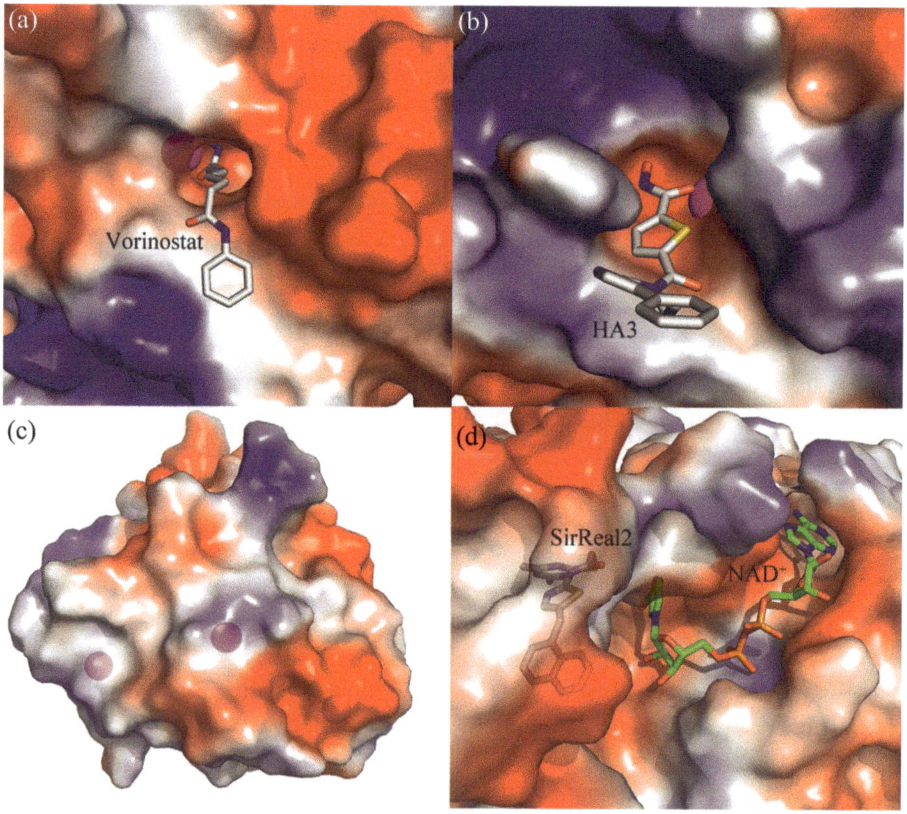

FIGURE 6 **Catalytic core surfaces of different HDAC subfamilies.**
(a) HDAC2 (PDB code: 4LXZ) of class I HDAC in complex with Vorinostat. (b) HDAC4 (PDB code: 2VQM) of class IIA HDAC in complex with HDAC4 inhibitor HA3. (c) HDAC6 (PDB code: 3C5K) of class IIB shown in electrostatic potential surface. (d) Sirt2 (PDB code: 4RMG) in complex with Sirtuin-rearranging ligand SirReal2 and NAD^+. The catalytic domains of HDACs are represented as an electrostatic potential surface and small molecules are shown as sticks, zinc ion in magenta sphere.

budding yeast (Kurdistani and Grunstein, 2003). They regulate the expression of thioredoxin binding protein-2 (TBP-2), myocyte enhancer factor-2 (MEF2), NF-κB, GATA (Kato et al., 2009) and have an effect on cell proliferation and cell survival. Class II HDACs (HDAC4-7, 9, and 10) contain additional regulatory domains and are larger than class I HDACs. Class II HDACs can be divided into class IIA (HDAC4, 5, 7, and 9) and class IIB (HDAC6 and HDAC10). Class IIA HDACs can shuttle between nucleus and cytoplasm depending on their phosphorus statutes while class IIB HDACs are mostly distributed in cytoplasm. Unlike other HDACs, class IIA HDACs are expressed highly in certain organs: HDAC5 and HDAC9 in muscles, heart, and brain; HDAC4 in brain and growth plates of the skeleton; HDAC7 in thymocytes and endothelial cells (Chang et al., 2004, 2006; Vega et al., 2004; Zhang et al., 2002). Class IIB HDACs possess higher catalytic activity than Class IIA HDACs. In mammalian cells, HDAC6 is expressed mainly in cytoplasma (Zhang et al., 2008) and it also functions as a tubulin deacetylase (Hubbert et al., 2002). To our knowledge, HDAC11 is the sole member of class IV (Gao et al., 2002) and located in the nucleus (Villagra et al., 2009). Class III HDACs (SIRT1-7) target a wide range of proteins in mitochondria, nucleus, and cytoplasm. Recent studies have elucidated that SIRT6 regulates lifespan in male mice and have important implications for age-related diseases (Kanfi et al., 2012). In addition, sirtuins regulate the cellular response to stress and make sure that damaged DNA does not propagate so that it is not accumulated (Saunders and Verdin, 2007).

HDACs have been proven to have key roles in the pathogenesis of cancers (Bolden et al., 2006; Minucci and Pelicci, 2006), neurological diseases (Butler and Bates, 2006; Kazantsev and Thompson, 2008), diabetes (Feng et al., 2008; Zhong et al., 2010), and immune disorders (Falkenberg and Johnstone, 2014) and have been the star target both in industrial and academic fields. Studies suggest that HDAC can regulate HIV transcription (Pagans et al., 2005; Richman et al., 2009; Williams et al., 2006) and HDAC inhibitors can be used to eradicate HIV infection in future clinical protocols (Ylisastigui et al., 2004). As the first epigenetics drug, HDAC inhibitor Vorinostat (Zolinza; Merck) (compound 18 in Figure 7) is approved by the US FDA for the treatment of cutaneous T cell lymphoma (CTCL) in 2006 (Grant et al., 2007) (structure shown in Figure 7). The approval of HDAC inhibitor Romidepsin (Istodax; Celgene) (compound 20 in Figure 7) in 2011 gives new hope for peripheral T-cell lymphoma (PTCL) patients (Kirk, 2012). Recently, Belinostat (compound 21 in Figure 7), developed by TopoTarget, has been approved for the treatment of PTCL (Mullard, 2015).

4.1.2 Computational Methods in HDAC Drug Design and Development

Apart from regular drug screen pipelines, computational methods can also benefit the discovery and development of HDAC inhibitors. In this section, we will review the applications of computational methods in HDACs drug

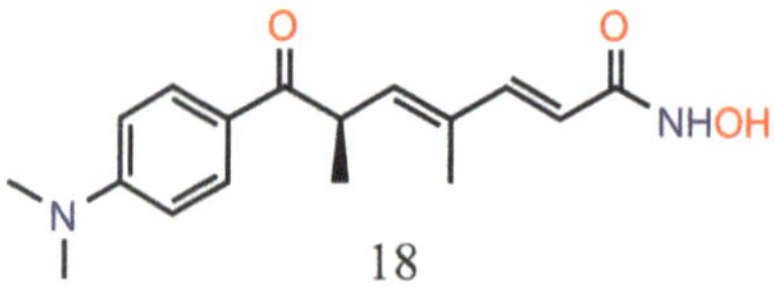

18
Trichostatin A (TSA)

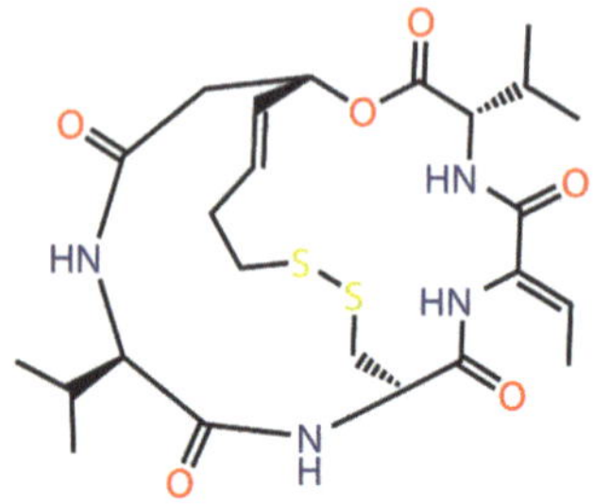

20
Romidepsin

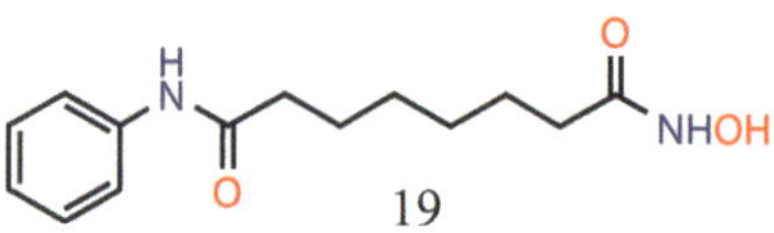

19
Vorinostat (SAHA)

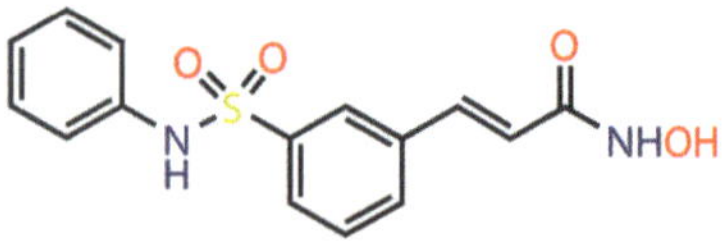

21
Belinostat

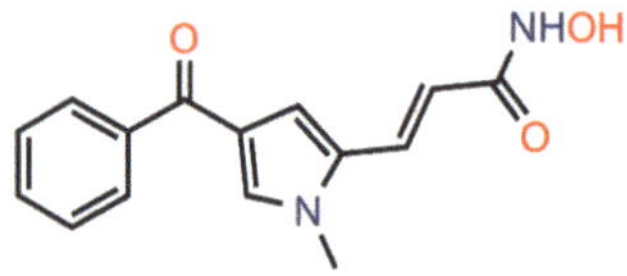

22

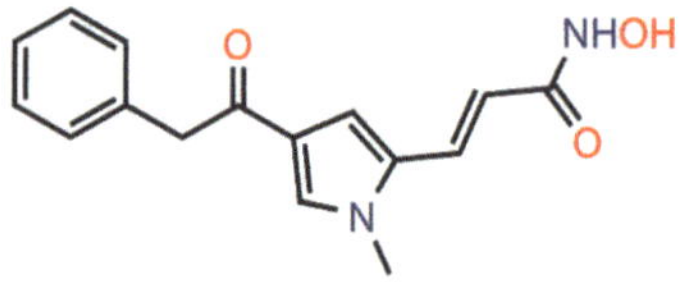

23

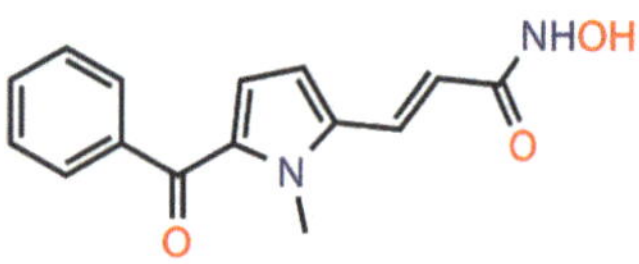

24

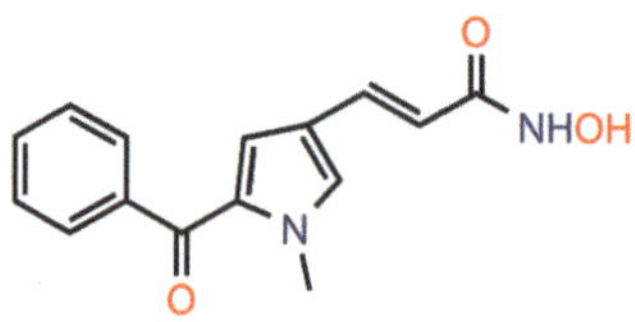

25

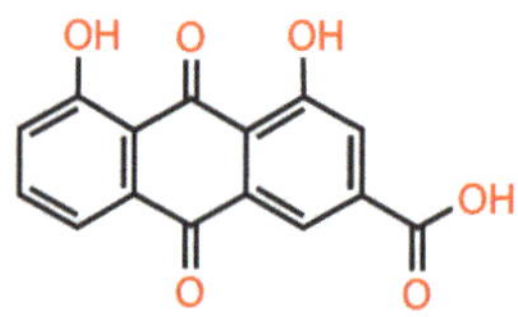

26
Rhein

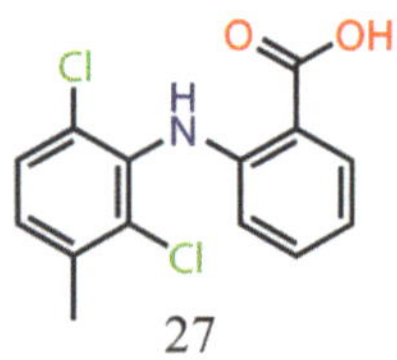

27
Meclofenamic Acid

FIGURE 7 Inhibitors of epigenetic erasers mentioned in this section.

discovery, including docking, pharmacophore, 3D-QSAR, MD, and so on. As of this writing, there are 56 HDAC X-ray structures in the Protein Data Bank, which enables the rational design of HDACs inhibitors. Meanwhile, a large set of active molecules targeting HDACs with diverse chemical structures make QSAR analysis and ligand-based pharmacophore possible. Advances in structural biology of HDACs and the deeper understanding of the acetylation mechanism with the help of MD and QM also benefit isoform selective HDAC inhibitor design. In this section, we will take the extensively studied HDAC1 as the study case and discuss how computational methods can accelerate HDAC1 inhibitor design.

4.1.3 High-Throughput Virtual Screening

Computational studies can be used to explore the binding modes of active HDAC inhibitors and we can gain insight into optimizing the activity of identified HDAC inhibitors. Mai et al. reported a series of 3-(4-benzoyl-1-methyl-1*H*-2-pyrrolyl)-*N*-hydroxy-2-propenamides (APHAs) and reported the binding modes of HDAC inhibitors compound 22 (Massa et al., 2001). The crystal structure of HDAC1 was not available and thus a homology modeling method was applied to build the HDAC1 structure using the X-ray structure of histone deacetylase-like protein (HDLP). Compound 22, which induced histone hyperacetylation, was chosen as the lead compound for further optimization (Mai et al., 2002). Proper substitution at the C4-position made compound 22, 3.8-fold more potent than compound 22 in vitro (IC_{50} = 0.1 μM), and docking stimulations revealed that the C4-substitution of compound 8 is accounted for a more considerable fitting into HDAC1 pocket, making an enhanced interaction with zinc ion (SAD, DOCK, Autodock) (Mai et al., 2003). Shortly after that, they docked compounds 24 and 25 into the modeled HDAC1 catalytic core and predicted the biological activities of 24 and 25 with three docked-bound conformations with the VALIDATE model (Ragno et al., 2004). Compounds 24 and 25 were synthesized and tested against mouse HDAC1 and maize HD2 enzymes. The mouse HDAC1 IC_{50} values of compounds 24 and 25 were 0.25 and 0.78 μM, respectively, while IC_{50} of compound 23 was 4.9 μM. Compound 25 also showed dose-dependent growth inhibition and hemoglobin accumulation effects on MEL cells. Park identified six novel HDAC1 inhibitors with IC_{50} values ranging from 1 to 100 μM by a structure-based virtual screening approach (Park et al., 2010). The compounds have various chelating groups for the active site zinc ion, including *N*-[1,3,4]thiadiazol-2-yl sulfonamide, *N*-thiazol-2-yl sulfonamide, and hydroxamic acid moieties. Quantum chemical computation at the B3LYP/6-31G** level suggest that these groups may serve as zinc chelators, an essential component of HDAC inhibitors.

Pharmacophore models can also be used to identify active HDAC inhibitors. Most current HDAC inhibitors share a common pharmacophore model, consisting of a zinc binding group (ZBG) or warhead to chelate the catalytic zinc, a capping group binding on the surface of active site, and a linker (Chen et al., 2013) (Figure 8). ZBG is significantly responsible for potency and sometimes

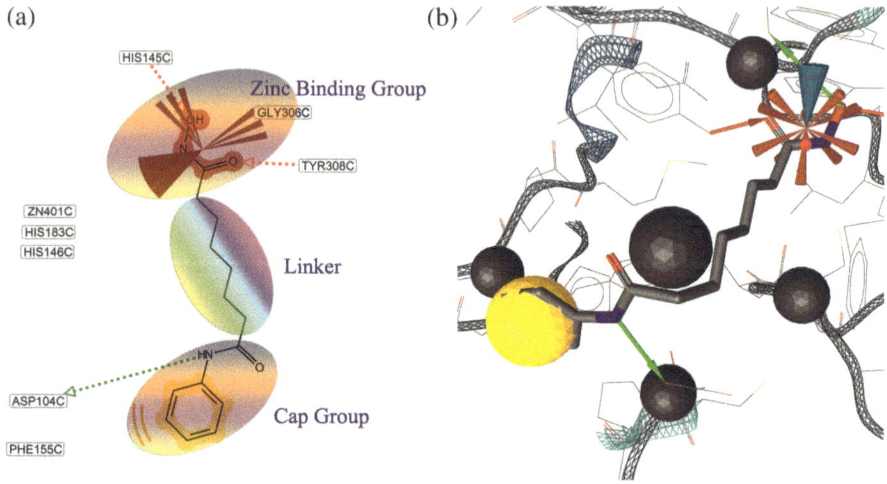

FIGURE 8 A common pharmacophore model, consisting of zinc binding group (ZBG). (a) Common pharmacophore for HDAC inhibitors. Most current HDAC inhibitors share a common pharmacophore model, consisted of zinc binding group (ZBG), linker, and cap group. (b) A 3D-view of the pharmacophore model. The pharmacophore is generated with LigandScout 2.02 (Wolber and Langer, 2005) with X-ray crystal structure of HDAC2 in complex with SAHA (vorinostat) (PDB code: 4LXZ). Pharmacophore features are defined as follows: hydrogen bond donor, red arrow; hydrogen bond acceptor, red arrow; excluded volumes, gray ball; hydrophobic interactions, yellow ball; zinc binding feature, cone in dark cyan; negative ionizable area, radial cone in red.

for isoform selectivity (Methot et al., 2008) and several computational studies of ZBGs have been performed. Vanommeslaeghe et al. ranked a series of ZBGs at the B3LYP/6-31G* level and proposed a general framework for chemical groups binding to HDACs (Vanommeslaeghe et al., 2005).

4.1.4 Quantitative Structure Activity Relationship

Guo explored the binding mode of indole amide analogues as potent HDAC1 inhibitors and analyzed the 3D-QSAR including comparative molecular field analysis (CoMFA) and comparative molecular similarity indices analysis (CoMSIA) (Guo et al., 2005). The docking studies revealed that the orientation and hydrogen-bonding interactions are similar to those in the crystal structure of the TSA–HDAC complex. A similar study applying docking and 3D-QSAR is also conducted to rationalize the activity trends and selectivity of hydroxamic acids.

4.1.5 Molecular Dynamics

Computational studies can be used to explore the binding modes of active HDAC inhibitors and we can gain insight into optimizing the activity of identified HDAC inhibitors. An alanine scan and molecular dynamics simulations were performed to assess the influence of residues in the 11 Å channel of HDAC1 catalytic site, which can facilitate HDAC1 drug design (Weerasinghe et al., 2008). According to this study, Phe150 is strictly required to maintain HDAC1 activity and a modification of the linker region may impart isoform selectivity. Molecular dynamic simulations and molecular/generalized-Born surface area (MM-GBSA) free energy calculations indicated that van der Waals and nonpolar interactions are the driving factors contributing to the activities of four hydroxamic acid inhibitors to HDLP (Yan et al., 2008). A computational protocol sequentially involving homology modeling, docking simulations, and MD was conducted to understand the differences between the binding modes (Wang et al., 2005). This study suggested that optimization of the interactions in the channel and on the surface considering the flexibility of the protein could weaken the metal binding of the ligands. The finding that the selectivity of reported isoform-specific inhibitors can be rationalized and reproduced indicting that this computational procedure can be applied to screen isoform-specific HDAC inhibitors.

4.1.6 Quantum Mechanics

DFT calculations uncovered that propitiate state of hydroxamic acids in the active site were of great importance for zinc binding and provided guidance for the design of novel, nonhydroxamic acid HDAC inhibitors (Vanommeslaeghe et al., 2005; Wang et al., 2007). Chen et al. developed a two-stage protocol to select potential bidentate zinc binders with M05-2X calculations followed by QM/MM hybrid calculations (Chen et al., 2013). This study correlated well with experimental data of ZBGs and predicted additional ZBGs such as α-aminoalcohols, guaiacolate, and 2-(aminomethyl)-pyridines for further research. Two hydrochloric features were found based on a 3D pharmacophore model, which might be responsible for the selectivity of HDAC1 inhibitors (Zhu et al., 2010). A hybrid approach combing docking and energy-optimized pharmacophore revealed that inhibitors with at least one aromatic ring in the linker region hold higher affinities and additional aromatic rings in the ZBG and the cap could enhance the binding efficiencies of inhibitors (Kalyaanamoorthy and Chen, 2013). Di Micco et al. carried out structural analysis by docking simulations and DFT calculation and obtained three selective ligands on the basis of structural guidelines (Di Micco et al., 2013).

A similar integrated computational study with newly developed force field parameters for zinc ion is also applied to elucidate the binding modes of active HDAC1 inhibitors and the strategy proved to be successful in ranking the 12 derivatives with indole amide, pyrrole, and sulfonamide moieties (Park and Lee, 2004). Elongatio Estiun or enlargement of the hydrophobic inhibitor head can contribute to the interactions with Pro29 and His28 and this can be used to optimize the activities of the HDAC1 inhibitor.

Computational analysis of HDAC structures can give insights into HDAC inhibitor design. Natural products have played a conspicuous role in HDAC drug design and development, so Zhang et al. evaluated the binding poses of natural products with docking and molecular dynamics methods (Zhang et al., 2013). Residue Asp155, Tyr273, Thr274, and Arg276 were discovered to have a key role in the binding, and water molecules should be taken into consideration owing to their contribution to the hydrogen bond network.

4.2 RNA Demethyltransferases

Nucleic acid methylation is one of the most important covalent modifications in epigenetics. The most abundant modification in RNA is the methylation of *N*6–methyladenine (m^6A), which was discovered in mRNA in 1974 (Desrosie et al., 1974). To our best knowledge, m^6A is the most general and conserved methylation modification on the residues of mRNA and lncRNA in a variety of eukaryotes such as mammals, flies, meiotic yeast, and plants (Ye et al., 2014). Regarded as a typical example in RNA methylation, m^6A has been featured as dynamic and reversible. The methyltransferases and demethyltransferases are designated as writers and erasers, respectively, and the level of methylation is determined by the balance of writers and erasers on a particular RNA site. The readers are responsible for processing and determining the subcellular behavior of RNAs with these sites, including transport, splicing, locating, turnover, translation, and stability (Fu et al., 2014). The human fat mass and obesity-associated protein, FTO, is the first eraser of RNA discovered in genome-wide association studies (Dina et al., 2007; Frayling et al., 2007; Scott et al., 2007), which belongs to the family of Fe^{2+}- and 2-oxoglutarate (2OG)-dependent AlkB dioxygenases. Studies have revealed the close relationship between FTO and various diseases including breast cancer (Kaklamani et al., 2011), Alzheimer disease (Keller et al., 2011), type II diabetes (Wehr et al., 2010), degenerative disc disease (Lao et al., 2014), and cardiovascular disease (Hotta et al., 2011; Pausova et al., 2009). It has been proposed that inhibition of FTO activity can be a potential treatment of extreme obesity (Thalhammer et al., 2012).

Discovery and development of FTO inhibitors not only can benefit the fundamental study of RNA demethylation, but can also provide opportunities for the treatment of cancer and obesity. Several crystals of FTO bound to small molecule inhibitors have been reported (Aik et al., 2013; Han et al., 2010; Huang et al., 2014), making structure-based drug design and development possible. The first identification of several FTO inhibitors was reported by our group (Chen et al., 2012), before which no inhibitor had been developed for FTO. We docked SPECS database against the crystal structure of FTO (PDB ID: 3LFM) and obtained two FTO inhibitors. Subsequently, a similarity search was performed through an in-house database and the natural product rhein (compound 26 in Figure 7) proved to be the most potent compound in vitro.

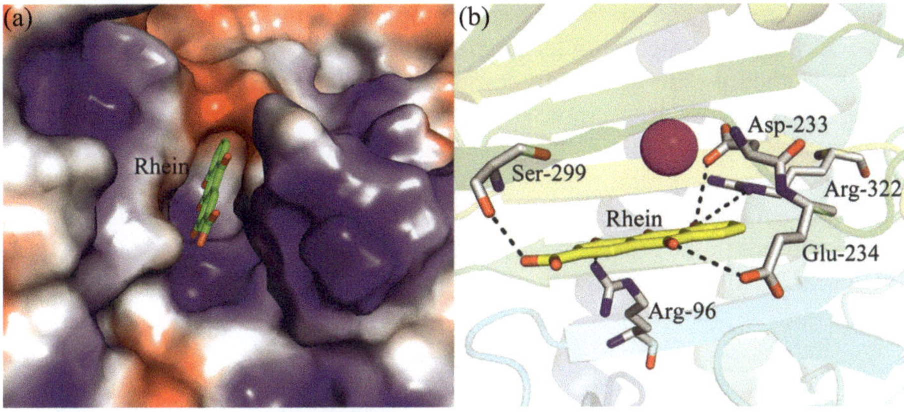

FIGURE 9 **Rhein in complex with FTO.**
(a) Rhein in the binding pocket of FTO. Binding pocket is shown in electrostatic potential surface and rhein is shown as sticks (PDB code: 4IE7). (b) Key residue interactions between rhein and FTO. Key residues are labeled and shown as sticks. The zinc ion is shown in magenta sphere while rhein is in yellow sticks.

Moreover, rhein can increase the modification level of m^{6}A in cells and showed low cytotoxicity. Aik and coworkers reported several cocrystal structures of FTO including rhein, which shed light on the development of more potent FOT inhibitors (Aik et al., 2013) (Figure 9). More recently, a selective FTO inhibitor meclofenamic acid (compound 27 in Figure 7) was identified and the crystal was determined, which elucidated the structural basis (Huang et al., 2015).

5. PROTEIN–PROTEIN INTERACTIONS

5.1 Overview of Epigenetic Protein–Protein Interactions

Histone modifying enzymes in epigenetics need to collaborate with cofactors, which involves plenty of protein–protein interactions (PPIs) (Figure 10). Undoubtedly, it is very beneficial to develop epigenetic small molecule probes targeting protein–protein interaction sites. Histone-modifying enzymes can interact with various nonhistone proteins to form functional complexes. In other words, there are significant therapeutic implications to design selective inhibitors against PPIs. Considering the fact that histone-modifying enzymes are highly homologous to their family members and share similar catalytic core pockets, it has been a difficult task to design selective inhibitors against them. PPIs are structurally more specific and it is an alternative strategy to design selective histone-modifying enzyme inhibitors targeting PPIs. Although it opens a new window in epigenetic drug discovery, it is still difficult to develop small molecule inhibitors against PPI interface, which involves extensive hydrophobic contacts (Wells and McClendon, 2007). There has been one study to characterize PPIs, illustrating fundamental properties, concerning the size, shape, complementarity, and residue propensities, which provides better understanding about this issue (Jones and Thornton, 1996). Higher-resolution protein complexes resolved, more effective computational tools developed, and more successful stories reported shows great potential to discover and develop small molecule epigenetic drugs and probes targeting PPIs that present less structural conservation. Apart from a histone reader like BRD4 that has been discussed previously, we will address the applications of computational methods in the design and discovery of PPI inhibitors against Menin-MLL1, WDR5-MLL1, and EZH2-EED interfaces (Figure 10).

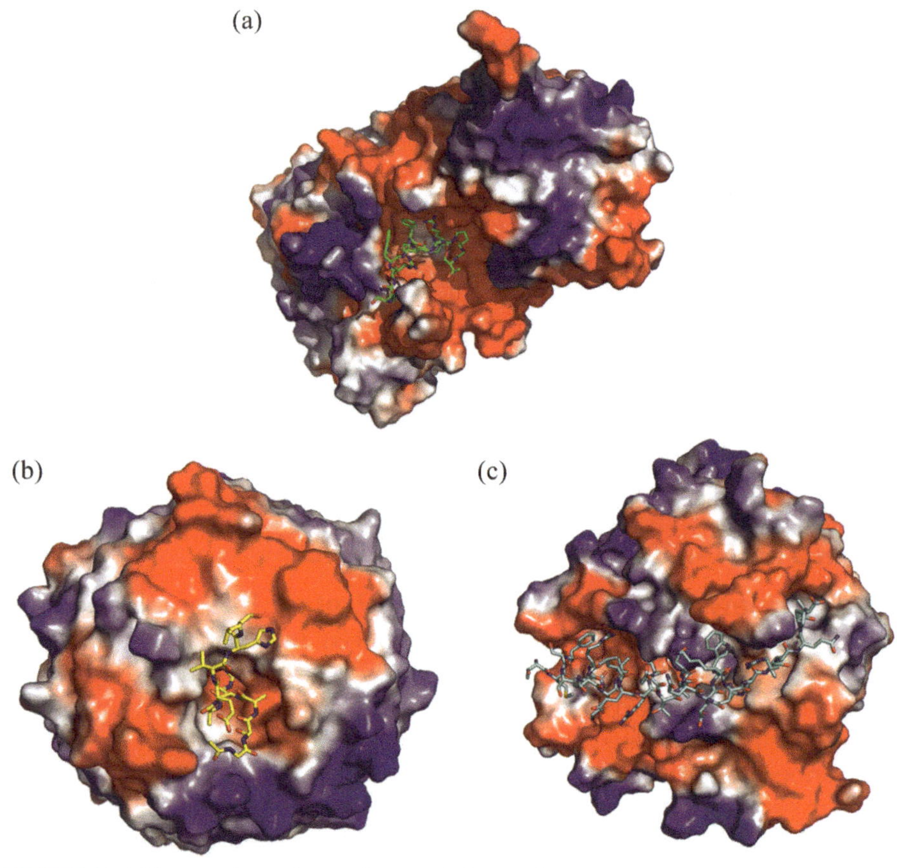

FIGURE 10 Different epigenetic protein–protein interaction modes.
(a) Menin–MLL1 complex (PDB code: 4GQ6). MLL1 is shown as sticks colored by element with carbon in yellow. Menin is used to generate the vacuum electrostatics model. (b) MLL1–WDR5 complex (PDB code: 3EG6) MLL1 is shown as sticks colored by element with carbon in cyan. WDR5 is used to generate the vacuum electrostatics model. (C) EZH2–EED complex (PDB code: 2QXV) EZH2 is shown as sticks colored by element with carbon in green. EED is used to generate the vacuum electrostatics surface. All figures are generated in pymol v.1.3r1.

5.2 Menin–MLL1

Menin has been discovered and validated as the oncogenic cofactor with the mixed lineage leukemia 1 (MLL1) fusion proteins with methyltransferase ability of histone H3 at lysine 4 and the protein–protein interaction required for development of leukemia has been validated in vivo (Yokoyama et al., 2005). MLL1 interacts with menin by menin-binding motif 1 (MBM1, K_d = 53 nM) and menin-binding motif 2 (MBM2, K_d = 1.4 μM) in a bivalent mode (Grembecka et al., 2010; Shi et al., 2012). The interface between menin and MLL1 shows great potential for drug discovery and development (Figure 11(c)), and disruption of menin–MLL interaction may play a significant role in reversing

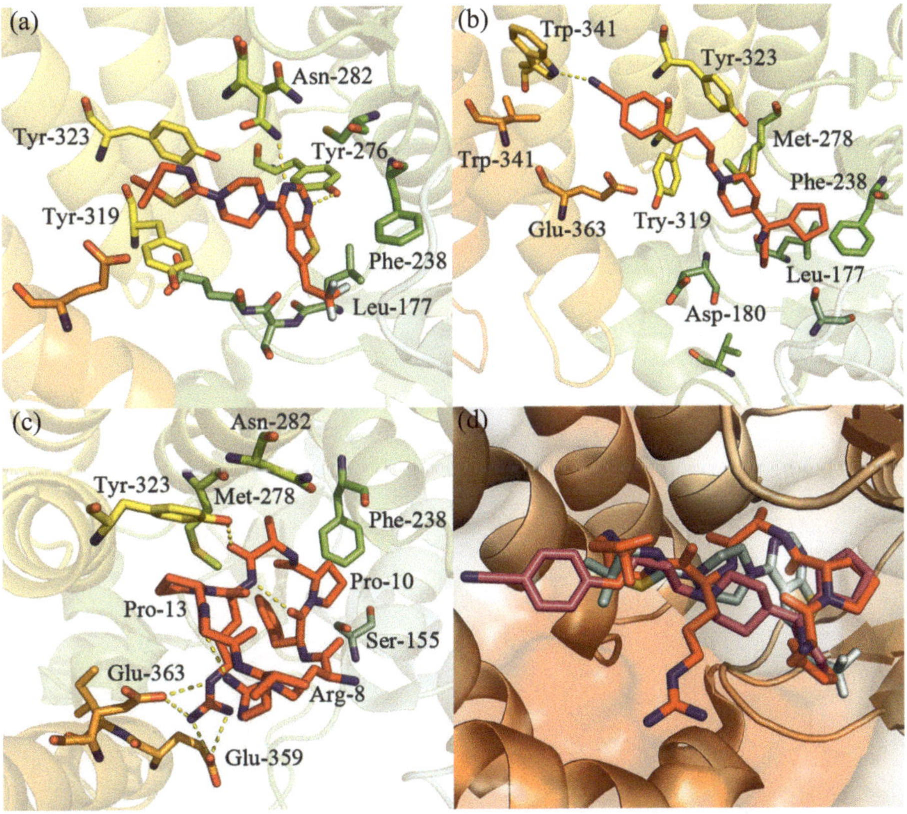

FIGURE 11 Protein–protein interactions of Menin–MLL1, WDR5–MLL1, and EZH2–EED interfaces.

(a) Menin interacts with MI-2-2 (PDB code: 4GQ4). Hydrogen bonds are shown in yellow dashed lines. (b) Menin interacts with MIV-6-R (PDB code: 4OG8). (c) The binding mode of Menin and MLL1 (PDB code: 4GQ6). Only residues 8–15 have been shown in the picture. (d) The alignment of MI-2-2, MIV-6-R, and native ligand MLL1. Both key residues involved in the pocket are properly labeled.

the leukemogenesis. The first generation of small-molecule epigenetic inhibitors have been developed by high-throughput screening based on fluorescence polarization assay with nanomolar affinity (MI-2: $K_d = 158\,\text{nM}$) (Grembecka et al., 2012). Based on the first-generation inhibitors, second-generation epigenetic inhibitors with the same scaffold and higher affinity present similar binding modes (Figure 11(a)) (Shi et al., 2012). Subsequently, novel classes of small molecules have also been developed. Through structure-based optimization, the most potent compound MIV-6-R (compound 28 in Figure 12) specifically interrupts the interaction between MLL-menin with $IC_{50} = 56\,\text{nM}$ and proves that developing small-molecule inhibitors against the PPI is absolutely feasible (Figure 11(b)). With regard to the cocrystallized structure available in the Protein Data Bank, small molecule probes compete with MLL1 in the same pocket, which provides a structural basis for the proof-of-concept that epigenetic protein–protein interaction is targetable (Figure 11(d)).

To use a computational approach to develop small inhibitors against the menin–MLL1 interface, pharmacophore models can be generated from the

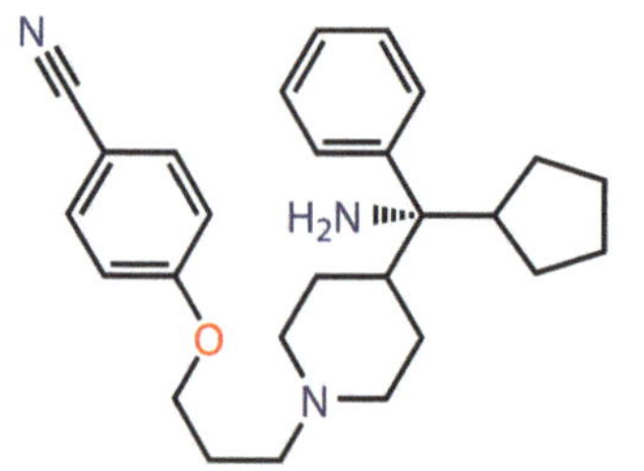

28
MIV-6-R

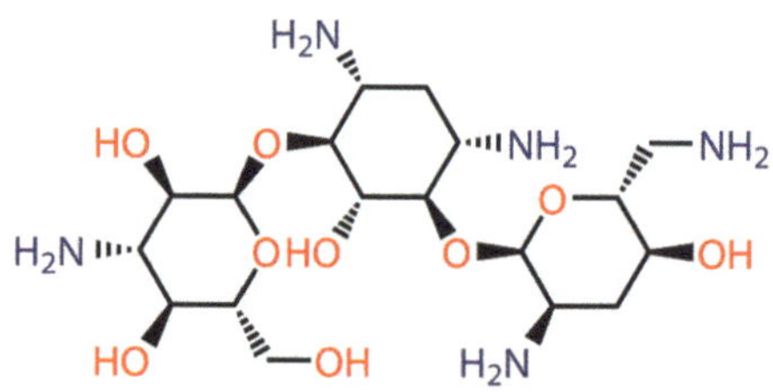

29
tobramycin

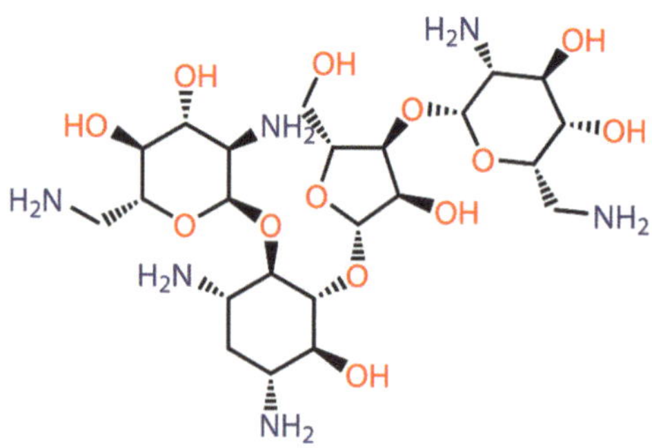

30
neomycin

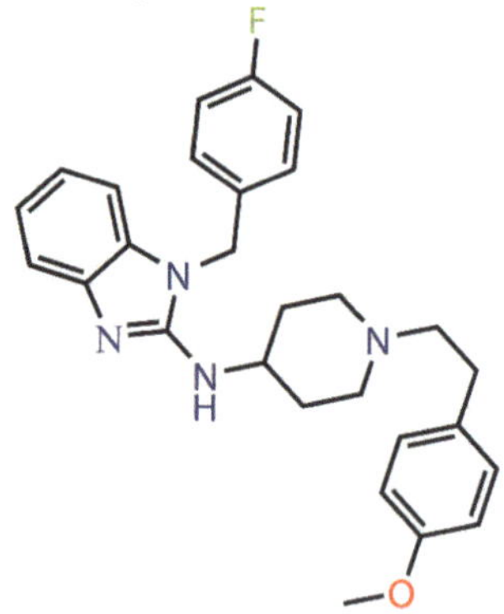

31
Astemizole

FIGURE 12 Small molecules disrupting protein–protein interactions.

complex structure (e.g., PDB code 4GQ6) available in the Protein Data Bank (Shi et al., 2012) with the predefined features (hydrogen bond acceptor, hydrogen bond donor, hydrophobic, and excluded volumes). Receptor-ligand-based pharmacophore explicitly describes the binding mode between the receptor and the native protein ligand or small-molecule inhibitors, which can be further utilized to improve the enrichment in the following virtual screening. Through cost-effective structure-based virtual screening and biochemical validation, the clinically used drugs tobramycin and neomycin (compound 29 and 30 in Figure 12) have been identified from the collection of clinical drugs in an in-house library, which shed light on the discovery and development of functional chemical epigenetic probes targeting PPIs (Li et al., 2014). The structure-based computational methods can also be applied in the discovery and development of macromolecule epigenetic probes to interrupt the menin-MLL1 interactions. A series of peptidomimetics have been synthesized and optimized that mimic the binding modes of native ligand protein. The success of the story laid solid foundation for further design and development of small-molecule epigenetic probes based on computational approaches (Zhou et al., 2013).

5.3 WDR5–MLL1

As described above, mixed lineage leukemia 1 (MLL1) is a methyltransferase for histone H3 lysine 4. MLL1 not only interacts with menin, but also forms a complex with Trp–Asp repeat domain 5 (WDR5), retinoblastoma binding protein 5 (RbBP5), and absent small or homeotic-2-like (ASH2L), which are indispensable in enzymatic activity (Dou and Hess, 2008; Dou et al., 2006). The disruption of the MLL1–WDR5 complex significantly inhibited the MLL1 enzymatic activity (Patel et al., 2008). Previous studies indicated that the minimal motif -CO-ARA-NH of MLL1 is sufficient in binding with WDR5 with $K_i = 120\,\text{nM}$, which provides a template for further study (Figure 13).

Based on the stepwise structure–activity relationship analysis, different N- (Ala1) and C- (Ala3) termini with different substituent groups have been designed and synthesized, which presented different physical properties including hydrophobicities, sizes, and shapes. The combination of most favorable substituent groups led to the discovery of MLL1 mimetic 1 with IC_{50} value 2.9 nM and $K_i < 1\,\text{nM}$. It's a successful story that combines the structure-based study with experimental validation for reaching the hanging fruits on epigenetic trees (Karatas et al., 2013).

5.4 EZH2–EED

There is widespread concern over the discovery and development of selective probes of polycomb repressive complex 2 (PRC2) in epigenetic drug discovery. PRC2 specifically catalyzes the lysine 27 of histone H3, which is involved in

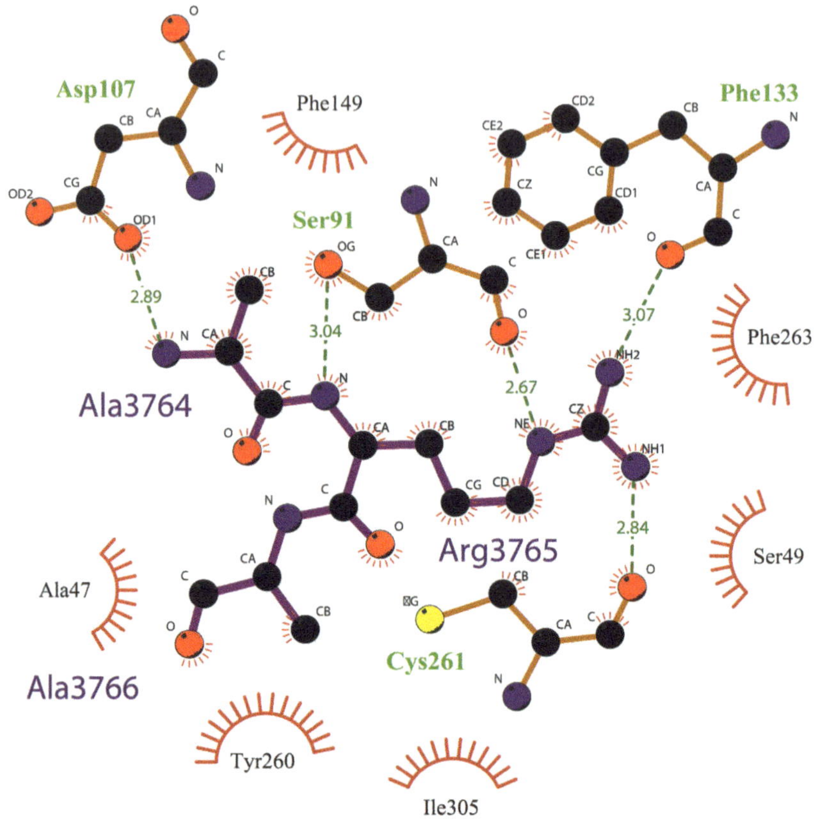

FIGURE 13 Key interactions between minimal motif ALA of MLL1 and WDR5.
This figure is generated with LigPlot+ v.1.4 (Laskowski and Swindells, 2011) making use of modified X-ray structure (PDB code: 3EG6) with default settings. Residues involved with hydrogen bonds are in green while hydrophobic interaction related residues are labeled in black.

various biological functions including the repression of genes in specific loci (Margueron and Reinberg, 2011). Recent discoveries show that the overexpression of the core component of PRC2 including the catalytic unit enhancer of zeste homolog 2 (EZH2), the cofactor embryonic ectoderm development (EED), together with suppressor of zeste 12 homolog (SUZ12) is tightly correlated with the development and invasion of malignant tumors (Chang and Hung, 2012; Chase and Cross, 2011). The development of EZH2 small-molecule inhibitors will provide a molecular basis understanding of the biology functions and accelerate the pace of medical therapy development, which has always been the main interest in industry and academia (Knutson et al., 2012; McCabe et al., 2012; Qi and Chana, 2012). However, EZH2 inhibitors directly compete with the binding site of the natural substrate SAM and there are more than 50 methyltransferases that share the similar binding pocket. So, it costs too much to determine the specificity of the identified active compounds (Copeland et al., 2009).

To bring the biological function into full play, the EZH2, EED, and SUZ12 need to form a complex stably, so that interference in the EZH2–EED interaction can make a real difference in repressing the catalytic activity (Kim et al., 2013). With the advances in the development of computational tools, the docking calculations have been widely applied in the discovery of the epigenetic small molecule probes. User-friendly software with graphical interface integrated with different modules has been widely used in drug discovery, which shows different speed and accuracy (Friesner, 2006).

Targeting the interface of EZH2-EED, an in-house database that included a collection of about 1000 clinical drugs was screened in silico ranked by Glide-score with the Glide program integrated in the Maestro program. It's noteworthy that through the procedure above, Astemizole (compound 31 in Figure 12) was identified as an inhibitor of EZH2-EED interaction, which destabilizes the PRC2 complex and shows the significant inhibition in its mehyltransferases in cell levels; this is another successful story of the discovery of epigenetic small-molecule probes targeting PPIs based on computational approaches (Kong et al., 2014).

6. FUTURE DIRECTIONS

During the last decades, computational methods have been routine in a rational design of epigenetic inhibitors and probes. With enriched knowledge of structure biology and molecular pathways of epigenetic modifying proteins, structure-based drug design like docking and complex-based pharmacophore has been widely used. When the crystal structure of epigenetic protein is not available, homology modeling makes docking-based virtual screens possible. Both MM and QM will facilitate a deeper understanding of molecular mechanisms, which can give insight into selective inhibitor design despite their time-consuming and CPU costing features. Once there are series of analogues of epigenetic inhibitors, 3D-QSAR or ligand-based pharmacophore is a useful tool for medicinal chemists.

Epigenetic modifying enzymes share similar catalytic mechanisms and conserved catalytic cores, making selective inhibitor design a challenging task. Epigenetic PPIs are more specific than catalytic domains and targeting PPI modulators can be an efficient strategy, and some protein–protein docking programs are available. Some epigenetic targets are metal dependent, such as HDACs, and thus zinc metalloprotein docking score has a potential role in HDAC inhibitor design. With rapid development of computer science, QM will be used more and more widely, which is more accurate but time-consuming. Also, data mining can be applied in inhibitor design and development.

REFERENCES

Ahmad, K., Henikoff, S., 2002. Epigenetic consequences of nucleosome dynamics. Cell 111, 281–284.

Aik, W., Demetriades, M., Hamdan, M.K., Bagg, E.A., Yeoh, K.K., Lejeune, C., Schofield, C.J., 2013. Structural basis for inhibition of the fat mass and obesity associated protein (FTO). J. Med. Chem. 56, 3680–3688.

Allis, C.D., Berger, S.L., Cote, J., Dent, S., Jenuwien, T., Kouzarides, T., Zhang, Y., 2007. New nomenclature for chromatin-modifying enzymes. Cell 131, 633–636.

Anglin, J.L., Deng, L., Yao, Y., Cai, G., Liu, Z., Jiang, H., Song, Y., 2012. Synthesis and structure-activity relationship investigation of adenosine-containing inhibitors of histone methyltransferase DOT1L. J. Med. Chem. 55, 8066–8074.

Arrowsmith, C.H., Bountra, C., Fish, P.V., Lee, K., Schapira, M., 2012. Epigenetic protein families: a new frontier for drug discovery. Nat. Rev. Drug Discov. 11, 384–400.

Auclair, G., Weber, M., 2012. Mechanisms of DNA methylation and demethylation in mammals. Biochimie 94, 2202–2211.

Bamborough, P., Diallo, H., Goodacre, J.D., Gordon, L., Lewis, A., Seal, J.T., Chung, C.W., 2012. Fragment-based discovery of bromodomain inhibitors part 2: optimization of phenylisoxazole sulfonamides. J. Med. Chem. 55, 587–596.

Bannister, A.J., Zegerman, P., Partridge, J.F., Miska, E.A., Thomas, J.O., Allshire, R.C., Kouzarides, T., 2001. Selective recognition of methylated lysine 9 on histone H3 by the HP1 chromo domain. Nature 410, 120–124.

Basavapathruni, A., Jin, L., Daigle, S.R., Majer, C.R., Therkelsen, C.A., Wigle, T.J., Olhava, E.J., 2012. Conformational adaptation drives potent, selective and durable inhibition of the human protein methyltransferase DOT1L. Chem. Biol. Drug Des. 80, 971–980.

Bello, M., Martinez-Archundia, M., Correa-Basurto, J., 2013. Automated docking for novel drug discovery. Expert Opin. Drug Discov. 8, 821–834.

Berger, S.L., 2007. The complex language of chromatin regulation during transcription. Nature 447, 407–412.

Berger, S.L., Kouzarides, T., Shiekhattar, R., Shilatifard, A., 2009. An operational definition of epigenetics. Genes Dev. 23, 781–783.

Bestor, T.H., 1988. Cloning of a mammalian DNA methyltransferase. Gene 74, 9–12.

Bestor, T.H., 2000. The DNA methyltransferases of mammals. Hum. Mol. Genet. 9, 2395–2402.

Blundell, T.L., Jhoti, H., Abell, C., 2002. High-throughput crystallography for lead discovery in drug design. Nat. Rev. Drug Discov. 1, 45–54.

Bode, A.M., Dong, Z., 2004. Post-translational modification of p53 in tumorigenesis. Nat. Rev. Cancer 4, 793–805.

Bolden, J.E., Peart, M.J., Johnstone, R.W., 2006. Anticancer activities of histone deacetylase inhibitors. Nat. Rev. Drug Discov. 5, 769–784.

Brand, M., Measures, A.M., Wilson, B.G., Cortopassi, W.A., Alexander, R., Hoss, M., Conway, S.J., 2015. Small molecule inhibitors of bromodomain-acetyl-lysine interactions. ACS Chem. Biol. 10, 22–39.

Brueckner, B., Boy, R.G., Siedlecki, P., Musch, T., Kliem, H.C., Zielenkiewicz, P., Lyko, F., 2005. Epigenetic reactivation of tumor suppressor genes by a novel small-molecule inhibitor of human DNA methyltransferases. Cancer Res. 65, 6305–6311.

Butler, R., Bates, G.P., 2006. Histone deacetylase inhibitors as therapeutics for polyglutamine disorders. Nat. Rev. Neurosci. 7, 784–796.

Carvalho, A.T., Gouveia, L., Kanna, C.R., Warmlander, S.K., Platts, J.A., Kamerlin, S.C., 2014. Understanding the structural and dynamic consequences of DNA epigenetic modifications: computational insights into cytosine methylation and hydroxymethylation. Epigenetics 9, 1604–1612.

Caulfield, T., Medina-Franco, J.L., 2011. Molecular dynamics simulations of human DNA methyltransferase 3B with selective inhibitor nanaomycin A. J. Struct. Bio. 176, 185–191.

Chang, C.J., Hung, M.C., 2012. The role of EZH2 in tumour progression. Br. J. Cancer 106, 243–247.

Chang, S., McKinsey, T.A., Zhang, C.L., Richardson, J.A., Hill, J.A., Olson, E.N., 2004. Histone deacetylases 5 and 9 govern responsiveness of the heart to a subset of stress signals and play redundant roles in heart development. Mol. Cell Biol. 24, 8467–8476.

Chang, S., Young, B.D., Li, S., Qi, X., Richardson, J.A., Olson, E.N., 2006. Histone deacetylase 7 maintains vascular integrity by repressing matrix metalloproteinase 10. Cell 126, 321–334.

Chase, A., Cross, N.C., 2011. Aberrations of EZH2 in cancer. Clin. Cancer Res. 17, 2613–2618.

Chen, B., Ye, F., Yu, L., Jia, G., Huang, X., Zhang, X., Yang, C.G., 2012. Development of cell-active N6-methyladenosine RNA demethylase FTO inhibitor. J. Am. Chem. Soc. 134, 17963–17971.

Chen, D.G., Ma, H., Hong, H., Koh, S.S., Huang, S.M., Schurter, B.T., Stallcup, M.R., 1999. Regulation of transcription by a protein methyltransferase. Science 284, 2174–2177.

Chen, K., Xu, L., Wiest, O., 2013. Computational exploration of zinc binding groups for HDAC inhibition. J. Org. Chem. 78, 5051–5055.

Chen, L.F., Greene, W.C., 2004. Shaping the nuclear action of NF-kappaB. Nat. Rev. Mol. Cell Biol. 5, 392–401.

Cheng, X., Blumenthal, R.M., 2008. Mammalian DNA methyltransferases: a structural perspective. Structure 16, 341–350.

Cheng, X., Collins, R.E., Zhang, X., 2005. Structural and sequence motifs of protein (Histone) methylation enzymes. Annu. Rev. Biophys. Biomol. Struct. 34, 267–294.

Cheung, N., Chan, L.C., Thompson, A., Cleary, M.L., So, C.W.E., 2007. Protein arginine-methyltransferase-dependent oncogenesis. Nat. Cell Biol. 9, 1208–1215.

Chung, C.-W., Tough, D.F., 2012. Bromodomains: a new target class for small molecule drug discovery. Drug Discov. Today Ther. Strateg. 9, e111–e120.

Chung, C.W., Dean, A.W., Woolven, J.M., Bamborough, P., 2012. Fragment-based discovery of bromodomain inhibitors part 1: inhibitor binding modes and implications for lead discovery. J. Med. Chem. 55, 576–586.

Copeland, R.A., Solomon, M.E., Richon, V.M., 2009. Protein methyltransferases as a target class for drug discovery. Nat. Rev. Drug Discov. 8, 724–732.

Daigle, S.R., Olhava, E.J., Therkelsen, C.A., Majer, C.R., Sneeringer, C.J., Song, J., Pollock, R.M., 2011. Selective killing of mixed lineage leukemia cells by a potent small-molecule DOT1L inhibitor. Cancer Cell 20, 53–65.

Deplus, R., Brenner, C., Burgers, W.A., Putmans, P., Kouzarides, T., de Launoit, Y., Fuks, F., 2002. Dnmt3L is a transcriptional repressor that recruits histone deacetylase. Nucleic Acids Res. 30, 3831–3838.

Desrosie, R., Frideric, K., Rottman, F., 1974. Identification of methylated nucleosides in messenger RNA from Novikoff hepatoma-cells. Proc. Natl. Acad. Sci. USA 71, 3971–3975.

Dhalluin, C., Carlson, J.E., Zeng, L., He, C., Aggarwal, A.K., Zhou, M.-M., Zhou, M.-M., 1999. Structure and ligand of a histone acetyltransferase bromodomain. Nature 399, 491–496.

Di Micco, S., Chini, M.G., Terracciano, S., Bruno, I., Riccio, R., Bifulco, G., 2013. Structural basis for the design and synthesis of selective HDAC inhibitors. Bioorg. Med. Chem. 21, 3795–3807.

Dina, C., Meyre, D., Gallina, S., Durand, E., Korner, A., Jacobson, P., Froguel, P., 2007. Variation in FTO contributes to childhood obesity and severe adult obesity. Nat. Genet. 39, 724–726.

Dou, Y., Hess, J.L., 2008. Mechanisms of transcriptional regulation by MLL and its disruption in acute leukemia. Int. J. Hematol. 87, 10–18.

Dou, Y., Milne, T.A., Ruthenburg, A.J., Lee, S., Lee, J.W., Verdine, G.L., Roeder, R.G., 2006. Regulation of MLL1 H3K4 methyltransferase activity by its core components. Nat. Struct. Mol. Biol. 13, 713–719.

Egger, G., Liang, G., Aparicio, A., Jones, P.A., 2004. Epigenetics in human disease and prospects for epigenetic therapy. Nature 429, 457–463.

Falkenberg, K.J., Johnstone, R.W., 2014. Histone deacetylases and their inhibitors in cancer, neurological diseases and immune disorders. Nat. Rev. Drug Discov. 13, 673–691.

Feinberg, A., Vogelsten, B., 1983. Hypomethylation distinguishes genes of some human cancers from their normal counterparts. Nature 301, 89–92.

Feinberg, A.P., Ohlsson, R., Henikoff, S., 2006. The epigenetic progenitor origin of human cancer. Nat. Rev. Genet. 7, 21–33.

Fellinger, K., Rothbauer, U., Felle, M., Langst, G., Leonhardt, H., 2009. Dimerization of DNA methyltransferase 1 is mediated by its regulatory domain. J. Cell Biochem. 106, 521–528.

Feng, B., Chen, S., Chiu, J., George, B., Chakrabarti, S., 2008. Regulation of cardiomyocyte hypertrophy in diabetes at the transcriptional level. Am. J. Physiol. Endocrinol. Metab. 294, E1119–E1126.

Feng, Q., Wang, H.B., Ng, H.H., Erdjument-Bromage, H., Tempst, P., Struhl, K., Zhang, Y., 2002. Methylation of H3-lysine 79 is mediated by a new family of HMTases without a SET domain. Curr. Biol. 12, 1052–1058.

Filippakopoulos, P., Picaud, S., Mangos, M., Keates, T., Lambert, J.P., Barsyte-Lovejoy, D., Knapp, S., 2012. Histone recognition and large-scale structural analysis of the human bromodomain family. Cell 149, 214–231.

Filippakopoulos, P., Qi, J., Picaud, S., Shen, Y., Smith, W.B., Fedorov, O., Bradner, J.E., 2010. Selective inhibition of BET bromodomains. Nature 468, 1067–1073.

Frayling, T.M., Timpson, N.J., Weedon, M.N., Zeggini, E., Freathy, R.M., Lindgren, C.M., McCarthy, M.I., 2007. A common variant in the FTO gene is associated with body mass Index and predisposes to childhood and adult obesity. Science 316, 889–894.

Friesner, R.A., 2006. Extra precision glide: docking and scoring incorporating a model of hydrophobic enclosure for protein-ligand complexes. J. Med. Chem. 49, 6177–6196.

Fu, Y., Dominissini, D., Rechavi, G., He, C., 2014. Gene expression regulation mediated through reversible m(6)A RNA methylation. Nat. Rev. Genet. 15, 293–306.

Gallenkamp, D., Gelato, K.A., Haendler, B., Weinmann, H., 2014. Bromodomains and their pharmacological inhibitors. Chem. Med. Chem. 9, 438–464.

Gao, L., Cueto, M.A., Asselbergs, F., Atadja, P., 2002. Cloning and functional characterization of HDAC11, a novel member of the human histone deacetylase family. J. Biol. Chem. 277, 25748–25755.

Garnier, J.-M., Sharp, P.P., Burns, C.J., 2014. BET bromodomain inhibitors: a patent review. Expert Opin. Ther. Pat. 24, 185–199.

Gerona-Navarro, G., Yoel, R., Mujtaba, S., Frasca, A., Patel, J., Zeng, L., Zhou, M.M., 2011. Rational design of cyclic peptide modulators of the transcriptional coactivator CBP: a new class of p53 inhibitors. J. Am. Chem. Soc. 133, 2040–2043.

Goll, M.G., Bestor, T.H., 2005. Eukaryotic cytosine methyltransferases. Annu. Rev. Biochem. 74, 481–514.

Grant, S., Easley, C., Kirkpatrick, P., 2007. Vorinostat. Nat. Rev. Drug Discov. 6 (1), 21–22.

Greer, E.L., Shi, Y., 2012. Histone methylation: a dynamic mark in health, disease and inheritance. Nat. Rev. Genet. 13, 343–357.

Gregoretti, I.V., Lee, Y.M., Goodson, H.V., 2004. Molecular evolution of the histone deacetylase family: functional implications of phylogenetic analysis. J. Mol. Biol. 338, 17–31.

Grembecka, J., Belcher, A.M., Hartley, T., Cierpicki, T., 2010. Molecular basis of the mixed lineage leukemia-menin interaction: implications for targeting mixed lineage leukemias. J. Biol. Chem. 285, 40690–40698.

Grembecka, J., He, S., Shi, A., Purohit, T., Muntean, A.G., Sorenson, R.J., Cierpicki, T., 2012. Menin-MLL inhibitors reverse oncogenic activity of MLL fusion proteins in leukemia. Nat. Chem. Biol. 8, 277–284.

Guo, Y.S., Xiao, J.F., Guo, Z.R., Chu, F.M., Cheng, Y.G., Wu, S., 2005. Exploration of a binding mode of indole amide analogues as potent histone deacetylase inhibitors and 3D-QSAR analyses. Bioorg. Med. Chem. 13, 5424–5434.

Han, Z., Niu, T., Chang, J., Lei, X., Zhao, M., Wang, Q., Chai, J., 2010. Crystal structure of the FTO protein reveals basis for its substrate specificity. Nature 464, 1205–1209.

Heil, C.S., 2014. No detectable effect of the DNA methyltransferase DNMT2 on Drosophila meiotic recombination. G3 (Bethesda) 4, 2095–2100.

Honarparvar, B., Govender, T., Maguire, G.E.M., Soliman, M.E.S., Kruger, H.G., 2014. Integrated approach to structure-based enzymatic drug design: molecular modeling, spectroscopy, and experimental bioactivity. Chem. Rev. 114, 493–537.

Hotta, K., Kitamoto, T., Kitamoto, A., Mizusawa, S., Matsuo, T., Nakata, Y., Sekine, A., 2011. Association of variations in the FTO, SCG3 and MTMR9 genes with metabolic syndrome in a Japanese population. J. Hum. Genet. 56, 647–651.

Hu, J., Chen, S., Kong, X., Zhu, K., Cheng, S., Zheng, M., Luo, C., 2015. Interaction between DNA/histone methyltransferases and their inhibitors. Curr. Med. Chem. 22, 360–372.

Huang, D., Rossini, E., Steiner, S., Caflisch, A., 2014. Structured water molecules in the binding site of bromodomains can be displaced by cosolvent. Chem. Med. Chem. 9, 573–579.

Huang, Y., Yan, J., Li, Q., Li, J., Gong, S., Zhou, H., Yang, C.-G., 2015. Meclofenamic acid selectively inhibits FTO demethylation of m6A over ALKBH5. Nucleic Acids Res. 43, 373–384.

Hubbert, C., Guardiola, A., Shao, R., Kawaguchi, Y., Ito, A., Nixon, A., Yao, T.-P., 2002. HDAC6 is a microtubule-associated deacetylase. Nature 417, 455–458.

Issa, J.P., Kantarjian, H.M., 2009. Targeting DNA methylation. Clin. Cancer Res. 15, 3938–3946.

Jakovcevski, M., Akbarian, S., 2012. Epigenetic mechanisms in neurological disease. Nat. Med. 18, 1194–1204.

Jones, S., Thornton, J.M., 1996. Principles of protein-protein interactions. Proc. Natl. Acad. Sci. USA 93, 13–20.

Kaklamani, V., Yi, N., Sadim, M., Siziopikou, K., Zhang, K., Xu, Y., Mantzoros, C., 2011. The role of the fat mass and obesity associated gene (FTO) in breast cancer risk. BMC Med. Genet. 12, 52.

Kalyaanamoorthy, S., Chen, Y.-P.P., 2013. Energy based pharmacophore mapping of HDAC inhibitors against class I HDAC enzymes. Biochim. Biophys. Acta Proteins Proteomics 1834, 317–328.

Kanfi, Y., Naiman, S., Amir, G., Peshti, V., Zinman, G., Nahum, L., Cohen, H.Y., 2012. The sirtuin SIRT6 regulates lifespan in male mice. Nature 483, 218–221.

Karatas, H., Townsend, E.C., Cao, F., Chen, Y., Bernard, D., Liu, L., Wang, S., 2013. High-affinity, small-molecule peptidomimetic inhibitors of MLL1/WDR5 protein-protein interaction. J. Am. Chem. Soc. 135, 669–682.

Kato, T., Shimono, Y., Hasegawa, M., Jijiwa, M., Enomoto, A., Asai, N., Takahashi, M., 2009. Characterization of the HDAC1 complex that regulates the sensitivity of cancer cells to oxidative stress. Cancer Res. 69, 3597–3604.

Kazantsev, A.G., Thompson, L.M., 2008. Therapeutic application of histone deacetylase inhibitors for central nervous system disorders. Nat. Rev. Drug Discov. 7, 854–868.

Keller, L., Xu, W., Wang, H.-X., Winblad, B., Fratiglioni, L., Graff, C., 2011. The obesity related gene, FTO, interacts with APOE, and is associated with Alzheimer's disease risk: a prospective cohort study. J. Alzheimers Dis. 23, 461–469.

Kim, J., Daniel, J., Espejo, A., Lake, A., Krishna, M., Xia, L., Bedford, M.T., 2006. Tudor, MBT and chromo domains gauge the degree of lysine methylation. EMBO Rep. 7, 397–403.

Kim, W., Bird, G.H., Neff, T., Guo, G., Kerenyi, M.A., Walensky, L.D., Orkin, S.H., 2013. Targeted disruption of the EZH2-EED complex inhibits EZH2-dependent cancer. Nat. Chem. Biol. 9, 643–650.

Kirk, R., 2012. Hematological cancer: a new option for patients with PTCL. Nat. Rev. Clin. Oncol. 9, 126.

Knutson, S.K., Wigle, T.J., Warholic, N.M., Sneeringer, C.J., Allain, C.J., Klaus, C.R., Kuntz, K.W., 2012. A selective inhibitor of EZH2 blocks H3K27 methylation and kills mutant lymphoma cells. Nat. Chem. Biol. 8, 890–896.

Kong, X., Chen, L., Jiao, L., Jiang, X., Lian, F., Lu, J., Luo, C., 2014. Astemizole arrests the proliferation of cancer cells by disrupting the EZH2-EED interaction of polycomb repressive complex 2. J. Med. Chem. 57, 9512–9521.

Konze, K.D., Ma, A., Li, F., Barsyte-Lovejoy, D., Parton, T., Macnevin, C.J., Jin, J., 2013. An orally bioavailable chemical probe of the lysine methyltransferases EZH2 and EZH1. ACS Chem. Biol. 8, 1324–1334.

Kovacs, J.J., Murphy, P.J.M., Gaillard, S., Zhao, X., Wu, J.-T., Nicchitta, C.V., Yao, T.-P., 2005. HDAC6 regulates Hsp90 acetylation and chaperone-dependent activation of glucocorticoid receptor. Mol. Cell. 18, 601–607.

Kuck, D., Caulfield, T., Lyko, F., Medina-Franco, J.L., 2010a. Nanaomycin A selectively inhibits DNMT3B and reactivates silenced tumor suppressor genes in human cancer cells. Mol. Cancer Ther. 9, 3015–3023.

Kuck, D., Singh, N., Lyko, F., Medina-Franco, J.L., 2010b. Novel and selective DNA methyltransferase inhibitors: docking-based virtual screening and experimental evaluation. Bioorg. Med. Chem. 18, 822–829.

Kurdistani, S.K., Grunstein, M., 2003. Histone acetylation and deacetylation in yeast. Nat. Rev. Mol. Cell Biol. 4, 276–284.

Lao, L., Zhong, G., Li, X., Liu, Z., 2014. A preliminary association study of fat mass and obesity associated gene polymorphisms and degenerative disc disease in a Chinese Han population. J. Int. Med. Res. 42, 205–212.

Larkin, M.A., Blackshields, G., Brown, N.P., Chenna, R., McGettigan, P.A., McWilliam, H., Valentin, F., Wallace, I.M., Wilm, A., Lopez, R., Thompson, J.D., Gibson, T.J., Higgins, D.G., 2007. Clustal W and Clustal X Version 2.0. Bioinformatics 23, 2947–2948.

Laskowski, R.A., Swindells, M.B., 2011. LigPlot$^+$: multiple ligand–protein interaction diagrams for drug discovery. J. Chem. Inf. Model 51, 2778–2786.

Lavecchia, A., Di Giovanni, C., 2013. Virtual screening strategies in drug discovery: a critical review. Curr. Med. Chem. 20, 2839–2860.

Li, B., Carey, M., Workman, J.L., 2007. The role of chromatin during transcription. Cell 128, 707–719.

Li, E., 2002. Chromatin modification and epigenetic reprogramming in mammalian development. Nat. Rev. Genet. 3, 662–673.

Li, K.K., Luo, C., Wang, D.X., Jiang, H.L., Zheng, Y.G., 2012. Chemical and biochemical approaches in the study of histone methylation and demethylation. Med. Res. Rev. 32, 815–867.

Li, L., Zhou, R., Geng, H., Yue, L., Ye, F., Xie, Y., Luo, C., 2014. Discovery of two aminoglycoside antibiotics as inhibitors targeting the menin-mixed lineage leukaemia interface. Bioorg. Med. Chem. Lett. 24, 2090–2093.

Lionta, E., Spyrou, G., Vassilatis, D.K., Cournia, Z., 2014. Structure-based virtual screening for drug discovery: principles, applications and recent advances. Curr. Top. Med. Chem. 14, 1923–1938.

Lucas, X., Wohlwend, D., Hugle, M., Schmidtkunz, K., Gerhardt, S., Schule, R., Gunther, S., 2013. 4-Acyl pyrroles: mimicking acetylated lysines in histone code reading. Angew. Chem. Int. Ed. Engl. 52, 14055–14059.

Magno, A., Steiner, S., Caflisch, A., 2013. Mechanism and kinetics of acetyl-lysine binding to bromodomains. J. Chem. Theory Comput. 9, 4225–4232.

Mai, A., Massa, S., Ragno, R., Cerbara, I., Jesacher, F., Loidl, P., Brosch, G., 2003. 3-(4-Aroyl-1-methyl-1H-2-pyrrolyl)-N-hydroxy-2-alkylamides as a new class of synthetic histone deacetylase inhibitors. 1. Design, synthesis, biological evaluation, and binding mode studies performed through three different docking procedures. J. Med. Chem. 46, 512–524.

Mai, A., Massa, S., Ragno, R., Esposito, M., Sbardella, G., Nocca, G., Brosch, G., 2002. Binding mode analysis of 3-(4-Benzoyl-1-methyl-1H-2-pyrrolyl)-N-hydroxy-2-propenamide: a new synthetic histone deacetylase inhibitor inducing histone hyperacetylation, growth inhibition, and terminal cell differentiation. J. Med. Chem. 45, 1778–1784.

Margueron, R., Reinberg, D., 2011. The polycomb complex PRC2 and its mark in life. Nature 469 (7330), 343–349.

Martin, C., Zhang, Y., 2005. The diverse functions of histone lysine methylation. Nat. Rev. Mol. Cell Biol. 6, 838–849.

Massa, S., Mai, A., Sbardella, G., Esposito, M., Ragno, R., Loidl, P., Brosch, G., 2001. 3-(4-Aroyl-1H-pyrrol-2-yl)-N-hydroxy-2-propenamides, a new class of synthetic histone deacetylase inhibitors. J. Med. Chem. 44, 2069–2072.

Maurer-Stroh, S., Dickens, N.J., Hughes-Davies, L., Kouzarides, T., Eisenhaber, F., Ponting, C.P., 2003. The tudor domain 'Royal family': tudor, plant agenet, chromo, PWWP and MBT domains. Trends Biochem. Sci. 28, 69–74.

McCabe, M.T., Ott, H.M., Ganji, G., Korenchuk, S., Thompson, C., Van Aller, G.S., Creasy, C.L., 2012. EZH2 inhibition as a therapeutic strategy for lymphoma with EZH2-activating mutations. Nature 492, 108–112.

Medina-Franco, J.L., Mendez-Lucio, O., Duenas-Gonzalez, A., Yoo, J., 2015. Discovery and development of DNA methyltransferase inhibitors using in silico approaches. Drug Discov. Today 20, 569–577.

Methot, J.L., Chakravarty, P.K., Chenard, M., Close, J., Cruz, J.C., Dahlberg, W.K., Miller, T.A., 2008. Exploration of the internal cavity of histone deacetylase (HDAC) with selective HDAC1/HDAC2 inhibitors (SHI-1:2). Bioorg. Med. Chem. Lett. 18, 973–978.

Min, J.R., Feng, Q., Li, Z.Z., Zhang, Y., Xu, R.M., 2003. Structure of the catalytic domain of human DOT1L, a Non-SET domain nucleosomal histone methyltransferase. Cell 112, 711–723.

Minucci, S., Pelicci, P.G., 2006. Histone deacetylase inhibitors and the promise of epigenetic (and more) treatments for cancer. Nat. Rev. Cancer 6, 38–51.

Mirguet, O., Lamotte, Y., Chung, C.W., Bamborough, P., Delannee, D., Bouillot, A., Nicodeme, E., 2014. Naphthyridines as novel BET family bromodomain inhibitors. Chem. Med. Chem. 9, 580–589.

Mujtaba, S., Zeng, L., Zhou, M.M., 2007. Structure and acetyl-lysine recognition of the bromodomain. Oncogene 26, 5521–5527.

Mullard, A., 2015. 2014 FDA drug approvals. Nat. Rev. Drug Discov. 14, 77–81.

Murray, C.W., Rees, D.C., 2009. The rise of fragment-based drug discovery. Nat. Chem. 1, 187–192.

Okada, Y., Feng, Q., Lin, Y.H., Jiang, Q., Li, Y.Q., Coffield, V.M., Zhang, Y., 2005. hDOT1L links histone methylation to leukemogenesis. Cell 121, 167–178.

Otani, J., Nankumo, T., Arita, K., Inamoto, S., Ariyoshi, M., Shirakawa, M., 2009. Structural basis for recognition of H3K4 methylation status by the DNA methyltransferase 3A ATRX–DNMT3–DNMT3L domain. EMBO Rep. 10, 1235–1241.

Pagans, S., Pedal, A., North, B.J., Kaehlcke, K., Marshall, B.L., Dorr, A., Ott, M., 2005. SIRT1 regulates HIV transcription via tat deacetylation. PLoS Biol. 3, 210–220.

Palazzo, A., Ackerman, B., Gundersen, G.G., 2003. Cell biology (communication arising): tubulin acetylation and cell motility. Nature 421, 230.

Park, H., Kim, S., Kim, Y.E., Lima, S.-J., 2010. A structure-based virtual screening approach toward the discovery of histone deacetylase inhibitors: identification of promising zinc-chelating groups. Chem. Med. Chem. 5, 591–597.

Park, H., Lee, S., 2004. Homology modeling, force field design, and free energy simulation studies to optimize the activities of histone deacetylase inhibitors. J. Comput. Aided Mol. Des. 18, 375–388.

Patel, A., Vought, V.E., Dharmarajan, V., Cosgrove, M.S., 2008. A conserved arginine-containing motif crucial for the assembly and enzymatic activity of the mixed lineage leukemia protein-1 core complex. J. Biol. Chem. 283, 32162–32175.

Pausova, Z., Syme, C., Abrahamowicz, M., Xiao, Y., Leonard, G.T., Perron, M., Paus, T., 2009. A common variant of the FTO gene is associated with not only increased adiposity but also elevated blood pressure in French Canadians. Circ. Cardiovasc. Genet. 2, 260–269.

Pellecchia, M., Sem, D.S., Wuthrich, K., 2002. NMR in drug discovery. Nat. Rev. Drug Discov. 1, 211–219.

Polo, S.E., Almouzni, G., 2005. Histone metabolic pathways and chromatin assembly factors as proliferation markers. Cancer Lett. 220, 1–9.

Pradhan, M., Esteve, P.O., Chin, H.G., Samaranayke, M., Kim, G.D., Pradhan, S., 2008. CXXC domain of human DNMT1 is essential for enzymatic activity. Biochemistry 47, 10000–10009.

Qi, W., Chana, H., 2012. Selective inhibition of EZH2 by a small molecule inhibitor blocks tumor cells proliferation. Proc. Natl. Acad. Sci. USA 109.

Ragno, R., Mai, A., Massa, S., Cerbara, I., Valente, S., Bottoni, P., Brosch, G., 2004. 3-(4-Aroyl-1-methyl-1*H*-pyrrol-2-yl)-*N*-hydroxy-2-propenamides as a new class of synthetic histone deacetylase inhibitors. 3. Discovery of novel lead compounds through structure-based drug design and docking studies. J. Med. Chem. 47, 1351–1359.

Rees, D.C., Congreve, M., Murray, C.W., Carr, R., 2004. Fragment-based lead discovery. Nat. Rev. Drug Discov. 3, 660–672.

Richman, D.D., Margolis, D.M., Delaney, M., Greene, W.C., Hazuda, D., Pomerantz, R.J., 2009. The Challenge of finding a cure for HIV infection. Science 323, 1304–1307.

Richon, V.M., Johnston, D., Sneeringer, C.J., Jin, L., Majer, C.R., Elliston, K., Copeland, R.A., 2011. Chemogenetic analysis of human protein methyltransferases. Chem. Biol. Drug Des. 78, 199–210.

Robert, X., Gouet, P., 2014. Deciphering key features in protein structures with the new ENDscript server. Nucl. Acids Res. 42, W320–W324.

Robertson, K.D., 2001. DNA methylation, methyltransferases, and cancer. Oncogene 20, 3139–3155.

Rooney, T.P., Filippakopoulos, P., Fedorov, O., Picaud, S., Cortopassi, W.A., Hay, D.A., Conway, S.J., 2014. A series of potent CREBBP bromodomain ligands reveals an induced-fit pocket stabilized by a cation-pi interaction. Angew. Chem. Int. Ed. Engl. 53, 6126–6130.

Sanchez, R., Zhou, M.M., 2011. The PHD finger: a versatile epigenome reader. Trends Biochem. Sci. 36, 364–372.

Saunders, L.R., Verdin, E., 2007. Sirtuins: critical regulators at the crossroads between cancer and aging. Oncogene 26, 5489–5504.

Schaefer, M., Lyko, F., 2010. Solving the Dnmt2 enigma. Chromosoma 119, 35–40.

Schrodinger, L.L.C., 2010. The PyMOL Molecular Graphics System. Version 1.3r1.

Scott, L.J., Mohlke, K.L., Bonnycastle, L.L., Willer, C.J., Li, Y., Duren, W.L., Boehnke, M., 2007. A genome-wide association study of type 2 diabetes in finns detects multiple susceptibility variants. Science 316, 1341–1345.

Shi, A., Murai, M.J., He, S., Lund, G., Hartley, T., Purohit, T., Cierpicki, T., 2012. Structural insights into inhibition of the bivalent menin-MLL interaction by small molecules in leukemia. Blood 120, 4461–4469.

Siedlecki, P., Boy, R.G., Musch, T., Brueckner, B., Suhai, S., Lyko, F., Zielenkiewicz, P., 2006. Discovery of two novel, small-molecule inhibitors of DNA methylation. J. Med. Chem. 49, 678–683.

Song, J.K., Rechkoblit, O., Bestor, T.H., Patel, D.J., 2011. Structure of DNMT1-DNA complex reveals a role for autoinhibition in maintenance DNA methylation. Science 331, 1036–1040.

Song, J.K., Teplova, M., Ishibe-Murakami, S., Patel, D.J., 2012. Structure-based mechanistic insights into DNMT1-mediated maintenance DNA methylation. Science 335, 709–712.

Spannhoff, A., Sippl, W., Jung, M., 2009. Cancer treatment of the future: inhibitors of histone methyltransferases. Int. J. Biochem. Cell 41, 4–11.

Suetake, I., Shinozaki, F., Miyagawa, J., Takeshima, H., Tajima, S., 2004. DNMT3L stimulates the DNA methylation activity of Dnmt3a and Dnmt3b through a direct interaction. J. Biol. Chem. 279, 27816–27823.

Suzuki, T., Tanaka, R., Hamada, S., Nakagawa, H., Miyata, N., 2010. Design, synthesis, inhibitory activity, and binding mode study of novel DNA methyltransferase 1 inhibitors. Bioorg. Med Chem. Lett. 20, 1124–1127.

Tao, T., Liu, D.C., Liu, C.H., Xu, B., Chen, S.Q., Yin, Y., Chen, M., 2014. Autoregulatory feedback loop of EZH2/miR-200c/E2F3 as a driving force for prostate cancer development. Biochim. Biophys. Acta Gene Regul. Mech. 1839, 858–865.

Thalhammer, A., Aik, W., Bagg, E.A.L., Schofield, C.J., 2012. The potential of 2-oxoglutarate oxygenases acting on nucleic acids as therapeutic targets. Drug Discov. Today Ther. Strateg. 9, e91–e100.

Trievel, R.C., 2004. Structure and function of histone methyltransferases. Crit. Rev. Eukaryot. Gene Expr. 14, 147–169.

Vallot, C., Herault, A., Boyle, S., Bickmore, W.A., Radvanyi, F., 2015. PRC2-independent chromatin compaction and transcriptional repression in cancer. Oncogene 34, 741–751.

Vanommeslaeghe, K., Loverix, S., Geerlings, P., Tourwe, D., 2005. DFT-based ranking of zinc-binding groups in histone deacetylase inhibitors. Bioorg. Med. Chem. 13, 6070–6082.

Vega, R.B., Matsuda, K., Oh, J., Barbosa, A.C., Yang, X., Meadows, E., Olson, E.N., 2004. Histone deacetylase 4 controls chondrocyte hypertrophy during skeletogenesis. Cell 119, 555–566.

Vidanes, G.M., Bonilla, C.Y., Toczyski, D.P., 2005. Complicated tails: histone modifications and the DNA damage response. Cell 121, 973–976.

Vidler, L.R., Brown, N., Knapp, S., Hoelder, S., 2012. Druggability analysis and structural classification of bromodomain acetyl-lysine binding sites. J. Med. Chem. 55, 7346–7359.

Vidler, L.R., Filippakopoulos, P., Fedorov, O., Picaud, S., Martin, S., Tomsett, M., Hoelder, S., 2013. Discovery of novel small-molecule inhibitors of BRD4 using structure based virtual screening. J. Med. Chem. 56, 8073–8088.

Villagra, A., Cheng, F., Wang, H.W., Suarez, I., Glozak, M., Maurin, M., Sotomayor, E.M., 2009. The histone deacetylase HDAC11 regulates the expression of interleukin 10 and immune tolerance. Nat. Immunol. 10, 92–100.

Wang, D., Helquist, P., Wiest, O., 2007. Zinc binding in HDAC Inhibitors: a DFT study. J. Org. Chem. 72, 5446–5449.

Wang, D.F., Helquist, P., Wiech, N.L., Wiest, O., 2005. Toward selective histone deacetylase inhibitor design: homology modeling, docking studies, and molecular dynamics simulations of human class I histone deacetylases. J. Med. Chem. 48, 6936–6947.

Wang, J., Chen, L., Sinha, S.H., Liang, Z., Chai, H., Muniyan, S., Luo, C., 2012. Pharmacophore-based virtual screening and biological evaluation of small molecule inhibitors for protein arginine methylation. J. Med. Chem. 55, 7978–7987.

Weerasinghe, S.V.W., Estiu, G., Wiest, O., Pflum, M.K.H., 2008. Residues in the 11 A channel of histone deacetylase 1 promote catalytic activity: implications for designing isoform-selective histone deacetylase inhibitors. J. Med. Chem. 51, 5542–5551.

Wehr, E., Schweighofer, N., Möller, R., Giuliani, A., Pieber, T.R., Obermayer-Pietsch, B., 2010. Association of FTO gene with hyperandrogenemia and metabolic parameters in women with polycystic ovary syndrome. Metabolism 59, 575–580.

Wells, J.A., McClendon, C.L., 2007. Reaching for high-hanging fruit in drug discovery at protein-protein interfaces. Nature 450, 1001–1009.

Williams, S.A., Chen, L.F., Kwon, H., Ruiz-Jarabo, C.M., Verdin, E., Greene, W.C., 2006. NF-kappa B p50 promotes HIV latency through HDAC recruitment and repression of transcriptional initiation. EMBO J. 25, 139–149.

Wolber, G., Langer, T., 2005. LigandScout: 3-D pharmacophores derived from protein-bound ligands and their use as virtual screening filters. J. Chem. Inf. Model 45, 160–169.

Yamaguchi, Y., Kurokawa, M., Imai, Y., Izutsu, K., Asai, T., Ichikawa, M., Hirai, H., 2004. AML1 is functionally regulated through p300-mediated acetylation on specific lysine residues. J. Biol. Chem. 279, 15630–15638.

Yan, C., Xiu, Z., Li, X., Li, S., Hao, C., Teng, H., 2008. Comparative molecular dynamics simulations of histone deacetylase-like protein: binding modes and free energy analysis to hydroxamic acid inhibitors. Proteins Struct. Funct. Bioinf. 73, 134–149.

Yang, X.J., Seto, E., 2007. HATs and HDACs: from structure, function and regulation to novel strategies for therapy and prevention. Oncogene 26, 5310–5318.

Ye, F., Zhang, L., Jin, L., Zheng, M., Jiang, H., Luo, C., 2014. Repair of methyl lesions in RNA by oxidative demethylation. Med. Chem. Comm. 5, 1797–1803.

Yen, R.W.C., Vertino, P.M., Nelkin, B.D., Yu, J.J., Eldeiry, W., Cumaraswamy, A., Baylin, S.B., 1992. Isolation and characterization of the cDNA-encoding human DNA methyltransferase. Nucleic Acids Res. 20, 2287–2291.

Ylisastigui, L., Archin, N.M., Lehrman, G., Bosch, R.J., Margolis, D.M., 2004. Coaxing HIV-1 from resting CD4 T cells: histone deacetylase inhibition allows latent viral expression. Aids 18, 1101–1108.

Yokoyama, A., Somervaille, T.C., Smith, K.S., Rozenblatt-Rosen, O., Meyerson, M., Cleary, M.L., 2005. The menin tumor suppressor protein is an essential oncogenic cofactor for MLL-associated leukemogenesis. Cell 123, 207–218.

Yue, W.W., Hassler, M., Roe, S.M., Thompson-Vale, V., Pearl, L.H., 2007. Insights into histone code syntax from structural and biochemical studies of CARM1 methyltransferase. EMBO J. 26, 4402–4412.

Yun, M., Wu, J., Workman, J.L., Li, B., 2011. Readers of histone modifications. Cell Res. 21, 564–578.

Zhang, C.L., McKinsey, T.A., Chang, S., Antos, C.L., Hill, J.A., Olson, E.N., 2002. Class II histone deacetylases act as signal-responsive repressors of cardiac hypertrophy. Cell 110, 479–488.

Zhang, L., Li, M., Zhang, L., Xu, W., 2013. Discovering the binding modes of natural products with histone deacetylase 1. Med. Chem. 9, 126–132.

Zhang, X., Cheng, X., 2003. Structure of the predominant protein arginine methyltransferase PRMT1 and analysis of its binding to substrate peptides. Structure 11, 509–520.

Zhang, X., Zhou, L., Cheng, X.D., 2000. Crystal structure of the conserved core of protein arginine methyltransferase PRMT3. EMBO J. 19, 3509–3519.

Zhang, Y., Kwon, S., Yamaguchi, T., Cubizolles, F., Rousseaux, S., Kneissel, M., Matthias, P., 2008. Mice lacking histone deacetylase 6 have hyperacetylated tubulin but are viable and develop normally. Mol. Cell Biol. 28, 1688–1701.

Zhao, L., Cao, D., Chen, T., Wang, Y., Miao, Z., Xu, Y., Shen, J., 2013. Fragment-based drug discovery of 2-thiazolidinones as inhibitors of the histone reader BRD4 bromodomain. J. Med. Chem. 56, 3833–3851.

Zhong, L., D'Urso, A., Toiber, D., Sebastian, C., Henry, R.E., Vadysirisack, D.D., Mostoslavsky, R., 2010. The histone deacetylase Sirt6 regulates glucose homeostasis via Hif1alpha. Cell 140, 280–293.

Zhou, H., Liu, L., Huang, J., Bernard, D., Karatas, H., Navarro, A., Wang, S., 2013. Structure-based design of high-affinity macrocyclic peptidomimetics to block the menin-mixed lineage leukemia 1 (MLL1) protein-protein interaction. J. Med. Chem. 56, 1113–1123.

Zhu, Y., Li, H.-F., Lu, S., Zheng, Y.-X., Wu, Z., Tang, W.-F., Lu, T., 2010. Investigation on the isoform selectivity of histone deacetylase inhibitors using chemical feature based pharmacophore and docking approaches. Eur. J. Med. Chem. 45, 1777–1791.

CHAPTER 5

In Silico Optimization of the First DNA-Independent Mechanism-Based Inhibitor of Mammalian DNA Methyltransferase DNMT1

Patrik Nikolić[2,*], Vedran Miletić[3,*], Ivica Odorcić[2], Željko M. Svedružić[1]

[1]Department of Biotechnology and Faculty of Medicine, University of Rijeka, Rijeka, Croatia; [2]Department of Biotechnology, University of Rijeka, Rijeka, Croatia; [3]Department of Informatics and Faculty of Engineering, University of Rijeka, Rijeka, Croatia

ABSTRACT

Background: Inhibitors of DNA methylation can be used to control functional organization of human genome in basic research and regenerative medicine.

Results: We describe a set of adenosyl-1-methyl-pyrimidin-2-one derivatives as novel mechanism-based suicide-inhibitors of mammalian DNA methyltransferase DNMT1. The inhibitors are designed to act as transition state analogs that can bind simultaneously to the cofactor binding site and the active site on DNMT1. The two binding sites and the mechanism-based suicide-inhibition are combined to provide highly potent and highly specific inhibition. The stability of presented DNMT1-inhibitor complex is described in detail using 58 modifications that affect flexibility and binding interactions at specific sites within the complex.

Conclusions: Presented results can guide synthesis and optimization of some highly specific mechanism-based inhibitors of mammalian DNA methylation with specific pharmacological properties.

KEYWORDS

Cell differentiation; CP2K; Enzymatic mechanism; Epigenetics; Epigenome; GAMESS; Gromacs; iPSC; Molecular docking; Molecular dynamics; Pharmacophoric groups; Quantum mechanics; Rational-drug-design.

CONTENTS

1. INTRODUCTION

DNA methylation is a fundamental mechanism in the functional organization of the human genome. DNA methylation is one of the first steps in epigenetic regulation and the most enduring epigenetic landmark (Covic et al., 2010). The inhibitors of DNA methylation can be used in studies of human epigenome

* Equal contributions.

Epi-Informatics. http://dx.doi.org/10.1016/B978-0-12-802808-7.00005-8

(Romanoski et al., 2015), or in creation of induced pluripotent stem cells and cellular reprogramming (Khavari et al., 2010; Pennarossa et al., 2013; Sen et al., 2010). Inhibitors of DNA methylation can be also used in clinics for treatment of oncogenic transformation, viral infections, immunological disorders, or neurological and psychiatric impairments (Babenko et al., 2012; Gros et al., 2012; Heerboth et al., 2014).

DNMT1 is the principal DNA methyltransferase in mammalian cells (Svedruzic, 2008, 2011). More than 20 different inhibitors of mammalian DNMT1 have been described in the last 30 years (Xu et al., 2010). Unfortunately none of those studies gave consistent results and sustainable progress (Heerboth et al., 2014; Xu et al., 2010). The observed inconsistencies and the lack of progress could be in large part attributed to assay design and regulation of DNMT1 activity in cells. DNMT1 in cells can interact with about 40 different proteins and with some RNA molecules (Di Ruscio et al., 2013; Qin et al., 2011; Svedruzic, 2011). Thus, DNA methylation in cells can be affected by any change in DNA metabolism, DNA repair, chromatin organization, and cell cycle control (Qin et al., 2011; Svedruzic, 2011). Cell-based studies of inhibition of DNA methylation cannot differentiate between compounds that target DNMT1 directly from the compounds that can stall DNA methylation by causing DNA damage or other changes in DNA structure and metabolism. Thus, the screenings for inhibitors have to start with purified DNMT1, and then well-characterized compounds can be used to study inhibition of DNA methylation in cells. The screenings with purified DNMT1 have to be designed to differentiate between compounds that bind to DNMT1 from the compounds that interfere with DNA methylation by binding to the DNA substrate. None of the published studies of inhibition of DNA methylation has included all the required precautions, and still none of the published studies found compounds with IC50 values significantly below 1 μM (Gros et al., 2012; Xu et al., 2010).

Mechanism-based inhibitors can give highest specificity and lowest toxicity (Fersht, 1998). Different cytosine derivatives are the only true mechanism-based inhibitors that have been developed in the last 40 years (Xu et al., 2010). The cytosine derivatives were introduced based on mechanistic similarities between DNA methyltransferase and tymidilate sythetase (Wu and Santi, 1987). The cytosine derivatives have been very useful in mechanistic studies of DNA methyltransferases (Song et al., 2012; Svedruzic and Reich, 2004, 2005a), however their full applicability as inhibitors of DNA methylation was very limited due to high toxicity in cells (Xu et al., 2010). Very few improvements have been introduced following the initial studies, mostly due to the unusually challenging mechanism of the target base attack in the process of DNA methylation (Song et al., 2011, 2012; Svedruzic, 2008; Svedruzic and Reich, 2004, 2005a,b). Consistent with earlier activity studies recent crystal structures showed that the target base attack depends on multiple flexible loops in the protein structure (Estabrook et al., 2009; Song et al., 2011, 2012; Svedruzic and Reich, 2004, 2005a). Regulation of protein function by flexible loops in protein structure can be described very effectively by combining experimental results and computational methods (Medina-Franco and Caulfield, 2011; Yoo et al., 2012; Yoo and Medina-Franco, 2012).

In this study we present a novel strategy to the design of a whole family of mechanism-based suicide-inhibitors. The inhibitors are designed to act as a transition-state analog by binding simultaneously to the AdoMet site and the active site on DNMT1 (Fersht, 1998). The combination of two binding sites and the mechanism-based suicide-inhibition was chosen to provide highly potent and highly specific inhibition of DNA methyltransferases (Fersht, 1998). The presented strategy is a combination of insights from extensive enzyme activity studies, recent crystal structures, and some of the most recent computational methods (Medina-Franco and Caulfield, 2011; Song et al., 2011, 2012; Svedruzic and Reich, 2004, 2005a).

2. LEAD STRUCTURE FOR THE MECHANISM-BASED TRANSITION-STATE ANALOG

The structures 4.1 and 4.2 in Figure 1 show the concept of mechanism-based suicide-inhibition, while the structure in Figure 2 shows a realistic functional derivative of the conceptual structure 4.1 in Figure 1 (Merino, 2013; Vorbrüggen and Ruh-Pohlenz, 2001). In essence our mechanism-based inhibitor can mimic the last step in the catalytic cycle of DNA methyltransferases (Figure 1). This concept was first presented in our earlier paper on catalytic mechanism of DNMT1 (Svedruzic, 2008), and subsequently a very similar concept was used to prepare a novel inhibitor of histone methyltransferase DOT1L (Daigle et al., 2013).

We will use the structure presented in Figure 2 as a lead compound in our in silico studies with a desire to design a whole family of novel improved mechanism-based inhibitors. The lead compound consists of four functional parts. The first two parts are adenosine and ribose rings that can anchor the inhibitor to the AdoMet/AdoHcys site on DNMT1 (Song et al., 2012). The third part of the inhibitor is a flexible linker that could mimic catalytic intermediates in the rate-limiting methyltransfer step (Svedruzic and Reich, 2004, 2005a; Yang et al., 2013). The linker is expected to bind in a 3.2 Å long cavity that represents a reaction coordinate (Song et al., 2012). The linker is expected to act as a bridge between the AdoMet site and the active site (Song et al., 2012). The fourth part of the lead compound is the target base ring (Figure 2). The target base ring can bind in the active site cavity, where it can form hydrogen bonds with Arg1310 and Arg1312 and a covalent adduct with active site Cys 1226 (Figure 1). The linker and the target base ring are expected to assure that the inhibitor will preferentially target DNA methyltransferases in cells rather than the other enzymes that can also bind adenine and ribose rings.

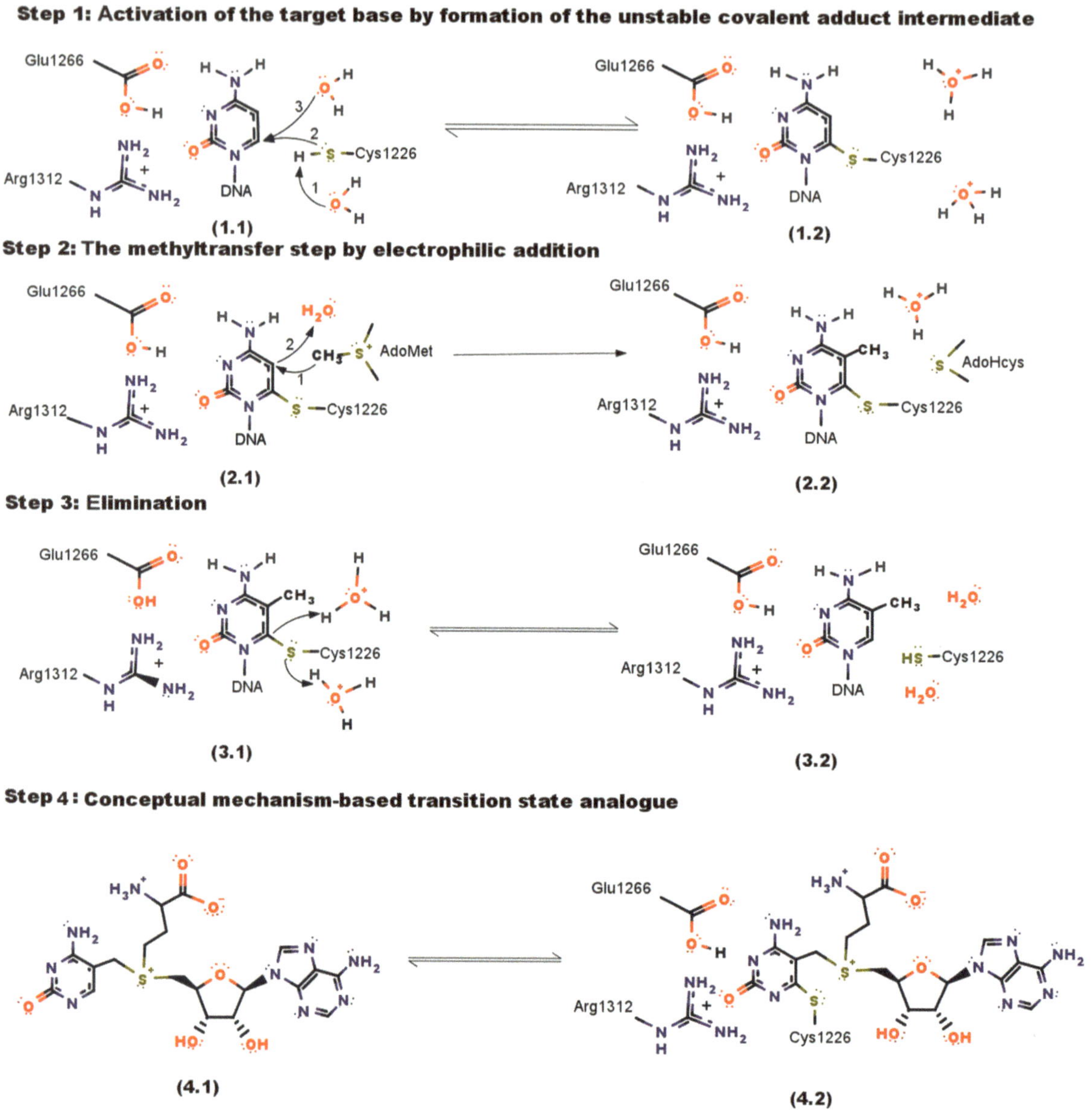

FIGURE 1 Catalytic mechanism of DNA methyltransferase DNMT1.

The catalytic mechanism of bacterial and mammalian cytosine-C5 DNA methyltransferase has been described in detail in the last 40 years in different crystal structures, enzyme kinetics studies, enzyme substrate binding studies, and QM/MM studies (Santi and Hardy, 1987; Song et al., 2011, 2012; Svedruzic and Reich, 2004, 2005a; Yang et al., 2013; Zhou et al., 2002). The catalysis consists of three main steps. The catalysis is initiated when the aromatic target base ring is posited in the active site cavity (step 1.1). In case of human and mouse enzymes the active site is composed of conserved Arg 1310, Arg 1312, Glu1266, and Cys 1126 residues (step 1.1). The enzyme can form unstable covalent adduct intermediate when conjugated aromatic bonds in the target base ring are positioned in asymmetric active site between polar amino acids (steps 1.1–1.2). Kinetic and QM/MM studies showed that the formation of the covalent adduct intermediate is a reversible step unless the reaction is driven forward by the rate-limiting irreversible methyltransfer step (Svedruzic and Reich, 2004, 2005a; Yang et al., 2013). Following the methyltransfer step (steps 2.1–2.2), there is a reversible target base elimination step (steps 3.1–3.2). We have decided to target this reversible step with a mechanism-based suicide inhibitor. The structure 4.1 is supposed to mimic the transition from structures 2.1 to 2.2. Thus, when structure 4.1 is positioned in the active site it should trigger formation of covalent adduct shown as the structure 4.2. This is in essence the same process as transition from structure 3.2 to structure 3.1. The figures were drawn in ChemAxon Marvin 15.0.1 (ChemAxon).

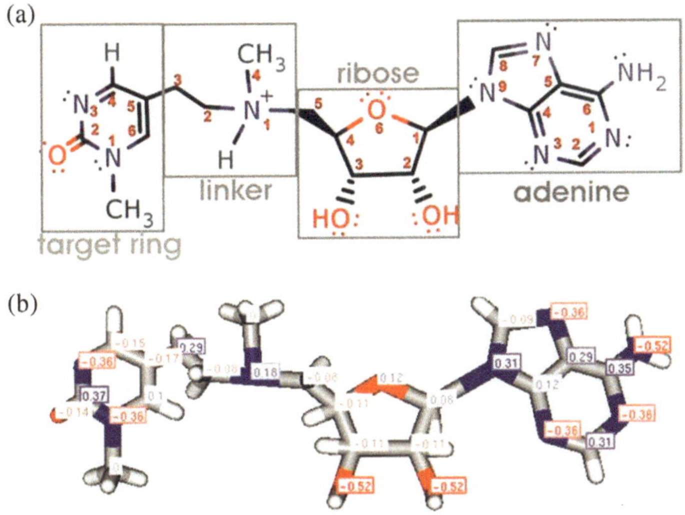

FIGURE 2 **(a and b). Lead structure for the mechanism-based suicide-inhibitor of DNMT1.** (a) The figure shows our first prototype for a mechanism-based inhibitor of DNMT1. The structure is derived from the structure 4.1 in Figure 1 as a reasonable challenge for organic synthesis with reasonable agreement with Lipinski's rules (Merino, 2013; Vorbrüggen and Ruh-Pohlenz, 2001). The structure is designed to act as a transition state analog and consists of four functional parts: adenine, ribose, tertiary amine linker, and 1-methyl-pyrimidin-2-one as the target base ring. The adenine and ribose parts are selected to assure that the lead compound can bind to AdoHcys/AdoMet sites. Tetrahedral tertiary amine linker is chosen as a flexible structure that can mimic the transition state structure (Figure 1). The 2-pyrmidine(2-one) was chosen as the target base ring instead of cytosine, to avoid problems with deamination at carbon 4 and to decrease polar surface on the target base ring. Different parts on the structure are labeled with numbers to help the reader follow different modifications throughout the text. (b) The present structure is in agreement with the Lipinski's rules except for its low LogD value. The relative molecular mass is 417.45 Da. There are six rotatable bonds, and a total of 58 bonds between 55 atoms. At pH 7.40 the polar surface area is 156.42 $Å^2$ while van der Waals surface area is 562.17 $Å^2$. There are five H bond donor sites and 13 acceptor sites. At pH 7.40, 94% of the ligand is protonated at the linker nitrogen and the compound is positively charged. The LogD value in protonated form is −2.63, and −1.88 in unprotonated form. To improve ADME properties the future modifications will be designed to increase the LogD value. This can be achieved by removing the positive charged nitrogen at the linker, and by removing hetero atoms at the ribose and adenine ring that give highest negative contribution to the LogD value. The numbers on the structure show contribution to the total LogD for each atom. All figures and calculations were prepared in ChemAxon Marvin suite version 15.3.16 (ChemAxon).

3. EVALUATION OF THE PROPOSED LEAD COMPOUND AS A MECHANISM-BASED INHIBITOR OF DNMT1

We show that the presented lead compound can bind to DNMT1 and act as a transition state analog and mechanism-based suicide-inhibitor using four different in silico approaches (Sections 3.1 to 3.4). First, rigid-body-docking with Chimera-Vina programs showed the lead compound can dock in its cleft on DNMT1 and form binding interactions (Figures 3 and 4; Pettersen et al., 2004; Trott and Olson, 2010). Second, Gromacs molecular mechanics/molecular dynamics (MM/MD) studies were used to analyze how molecular flexibility and bulk solvent molecules can affect stability of DNMT1-ligand complex (Table 1, lead compound entry). Third, specific modifications in the lead compound structure were chosen to evaluate binding interactions and flexibility at different positions within the DNMT1-ligand complex (Tables 1–5). In all three steps the binding was described as successful if the ligand in the binding site can overlap with AdoHcys and target cytosine in the corresponding crystal structures (Figure 3). Such overlap can be expected if the inhibitor can act as a transition state analog (Figure 3). In the final fourth step, QM/MM studies were used to test whether the observed complex can support formation of a covalent adduct between the target base ring and the active site Cys1226 (Section 3.4). Formation of the covalent adduct is the key step in the process of suicide inhibition that can assure a high potency and a high specificity for DNA methyltransferase (Figure 1).

In the first step, we used rigid-body-docking studies with programs Vina-Chimera as a fast and computationally light approach that can test if the lead compound can dock in its binding site and form supportive interactions (Pettersen et al., 2004; Trott and Olson, 2010). The corresponding search box included the whole catalytic domain, the whole DNA binding domain and about 15 Å of the surrounding space. The protein binding site was treated as a rigid body, and the docking was restricted to a maximum of nine conformers. The docking was successful if at least one of the nine binding conformers could overlap with AdoHcys and target cytosine in the corresponding crystal structures (Figure 3; Table 1(a), the lead compound entry).

In the second step, we subjected selected complex from the rigid-body-docking studies to MM/MD simulations using program Gromacs (Hess et al., 2008; Van Der Spoel et al., 2005). The selected complex was first minimized using molecular mechanics protocols in the presence of bulk solvent molecules. The molecular mechanics minimizations in essence represent evaluation of the initial rigid-body-docking complex as a flexible-docking problem. The minimizations resulted in only minor changes in the DNMT1 complex with the lead compound, indicating the initial rigid-body-docking was suitable for our application. Next, molecular dynamics simulations were used to analyze how bulk solvent molecules and molecular flexibility can affect interactions within the complex.

First, we find that the adenine and the ribose rings are firmly anchored in their binding site within 20 nsec simulations (Figures 3 and 4). The solvent can break

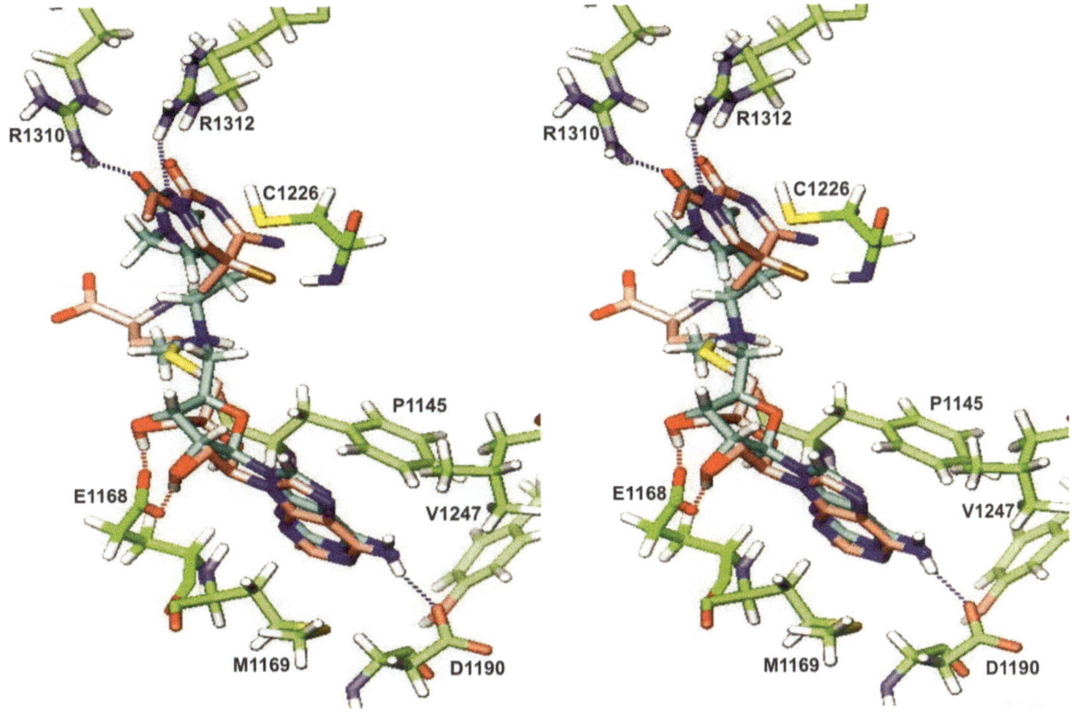

FIGURE 3 The lead compound can bind to the crystal structure of mouse DNMT1 (PDB: 4DA4 (Song et al., 2012)). The stereo image was prepared in VMD (Humphrey et al., 1996) and shows the lead compound (cyan), and the crystal structure of interacting amino acids (green), AdoHcys (pink), and the target cytosine (pink) (PDB: 4DA4 (Song et al., 2012)). Hydrogen atoms are shown in white, oxygen in red, nitrogen in blue, and sulfur in yellow. The figure shows that our lead compound can dock to DNMT1 in the same position as AdoHcy and the target cytosine. The figure was prepared by superimposing one of the MM/MD simulation frames with the lead compound in complex with DNMT1 to the crystal structure of DNMT1 in complex with AdoHcys and the target cytosine (PDB: 4DA4 (Song et al., 2012)). First, we find that the adenine and ribose part of the lead compound can overlap with AdoHcys in crystal structure in all simulation frames. The ribose rings are always anchored to their binding site by two hydrogen bonds with Glu 1168. The adenine rings are buried in a hydrophobic pocket and anchored with several hydrogen bonds to Asp1190 and the corresponding section of peptide backbone. MM/MD simulations show that all the hydrogen bonds are dynamic and sometimes interrupted by the bulk solvent molecules. However, adenine and ribose rings stay in place since the binding is always supported by multiple interactions. Second, we find that linker on the lead compound can fully mimic the transition state structure. The ethyl part of the linker is parallel to the methyl group on carbon-5 of target cytosine in the place of expected reaction coordinate (Svedruzic and Reich, 2004, 2005a; Yang et al., 2013). The positively charged tertiary amine is in position of the sulfur atom on AdoHcys, while the methyl-tail of the linker overlaps with the homocysteine tail. Third, the target base ring on the lead compound is positioned in same orientation as the target cytosine (pink) so that the two structures can practically overlap. Both rings form dynamic hydrogen bonds between the carbonyl oxygen and Arg 1310 or Arg 1312. The interactions with the two Arg residues are crucial for positioning the carbon 6 atom on the target base ring within 4 Å from the active site Cys1226. This proximity is crucial for formation of covalent adduct between the enzyme and the target base (Figure 1; Yang et al., 2013).

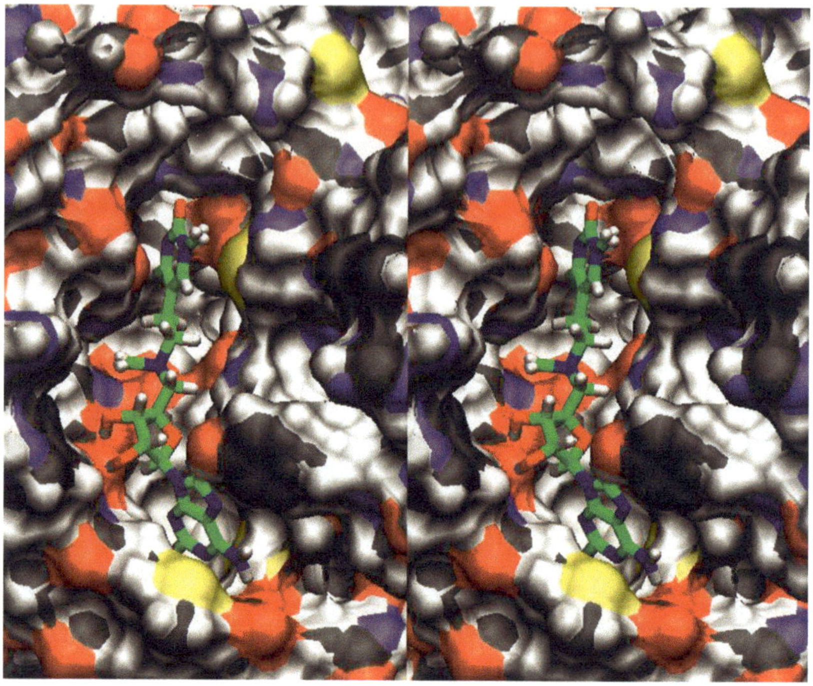

FIGURE 4 **Stereo image of the lead compound in its binding cleft on mouse DNMT1 (PDB: 4DA4 (Song et al., 2012)).**
The image shows the lead compound bound to mouse DNMT1 just as in Figure 3, but this time in surface mode to highlight the depth and the structure of binding site cleft. The image was prepared in VMD (Humphrey et al., 1996), and shows the surface hydrogen in white, carbon in black, nitrogen in blue, oxygen in red, and sulfur in yellow. The lead compound is shown in stick mode and colored green. The image shows one of the most open Gromacs MM/MD simulation frames to show depth and surface of the active site cavity. All the amino acids shown in Figure 3 can be easily recognized. The surrounding solvent molecules were removed for clarity. The reader is looking at the binding site cleft from the position of the DNA substrate, therefore the methyl group on the nitrogen 1 of the target base ring is positioned almost perpendicular to the plane of paper. The adenine ring is positioned in a tight surface pocket enclosed by Met1169, while the amino group on carbon 6 is positioned toward Asp 1190. The ribose ring is anchored in a wide ridge with its two OH groups atop of Glu1168. The linker is positioned in the widest part of the binding site cleft just above AdoHcys/AdoMet binding cavity. The target base ring is positioned in the active site cavity where it is stabilized by two dynamic hydrogen bonds between its carbonyl group and Arg1310 and Arg1312. The carbon 6 of the target base ring is within 4 Å from the active Cys1226, which is close to the optimal distance for formation of the covalent adduct (Yang et al., 2013).

the hydrogen bonds between DNMT1 and ribose or adenine ring (Figure 3), however such interruptions have very little effect on binding since the two rings are held in place by multiple interactions (Table 1(a), RMSD the lead compound entry). Second, we find that unlike adenosine and ribose rings, the linker and the target base constantly swivel within the confines of the binding site cavity (Figure 4). The swiveling results in two major conformations, which show as two distinct steps in the plots of average RMSD values (Table 1(a), the lead compound entry). MM/MD simulations plots also show that the distances between active site Cys 1226 and carbon 6 on the target base oscillate between 4 and 6 Å, with an average distance of 4.7 Å. Such oscillations can significantly affect the rate of formation of the covalent adduct and the suicide inhibition (Figure 6). MM/MD simulations show that the oscillations are driven by two sets of motions. First, the mobility of the active site Cys 1226 is driven by conformational changes in the active site loop (amino acids 1220 to 1236 (Song et al., 2012; Svedruzic and Reich, 2004)). The loop forms the edges of the active site cavity (Figure 4), that is continually closing and opening with the oscillation amplitude of about 4 Å at the tip of the loop. The second set of motions within the active site is wobbling of the target base ring around its two hydrogen bonds with Arg 1310 and Arg 1312 (Figure 3). In some frames the ring can rotate by as much as 360°, and such rotations can be detected as the extremes on Cys1226-carbon-6-distance plots (Table 1(a), the lead compound entry).

In the third step we prepared 58 modifications in the lead compound to analyze how specific changes in functional groups and ligand flexibility can affect the binding interactions (Tables 1–5). This ligand modification approach was a very rigorous test of the reliability of presented in silico studies. The modifications also gave valuable insights about interaction forces and showed a number of possible improvements in the lead compound structure. For example, we found a number of apparently subtle modifications in the lead structure that could prevent the initial rigid body docking (Tables 1–5). Some modifications did not significantly affect the initial rigid-body docking, but the subsequent MM/MD simulations showed differences in the stability of DNMT1-ligand complex (Tables 1–5). The observed differences in binding interactions are illustrated by changes in the average-ligand-RMSD plots and by changes in Cys1226-carbon-6-distance plots relative to the lead compound (Tables 1–5). Ligand RMSD plots were used to track MM/MD frames that show conformations with significant changes in DNMT1-ligand interaction. Distinct steps in RMSD plots show different conformations of the ligand within the complex, while uniform RMSD plots with low RMSD values indicate a stabile complex. Similarly Cys1226-carbon-6-distance plots were used to select MM/MD frames for QM/MM studies of covalent bond formation and mechanism-based suicide inhibition step (Figure 6). Covalent bond formation and the suicide inhibition step can be expected if the distance between the active site Cys 1226 and carbon 6 on the target base is within 3–4 Å, and if the two groups can meet within a reactive angle (Figure 6). It is also necessary to add that many of the presented modifications gave higher more favorable LogD values for the ligand relative to the lead compound (Figure 2; Tables 1–4).

3.1 Evaluations of Binding Interactions at the Adenine Part of the Lead Compound

All MM/MD simulation frames showed that the adenine ring sits tightly in its hydrophobic pocket behind Met 1169 and fully overlaps with adenine ring of AdoHcys in the crystal structure (Figure 3; Song et al., 2012). In addition to the tight binding pocket, the binding is stabilized by three highly dynamic hydrogen bonds between the ring nitrogen atoms and the protein backbone (Figure 3). MM/MD simulations showed that each of the hydrogen bonds is frequently broken or interrupted by bulk solvent molecules, but that does not affect the position of the ring in its binding cavity (Figures 4 and 5). We have analyzed two different modification sets. First modifications at the carbon 6 amino group and second modifications within the ring structure (Tables 1(a) and (b)). Different modifications in the adenine ring that can affect binding to DNMT1 have been experimentally described (Isakovic et al., 2009).

The amino group on carbon 6 is positioned on a hydrophobic ridge out of the adenosine binding pocket where it makes maximum of three dynamic hydrogen bonds with protein backbone and bulk solvent molecules (Figures 3–5). We have replaced the amino group with a hydrogen atom, nonpolar methyl group, or carbonyl group (Table 1(a)). The binding was not observed only when a hydrogen atom was placed on carbon 6, suggesting that a group of a certain size is needed to guide the adenosine ring in its binding pocket (Figures 4 and 5). Both methyl group and carbonyl groups support binding, which indicates that the hydrogen bonds with the amino group on carbon 6 are not crucial for the DNMT1-ligand interaction. Precisely, MM/MD simulations showed that the methyl and carbonyl groups can reposition the modified ring in the ligand binding cavity so that optimal interactions with other amino acids can be achieved. The repositioning can also increase the ligand RMSD values, yet still Cys1226-carbon-6 distance plots show that both modifications can support the suicide inhibition step (Table 1(a)).

We next introduced modifications within the ring structure (Table 1(b)). We made the ring more hydrophobic by replacing the polar nitrogen atoms with carbon atoms or by replacing the aromatic ring with the saturated ring (Table 1(b)). The increase in hydrophobic surface can bury the modified ring in its binding pocket and decrease the ligand mobility in its binding site groove (Figures 4 and 5; Table 1(b)). This results in a decrease in RMSD values, and in a very favorable decrease in Cys1226-carbon-6 distance plots (Table 1(b)). The only exception is a fully saturated ring, which is too thick to slip in the binding site pocket that is suited for a flat aromatic ring. The ligand with the saturated ring can dock in the binding site groove, but the corresponding RMSD and Cys1226-carbon-6 distance plots show that the ligand with saturated ring is

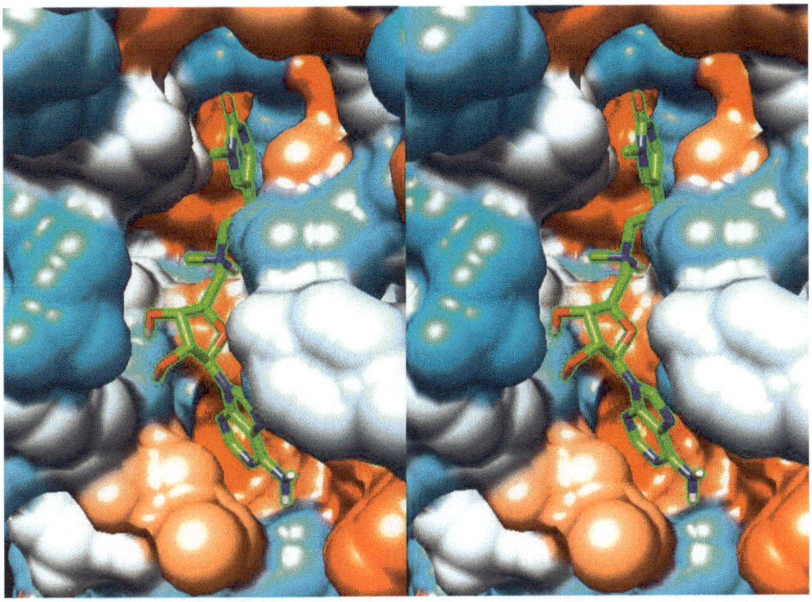

FIGURE 5 The lead structure binds to both hydrophobic and hydrophilic surface in its binding site cleft on mouse DNMT1 (PDB: 4DA4 (Song et al., 2012)).
The stereo image shows the lead compound bound to mouse DNMT1 just as in Figures 3 and 4, but this time surface in the binding cavity is colored according to its hydrophobic (brown) and hydrophilic (blue) properties in program Chimera (Pettersen et al., 2004). The figure shows that the adenine ring is positioned in a deep hydrophobic pocket, except for its amino group on carbon 6, which makes hydrogen bonds with hydrophilic Asp 1190. The ribose ring is positioned on a tight ridge so that its two OH groups sit atop of the hydrophilic surface of Glu 1168 while its acetal oxygen is buried in the slightly hydrophobic cavity. The nitrogen and the methyl tail part of the linker sit in the hydrophilic AdoMet/AdoCys cavity. The ethyl part of the linker that is positioned in the reaction coordinate grove sits on a hydrophobic surface. Finally, the target base 1-methyl-pyrimid-2-one ring is positioned in a hydrophobic active site cavity where it makes dynamic hydrogen bonds through its carbonyl oxygen and the hydrophilic surface of the active site Arg 1310 and 1312.

only loosely bound in the binding site groove and cannot support the suicide inhibition step (Figure 6).

In conclusion, we found several modifications in the adenine part of the ligand that can support binding of the ligand to DNMT1 (Tables 1(a) and (b)). These modifications can be used to decrease the large negative contribution of the adenine ring to the LogD value (Figure 2(b)), and to possibly eliminate some of the nonspecific interaction with numerous other proteins that can bind the adenine ring in cells (Aravind, 2011). Preparation of different modifications in the adenine ring have been described in the literature (Merino, 2013; Vorbrüggen and Ruh-Pohlenz, 2001), and many of the modified structures are commercially available.

3.2 Evaluations of Binding Interactions at the Ribose Part of the Lead Compound

In all DNMT1 crystal structures the two OH groups on the ribose ring of AdoHcys and AdoMet make hydrogen bonds with the negatively charged Glu1168 (Song et al., 2011, 2012; Takeshita et al., 2011). The same hydrogen bonds can be observed between Glu1168 and our lead compound (Figure 3). The Gromacs MM/MD simulations showed that the ribose part of the inhibitor overlaps with the ribose part of crystalline AdoHcys in all simulation frames (Figure 3). The crystal structures and MM/MD simulations showed that both —OH groups and Glu1168 can make dynamic hydrogen bonds with the bulk water molecules (Song et al., 2011, 2012; Takeshita et al., 2011). Despite those fluctuations the ribose is firmly anchored to its place. We have modified all six atoms in the ribose ring to test how each atom can contribute to binding the lead compound (Table 2). We designed modifications in the ribose ring with two aims. The first aim is to increase specificity of the lead compound, since numerous molecules in cells can bind ribose. The second aim is to increase the LogD value, and thus enhance ADME properties of the inhibitor. The two —OH group give the highest contribution to the LogD value of the presented lead compound (−0.52 per one OH group, Figure 2).

We find that our lead compound cannot bind to DNMT1 if both OH groups are replaced with H atoms (Table 2). The ligand can bind if the OH group on carbon 3 is kept and the OH group on carbon 2 is replaced with H atoms. However, there is no binding if the OH group on carbon 2 is kept and the OH group on carbon 3 is replaced with H atoms. Gromacs MM/MD simulations show that the OH group on the position 3 has more influence on binding than on position 2, since it has more favorable interaction geometry with Glu 1168 and tighter interactions with the walls of the binding site grove. The binding will be also abolished when a double bond is introduced between carbons 2 and 3 (Table 2). In that case the two OH groups are lifted in the plane of the ribose ring, and thus cannot satisfy the optimal geometry for the hydrogen bond formation. We also prepared several positive controls to prove that interaction with Glu 1168 is crucial for binding (Table 2). In all cases the ligand can bind to its site when the positive group is in the position of the OH groups, while the binding is not observed when the OH groups are replaced with negative groups.

Apart from the interactions with Glu 1168, the binding of the lead compound depends on a tight fitting of its ribose ring in its binding cavity. Precisely, all modifications at the acetal oxygen on position 6 can abolish binding by displacing the ring from it position. The acetal oxygen is in direct van der Waals contact with Phe1145. Thus any group at position 6 that is bigger than the oxygen atom can push the ribose ring out of its binding grove. The only modification that can be tolerated at that position is replacement of oxygen with the secondary nitrogen that has one hydrogen atom directly below the plane of the ring. We also found several surprises. First, the ligand binding is observed when both OH groups are removed and carbon 3 is replaced with an oxygen atom (Table 2). MM/MD simulation showed that such protein-ligand complex is stabilized in its position by a water molecule that can wedge between Glu 1168 and the oxygen atom at position 3. The second surprise was observed when the two OH groups were replaced with chlorine atoms. The resulting compound has optimal size for wedging into the binding site cleft in protein in rigid-body docking (Table 2). However such ligand lacks the binding interactions so it can quickly escape from the bind site cleft in MM/MD simulations.

In conclusion just like the adenosine ring, the ribose ring in the lead compound can tightly overlap with the corresponding parts in crystal structures of AdoHcys (Figure 3). In contrast with the adenosine ring with the ribose ring we could not find modifications that can give the desired stability to the DNMT1-ligand complex.

3.3 Evaluation of Binding Contribution from the Linker Part of the Lead Compound

Both docking and MM/MD studies showed that a tertiary amine-linker can mimic the transition state intermediate of DNA methyltransferases (Figure 3). First, the ethyl branch on the linker is positioned in the same place as the methyl group on the target cytosine in the crystal structure of DNMT1 (Figure 3; Song et al., 2012). Second, the tertiary nitrogen is positioned in the same place as sulfur in AdoHcys, while its methyl-tail is positioned in the same place as the Hcys tail of AdoHcys (Figure 3). MM/MD simulations showed that the linker can swivel out of the binding site cavity together with the target base while the adenosyl part of the ligand is anchored in its place. Such swiveling can be explained with the binding interactions. The linker is positioned between three binding pockets on the DNMT1 surface where it can make only loose hydrogen bonds or hydrophobic interactions with DNMT1. The linker is wedged in its place by the firmly anchored adenosily part of the ligand and by the target base ring that makes two dynamic hydrogen bonds with Arg 1310 and 1312 in the active site cavity (Figures 3 and 4). Thus we prepared modifications in the linker with three specific aims. The first aim was to test how changes in the linker flexibility can affect the binding of the ligand (Table 3(b)). The second aim was to replace the positively charged nitrogen atom with some group that has more desirable ADME properties (Table 3(a)). The third aim was to wedge in the linker in the binding site grove by adding strategically placed methyl groups that can decrease flexibility and mobility of the linker in its binding site (Table 3(c)).

First we show that the lead structure can bind to DNMT1 with both protonated and unprotonated nitrogen atoms on the linker (Table 3(a)). This is not surprising, since we have found in earlier experiments that a charge at that position does not significantly affect binding of any ligand to DNMT1 (Svedruzic and Reich, 2004, 2005a). Second, we tested different modifications in the tail section of the linker (Figure 2, position 4). The tail end of the linker is expected to fit in the same site as the AdoHcys tail (Figure 3) where it can anchor the linker in its position by acting as a three-legged stool. Indeed a linker without its methyl-tail cannot support binding of the target base ring to the active site cavity (Figure 2; Table 3(a)). Next we replaced the methyl-tail with an ethyl-tail, to see if a longer tail can be a better anchor. Surprisingly, docking studies showed that the longer tail supports ligand binding much less than the methyl-tail, and only when the ligand nitrogen is in its unprotonated position. The corresponding MM/MD studies showed that the long tail has more collisions with the protein surface than the short methyl-tail. Those collisions tend to push the linker out of its binding site and thus decrease the chances for the tail to fit in the Hcys binding cavity. In the last attempt to find the best anchor in the tail section, we analyzed binding of the lead compound with hydroxy-methyl and hydroxy-ethyl tails. The hydroxy-methyl tail is attractive since it can lower the pKa of the tertiary amine so that the ligand will be mostly neutral at the physiological pH (Table 3(a)). The docking studies show that hydroxy-methyl will support binding while the hydroxy-ethyl tail will not. The corresponding MM/MD studies show that the hydroxy-methyl tail does not make significant hydrogen bonds with the protein surface, rather it supports the binding by acting as a better steric anchor than the hydroxy-ethyl tail.

Apart from its function as a tetrahedral steric anchor, the tertiary amine linker was selected as the most versatile structure for organic synthesis. Four atoms could be considered for the branch point of the linker: nitrogen, carbon, oxygen, and sulfur. Linkers with a bivalent oxygen or sulfur atom in position 1 cannot have a tail that might be needed to anchor the linker in the required position (Figure 2). A linker with methyl-sulfonium at position 1 does not have the required chemical stability. Docking analysis showed that linkers with oxygen or sulfur at position 1 can support docking of the target base in the binding site cleft, but the linker with the —CH_2— group cannot. MM/MD analysis showed that this surprising interaction can be attributed to a very tight van der Waals interaction between oxygen and Pro 1224, which can accommodate an oxygen atom but not the slightly bigger —CH_2— group. The bulky sulfur can wedge in the binding site cleft and force the target base toward the active site cavity. Finally, a linker with a tertiary carbon at position 1 can anchor the lead compound to its binding site just as the tertiary nitrogen linker. The carbon linker has also a more favorable LogD value (Figure 2), however the compounds with tertiary carbon at the branch point are more difficult to synthesize.

We have also attempted several strategies to stiffen the long arm of the linker and force the target base ring toward Cys 1226 in the active site. We found that introduction of E or Z double bond to any part of the linker cannot force the target base ring toward the active site Cys 1226. Similarly the target base cannot be forced toward Cys 1226 by replacing a flexible linker with some of the ring structures, since different rings structures were too big to fit in the required position. At the end, we found that the target base ring can be forced toward active site Cys 1226 when an *entgegen* double bond is introduced between carbons 2 and 3 and with another *entgegen* double bond or an oxygen atom placed at position 1 on the linker (Figure 2; Table 3(b)). The binding of the ligand can also be improved by adding methyl groups to the flexible linker (Table 3(c)). Two methyl groups on the linker can decrease the ligand mobility in the active site groove by wedging the flexible linker against the sides of the binding site cleft (Table 3(c)).

In conclusion, we found several modifications that can significantly improve our lead compound. These modifications can simplify preparation of the linker, increase the LogD value of the ligand, and increase stability of the DNMT1-ligand complex. Some modifications can decrease the linker flexibility with specifically placed double bonds or methyl-groups that can wedge the ligand in its binding site. Some modifications can increase the linker flexibility and still give the desired stability to the DNMT1-ligand complex. The tail section of the ligand can stabilize the DNMT1-ligand complex to some degree, however these effects are very limited since the tail can also destabilize the complex by collisions with the binding site surface. The presented modifications in the linker can be prepared together with the target base using the Biginelli reaction with different commercially available aldehydes, and subsequent modification on the nascent chain using pyridinium chlorochromate oxidations (Merino, 2013; Vorbrüggen and Ruh-Pohlenz, 2001).

3.4 QM/MM Analysis of the Target Base Ring in Proximity of the Active Site Cys 1226

We used QM/MM studies to show that the active site Cys 1226 can make a covalent bond with carbon 6 on the target base ring when the lead compound is bound to DNMT1 (Figure 6). Formation of the covalent bond is the key step in mechanism-based suicide-inhibition that can make our inhibitor highly potent and highly specific for DNA methyltransferase in cells (Figures 1 and 6). MM/MD plots of distances between cysteine 1226 and carbon 6 on the target base showed that those two atoms are separated by 3–4 Å when the lead compound or some of its derivatives are bonded to DNMT1 (Tables 1–4). Earlier QM/MM studies of bacterial DNA methyltransferases showed that active site cysteine can form a covalent bond with the target base ring when the two groups are about 3 to 4 Å apart (Yang et al., 2013).

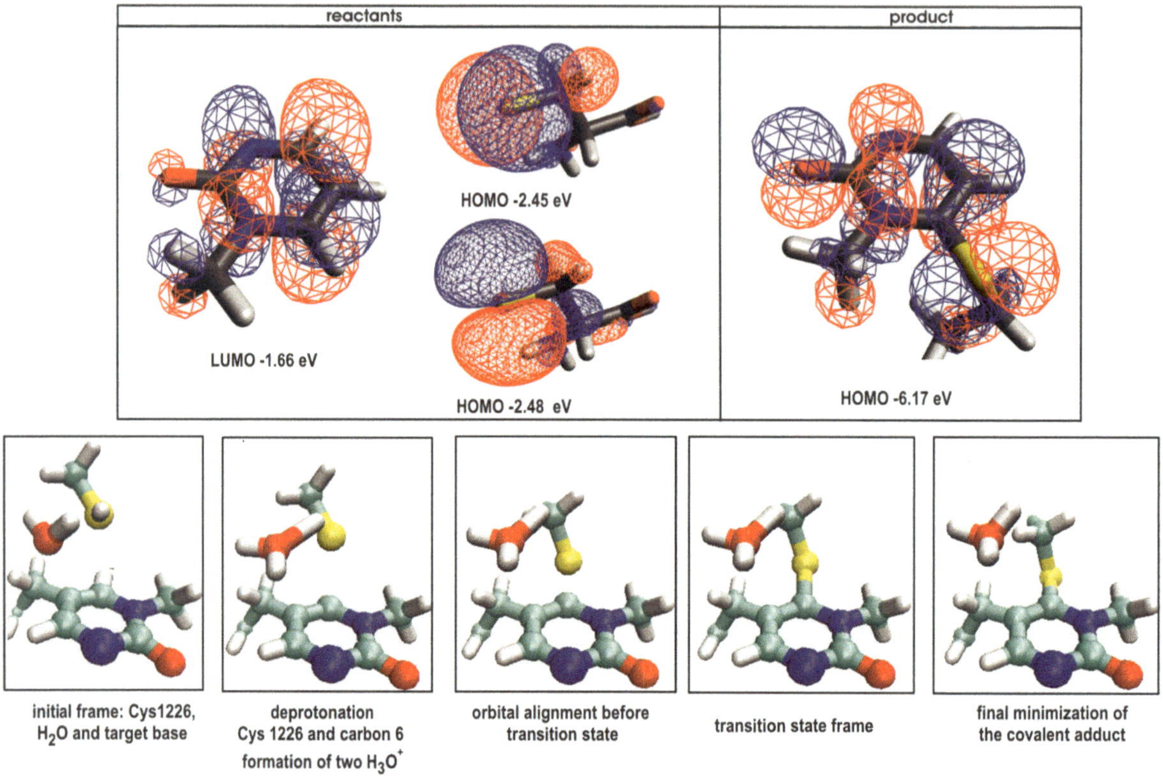

FIGURE 6 **Formation of a covalent bond between the lead compound and active site cysteine 1226.** ***(All figures were prepared in VMD (Humphrey et al., 1996).)***

The target base ring has LUMO orbitals at carbon 6 that are perpendicular to the plane of the ring. The nucleophilic sulfur anion has two HOMO orbitals with almost identical energy that are perpendicular to each other. The reaction can start when one of the two HOMO orbitals on the sulfur atoms come in the same plane as the target base ring. The orbitals and their energies were calculated in program Gamess (Gordon and Schmidt, 2005). In the bottom panels, selected QM/MM simulation frames show how nucleophilic orbitals on a sulfur anion can come in the plane with the electrophilic orbitals on carbon 6 of the target base. Hydrogen atoms are shown in white, carbon in cyan, oxygen in red, nitrogen in blue, and sulfur in yellow. The first panel shows one of the Gromacs MM/MD simulation frames that was used as a starting point for QM/MM simulations with program CP2K (VandeVondele et al., 2005). The active site cysteine 1226 is directly above the plane of the target base ring 3.8 Å from carbon 6. Both Cys 1226 and carbon 6 on the target base are protonated (first frame). The reaction starts with deprotonation of the active site Cys by a water molecule (second frame). When Cys 1226 anion collides with carbon 6 and its hydrogen atom, the carbon 6 can get deprotonated (third frame, deprotonation of carbon 6 takes place below the ring so it is not visible in the presented angle). Deprotonation is completed when the Cys 1226 anion gets in the same plane as the target base ring (third frame) so that the transition state complex and the covalent bond can be formed (fourth frame). At the end the new bonds are minimized to achieve the optimal bond angles. Both MM/MD and QM/MM simulations show that the bond formation is driven by the wobbling of the target base ring around the hydrogen bond axis between carbonyl oxygen and Arg 1310 and 1312. Up-and-down swiveling motion of Cys 1226 is driven by wobbling in the active site loop (Song et al., 2012). Looking at Figure 4, the bond forming motions can be driven by the fluctuations in the width of the binding site cavity, and by the rotation of the target base ring around its hydrogen bonds axis with Arg 1310 and Arg 1312.

The two highest occupied orbitals on nucleophilic cysteine are perpendicular to each other and have almost equal energy (Figure 6). The lowest unoccupied orbitals on the target base are positioned on carbon 6 and perpendicular to the plane of the ring. Our QM/MM simulations showed how the covalent bond can form when the active site Cys 1226 moves in the plane of the target base ring (Figure 6). Cys 1226 can be driven in the reactive position by closing the active site loop, which makes the sides of bind site cleft (Figure 4 vs Figure 7). The target base ring can turn toward the active site cysteine by wobbling around its two hydrogen bonds with Arg1310 and Arg 1312 (Figures 3 and 6). The formation of the covalent bond also requires two water molecules within an active site cavity (Figure 6). Interestingly, MM/MD studies showed that the methyl group on positions 1 or 4 of the target base ring can interfere with productive collisions between Cys 1226 and carbon 6 on the target base ring. Thus, we could regulate the rate of the suicide inhibition step by placing the methyl group on nitrogen 1 or carbon 4.

We analyzed lead compounds with both 1-methyl-pirimidin-2-one and cytosine as the target base ring. 1-methyl-pirimidin-2-one has several advantages. First, 1-methyl-pirimidin-2-one has smaller polar surface and larger LogD value than cytosine. Second, 1-methyl-pirimidin-2-one lacks the electron-rich amino group, so it is a better electrophile than cytosine (its LUMO is −1.4 eV, vs LUMO for cytosine is −0.9 eV). Third, derivatives of 1-methyl-pirimidin-2-one are easier to prepare than cytosine derivatives using the Biginelli reaction. Fourth, 1-methyl-pirimidin-2-one makes a stabile covalent adduct that can be observed in crystal structures, while cytosine makes a reversible covalent adduct (Yang et al., 2013; Zhou et al., 2002). Cytosine as a target base can also have some specific advantages. The lead compound with cytosine as the target base can bind tighter to DNMT1 since the amino group on the cytosine carbon 4 can form hydrogen bonds with Glu1266 in the active site (Song et al., 2012; Table 4). However, the amino group on carbon 4 is known to hydrolyze and turn into carbonyl group that has a repulsive interaction with Glu1266 (Svedruzic and Reich, 2005a; Table 4). Taken together these results indicate that 1-methyl-pirimidin-2-one is a better choice when irreversible suicide inhibition is desired, while cytosine can be a better choice when short-lasting reversible inhibition is desired. The reactivity of the target base can be also modulated by the methyl groups on carbons 1 and 4 (Table 4). The Cys1226-to-carbon-6 distance plots show that the methyl-groups can affect mobility of the target base ring and thus rate of formation of the covalent adduct (Figure 6).

Different modifications on the target base with different linkers at carbon 5 can be prepared in two or three synthetic steps using Biginelli reaction with different commercially available aldehydes and methyl-urea (Merino, 2013; Vorbrüggen and Ruh-Pohlenz, 2001).

4. OPTIMIZATION OF BINDING INTERACTIONS AT MULTIPLE SITES ON THE LEAD COMPOUND

The insights from different modifications at a single site can be combined to create improved compounds with modifications at multiple sites. MM/MD and docking studies showed that the binding interactions at the ribose ring and the linker region are the two most challenging sites for modification (Tables 1–4). MM/MD studies showed that the loss of the bonding interaction between the —OH group and Glu1168 will result in wobbling of the ribose ring in the binding site cleft and facilitate dissociation of the ligand. We added methyl-groups at the C2 and C3 opposite to the two —OH groups to wedge the ligand in its binding site and make the DNMT1-ligand complex less dependent on interaction with Glu1168 (Table 5). Docking and MM/MD studies showed that such methyl groups can stabilize the ribose ring and DNMT1-ligand complex even with some modifications that otherwise do not support formation of the complex (Table 2). Ribose molecules with methyl groups on carbon 2 and carbon 3 are commercially available from several suppliers.

5. BINDING OF THE MECHANISM-BASED INHIBITOR TO DIFFERENT CONFORMATIONS OF DNMT1

We next show that the presented lead compound can bind to different conformations of DNMT1 (Figure 4 vs Figure 7; Song et al., 2011, 2012; Takeshita et al., 2011). Different crystal structures show different conformations of DNMT1 that represent different stages in process of DNA methylation in cells. We are specifically interested in function of the active site loop (amino acids 1220 to 1236 (Song et al., 2011, 2012; Takeshita et al., 2011)). The active site loop is in its closed position when full-methylated DNA is bonded to DNMT1 (PDB code: 4DA4 (Song et al., 2012)). That structure represents the last step in the process of maintenance methylation when DNMT1 is in complex with the fully-methylated DNA substrate. The active site loop is in the open position when mouse DNMT1 is in complex with unmethylated DNA (PDB code: 3PT6 (Song et al., 2011)), or when DNA is not bonded to DNMT1 (PDB code: 3AV6 (Takeshita et al., 2011)). We would like to exploit the function of the active site loop and its influence on the suicide inhibition to design inhibitors that could preferentially target different metabolic functions of DNMT1 and thus provide specific pharmacological properties (Svedruzic, 2011).

The simulations show that RMSD values for the lead compound bound to DNMT1 are almost identical when the active site loop is open and closed (Table 6,

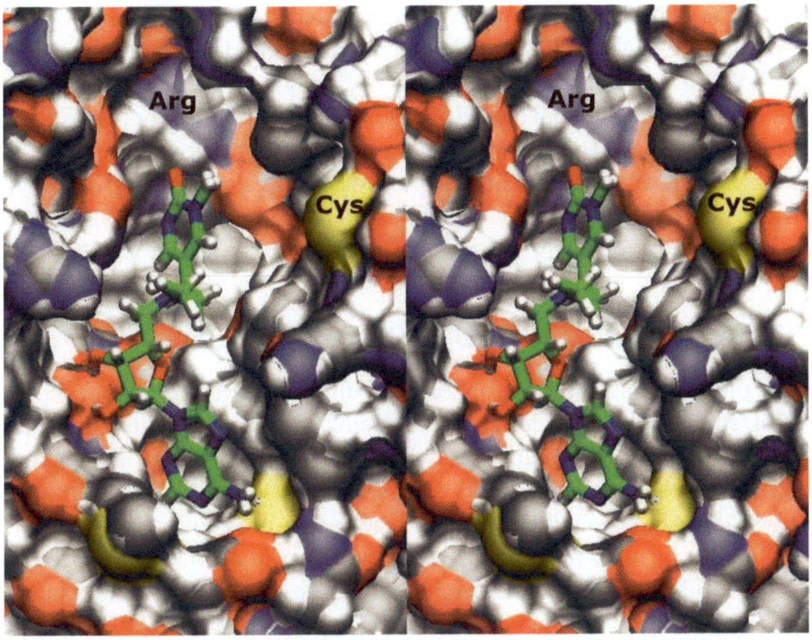

FIGURE 7 Stereo image of the lead structure bound to mouse DNMT1 with the active site loop open (PDB code:3AV6 (Takeshita et al., 2011)).

The image shows that the binding site for the lead compound is much wider when the active site loop is open (amino acids 1220–1236). Just as in previous figures, the inhibitor is held in its binding cavity by its three main interactions. The two —OH groups on the ribose ring, the adenine ring that is buried in the hydrophobic pocket, and the hydrogen bonding with Arg 1310 and Arg 1312. Closing the active site loop drives the active site cytosine Cys1226 to its catalytic position (Song et al., 2012). Taken together these observations indicate that the lead compound can bind to DNMT1 when its active site loop is in its open and closed positions. However, the suicide inhibition is much more likely when the active site loop is closed. The closing of the loop can be driven by DNA binding or by slow thermodynamic swiveling in the loop structure from closed to open conformation. Image was prepared in VMD (Humphrey et al., 1996).

average RMSD value 11.4 vs 12.4 Å). This is not surprising since the active site loop has little effect on AdoMet binding sites, which provide the majority of binding interactions for the lead compound (Song et al., 2011, 2012; Takeshita et al., 2011; Figure 3). However, there are significant differences on plots of Cys 1226 and carbon 6 distances. The average values for 4DA4 structure is 4.7 Å, while the average for 3AV6 is 9.1 Å (Table 6). Those differences can be attributed to the different positions in the active site loop (Song et al., 2011, 2012; Takeshita et al., 2011; Figure 4 vs Figure 7). The distance between Cys 1226 and carbon 6 on the target base can affect kinetics of the suicide inhibition step (Figure 6). Taken together these results indicate that the suicide inhibition can happen fast when DNMT1 is bonded to hemimethylated substrate in the process of maintenance methylation. Much slower suicide inhibition can be expected when DNMT1 is bonded to unmethylated DNA or when DNMT1 is in its DNA free form. In those cases the slow suicide inhibition step will depend on the slow thermal swiveling of the active site loop between the open and closed positions. The differences between fast and slow target base attack and the function of the active site loop have been described experimentally for DNMT1 and HhaI DNA methyltransferases (Estabrook et al., 2009; Svedruzic and Reich, 2004, 2005a).

6. CONCLUSIONS AND FUTURE DIRECTIONS

We have presented a combination of four different in silico approaches as an insightful strategy for the design of different mechanism-based inhibitors of DNMT1. The presented results show that it is possible to prepare a family of mechanism-based suicide inhibitors of DNMT1 that can bind to the enzyme-like transition state analogs. Different modifications in the original lead compound showed how to increase the stability of the complex and how to affect the rate of the suicide inhibition. Presented insights can guide different synthetic strategies to prepare a whole family of mechanism-based inhibitors of DNMT1 with optimal ADME properties for desired treatments. Many of the presented compounds can be prepared for less than 6000 USD in material costs by combining the different products of Biginelli reaction with different adenosine derivatives in total of three to five synthetic steps (Merino, 2013; Vorbrüggen and Ruh-Pohlenz, 2001).

The presented in silico approaches can also be extended in the future to prepare compounds specific for DNMT1, DNMT2, DNMT3, or some bacterial DNA methyltransferases (Guo et al., 2015; Li et al., 2013; Zhou et al., 2002). Selective compounds can be prepared by exploiting subtle differences between different DNA methyltransferase in their active site and interaction with the cofactor (Svedruzic and Reich, 2005a). The hope is to design inhibitors that will have preference for specific DNA methylation processes in cells (Svedruzic, 2011).

7. METHODOLOGY

7.1 Docking Analysis

The initial "rigid-body-docking" calculations used programs Autodock Vina as a plugin application in UCSF-Chimera (Pettersen et al., 2004; Trott and Olson, 2010). The protein coordinates were acquired from the Protein Data Bank, while the ligand structures were drawn in ChemAxon Marvin (ChemAxon). Prior to docking the ligand structures were tested with several programs to make sure that conversion between file formats did not distort chiral centers or aromatic bonds. In first step, both the protein and the ligand were protonated at pH = 7.2 using Amber99SB force-field tool in Chimera program. Than the search volume box 25.0 × 23.5 × 19.0 Å was prepared, which included the AdoHcy binding site, the active site where the target cytosine binds, and the interface between the catalytic domains (Song et al., 2012). The crystal water molecules have been deleted, except for the molecules that had hydrogen bonds with the residues at the AdoMet sites. In the latter case, we did docking with and without the water molecules. Docking was judged as successful if at least one of the nine conformers showed overlap with adenine rings and the target cytosine in the corresponding crystal structure (Figure 3). The hydrogen bonds and the interaction energies reported by Vina were used only as reference points and have to be taken with caution.

7.2 MD Simulation and Analysis

Classical molecular mechanics and molecular dynamic (MM/MMD) simulations were calculated using the program package GROMACS (Hess et al., 2008; Van Der Spoel et al., 2005). Position-restrained equilibrated MD simulations for large biomolecular systems (>200,000 atoms) were running on heterogeneous CPU/GPU workstations with CUDA-enabled NVIDIA GeForce and Tesla GPUs. Our systems were composed from protein (standard amino acid residues), ligand (nonstandard residue), and bulk solvent molecules. Each of these elements was prepared separately. The protein was processed with pdb2gmx, while the ligand was processed with ACPYPE10 (Sousa da Silva and Vranken, 2012). ACPYPE is an interface for Antechamber (part of AmberTools11), which is used to generate various topology types for various force fields (CHARMM, GROMOS, AMBER, OPLS) (Sousa da Silva and Vranken, 2012). The Amber99SB force field was used for MD simulations that have nonstandard residues such as ligands. We merged all coordinate files into a single coordinate file, and all topology files into single topology file after parameterization of the protein and the ligand.

The next step was preparation of solvent box and solvation of the system. We used rhombic dodecahedron solvent box type optimized to minimize the simulation wall-clock time. The TIP3P model was used for water since the solvent can significantly affect protein–ligand interactions. The prepared system was next minimized and equilibrated. The minimizations used a combination of steepest descent and conjugate gradient algorithms. When the most stabile state was achieved the temperature was introduced and the system was equilibrated to 310 K (NVT equilibration, V-rescale). Next, the pressure was equilibrated to 1 atm (NPT equilibration, Parrinello-Rahman). No restraints were used for the protein or the ligand when the system was minimized, but in NPT and NVT equilibration protein and ligand were restrained. The outputs of minimization and equilibration MD simulations provide insight into the potential energy of the system based on minimization of the temperature in NVT simulation, and minimization of the pressure in NPT simulation.

We calculated position-restrained MD simulations for mammalian DNMT1 in total duration of 20 ns, 10 million steps, 2 fs step-time, with a total of 100 recorded frames. Position-restrained MD simulations were chosen since a large fraction of amino acids in DNMT1 does not take part when ligands bind to the AdoMet site and the active site. Position-restrained MD runs did not restrain the amino acids in catalytic domains (residues 1135–1295), and amino acids in the adjacent DNA binding domain (residues 1455–1597) (Song et al., 2012). Such a strategy was justified by the observed b-factor values and by our earlier experiences with smaller bacterial DNA methyltransferase HhaI (data not published). Restraining could save us about 60% of the wall clock time in each simulation run. Following MM/MD simulations RMSD plots and Cys 1226 and carbon six distance plots were calculated using built-in GROMACS and VMD utilities (Humphrey et al., 1996).

7.3 QM and QM/MM Simulations

The frontier molecular orbitals were calculated using program Gamess (Gordon and Schmidt, 2005). The molecules were drawn in Avogadro (Hanwell et al., 2012), the calculations were set up using SCFTYP = RHF RUNTYP = ENERGY DFTTYP = B3LYP protocols with 6-31G basis set, water medium, and initial MO guess set as Huckel. The resulting molecular orbitals and the calculated energies have been visualized using VMD (Humphrey et al., 1996).

The CP2K program was used for quantum mechanics (QM) and hybrid quantum mechanics and molecular mechanic (QM/MM) simulation. QM simulations used the density functional theory (DFT) that was based on hybrid Gaussian and plane waves density functional (VandeVondele et al., 2005). TZV2P-GTH basis sets and BLYP exchange-correlation energy functional were used for 19 QM atoms in the active site. CHARMM force field parameters were used for MM atoms. Force field parameters were generated by Acpype for ligand (Sousa da Silva and Vranken, 2012), and taken from VMD for amino acids (Humphrey et al., 1996). PSF topology used in CP2K is generated by VMD using an automatic PSF builder plugin (Humphrey et al., 1996).

Nudged elastic band method was used for transition state search (Henkelman and Jónsson, 2000; Henkelman et al., 2000). PyMOL is used for selection of QM atoms in QM/MM simulation, and PyMOL plugin was used for generation of QM/MM section in CP2K input file. Replicas used in the nudged elastic band were drawn in Avogadro (Hanwell et al., 2012).

APPENDIX TABLES

Table 1A Optimization at Position 6 in Adenine Ring of the Presented Lead Compound

Structure Modification	LogD @ pH=7.4	Chimera-Vina Docking Interaction Energy Range H Bonds with DNMT1	Gromacs MM/MD Frames Cys1226 Ring C6 Distance Relative to the First Frame	Gromacs MM/MD Frames RMSD Ligand Relative to the First Frame
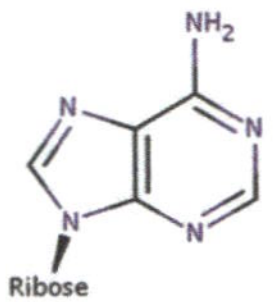 Lead-compound as a reference	−3.07	−9.7 to −7.9 kcal/mol 7 Hydrogen bonds: CYS 1191—N1 (adenosine) CYS 1191—N6 (adenosine) MET 1169—N3 (adenosine) GLU 1168—O3 (ribose) GLU 1168—O2 (ribose) PHE 1145—N1 (linker) ARG 1312—O2 (cytosine)	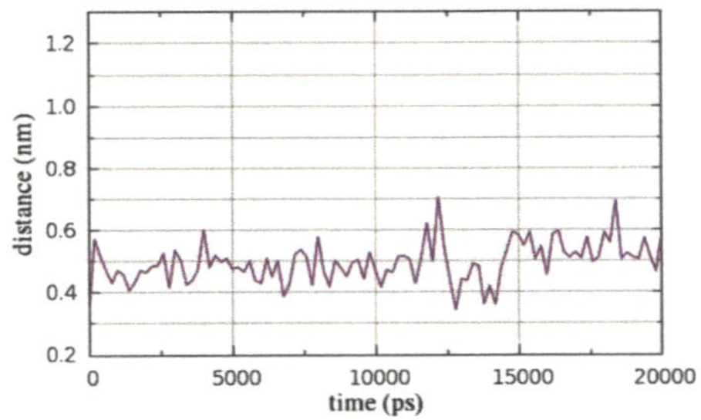	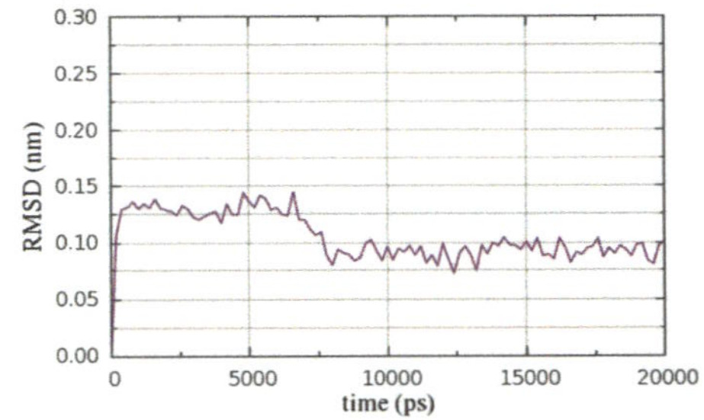
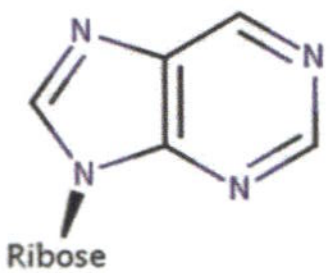 Nebularine	−2.84	Did not dock to the binding site cavity	N.A.	N.A.
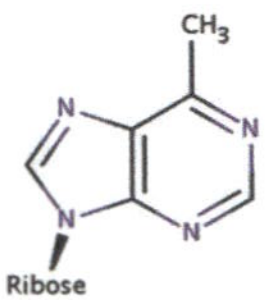	−2.60	−9.2 to −8.0 kcal/mol 3 Hydrogen bonds: MET 1169—N3 (adenosine) CYS 1191—N1 (adenosine) HOH 1704—O2 (linker)	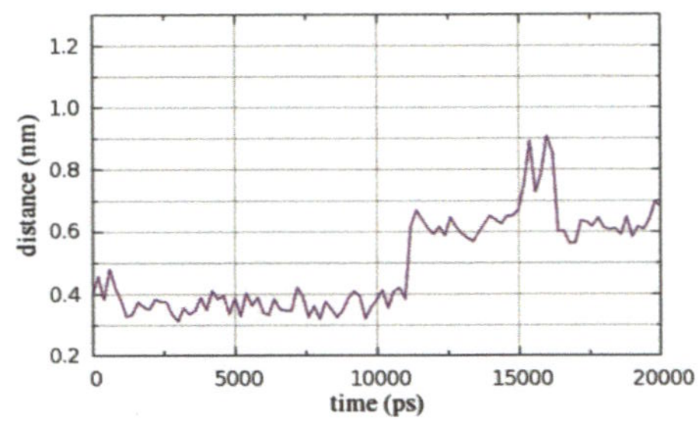	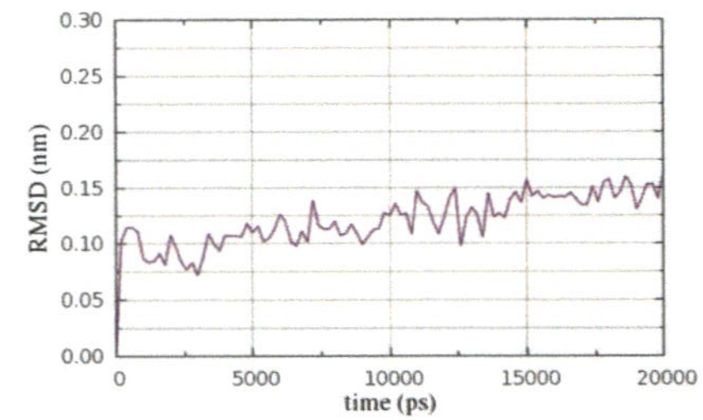
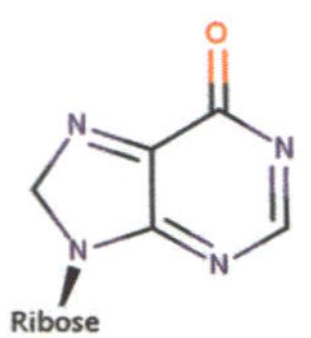	−2.44	10.1 to −8.1 kcal/mol 7 Hydrogen bonds MET 1169—N3 (adenosine) CYS 1191—O6 (adenosine) CYS 1191—N2 (adenosine) GLU 1168—O3 (ribose) GLY 1223—N1 (linker) ARG 1312—N3 (cytosine) ARG 1312—O2 (cytosine)	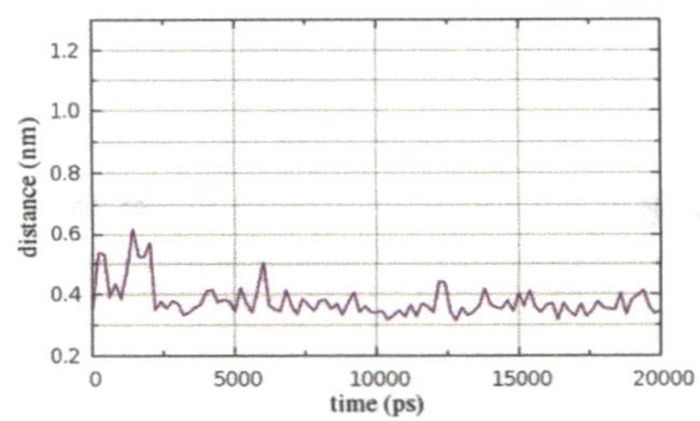	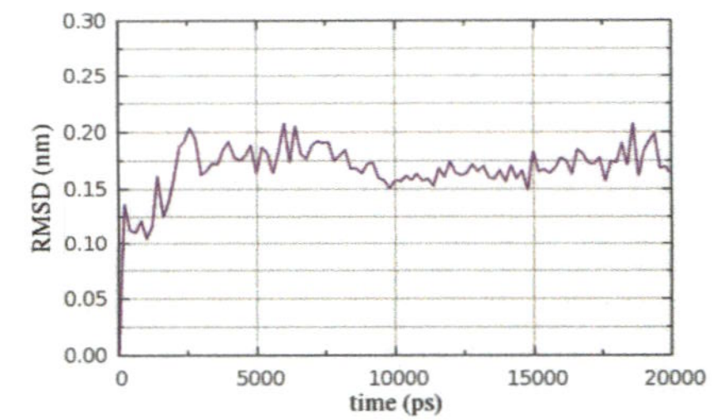

The LogD values represent calculated values for the lead compound with presented modification using weighted option in ChemAxon Marvin (ChemAxon). The LogD value for the lead compound is −3.07, thus any modification that has a LogD value higher than −3.07 represents a favorable LogD modification.

Table 1B Optimization of Heteroatoms in Adenosine Ring of the Presented Lead Compound

Structure Modification	LogD @ pH=7.4	Chimera-Vina Docking Interaction Energy Range H Bonds with DNMT1	Gromacs MM/MD Frames Cys1226 Ring C6 Distance Relative to the First Frame	Gromacs MM/MD Frames RMSD Ligand Relative to the First Frame
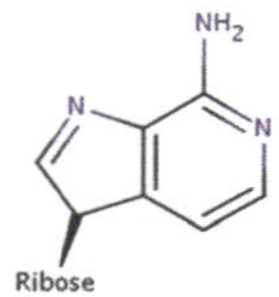	−2.64	Did not dock to the binding site cavity	N.A.	N.A.
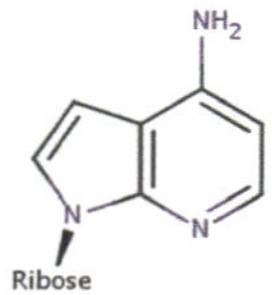	−2.84	−8.8 to −8.1 kcal/mol 4 Hydrogen bonds CYS 1191—N6 (adenosine) GLU 1168—O2 (ribose) ARG 1310—O2 (cytosine) ARG 1312—O2 (cytosine)	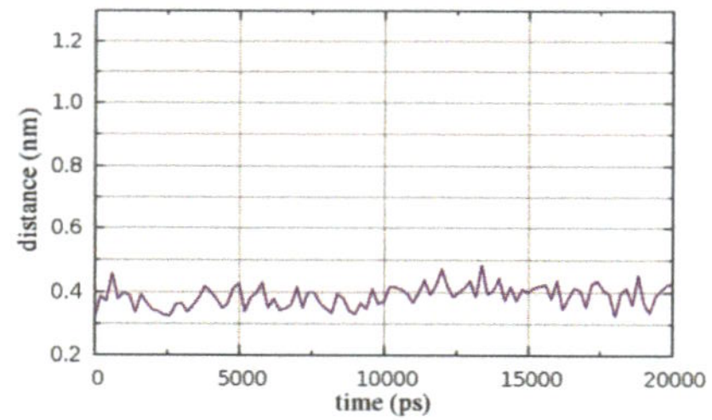	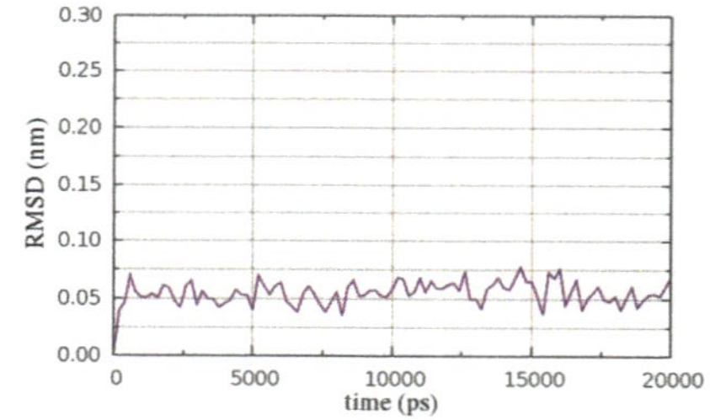
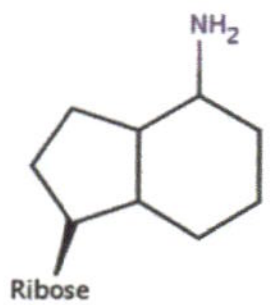	−1.20	−8.5 to −7.4 kcal/mol 5 Hydrogen bonds HOH 1773—N6 (adenosine) HOH 1704—O6 (ribose) GLN 1227—O2 (ribose) ARG 1310—O2 (cytosine) GLU 1312—O2 (cytosine)	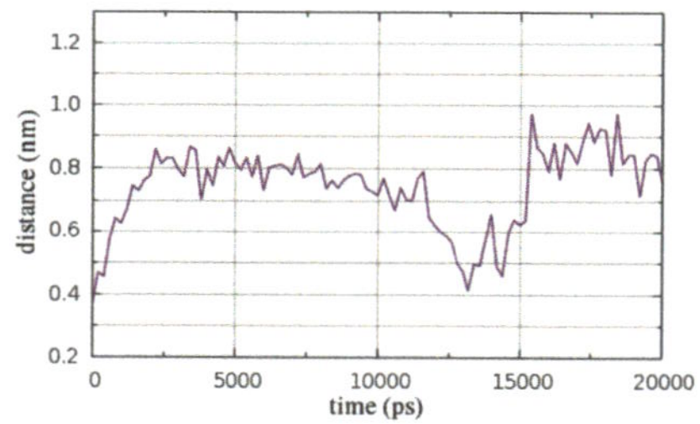	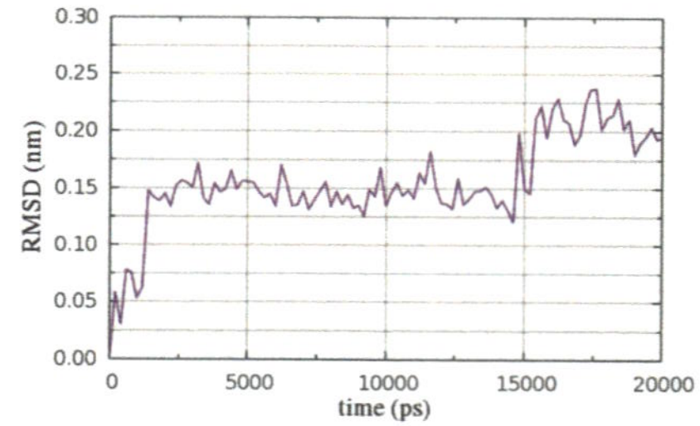
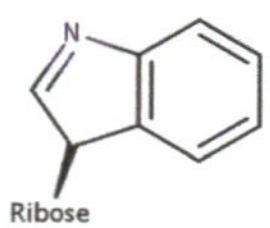	−1.31	−9.6 to −8.1 kcal/mol 5 Hydrogen bonds HOH 1704—O2 (ribose) HOH 1704—O3 (ribose) GLU 1168—O2 (ribose) ARG 1310—O2 (cytosine) ARG 1312—O2 (cytosine)	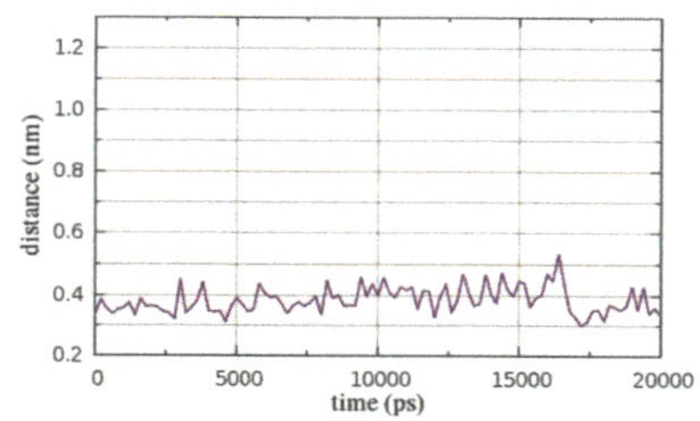	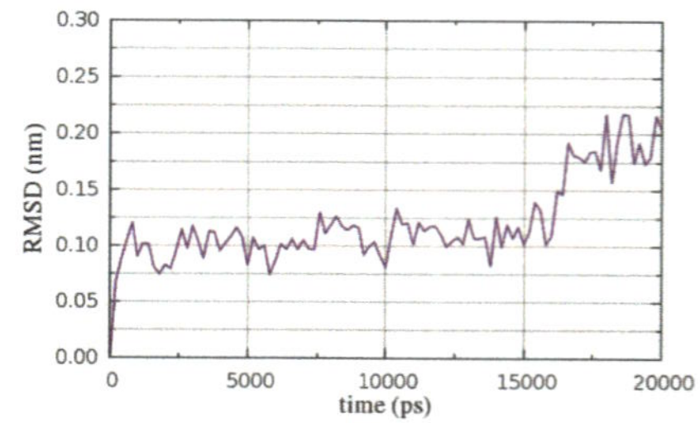

Table 2A Optimization of the Hydroxyl Groups on Ribose Ring of the Presented Lead Compound

Structure Modification	LogD @ pH = 7.4	Chimera-Vina Docking Interaction Energy Range H Bonds with DNMT1	Gromacs MM/MD Frames Cys1226 Ring C6 Distance Relative to the First Frame	Gromacs MM/MD Frames RMSD Ligand Relative to the First Frame
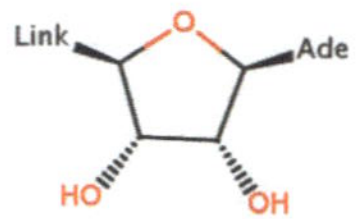 lead compound as reference	−3.07	−9.7 to −7.9 kcal/mol 7 Hydrogen bonds: CYS 1191—N1 (adenosine) CYS 1191—N6 (adenosine) MET 1169—N3 (adenosine) GLU 1168—O2 (ribose) GLU 1168—O3 (ribose) PHE 1145—N1 (linker) ARG 1312—O2 (cytosine)	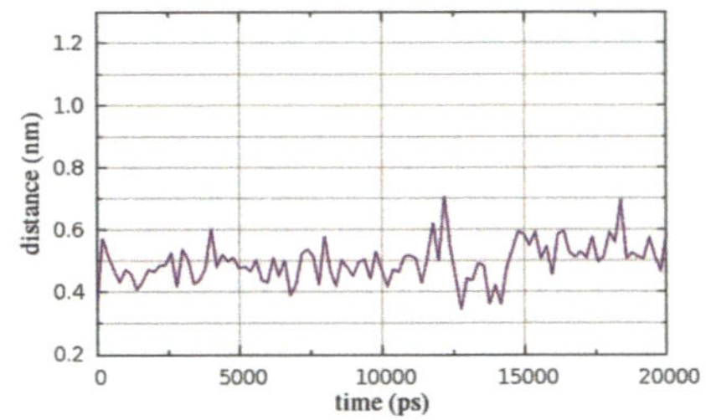	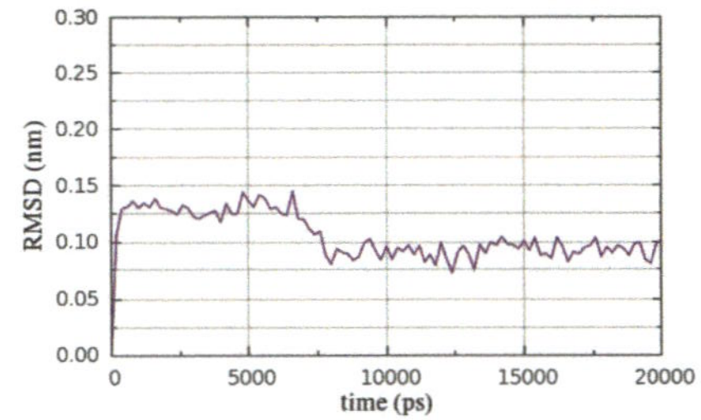
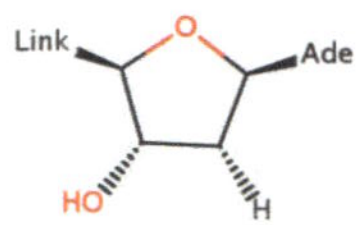	−2.46	−9.7 to −8.2 kcal/mol 5 Hydrogen bonds: CYS 1191—N1 (adenosine) CYS 1191—N6 (adenosine) MET 1169—N3 (adenosine) ARG 1312—O2 (cytosine) ARG 1312—N3 (cytosine)	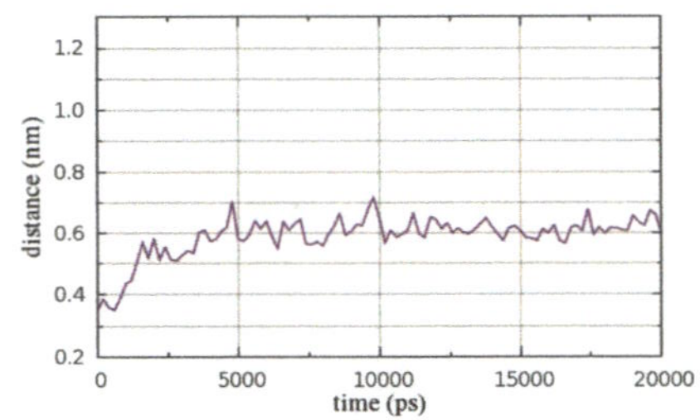	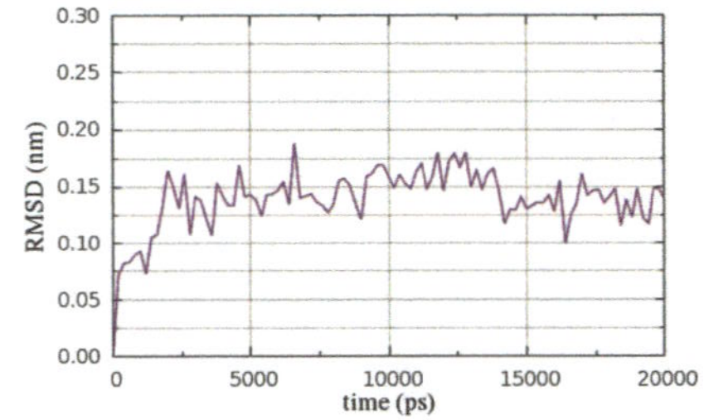
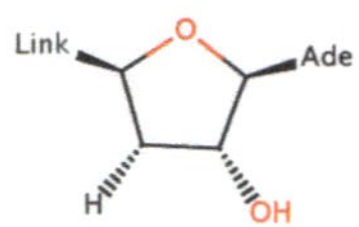	−2.87	Did not dock to the binding site cavity	N.A.	N.A.
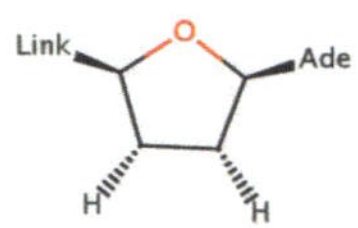	−2.25	Did not dock to the binding site cavity	N.A.	N.A.
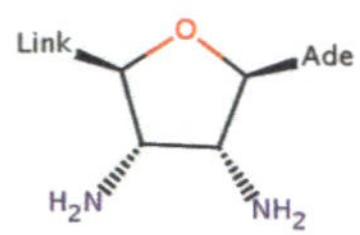	−5.03	−9.0 to −7.9 kcal/mol 4 Hydrogen bonds CYS 1191—N6 (adenosine) GLU 1168—N3 (ribose) ARG 1312—O2 (cytosine) ARG 1312—N3 (cytosine)	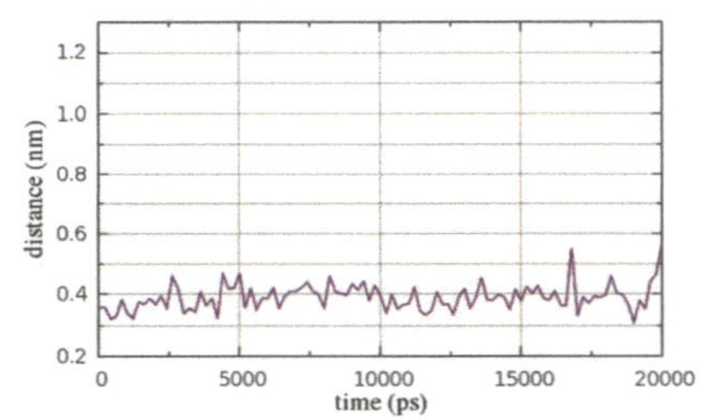	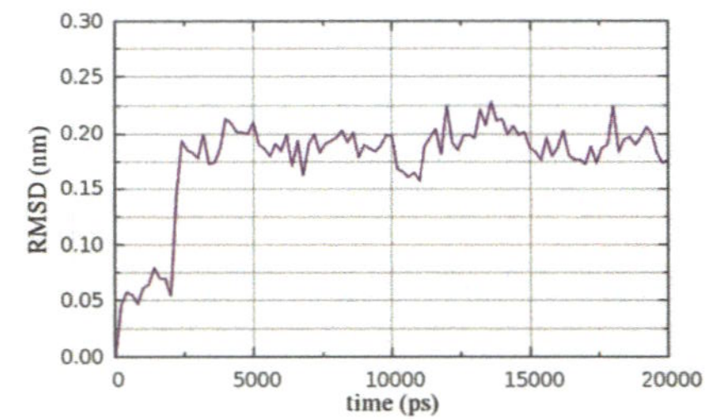

Continued...

Table 2A Optimization of the Hydroxyl Groups on Ribose Ring of the Presented Lead Compound (*Continued*)

Structure Modification	LogD @ pH = 7.4	Chimera-Vina Docking Interaction Energy Range H Bonds with DNMT1	Gromacs MM/MD Frames Cys1226 Ring C6 Distance Relative to the First Frame	Gromacs MM/MD Frames RMSD Ligand Relative to the First Frame
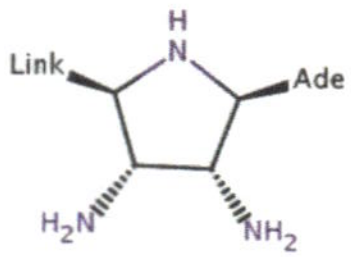	−5.07	Did not dock to the binding site cavity	N.A.	N.A.
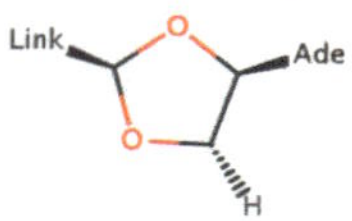	−1.23	−9.9 to −8.3 kcal/mol 5 Hydrogen bonds: CYS 1191—N1 (adenosine) CYS 1191—N6 (adenosine) MET 1169—N3 (adenosine) HOH 1704—O3 (ribose) ARG 1312—O2 (cytosine)	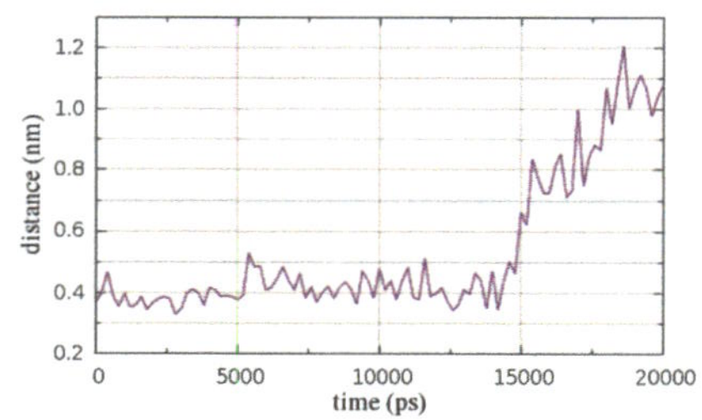	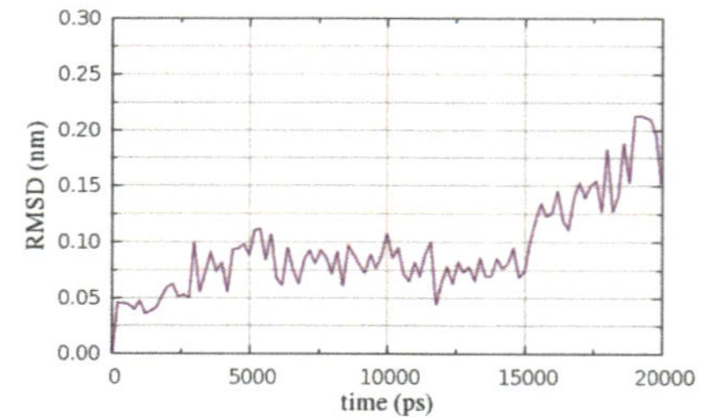
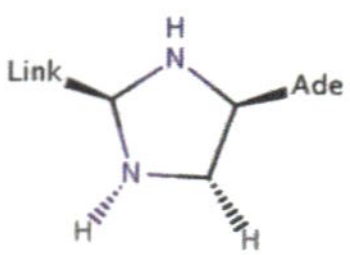	−2.03	−9.9 to −8.5 kcal/mol 4 Hydrogen bonds: CYS 1191—N1 (adenosine) CYS 1191—N6 (adenosine) MET 1169—N3 (adenosine) ARG 1312—O2 (cytosine)	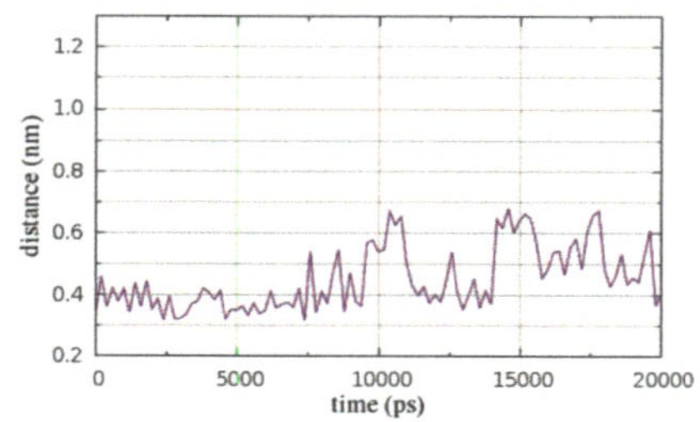	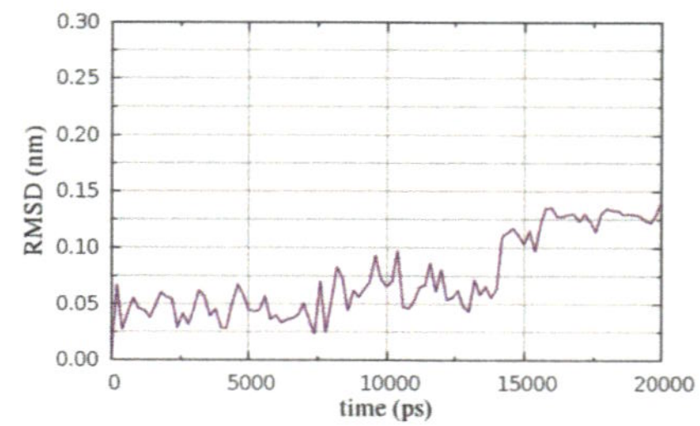
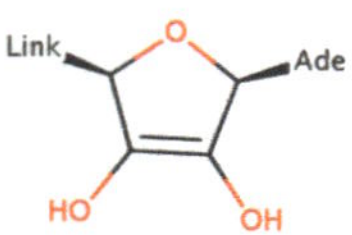	−4.16	Did not dock to the binding site cavity	N.A.	N.A.

The LogD values represent calculated values for the lead compound with presented modification using weighted option in ChemAxon Marvin (ChemAxon). The LogD value for the lead compound is −3.07, thus any modification that has a LogD value higher than −3.07 represents a favorable modification.

Table 2B Optimization of Acetal-Oxygen at Position 6 in Ribose Ring of the Presented Lead Compound

Structure Modification	LogD @ pH=7.4	Chimera-Vina Docking Interaction Energy Range H Bonds with DNMT1	Gromacs MM/MD Frames Cys1226 Ring C6 Distance Relative to the First Frame	Gromacs MM/MD Frames RMSD Ligand Relative to the First Frame
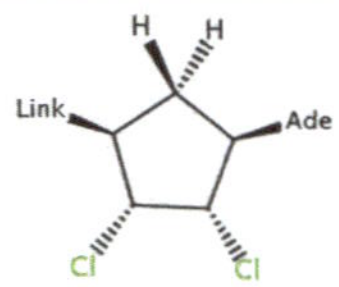	−1.53	−10.1 to −7.5 kcal/mol 2 Hydrogen bonds: HOH 1773—N1 (adenosine) ARG 1312—O2 (cytosine)	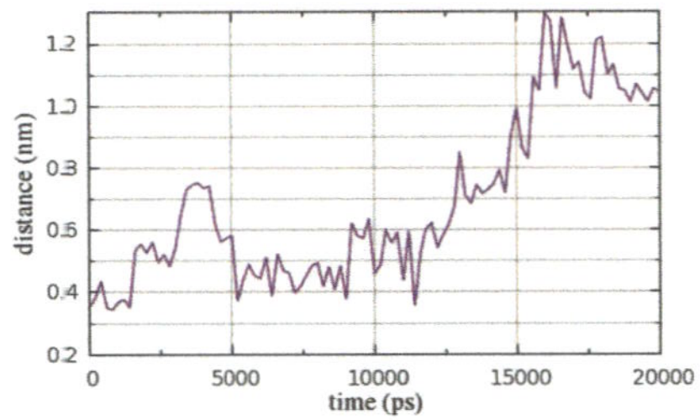	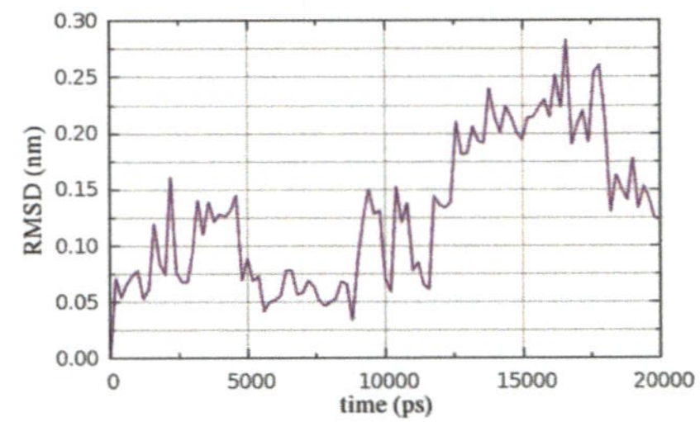

The following modifications at the acetal-oxygen could not dock into the binding cleft cavity.

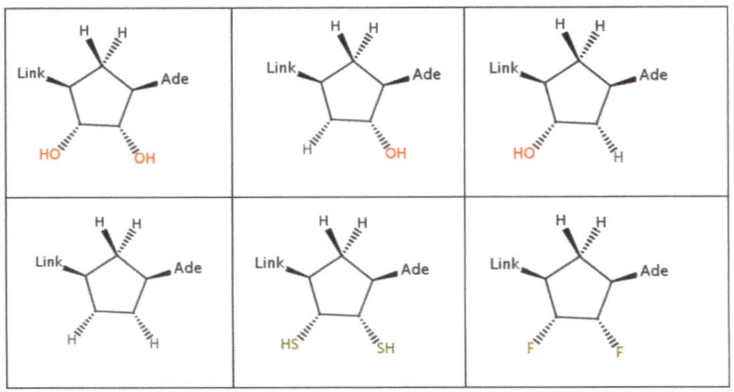

The LogD values represent calculated values for the lead compound with presented modification using weighted option in ChemAxon Marvin (ChemAxon). The LogD value for the lead compound is −3.07, thus any modification that has a LogD value higher than −3.07 represents a favorable modification.

Table 3A Optimization of the Linker Part of the Presented Lead Compound: Changes in the Linker Tail

Structure Modification	LogD @ pH = 7.4	Chimera-Vina Docking Interaction Energy Range H Bonds with DNMT1	Gromacs MM/MD Frames Cys1226 Ring C6 Distance Relative to the First Frame	Gromacs MM/MD Frames RMSD Ligand Relative to the First Frame
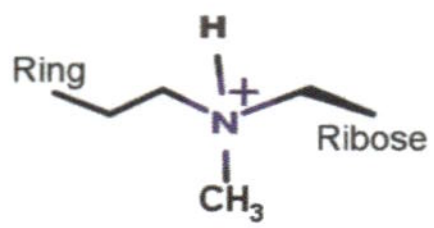 Lead compound as reference	−3.07	−9.7 to −7.9 kcal/mol 7 Hydrogen bonds: CYS 1191—N1 (adenosine) CYS 1191—N6 (adenosine) MET 1169—N3 (adenosine) HOH 1704—O3 (ribose) HOH 1704—O2 (ribose) PHE 1145—N1 (linker) ARG 1312—O2 (cytosine)	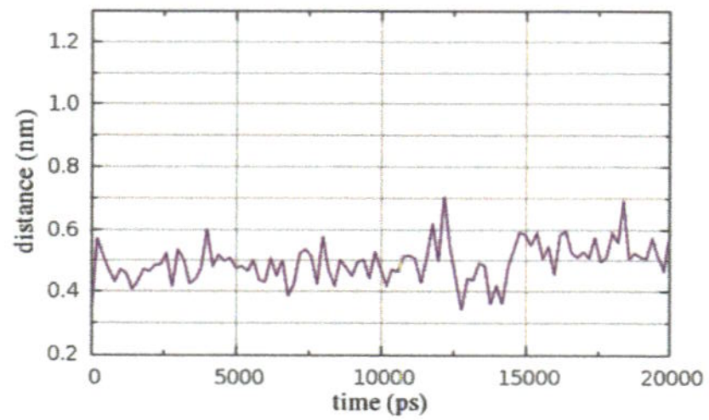	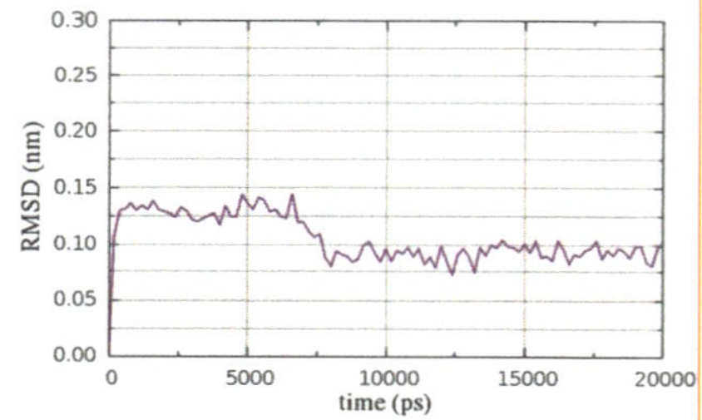
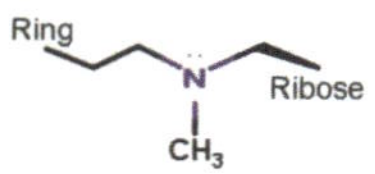	−3.07	−9.8 to −8.7 kcal/mol 5 Hydrogen bonds: CYS 1191—N6 (adenosine) CYS 1191—N1 (adenosine) MET 1169—N3 (adenosine) ARG 1310—O2 (cytosine) ARG 1312—O2 (cytosine)	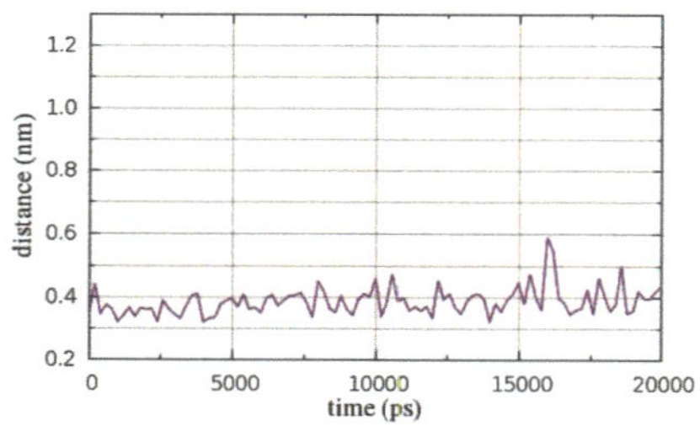	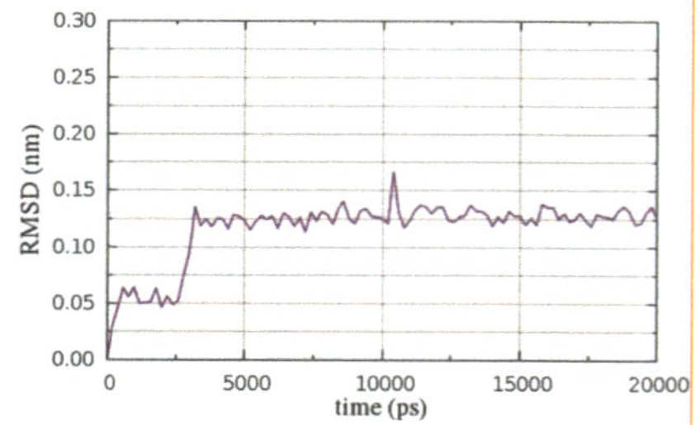
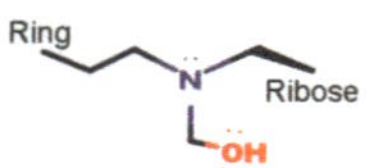	−2.67	−10.4 to −8.5 kcal/mol 6 Hydrogen bonds: CYS 1191—N1 (adenosine) CYS 1191—N6 (adenosine) MET 1169—N3 (adenosine) HOH 1704—O2 (ribose) GLU 1168—O3 (ribose) ARG 1312—O2 (cytosine)	N.A.	N.A.
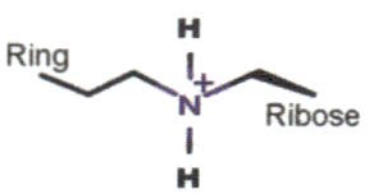	−2.17	Did not dock to the binding site cavity	N.A.	N.A.
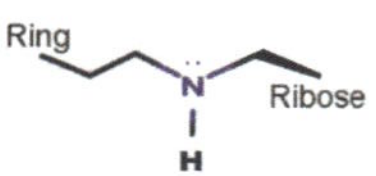	−2.17	Did not dock to the binding site cavity	N.A.	N.A.

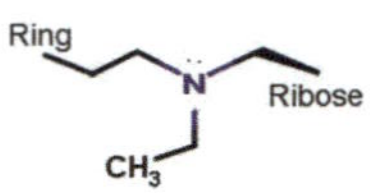	−1.34	−10.2 to −8.1 kcal/mol 6 Hydrogen bonds: CYS 1191—N1 (adenosine) CYS 1191—N6 (adenosine) MET 1169—N3 (adenosine) GLU 1168—O3 (ribose) GLU 1168—O2 (ribose) ARG 1312—O2 (cytosine)	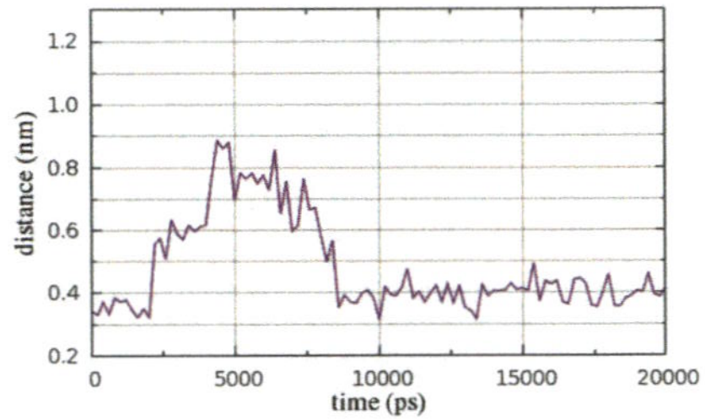	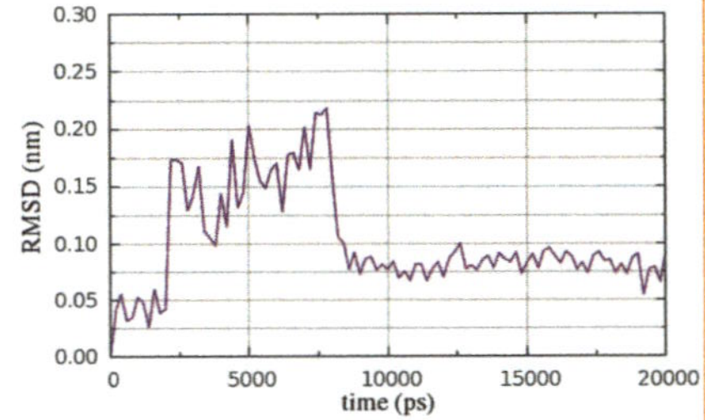
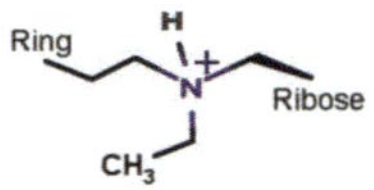	−1.34	Did not dock to the binding site cavity	N.A.	N.A.
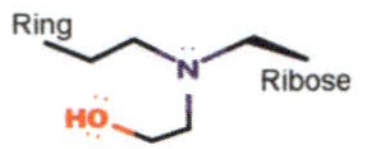	−2.08	Did not dock to the binding site cavity	N.A.	N.A.

The LogD values represent calculated values for the lead compound with presented modification using weighted option in ChemAxon Marvin (ChemAxon). The LogD value for the lead compound is −3.07, thus any modification that has LogD value higher than −3.07 represents a favorable modification.

Table 3B Evaluation of the Linker Part of the Lead Compound: Changes in Linker Flexibility

Structure Modification	LogD @ pH=7.4	Chimera-Vina Docking Interaction Energy Range H Bonds with DNMT1	Gromacs MM/MD Frames Cys1226 Ring C6 Distance Relative to the First Frame	Gromacs MM/MD Frames RMSD Ligand Relative to the First Frame
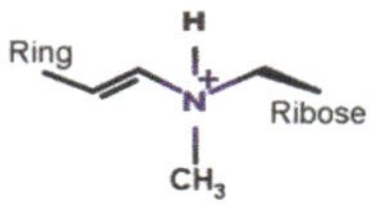	−2.16	−10.5 to −8.2 kcal/mol 7 Hydrogen bonds: CYS 1191—N1 (adenosine) CYS 1191—N6 (adenosine) MET 1169—N3 (adenosine) GLU 1168—O3 (ribose) GLU 1168—O2 (ribose) PHE 1145—N1 (linker) ARG 1312—O2 (cytosine)	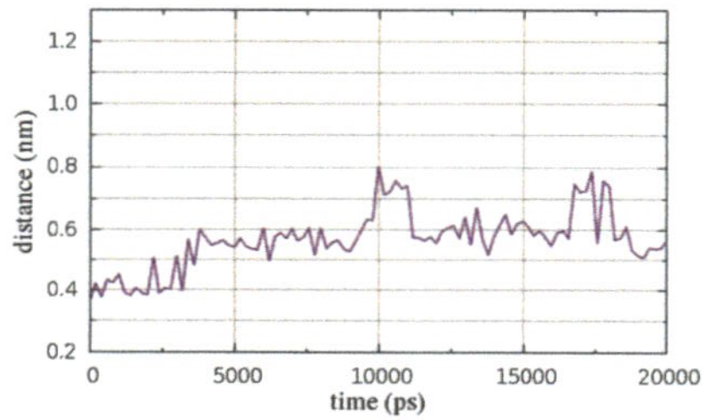	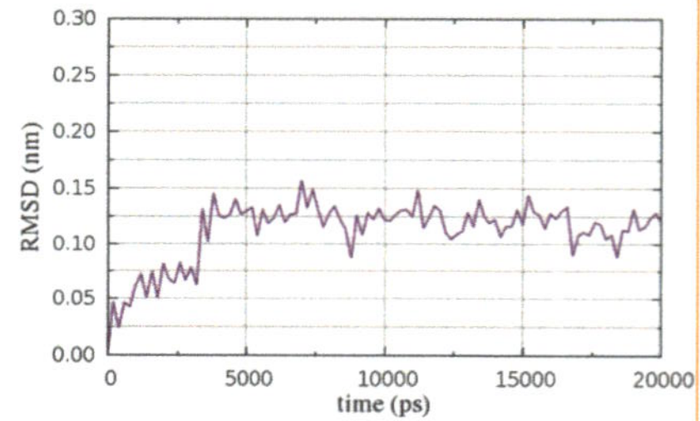
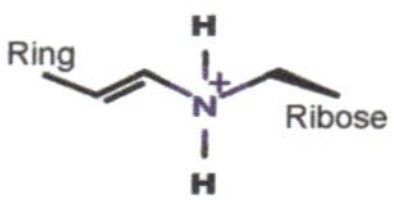	−2.41	−11.1 to −9.1 kcal/mol 7 Hydrogen bonds: CYS 1169—N1 (adenosine) CYS 1169—N6 (adenosine) MET 1169—N3 (adenosine) PHE 1145—N1 (linker) ARG 1310—O2 (cytosine) ARG 1312—O2 (cytosine) ARG 1312—N3 (cytosine)	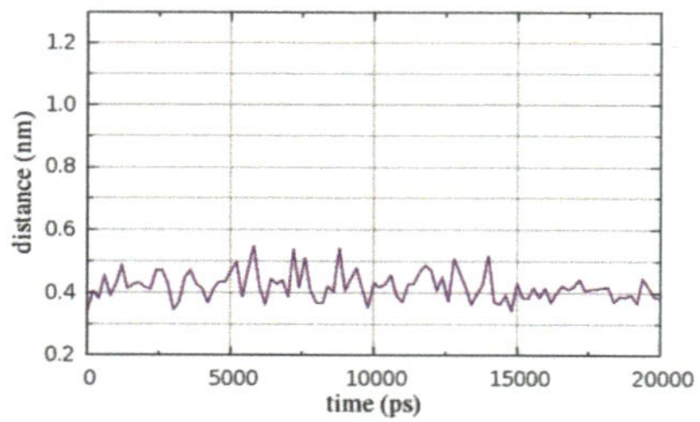	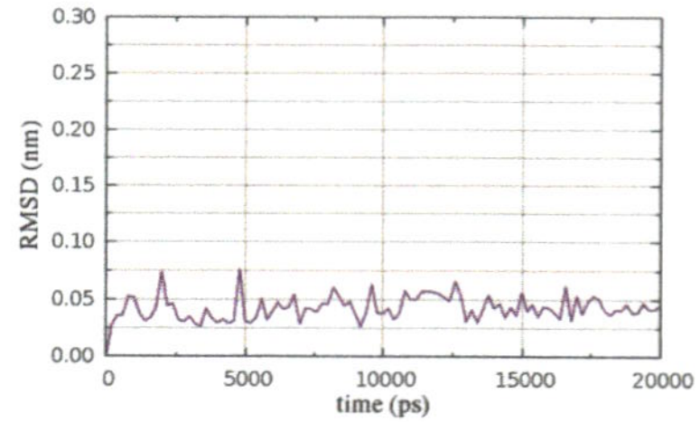
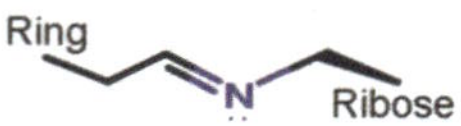	−0.42	did not dock to the binding site cavity	N.A.	N.A.
	1.44	did not dock to the binding site cavity	N.A.	N.A.
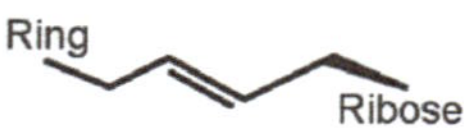	1.08	Did not dock to the binding site cavity	N.A.	N.A.
	1.08	Did not dock to the binding site cavity	N.A.	N.A.
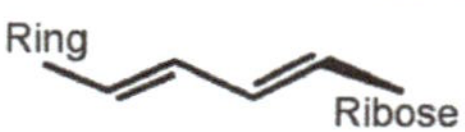	0.87	−10.2 to −8.3 kcal/mol 4 Hydrogen bonds: CYS 1191—N1 (adenosine) CYS 1191—N6 (adenosine) MET 1169—N3 (adenosine) ARG 1312—O2 (cytosine)	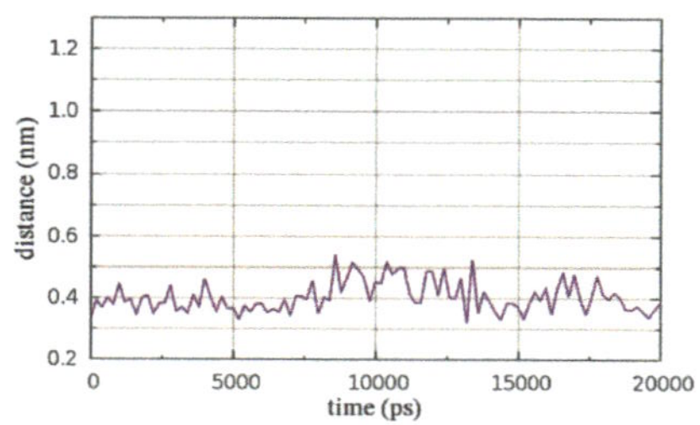	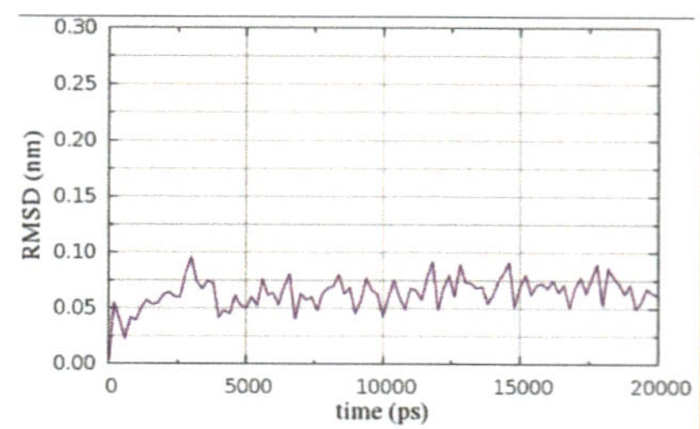

	0.06	−10.2 to −8.9 kcal/mol 8 Hydrogen bonds: CYS 1191—N1 (adenosine) CYS 1191—N6 (adenosine) MET 1169—N3 (adenosine) GLU 1168—O3 (ribose) GLU 1168—O2 (ribose) HOH 1704—O3 (ribose) HOH 1704—O2 (ribose) ARG 1312—O2 (cytosine)	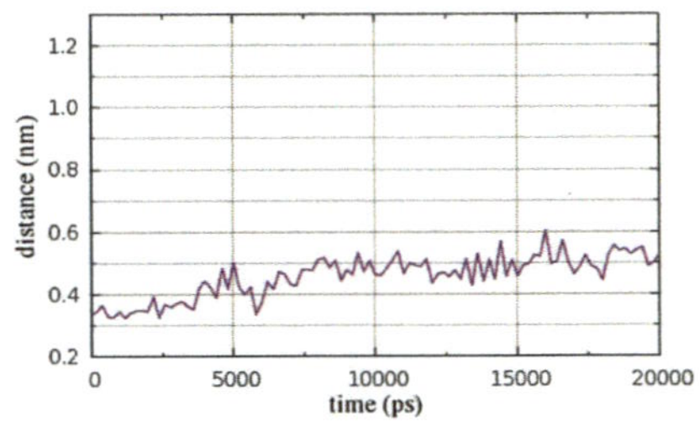	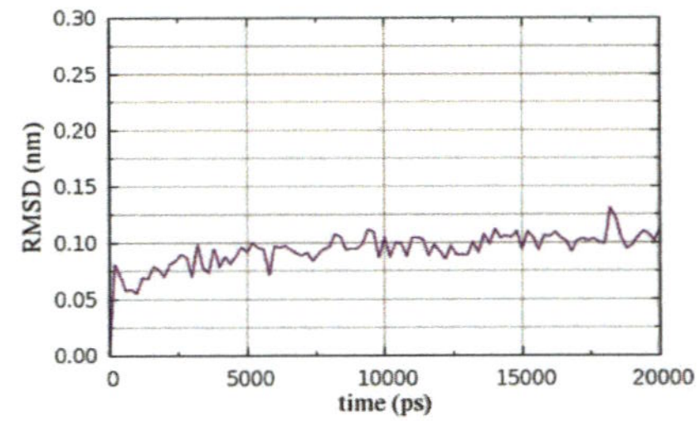
	−1.21	−10.4 to −8.2 kcal/mol 5 Hydrogen bonds: CYS 1191—N1 (adenosine) CYS 1191—N6 (adenosine) MET 1169—N3 (adenosine) ARG 1310—O2 (cytosine) ARG 1312—O2 (cytosine)	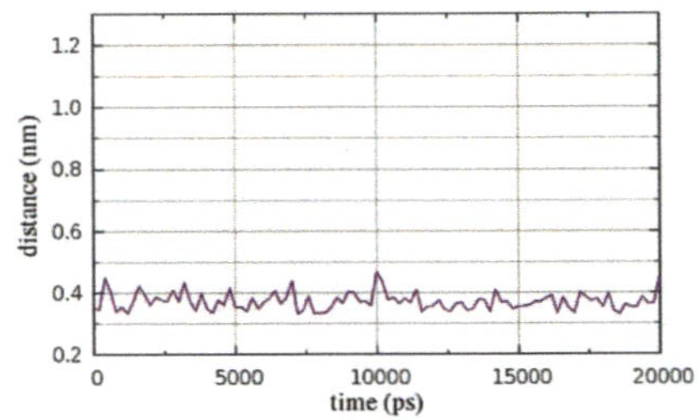	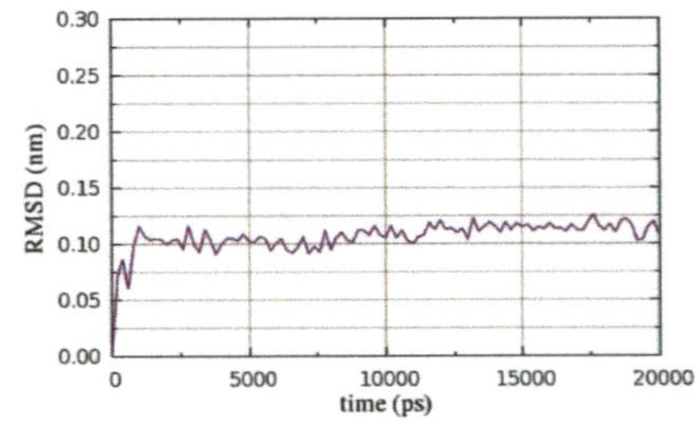
	0.07	−10.9 to −8.3 kcal/mol 7 Hydrogen bonds: CYS 1191—N1 (adenosine) CYS 1191—N6 (adenosine) MET 1169—N3 (adenosine) GLU 1168—O3 (ribose) GLU 1168—O2 (ribose) ARG 1310—O2 (cytosine) ARG 1312—O2 (cytosine)	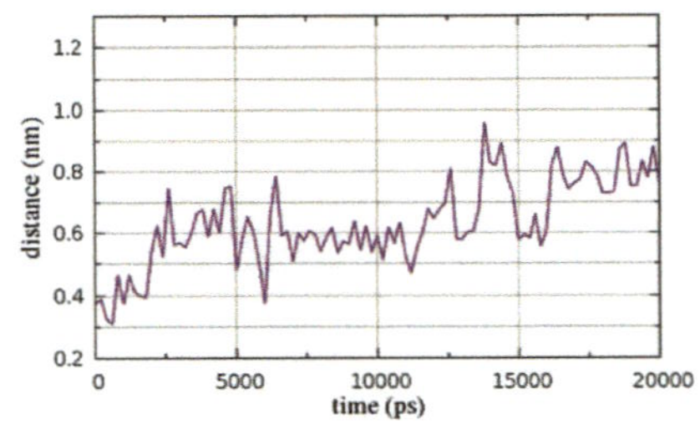	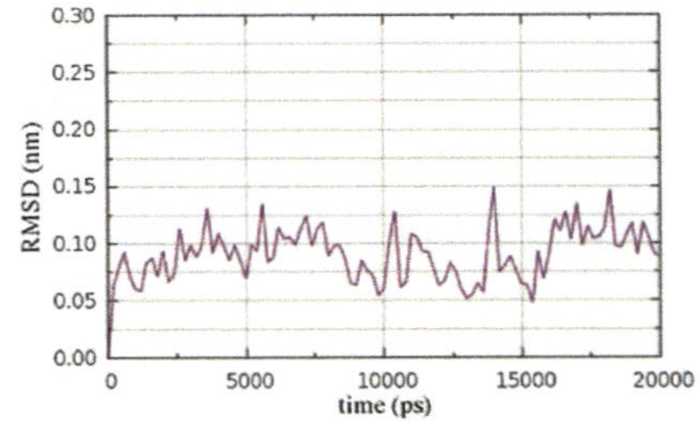

Table 3C Evaluation of the Linker Part of the Lead Compound: Changes in the Linker Mobility

Structure Modification	LogD @ pH=7.4	Chimera-Vina Docking Interaction Energy Range H Bonds with DNMT1	Gromacs MM/MD Frames Cys1226 Ring C6 Distance Relative to the First Frame	Gromacs MM/MD Frames RMSD Ligand Relative to the First Frame
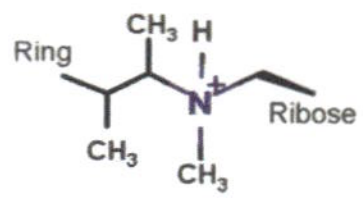	−2.84	−9.4 to −7.8 kcal/mol 5 Hydrogen bonds: CYS 1191—N1 (adenosine) CYS 1191—N6 (adenosine) MET 1169—N3 (adenosine) GLU 1168—O3 (ribose) GLU 1168—O2 (ribose)	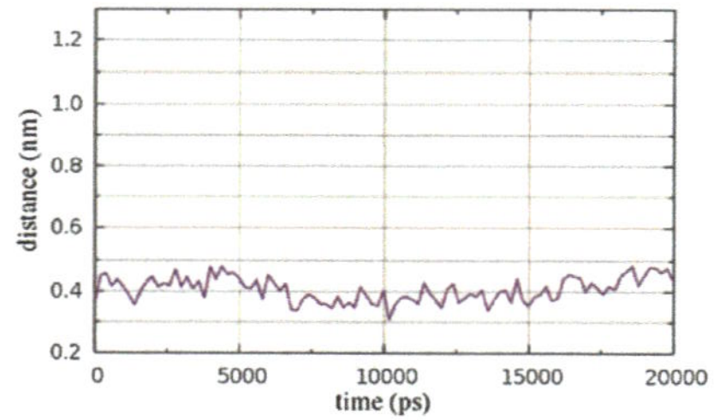	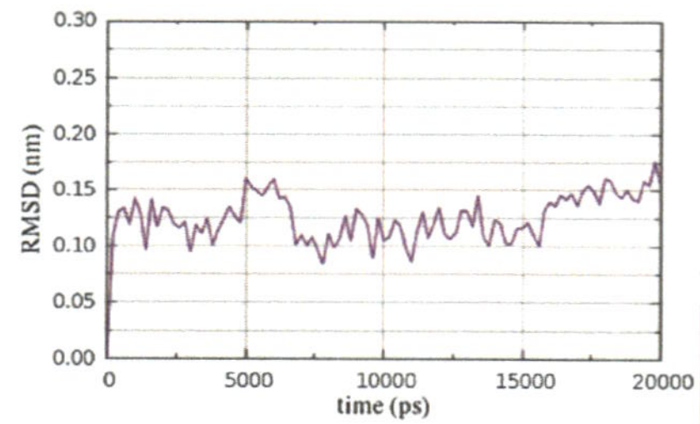
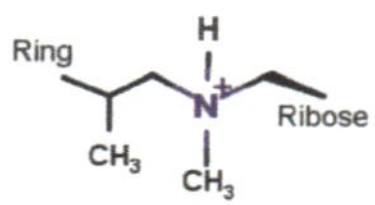	−2.99	−9.8 to −8.0 kcal/mol 6 Hydrogen bonds: CYS 1191—N1 (adenosine) CYS 1191—N6 (adenosine) MET 1169—N3 (adenosine) GLU 1168—O3 (ribose) GLU 1168—O2 (ribose) ARG 1312—O2 (cytosine)	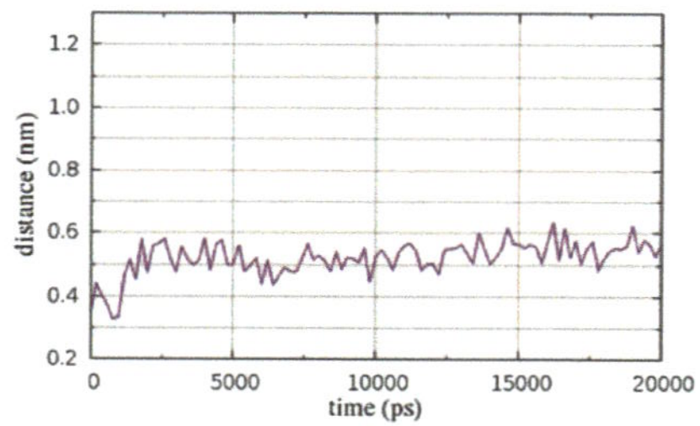	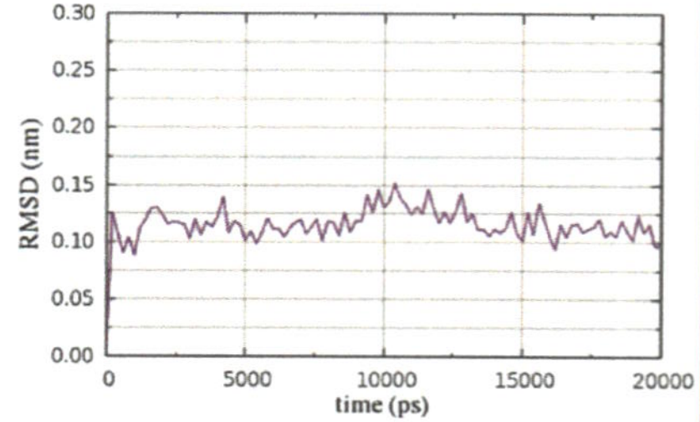
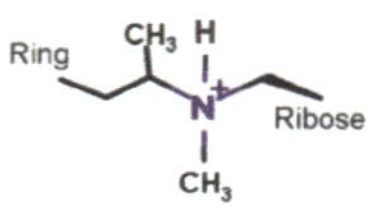	−2.92	−9.3 to −8.2 kcal/mol 8 Hydrogen bonds: CYS 463—N1 (adenosine) CYS 463—N6 (adenosine) MET 441—N3 (adenosine) GLU 1168—O3 (ribose) GLU 1168—O2 (ribose) ARG 1310—O2 cytosine) ARG 1310—O2 (cytosine) ARG 1312—N3 (cytosine)	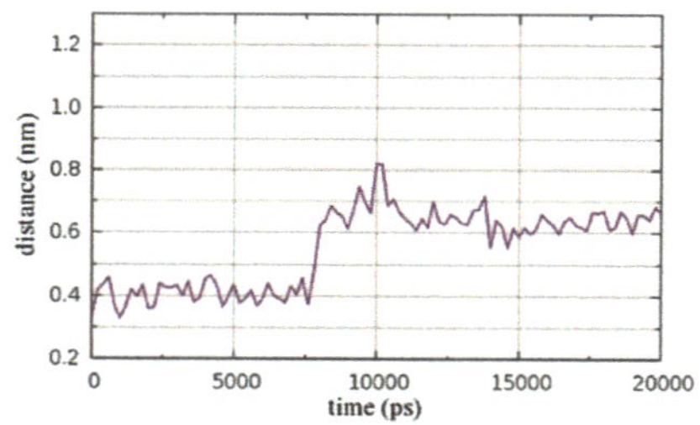	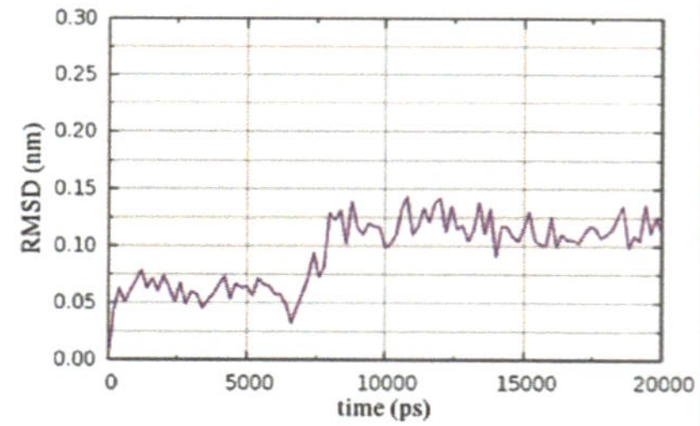

Table 4 Evaluation of Binding Interactions at the Target Base Ring

Structure	LUMO eV	Chimera-Vina Docking Interaction Energy Range H Bonds with DNMT1	Gromacs MM/MD Frames Cys1226 Ring C6 Distance Relative to the First Frame	Gromacs MM/MD Frames RMSD Ligand Relative to the First Frame
 1-methyl-pirimidin-2-one (lead compound)	−1.4	−9.7 to −7.9 kcal/mol 7 Hydrogen bonds: CYS 1191—N1 (adenosine) CYS 1191—N6 (adenosine) MET 1169—N3 (adenosine) GLU 1168—O3 (ribose) GLU 1168—O2 (ribose) PHE 1145—N1 (linker) ARG 1312—O2 (cytosine)	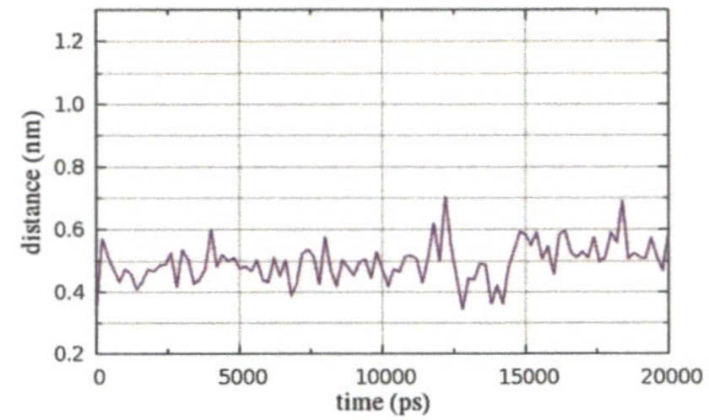	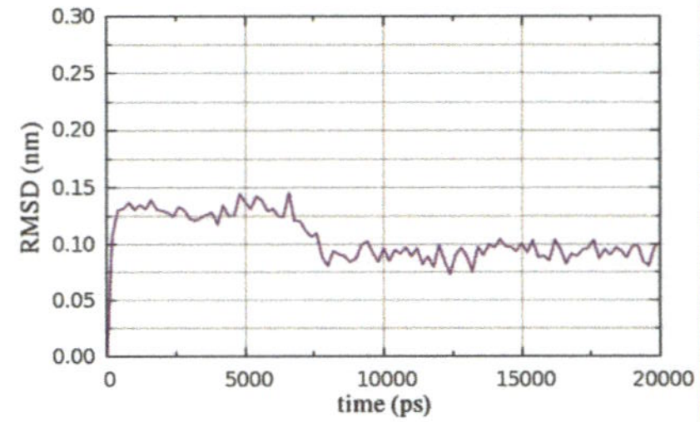
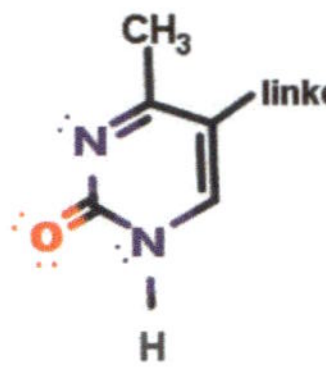	−1.4	Did not dock to the binding site cavity	N.A.	N.A.
	−1.5	−9.2 to −6.5 kcal/mol 6 Hydrogen bonds: CYS 1191—N1 (adenosine) MET 1169—N3 (adenosine) GLU 1168—O3 (ribose) GLU 1168—O2 (ribose) PHE 1145—N1 (linker) ARG 1312—O2 (cytosine)	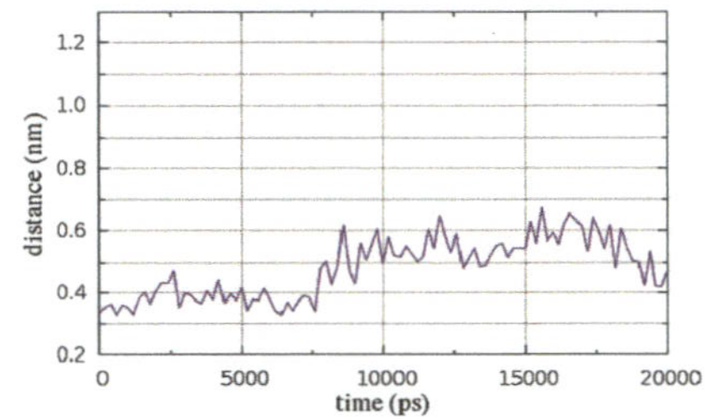	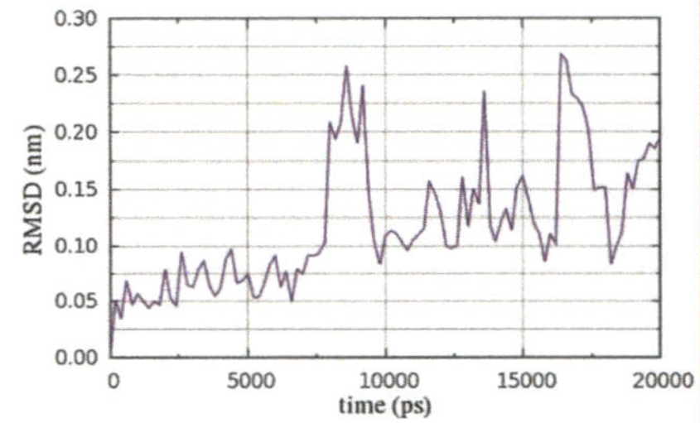
 1-methyl-cytosine	−0.9	−10.1 to −8.6 kcal/mol 12 Hydrogen bonds: CYS 1191—N1 (adenosine) CYS 1191—N6 (adenosine) ASP 1190—N6 (adenosine) MET 1169—N3 (adenosine) HOH 1704—O2 (ribose) HOH 1704—O3 (ribose) GLU 1168—O2 (ribose) PRO 1224—N4 (cytosine) GLU 1266—N4 (cytosine) ARG 1312—N3 (cytosine) ARG 1312—O2 (cytosine) ARG 1310—O2 (cytosine)	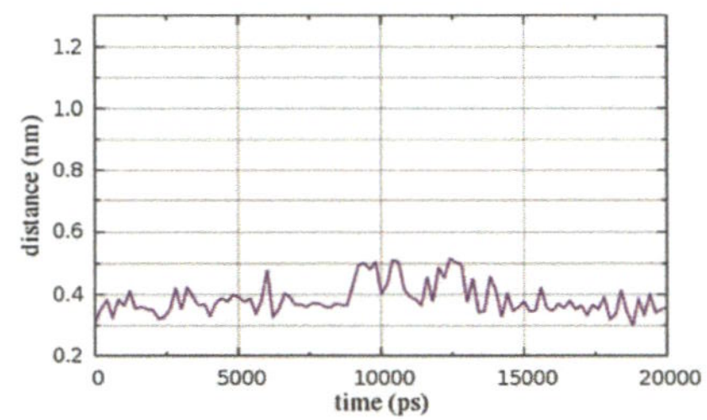	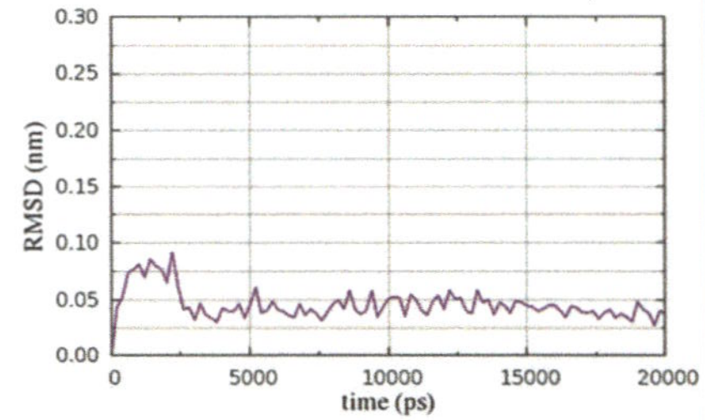

Continued...

Table 4 Evaluation of Binding Interactions at the Target Base Ring (*Continued*)

Structure	LUMO eV	Chimera-Vina Docking Interaction Energy Range H Bonds with DNMT1	Gromacs MM/MD Frames Cys1226 Ring C6 Distance Relative to the First Frame	Gromacs MM/MD Frames RMSD Ligand Relative to the First Frame
 1-methyl-uracil	−0.6	−10.0 to −8.3 kcal/mol 7 Hydrogen bonds: CYS 1191—N1 (adenosine) CYS 1191—N6 (adenosine) MET 1169—N3 (adenosine) GLU 1168—O2 (ribose) CYS 1226—O4 (cytosine) GLU 1266—N3 (cytosine) VAL 1579—O2 (cytosine)	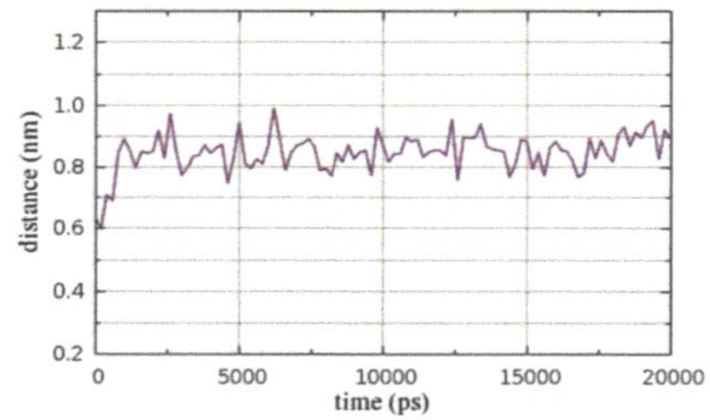	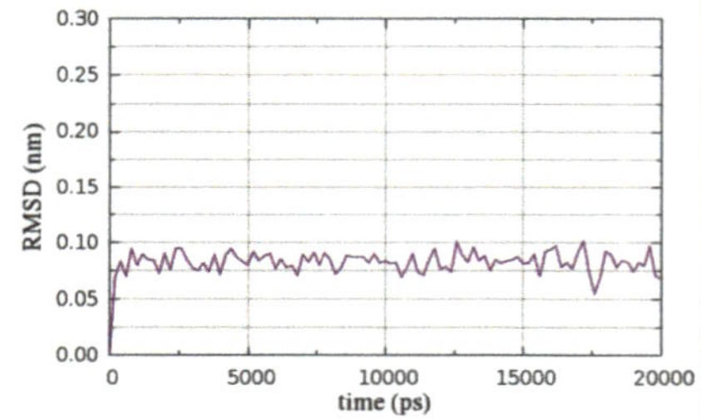
 Pirimidine-2-one	−1.4	−10.6 to −8.4 kcal/mol 10 Hydrogen bonds: CYS 1191—N1 (adenosine) CYS 1191—N6 (adenosine) MET 1169—N3 (adenosine) GLU 1168—O2 (ribose) HOH 1704—O2 (ribose) GLU 1168—O3 (ribose) HOH 1704—O3 (ribose) PHE 1145—N1 (linker) ARG 1312—N3 (cytosine) ARG 1312—O2 (cytosine)	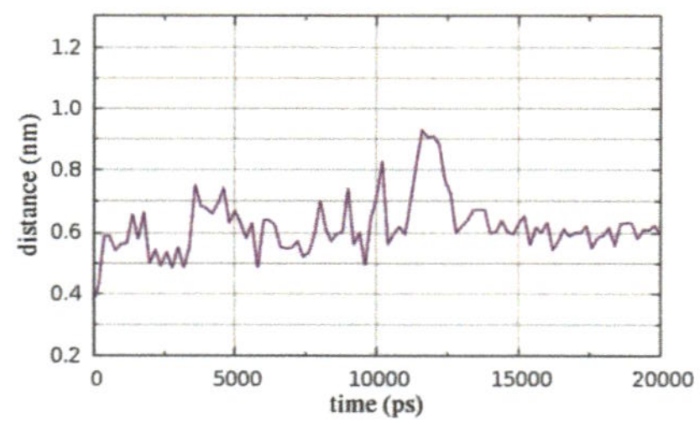	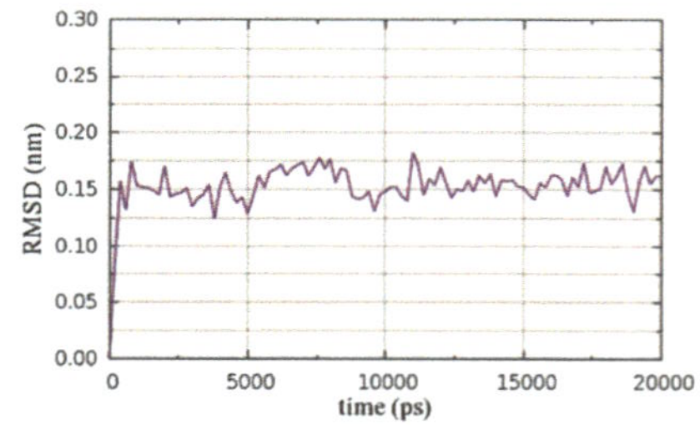

LUMO Energies Were Calculated Using Program Gamess (Gordon and Schmidt, 2005).

Table 5 Evaluation of the Rigidity of the DNMT1-Ligand Complex

Structure Modification	LogD @ pH=7.4	Chimera-Vina Docking Interaction Energy Range H Bonds with DNMT1	Gromacs MM/MD Frames Cys1226 Ring C6 Distance Relative to the First Frame	Gromacs MM/MD Frames RMSD Ligand Relative to the First Frame
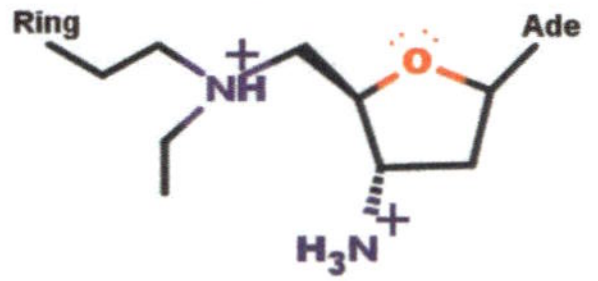	−4.24	−9.3 to −8.3 kcal/mol 6 Hydrogen bonds: CYS 1191—N6 (adenosine) CYS 1191—N1 (adenosine) GLU 1168—N3 (ribose) GLY 1223—N (linker) ARG 1312—N3 (citosine) ARG 1312—O (citcsine)	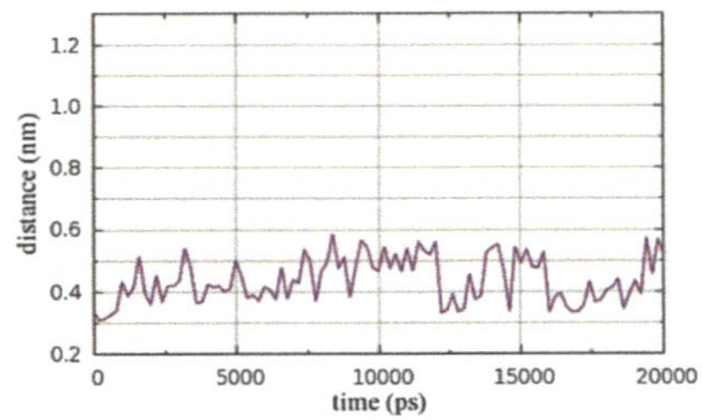	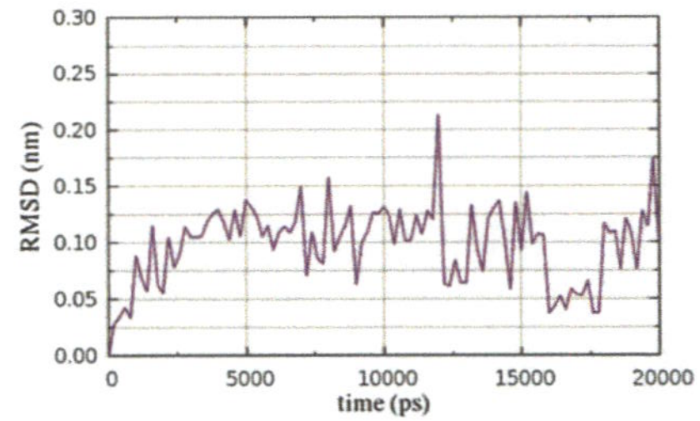
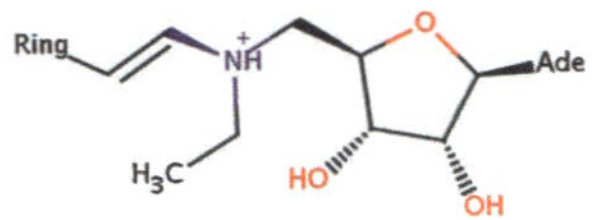	−1.96	−10.1 to −8.5 kcal/mol 3 Hydrogen bonds: CYS 1191—N6 (acenosine) CYS 1191—N1 (acenosine) HOH 1704—O3 (ribose)	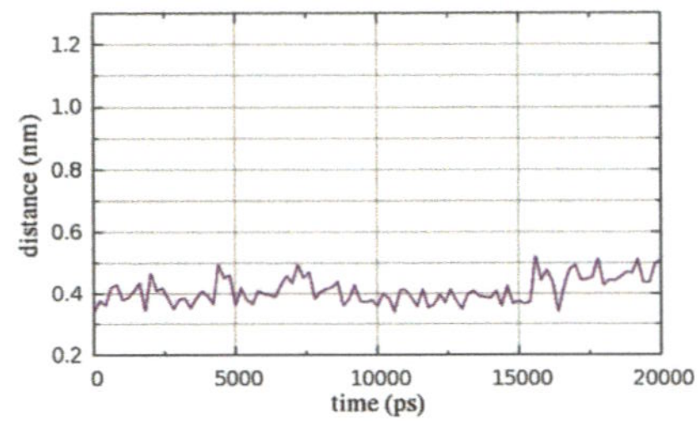	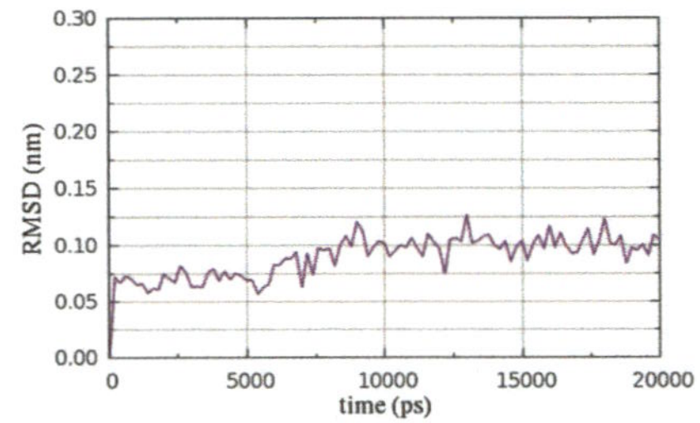
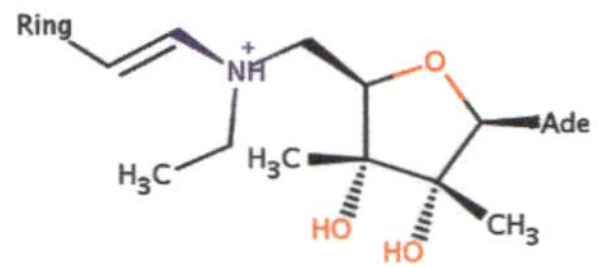	−1.43	−9.0 to −7.4 kcal/mol 6 Hydrogen bonds: CYS 1191—N6 (adenosine) CYS 1191—N1 (adenosine) MET 1169—N3 (adenosine) HOH 1704—O2 (ribose) HOH 1704—O3 (ribose) ARG 1312—O2 (cytosine)	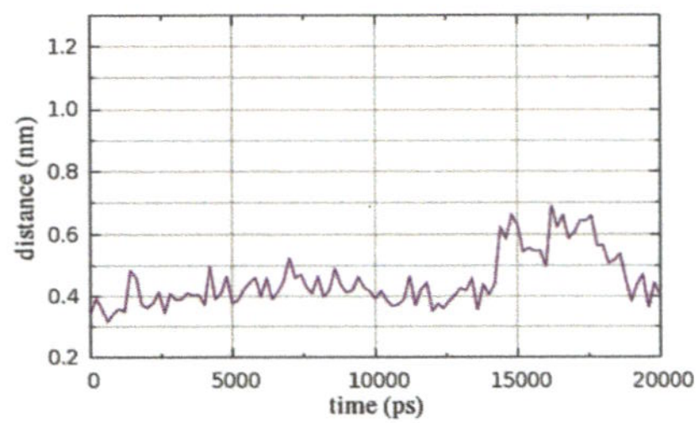	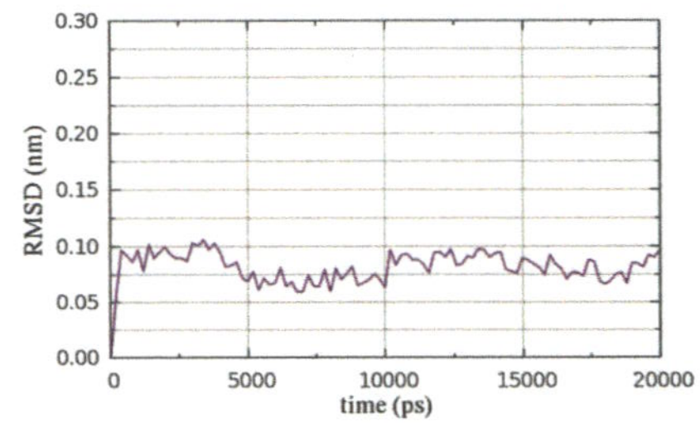
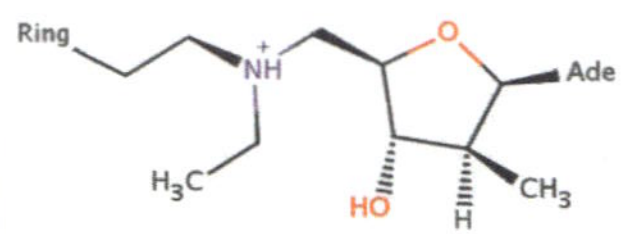	−2.46	−9.3 to −7.9 kcal/mol 5 Hydrogen bonds: CYS 1191—N6 (adenosine) CYS 1191—N1 (adenosine) MET 1169—N3 (adenosine) ARG 1310—O2 (citosine) ARG 1312—O2 (citosine)	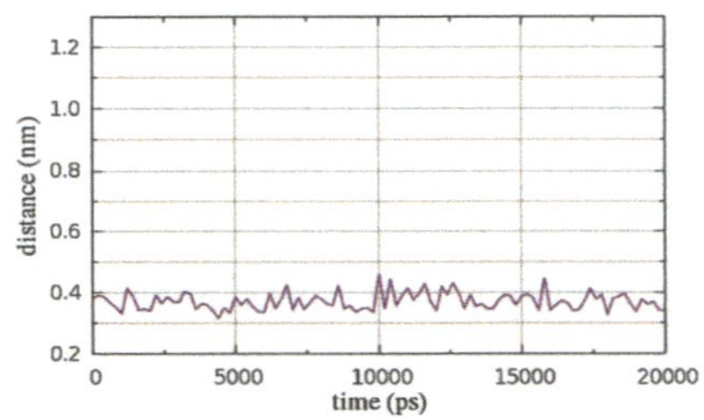	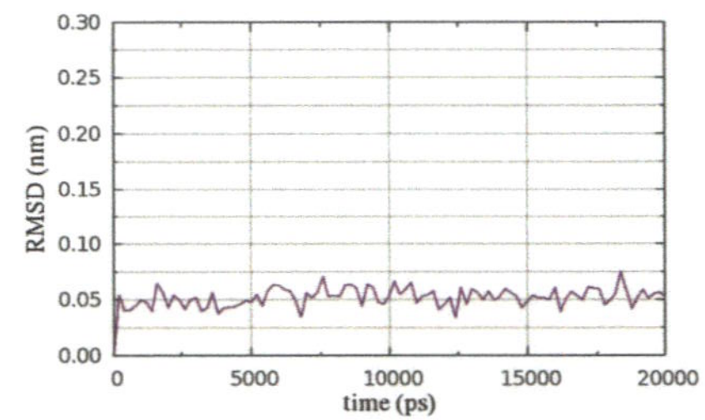

Continued...

Table 5 Evaluation of the Rigidity of the DNMT1-Ligand Complex (*Continued*)

Structure Modification	LogD @ pH = 7.4	Chimera-Vina Docking Interaction Energy Range H Bonds with DNMT1	Gromacs MM/MD Frames Cys1226 Ring C6 Distance Relative to the First Frame	Gromacs MM/MD Frames RMSD Ligand Relative to the First Frame
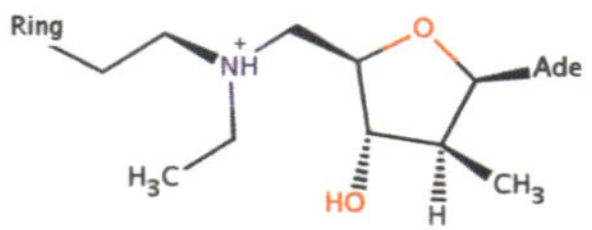	−2.08	−9.9 to −7.8 kcal/mol 5 Hydrogen bonds: CYS 1191—N6 (adenosine) CYS 1191—N1 (adenosine) MET 1169—N3 (adenosine) GLU 1168—O3 (ribose) ARG 1312—O2 (citosine)	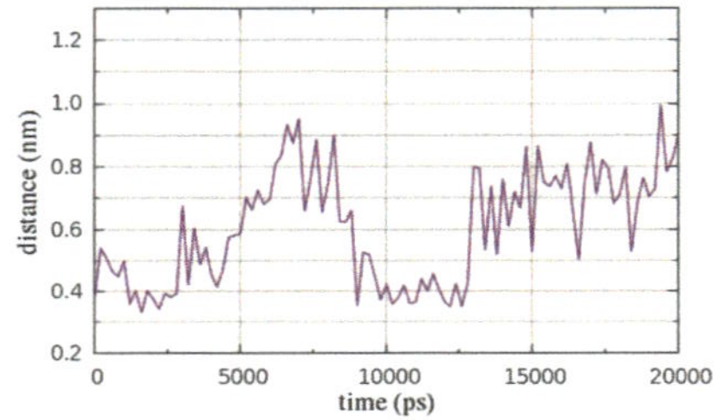	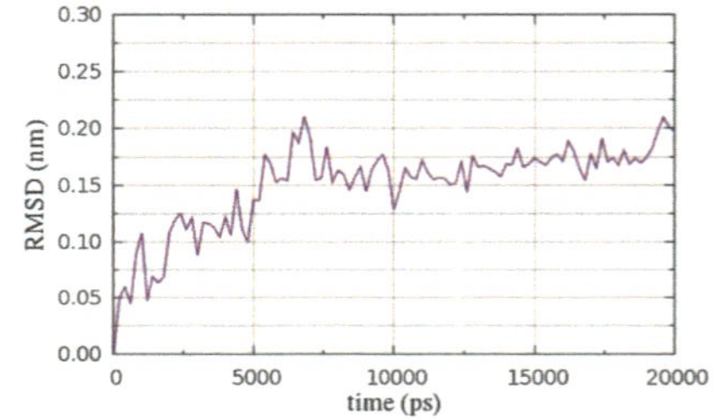
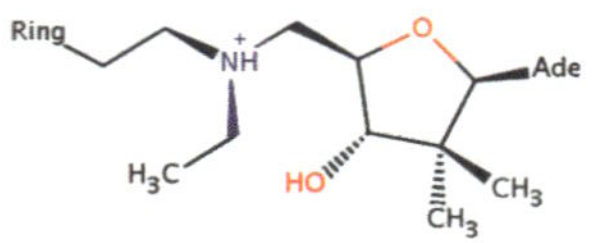	−1.70	−8.7 to −7.3 kcal/mol 5 Hydrogen bonds: CYS 1191—N6 (adenosine) CYS 1191—N1 (adenosine) MET 1169—N3 (adenosine) GLU 1168—O3 (ribose) ARG 1312—O2 (citosine)	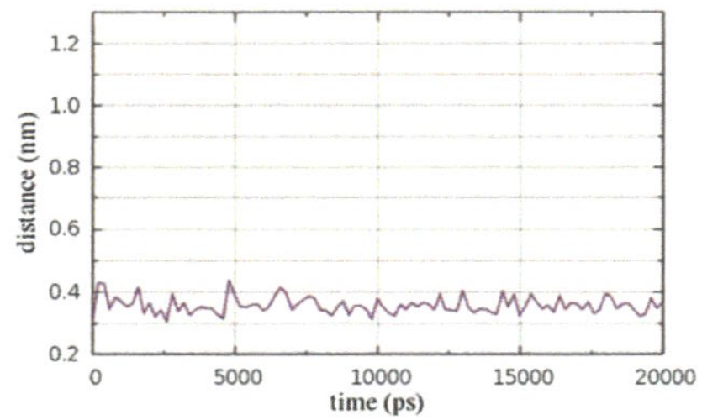	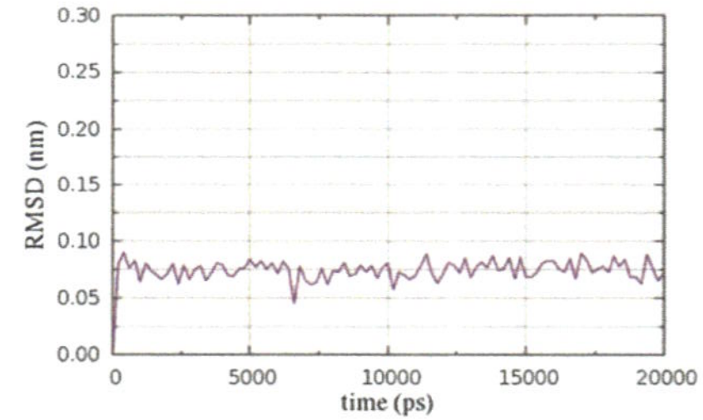
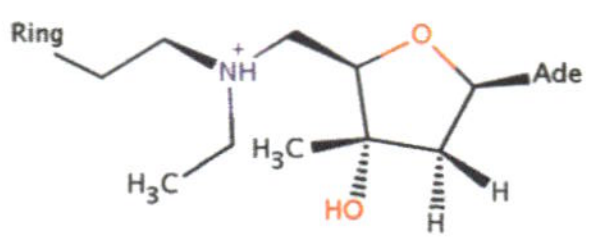	−2.13	Did not dock to the binding site cavity	N.A.	N.A.
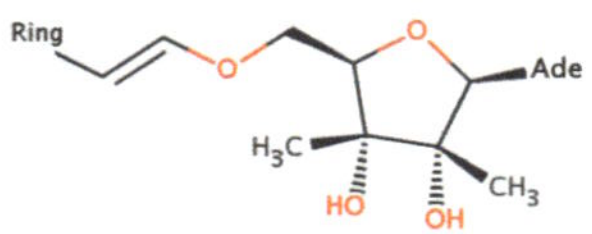	−1.38	−10.1 to −7.9 kcal/mol 6 Hydrogen bonds: CYS 1191—N6 (adenosine) CYS 1191—N1 (adenosine) MET 1169—N3 (adenosine) HOH 1704—O3 (ribose) ARG 1312—O2 (cytosine) ARG 1312—N3 (cytosine)	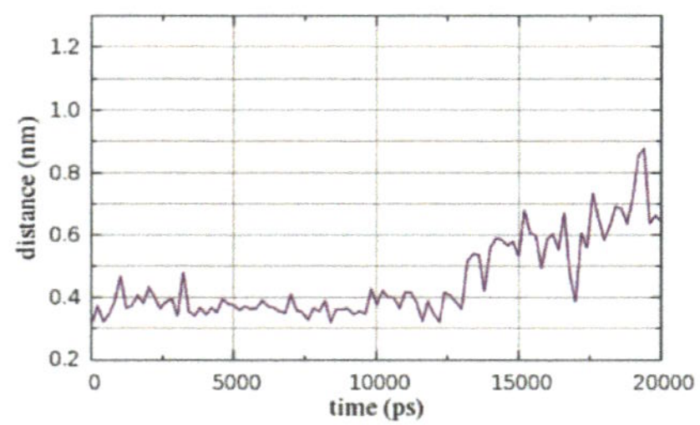	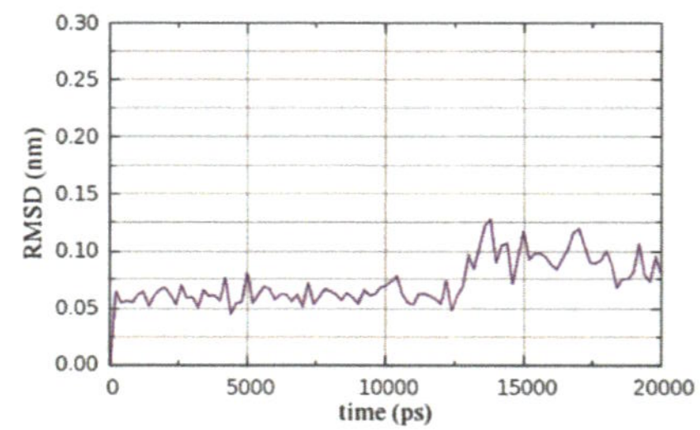

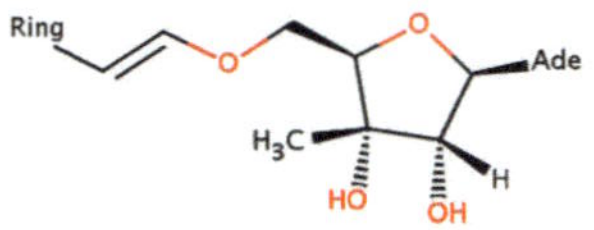	−1.66	−10.0 to −7.5 kcal/mol 7 Hydrogen bonds: CYS 1191—N6 (adenosine) CYS 1191—N1 (adenosine) MET 1169—N3 (adenosine) HOH 1704—O2 (ribose) GLU 1168—O3 (ribose) ARG 1310—O2 (citosine) ARG 1312—O2 (citosine)	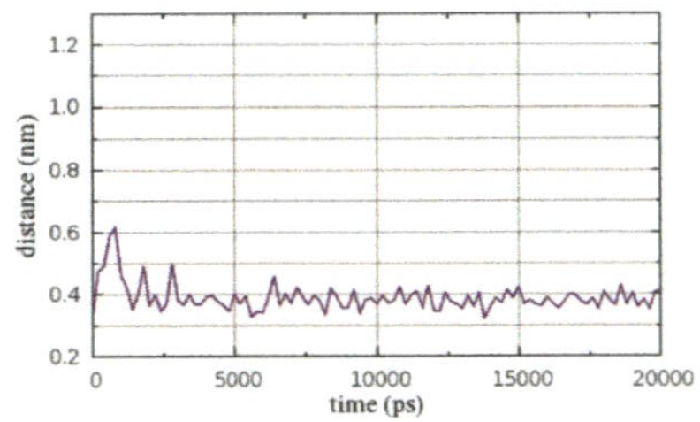	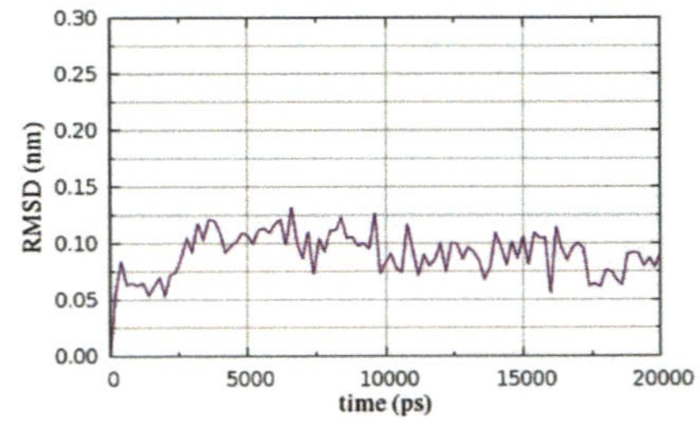
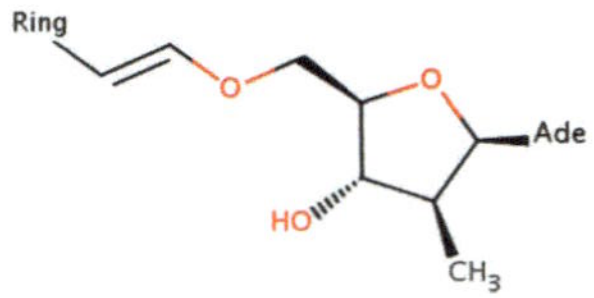	−0.66	−10.6 to −8.2 kcal/mol 5 Hydrogen bonds: CYS 1191—N6 (adenosine) CYS 1191—N1 (adenosine) MET 1169—N3 (adenosine) GLU 1168—O3 (ribose) ARG 1312—O2 (citosine)	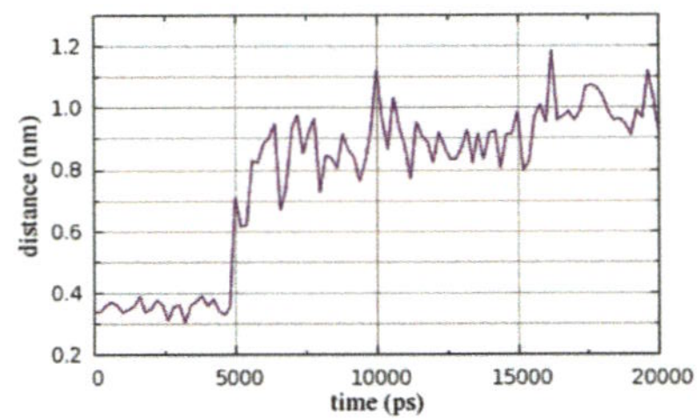	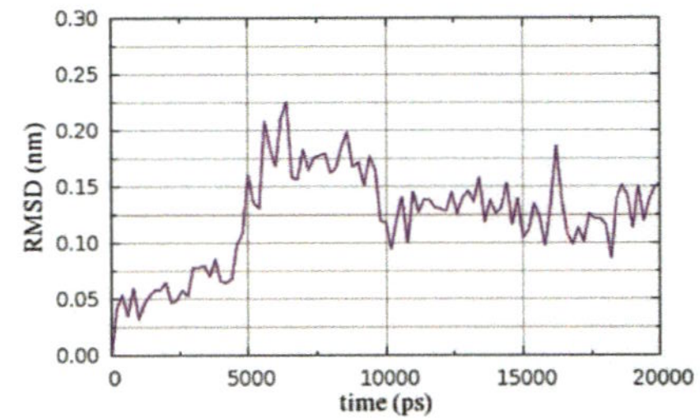
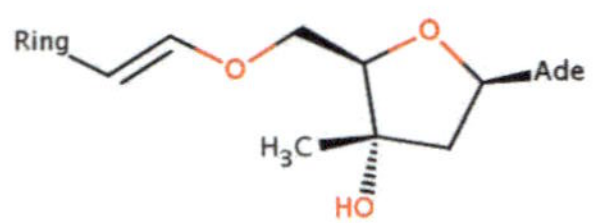	−0.76	Did not dock to the binding site cavity	N.A.	N.A.
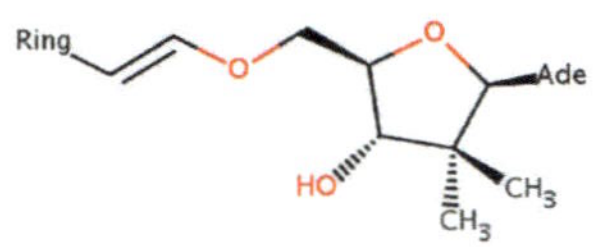	−0.28	Did not dock to the binding site cavity	N.A.	N.A.

The LogD values represent calculated values for the lead compound with presented modification that were calculated using weighted option in ChemAxon Marvin (ChemAxon). The LogD value for the lead compound is −3.07, thus any modification that has a LogD value higher than −3.07 represents a favorable modification.

Table 6 Binding the Lead Compound to DNMT1 with the Active Site Loop in Closed and Open Positions

PDB Structure	Chimera-Vina Docking Interaction Energy Range H Bonds with DNMT1	Gromacs MM/MD Frames Cys1226 Ring C6 Distance Relative to the First Frame	Gromacs MM/MD Frames RMSD Ligand Relative to the First Frame
4DA4 (Song et al., 2012) with the lead compound	−9.7 to −7.9 kcal/mol 7 Hydrogen bonds: CYS 1191—N1 (adenosine) CYS 1191—N6 (adenosine) MET 1169—N3 (adenosine) GLU 1168—O3 (ribose) GLU 1168—O2 (ribose) PHE 1145—N1 (linker) ARG 1312—O2 (cytosine)	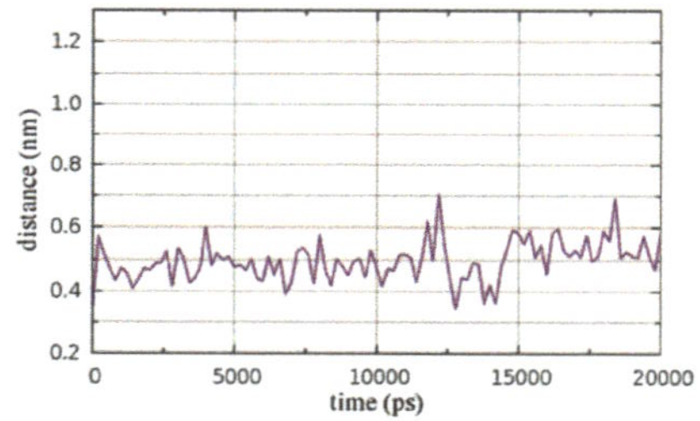	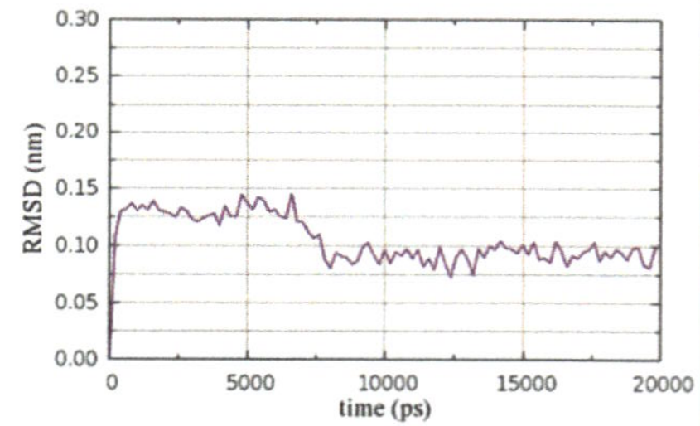
3AV6 (Takeshita et al., 2011) with the lead compound	−9.5 to −8.1 kcal/mol 8 Hydrogen bonds: CYS 838—N1 (adenosine) CYS 838—N6 (adenosine) MET 816—N3 (adenosine) GLU 815—O2 (ribose) ASN 1224—O3 (ribose) GLY 870—N1 (linker) ARG 957—O2 (cytosine) ARG 957—N3 (cytosine)	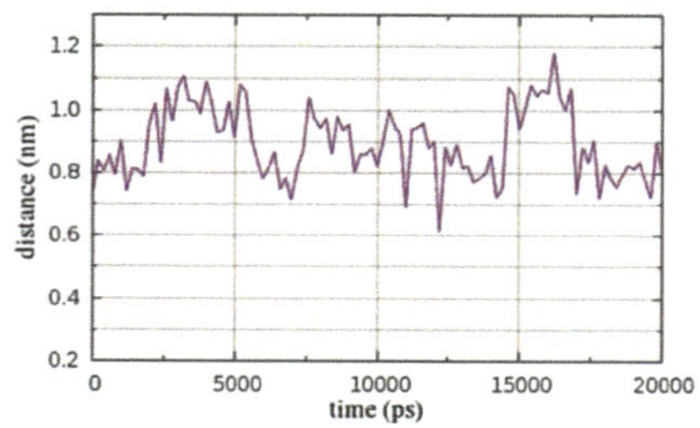	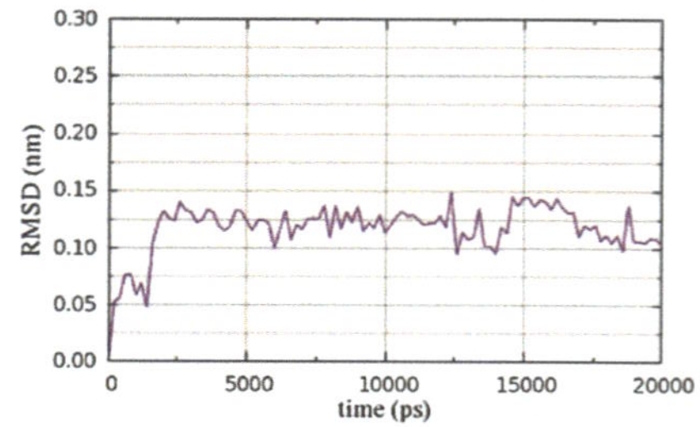

ACKNOWLEDGMENTS

This work has been supported by the Croatian Science Foundation's project number O-1505-2015; University of Rijeka project number 511-12; and the Croatian Ministry of Science. We gratefully acknowledge services of NVIDA CUDA Teaching Center and HPC facility at the Center for Advanced Computing and Modeling provided by the University of Rijeka. ZS is a paid consultant for Jiva Pharmaceuticals. None of the funding sources had any influence on design, interpretation, or dissemination of the results presented in these studies. We are proud supporters and contributors to the versatility and convenience of the open-source software.

REFERENCES

Aravind, L., 2011. Natural history of eukaryotic DNA methylation systems. In: Iyer, L.M., Abhiman, S. (Eds.), Modifications of Nuclear DNA and Its Regulatory Proteins. 101, 25–104.

Babenko, O., Kovalchuk, I., Metz, G.A., 2012. Epigenetic programming of neurodegenerative diseases by an adverse environment. Brain Res. 1444, 96–111.

ChemAxon, 2015. Marvin Was Used for Drawing, Displaying and Characterizing Chemical Structures, Substructures and Reactions, Marvin 15.0.1 (Version Number). ChemAxon.

Covic, M., Karaca, E., Lie, D., 2010. Epigenetic regulation of neurogenesis in the adult hippocampus. Heredity 105, 122–134.

Daigle, S.R., Olhava, E.J., Therkelsen, C.A., Basavapathruni, A., Jin, L., Boriack-Sjodin, P.A., Allain, C.J., Klaus, C.R., Raimondi, A., Scott, M.P., Waters, N.J., Chesworth, R., Moyer, M.P., Copeland, R.A., Richon, V.M., Pollock, R.M., 2013. Potent inhibition of DOT1L as treatment of MLL-fusion leukemia. Blood 122, 1017–1025.

Di Ruscio, A., Ebralidze, A.K., Benoukraf, T., Amabile, G., Goff, L.A., Terragni, J., Figueroa, M.E., Pontes, L.L.D.F., Alberich-Jorda, M., Zhang, P., 2013. DNMT1-interacting RNAs block gene-specific DNA methylation. Nature 503, 371–376.

Estabrook, R.A., Nguyen, T.T., Fera, N., Reich, N.O., 2009. Coupling sequence-specific recognition to DNA modification. J. Biol. Chem. 284, 22690–22696.

Fersht, A., 1998. Structure and Mechanism in Protein Science: A Guide to Enzyme Catalysis and Protein Folding (Hardcover), first ed. W. H. Freeman.

Gordon, M.S., Schmidt, M.W., 2005. Advances in electronic structure theory: GAMESS a decade later. In: Theory and Applications of Computational Chemistry: The First Forty Years, pp. 1167–1189.

Gros, C., Fahy, J., Halby, L., Dufau, I., Erdmann, A., Gregoire, J.-M., Ausseil, F., Vispé, S., Arimondo, P.B., 2012. DNA methylation inhibitors in cancer: recent and future approaches. Biochimie 94, 2280–2296.

Guo, X., Wang, L., Li, J., Ding, Z., Xiao, J., Yin, X., He, S., Shi, P., Dong, L., Li, G., Tian, C., Wang, J., Cong, Y., Xu, Y., 2015. Structural insight into autoinhibition and histone H3-induced activation of DNMT3A. Nature 517, 640–644.

Hanwell, M.D., Curtis, D.E., Lonie, D.C., Vandermeersch, T., Zurek, E., Hutchison, G.R., 2012. Avogadro: an advanced semantic chemical editor, visualization, and analysis platform. J. Cheminform. 4, 17.

Heerboth, S., Lapinska, K., Snyder, N., Leary, M., Rollinson, S., Sarkar, S., 2014. Use of epigenetic drugs in disease: an overview. Genet. Epigenet. 6, 9.

Henkelman, G., Jónsson, H., 2000. Improved tangent estimate in the nudged elastic band method for finding minimum energy paths and saddle points. J. Chem. Phys. 113, 9978–9985.

Henkelman, G., Uberuaga, B.P., Jónsson, H., 2000. A climbing image nudged elastic band method for finding saddle points and minimum energy paths. J. Chem. Phys. 113, 9901–9904.

Hess, B., Kutzner, C., van der Spoel, D., Lindahl, E., 2008. GROMACS 4: algorithms for highly efficient, load-balanced, and scalable molecular simulation. J. Chem. Theory Comput. 4, 435–447.

Humphrey, W., Dalke, A., Schulten, K., 1996. VMD – visual molecular dynamics. J. Mol. Graph. 14, 33–38.

Isakovic, L., Saavedra, O.M., Llewellyn, D.B., Claridge, S., Zhan, L., Bernstein, N., Vaisburg, A., Elowe, N., Petschner, A.J., Rahil, J., 2009. Constrained (l-)-S-adenosyl-l-homocysteine (SAH) analogues as DNA methyltransferase inhibitors. Bioorg. Med. Chem. Lett. 19, 2742–2746.

Khavari, D.A., Sen, G.L., Rinn, J.L., 2010. DNA methylation and epigenetic control of cellular differentiation. Cell Cycle 9, 3880–3883.

Li, S., Du, J., Yang, H., Yin, J., Ding, J., Zhong, J., 2013. Functional and structural characterization of DNMT2 from *Spodoptera frugiperda*. J. Mol. Cell Biol. 5, 64–66.

Medina-Franco, J.L., Caulfield, T., 2011. Advances in the computational development of DNA methyltransferase inhibitors. Drug Discov. Today 16, 418–425.

Merino, P., 2013. Chemical Synthesis of Nucleoside Analogues. John Wiley & Sons.

Pennarossa, G., Maffei, S., Campagnol, M., Tarantini, L., Gandolfi, F., Brevini, T.A., 2013. Brief demethylation step allows the conversion of adult human skin fibroblasts into insulin-secreting cells. Proc. Natl. Acad. Sci. U.S.A. 110, 8948–8953.

Pettersen, E.F., Goddard, T.D., Huang, C.C., Couch, G.S., Greenblatt, D.M., Meng, E.C., Ferrin, T.E., 2004. UCSF Chimera–a visualization system for exploratory research and analysis version. J. Comput. Chem. 25, 1605–1612.

Qin, W., Leonhardt, H., Pichler, G., 2011. Regulation of DNA methyltransferase 1 by interactions and modifications. Nucleus 2, 392–402.

Romanoski, C.E., Glass, C.K., Stunnenberg, H.G., Wilson, L., Almouzni, G., 2015. Epigenomics: Roadmap for regulation. Nature 518, 314–316.

Santi, D.V., Hardy, L.W., 1987. Catalytic mechanism and inhibition of tRNA (uracil-5-)methyltransferase: evidence for covalent catalysis. Biochemistry 26, 8599–8606.

Sen, G.L., Reuter, J.A., Webster, D.E., Zhu, L., Khavari, P.A., 2010. DNMT1 maintains progenitor function in self-renewing somatic tissue. Nature 463, 563–567.

Song, J., Rechkoblit, O., Bestor, T.H., Patel, D.J., 2011. Structure of DNMT1-DNA complex reveals a role for autoinhibition in maintenance DNA methylation. Science 331, 1036–1040.

Song, J., Teplova, M., Ishibe-Murakami, S., Patel, D.J., 2012. Structure-based mechanistic insights into DNMT1-mediated maintenance DNA methylation. Science 335, 709–712.

Sousa da Silva, A.W., Vranken, W.F., 2012. ACPYPE - AnteChamber PYthon Parser interfacE. BMC Res. Notes 5, 367.

Svedruzic, Z.M., 2008. Mammalian cytosine DNA methyltransferase Dnmt1: enzymatic mechanism, novel mechanism-based inhibitors, and RNA-directed DNA methylation. Curr. Med. Chem. 15, 92–106.

Svedruzic, Z.M., 2011. Dnmt1 structure and function. Prog. Mol. Biol. Transl. Sci. 101, 221–254.

Svedruzic, Z.M., Reich, N.O., 2004. The mechanism of target base attack in DNA cytosine carbon 5 methylation. Biochem 43, 11460–11473.

Svedruzic, Z.M., Reich, N.O., 2005a. DNA cytosine C5 methyltransferase Dnmt1: catalysis-dependent release of allosteric inhibition. Biochemistry 44, 9472–9485.

Svedruzic, Z.M., Reich, N.O., 2005b. Mechanism of allosteric regulation of Dnmt1's processivity. Biochemistry 44, 14977–14988.

Takeshita, K., Suetake, I., Yamashita, E., Suga, M., Narita, H., Nakagawa, A., Tajima, S., 2011. Structural insight into maintenance methylation by mouse DNA methyltransferase 1 (Dnmt1). Proc. Natl. Acad. Sci. U.S.A. 108, 9055–9059.

Trott, O., Olson, A.J., 2010. AutoDock Vina: improving the speed and accuracy of docking with a new scoring function, efficient optimization, and multithreading. J. Comput. Chem. 31, 455–461.

Van Der Spoel, D., Lindahl, E., Hess, B., Groenhof, G., Mark, A.E., Berendsen, H.J., 2005. GROMACS: fast, flexible, and free. J. Comput. Chem. 26, 1701–1718.

VandeVondele, J., Krack, M., Mohamed, F., Parrinello, M., Chassaing, T., Hutter, J., 2005. Quickstep: fast and accurate density functional calculations using a mixed Gaussian and plane waves approach. Comput. Phys. Commun. 167, 103–128.

Vorbrüggen, H., Ruh-Pohlenz, C., 2001. Handbook of Nucleoside Synthesis. Wiley.

Wu, J.C., Santi, D.V., 1987. Kinetic and catalytic mechanism of HhaI methyltransferase. J. Biol. Chem. 262, 4778–4786.

Xu, F., Mao, C., Ding, Y., Rui, C., Wu, L., Shi, A., Zhang, H., Zhang, L., Xu, Z., 2010. Molecular and enzymatic profiles of mammalian DNA methyltransferases: structures and targets for drugs. Curr. Med. Chem. 17, 4052.

Yang, J., Lior-Hoffmann, L., Wang, S., Zhang, Y., Broyde, S., 2013. DNA cytosine methylation: structural and thermodynamic characterization of the epigenetic marking mechanism. Biochemistry 52, 2828–2838.

Yoo, J., Kim, J.H., Robertson, K.D., Medina-Franco, J.L., 2012. Molecular modeling of inhibitors of human DNA methyltransferase with a crystal structure: discovery of a novel DNMT1 inhibitor. Adv. Protein Chem. Struct. Biol. 87.

Yoo, J.L., Medina-Franco, J., 2012. Inhibitors of DNA methyltransferases: insights from computational studies. Curr. Med. Chem. 19, 3475–3487.

Zhou, L., Cheng, X., Connolly, B.A., Dickman, M.J., Hurd, P.J., Hornby, D.P., 2002. Zebularine: a novel DNA methylation inhibitor that forms a covalent complex with DNA methyltransferases. J. Mol. Biol. 321, 591–599.

CHAPTER 6

Structure-Based Modeling of Histone Deacetylases Inhibitors

Rino Ragno
Rome Center for Molecular Design, Sapienza University of Rome, Rome, Italy

1. INTRODUCTION

Macromolecules (proteins, nucleic acids, and polysaccharides) are three-dimensional (3-D) structures obtained by either X-ray crystallography or nuclear magnetic resonance (NMR) spectroscopy and play a prominent role in the drug discovery (DD) process (Figure 1). In the latter, some boost was received from the inclusion of rapid synthetic procedures of combinatorial chemistry (Mjalli and Harris, 1996) and the elegance of rational library design (Matter et al., 2001). The prosperity of computational methods that explore

CONTENTS

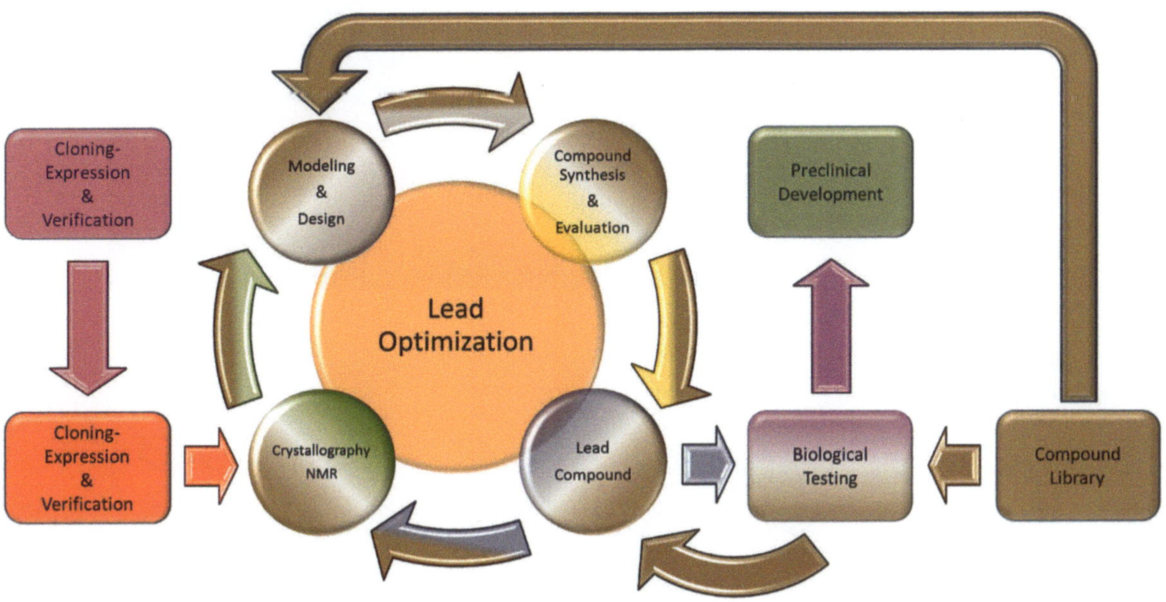

FIGURE 1
The drug discovery process scheme.

Epi-Informatics. http://dx.doi.org/10.1016/B978-0-12-802808-7.00006-X

both the receptor structure (structure-based (SB) approaches, ligand-based (LB) approaches like pharmacophore modeling (Langer, 2010), and 3D quantitative structure–activity relationships (3D QSAR) (Doweyko, 2006) provide novel opportunities for medicinal chemists to discover new lead compounds by either synthetic efforts stage (gray lane) or high throughput virtual screening of huge compound libraries using SB or LB approaches (brown flow, Figure 1). In the lead optimization iterative protocol, SBDD represents the top technique that during the drug discovery processes is pursued. In this scenario, the design process can have different starting points upon the availability of either the target structure or known ligands (Figure 2).

In 1999, Finnin et al. reported the first structure of a histone deacetylase-like protein (HDLP) cocrystallized with suberoylanilide hydroxamic acid (SAHA) and trichostatin A (TSA) that prompted numerous researchers to start SBDD studies in the epigenetic field (Chart 1). Rome Center for Molecular Design (RCMD) lab was among the first to report those kind of studies and many more are still ongoing after more than 15 years due to the enormous amount of data continuously

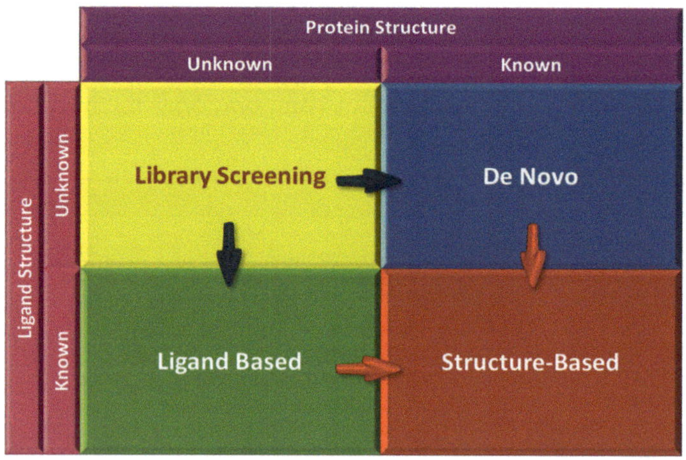

FIGURE 2
Different phases of drug design.

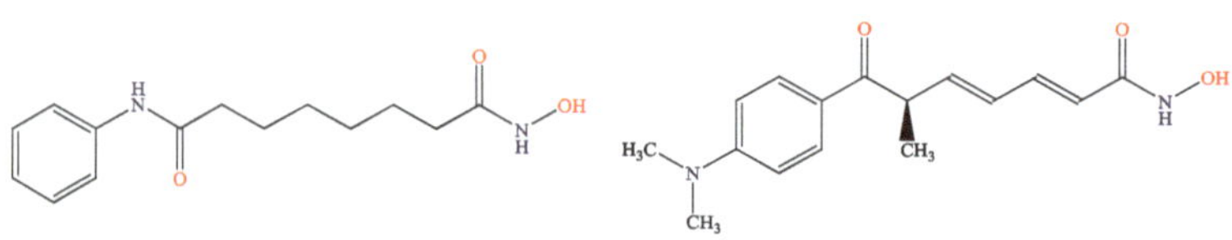

SAHA (**1**)

TSA (**2**)

CHART 1
TSA and SAHA chemical structures.

released on HDACs. This chapter focuses on SBDD approaches on histone deacetylase (HDAC) inhibitors (HDACi) that were applied by the RCMD lab. This report covers mainly Zn-based HDACi, although some studies on NAD-based HDAC (Sirtuin) inhibitors are also included at the end of the chapter.

2. SBDD STUDIES ON ZN-BASED HDAC INHIBITORS

2.1 Discovery of APHA Derivatives

Following the discovery of TSA as an antimycotic agent, pyrrole-based analogs **3** (Chart 2) were reported by Professor Artico's group as antifungal, antibacterial, and antiviral agents (Corelli et al., 1987; Massa et al., 1990). Almost 10 years later, prompted by the data from Finnin et al., SBDD studies were engaged on aroylpyrrolyl hydroxamic acid (APHA) derivatives to investigate their possibility as potential HDACi.

The newly available structural data were used to assess the predictability of an optimized VALIDATE scoring function (Head et al., 1996; Marshall et al., 2000a; Marshall et al., 2000b) obtained applying semiempirical AM1 calculated charges to a training set of 86 ligand/protein complexes. The use of quantum mechanic charges improved the predictability of the previous VALIDATE version toward those complexes constituted by proteins or enzymes containing metal cations such as those of HDLP with TSA and SAHA.

Conformational searches on a series of APHAs derivatives were done into the HDLP protein structure by means of macromodel software (Mohamadi et al., 1990) using a semiautomatic docking (SAD) procedure to provide a number of APHA/HDLP complex models. APHA derivatives pK_is were predicted in the submicromolar range by application of the reexamined VALIDATE model on the most energy-stable complexes (Table 1).

The encouraging activity values predictions pushed the design, synthesis, and biological evaluation of further APHA analogues that showed experimental IC_{50}s in the single-digit μM range against maize histone deacetylase HD2 used as enzyme source: *A new class of HDACi was disclosed by application of nonordinary SBDD protocol; very likely this was the first SBDD study on HDACi* (Massa et al., 2001) (Figure 3).

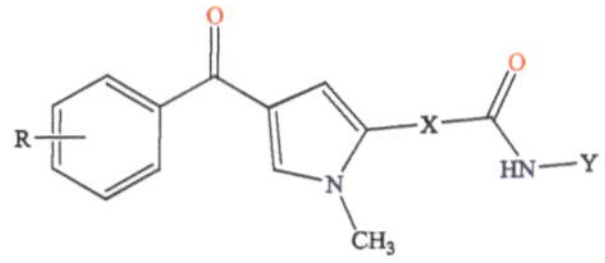

3

CHART 2

General chemical structure of APHA derivatives **3**.

Table 1 Predicted and Experimental Anti-HDAC pK_is of the Designed APHAs **3**, TSA, and SAHA

Compd	R	X	Y	Predicted pK_i	Exprmt pK_i
3a	H	CH=CH	OH	6.52	5.64
3b	Cl	CH=CH	OH	7.06	5.84
3c	F	CH=CH	OH	6.75	5.64
3d	NO_2	CH=CH	OH	6.02	5.63
3e	CH_3	CH=CH	OH	7.02	5.95
3f	OCH_3	CH=CH	OH	6.92	5.76
3g	$N(CH_3)_2$	CH=CH	OH	6.89	5.84
3h	H	CH_2—CH_2	OH	6.62	8.36
1 (TSA)				8.66	8.36
2 (SAHA)				6.27	6.88

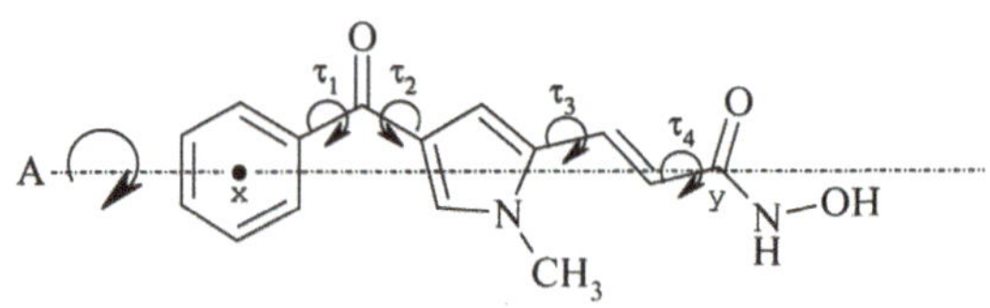

FIGURE 3
Semiautomatic docking procedure generating a family of 216 conformations in case of **3a**. (A) represents the axis of rotation passing through the center (x) of the benzene ring and the hydroxamic carbon atom (y). τ_1–τ_4 indicate the four rotatable bonds used for the grid searches.

Further biological investigations, using partially purified HDAC1 from mouse A20 cells [3H]acetate prelabeled chicken reticulocyte histones as substrate indeed confirmed APHAs as HDACi (Mai et al., 2002). Computational studies were also conducted in a surrogate HDAC1 structure derived from HDLP, which shared 36.5% identity. In these calculations the HDLP catalytic core was transformed into that of HDAC1 by single-point mutations of residues within 12 Å (Figure 4) from the co-crystallized TS- bound structure (pdb entry code 1C3R).

SAD calculations were reiterated for APHAs into the HDAC1 model and application of VALIDATE revealed that a lower error of prediction could be obtained when comparing experimental and calculated data using homogenous conditions as much as possible (Table 2). As a matter of fact an average absolute error of prediction (AAEP) of 0.78 was associated with HDLP predicted versus experimental HD2 pK_is, while comparing mouse HDAC1 expt pK_is with those predicted by the VALIDATE model into the modeled HDAC1 where an AAEP value of 0.28 was obtained. Inspection of APHA derivative **3a** binding mode into HDAC1 indicated that its potency did not reach that of TSA and SAHA

```
        10        20        30        40        50        60        70        80  Residue's number
                 |||  | |
MRKVCYYYDGDVGNYYYGQGHPMKPHRIRMTHNLLLNYGLYRKMEIYRPHKANAEEMTKYHSDDYIK FLRSIRPDNMSE  HDAC1
MKKVKLIGTLDYGKYRYPKNHPLKIPRVSLLLRFKDAMNLIDEKELIKSRPATKEELLLFHTEDYINTLMEAERSQSVPK  HDLP
                 YPKNHPLKI                                                        HDLP 12 Å core

        90       100       110       120       130       140       150       160  Residue's number
           ||||                                               |      ||
YSKQMQRFNVGE DCPVFDGLFEFCQLSTGGSVASAVKLNKQQTDIAVNWAGGLHHAKKSEASGFCYVNDIVLAILELLK  HDAC1
GAR   EKYNIGGYENPVSYAMFTGSSLATGSTVQAIEEFLK GNVAFNPAGGMHHAFKSRANGFCYINNPAVGIEYLRK  HDLP
           GGYENP                                            ANGFCY NNP           HDLP 12 Å core

       170       180       190       200       210       220       230       240  Residue's number
          | |  |                  | |   |   ||
Y HQRVLYIDIDIHHGDGVEEAFYTTDRVMTVSFHKYGEY FP GTGDLRDIGAGKGKYYAVNYPLRDGIDDESYEAIFK  HDAC1
KGFKRILYIDLDAHHCDGVQEAFYDTDQVFVLSLHQSPEYAFPFEKGFLEEIGEGKGKGYNLNIPLPKGLNDNEFLFALE  HDLP
         DLDAHHCDGV             SLHQ   YAFPFEK                                  HDLP 12 Å core

       250       260       270       280       290       300       310       320  Residue's number
                  | | |    | ||                              |
PVMSKVMEMFQPSAVVLQCGSDSLSGDRLGCFNLTIKGHAKCVEFVKSFNLPMLMLGGGGYTIRNVARCWTYETAVALDT  HDAC1
KSLEIVKEVFEPEVYLLQLGTDPLLEDYLSKFNLSNVAFLKAFNIVREVFGEGVYLGGGGYHPYALARAWTLIWCELSGR  HDLP
                 QLGTDPL  DYLSK                         GGGGYH                    HDLP 12 Å core
```

FIGURE 4

Sequence alignment of HDAC1, HDLP, and HDLP-12 Å core.

Table 2 Experimental versus VALIDATE-Predicted Anti-HDAC Activities of **3a–c** and for the References TSA and SAHA

Compd	HD2 Expt pK_i	Mouse HDAC1 Expt pK_i	HDAC1 Predicted pK_i	HDLP Predicted pK_i
3a	5.64	5.53	5.76	6.52
3b	5.95	5.57	5.83	7.02
3c	5.84	5.60	5.68	6.89
TSA	8.36	8.97	8.61	8.66
SAHA	6.88	7.17	6.69	6.27

due to a steric limitation exerted by the pyrrole N1-methyl group interacting with Phe198 and Leu265 residues, preventing the **3a** hydroxamate moiety to fully chelate the catalytic Zn ion (Figure 5). Furthermore, a determination correlation coefficient of r^2 of 0.99 confirmed the maize HD2 determined activity as predictive for the HDAC1, while in vivo hyperacetylation assays, growth inhibition, and cell differentiation confirmed APHAs to be specific inhibitors of HDAC activity in vivo endowed with potential antitumoral activities.

2.2 SBDD by Multidocking Protocol as Aid for Lead Optimization

The previously summarized studies introduced APHA as a new interesting HDACi chemical class, consequently, with the purpose of optimizing the lead compound **3a**, intensive medicinal chemistry efforts led to the synthesis and biological evaluation of new derivatives (Mai et al., 2003b). Accordingly, to sustain the drug design process, further computational studies were

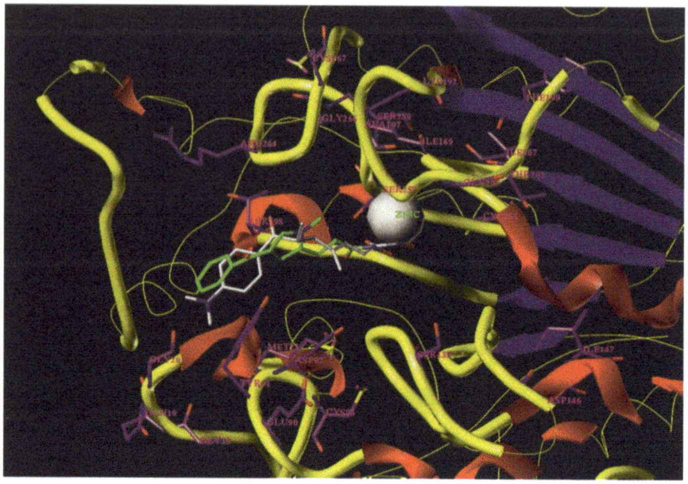

FIGURE 5

Three-dimensional model of **3a** (green) docked into the HDAC1 catalytic core. TSA (white), and the mutated HDLP→HDAC1 residues (purple) are also displayed. α-carbon atom trace of the HDAC/HDLP structure is in yellow.

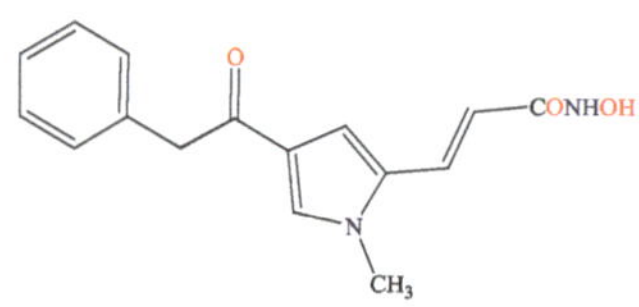

4

CHART 3

Structure of the second APHA lead compound **4**.

carried out. In particular the binding mode of the new lead compound 4 (Chart 3) was investigated by an unprecedented multidocking protocol, in which three different docking approaches, namely SAD, DOCK (Kuntz et al., 1982), and AutoDock (Morris et al., 2009), were used in a consensus docking procedure.

In the multidocking protocol all the ligand/HDAC1 complexes obtained were minimized by means of macromodel; the most stable ones were those proposed by AutoDock (Table 3). For comparison and procedure assessment, the same calculations were applied to **3a**, TSA, and SAHA. To compare the four docked molecules (1, 8, TSA, and SAHA), Figure 7 shows their binding conformations proposed by AutoDock. Similar to that observed for the complexes, the AutoDock conformations also showed to be more stable than those obtained with DOCK or SAD procedures (ligand bound steric energies, Table 3). Although comparing IC_{50} with predicted pK_is, application of the revisited

Table 3 Minimized Complex and Ligand-Bound Steric Energies (AMBER Force Field) Relative to **3**, **4**, SAHA, and TSA Obtained by DOCK, AutoDock, and SAD Procedures (AAEP was Calculated against HDAC1 Experimental Activities)

Docking Method	3a	4	SAHA	TSA
Minimized Complex Steric Energy (kJ/mol)				
DOCK	−7174.81	−9077.85	−7369.18	−8909.93
AutoDock	−8913.90	−9405.68	−10363.13	−9976.13
SAD	−7828.92	−9142.08	ND	ND
Ligand-Bound Steric Energy (kJ/mol)				
DOCK	−12.25	−49.03	−0.53	−48.20
AutoDock	−25.11	−53.02	−13.38	−52.74
SAD	−11.50	−54.90	ND	ND

Table 4 Experimental versus VALIDATE and AutoDock Predicted HDAC1 Inhibitory Activities of **3**, **4**, SAHA, and TSA

	Expt pIC_{50}		VALIDATE Predicted pK_i			
Compd	**HDAC1**	**HD2**	**SAD**	**DOCK**	**AutoDock**	**AutoDock Predicted pK_i**
TSA	8.70	8.15	8.61	7.26	7.92	7.41
SAHA	6.95	7.30	6.69	6.79	7.15	6.21
1	5.31	5.42	5.76	4.76	7.31	5.65
8	6.29	7.00	6.30	6.98	6.37	5.98
AAEP			0.20	0.71	0.76	0.67

VALIDATE model to the three docking methods revealed the superior predictive ability of the SAD method; nevertheless the internal AutoDock scoring function showed a good level of reliability (Table 4). The latter represented interesting results; in fact, considering the slowness in the VALIDATE procedure (all the complexes needed to be minimized and their AM1 charges calculated), the use of AutoDock without further calculation represented a step forward in speeding up the SBDD applied to HDACi with a minor increase in error prediction. This compromise is particularly true in the medicinal chemistry field where many times hundreds or even thousands of compounds need to be prioritized for the synthesis of biological assays.

On the other hand the good predicted activities could reflect some chance results, therefore analyses of binding modes was mandatory. Interestingly, docking results were all in good agreement each to other (compare **3a** and **4** in Figure 6) and with experimental values (compare TSA and SAHA in Figure 6). They found that different binding conformations of **3a** (Figure 6(a)) likely represent three different bound states of the ligand, confirming the low tightness of the

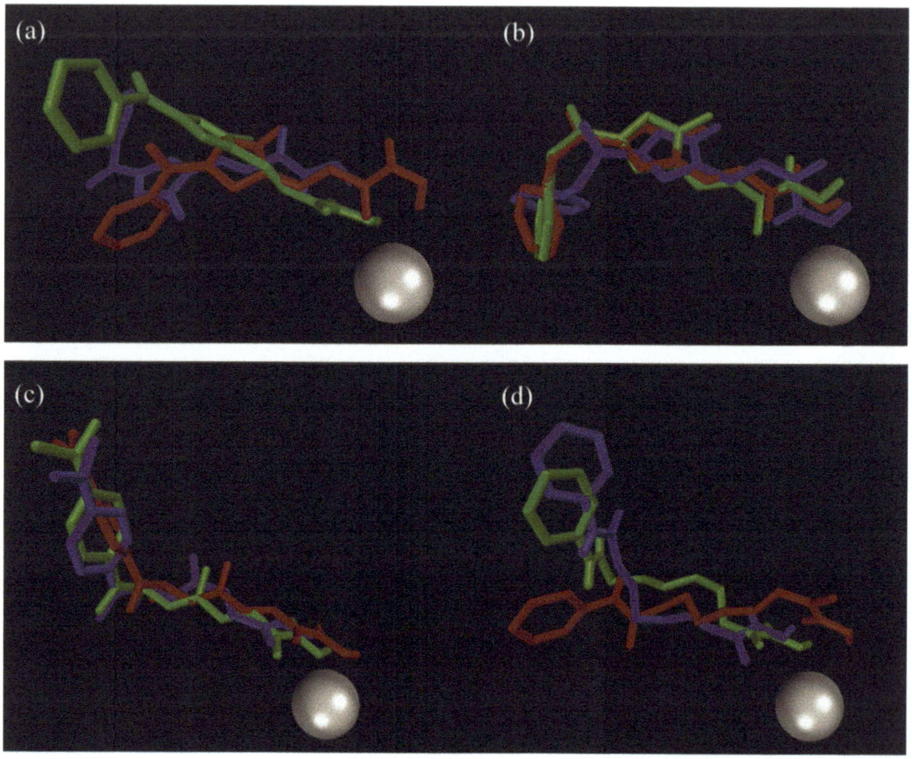

FIGURE 6

Superimposition of the docked conformations of **3a** (a), **4** (b), TSA (c), and SAHA (d) obtained from three docking procedures. The minimized experimental (TSA and SAHA) and the semiautomatic docked (SAD) (**3a** and **4**) conformations are in green. AutoDock conformations are in red. DOCK conformations are in blue. The gray sphere indicates the Zn ion position into the HDAC1. Figures were generated by means of AutoDockTools 1.5.6 (Morris et al., 2009).

HDAC1/**3a** complex. On the other hand, for structures endowed with higher activities, the docking methodologies led to almost the same results, reflecting that more stable and unique complexes are achieved for **4**, TSA, and SAHA (Figure 6(b)–(d)).

Inspection of **4** binding mode into the HDAC1 catalytic core, compared with that described for **3a**, revealed that the enhancement of inhibitory activity could be attributed to the higher flexibility of the pyrrole C4-substituent of **4**. Introduction of a methylene connection between the phenyl and the carbonyl group of the **3a** benzoyl moiety disrupts the extended phenyl-CO-pyrrole π-conjugation. As a consequence of this flexibility, **4** fits considerably better than **3a** into the HDAC1 pocket, showing a higher average steric fit ($\mathbf{3a}_{\text{average steric fit}} = 1.7$; $\mathbf{4}_{\text{average steric fit}} = 2.0$; Table 5) and providing a more favorable enthalpy of binding that offsets the slightly more unfavorable entropy

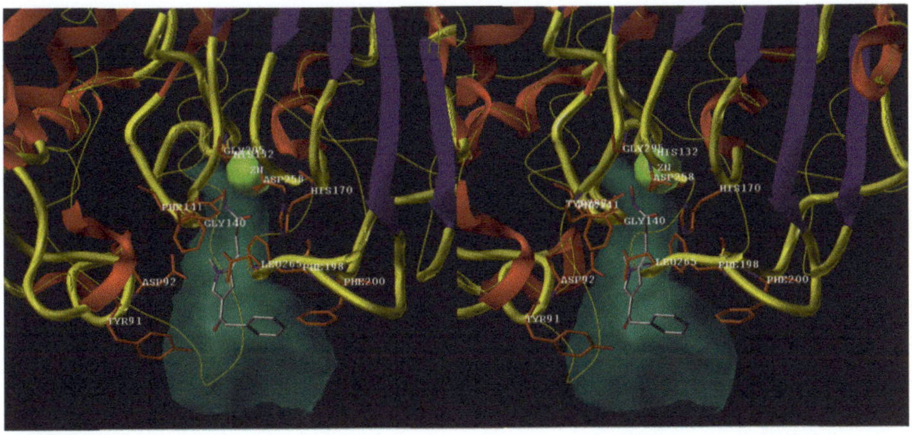

FIGURE 7

Stereoview of the SAD derived **4** conformation (white) into HDAC1 catalytic core. Residue side chains within 4 Å from the ligand are also displayed (orange). The volume binding site is highlighted in transparent cyan.

Table 5 Calculated Steric Fit as in VALIDATE method (Head et al., 1996) and Hydroxamate/Zn Distances

Comp	Docking Method	Steric Fit	CO···Zn	OH···Zn
3a	SAD	1.3	4.5	4.8
3a	AutoDock	2.3	2.8	4.4
3a	DOCK	1.5	4.7	4.8
4	SAD	1.8	3.3	3.1
4	AutoDock	1.9	3.0	3.1
4	DOCK	2.2	3.0	4.4

(Table 3). Furthermore, this enhanced fitting allows a closer positioning of 4 hydroxamate moiety to the zinc ion (**3a**: average CO···Zn = 4.0 Å, average OH···Zn = 4.7 Å; **4**: average CO···Zn = 3.1 Å, average OH···Zn = 3.5 Å).

Inspection of consensus-docking proposed **4** conformation into HDAC1 catalytic core (Figure 7) revealed Tyr297 hydroxyl group at hydrogen-bond distance from the **4** hydroxamate moiety (Tyr297-OH · · · $CO_{hydroxamate}$ = 3.25 Å; Tyr297-OH · · · $OH_{hydroxamate}$ = 2.42 Å). Moreover, favorable π-stacking interactions can be observed between the pyrrolylethylene chain and the Phe141 and Phe198 side chains. Regarding pyrrole N1-methyl group, differently from **3a**, it makes interactions with the α-carbon atom of Gly140 and partially with the Phe141 side chain. This switched position of **4** N1-methyl group allows the hydroxamate group to get closer to the catalytic zinc ion. Finally, in the pyrrole C4-phenylacetyl portion, the ketone points toward the Tyr91 hydroxyl group, hypothesizing that a water molecule could insert and make a hydrogen bond bridge, while the benzene ring points to a hydrophobic region of the enzyme delimited by the Phe200 side chain.

This was the first application of a multidocking and consensus-docking procedure ever reported on HDACi; later, many others reported single-docking applications on different molecules and HDAC structures (Lu et al., 2005; Wang et al., 2005, 2004; Xie et al., 2004).

2.3 SBDD/Synthetic Chemistry Tandem as Strategy for New HDACi

In view of the obtained positive results, a fruitful molecular modeling and synthetic chemistry tandem led to the preparation of a number of new APHA-related HDACi. In 2004, the replacement of the pyrrole-C_2 ethene moieties of APHA **3** with different alkyl/alkenyl chains and homologs with various hydrocarbon linker lengths between the terminal phenyl ring and the carbonyl adjacent to pyrrole ring led to the synthesis and evaluation of a new APHA series as anti-HD2 agents (Chart 4). With the aim of assessing the reliability of the synthetic effort, the binding modes of designed APHA derivatives (**5**) into the modeled HDAC1 catalytic pocket were investigated by molecular docking studies.

From docking studies, for each docked APHA derivative three different binding conformations were selected: (1) best docked (BD), the first AutoDock scored conformation; (2) best cluster (BC), the representative conformation of the most populated cluster; and (3) best energy (BE), the conformation extracted from the lowest energy complex obtained after a single-point minimization of all proposed complexes. AutoDock BE conformations predicted pK_i values and experimental pIC_{50}s resulted in good agreement with AAEP values as low as 0.91. Analysis of new APHA derivatives binding modes were inspected and focused on the most active (Chart 4: **5a**: R = phenyl, n = 2, Y = CH_2=CH_2, Z = CONHOH) while comparing with APHA lead compound **4**. Inspection showed the following common ligand/enzyme interactions: Gly140 carbonyl group at hydrogen-bond distance from NH hydroxamate moieties; favorable π-stacking interactions between the

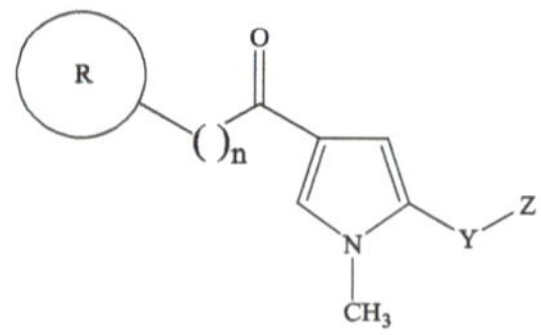

5

CHART 4
Designed APHA derivatives.

pyrrolylethylene groups and Phe141/Phe198 side chains; pyrrole N1-methyl groups made favorable interactions with the α-carbon atom of Gly140 and partially with the Phe198 side chain (white area of Figure 8). The greater difference between **4** and **5a** was found to be their pyrrole C4-acyl substituents interactions. The presence of the extra methylene in **5a** enhanced the flexibility, allowing the placement of the benzene ring into a hydrophobic cleft made by portions of the side chains of Pro22, Tyr91, Phe141, and Leu265 (orange area of Figure 8), while the carbonyl group points outside the HDAC1 binding pocket, able to make favorable hydrogen bonds with the water environment.

In a parallel study, we applied a different approach on the **3a** structure to discover novel leads, using SBDD and docking procedures. Two isomers of **3a**, 3-(2-benzoyl-1-methyl-1H-pyrrol-5-yl)-N-hydroxy-2-propenamide **6** and 3-(2-benzoyl-1-methyl-1H-pyrrol-4-yl)-N-hydroxy-2-propenamide **7** (Chart 5) were designed and their HDAC inhibiting activity was predicted by computational studies.

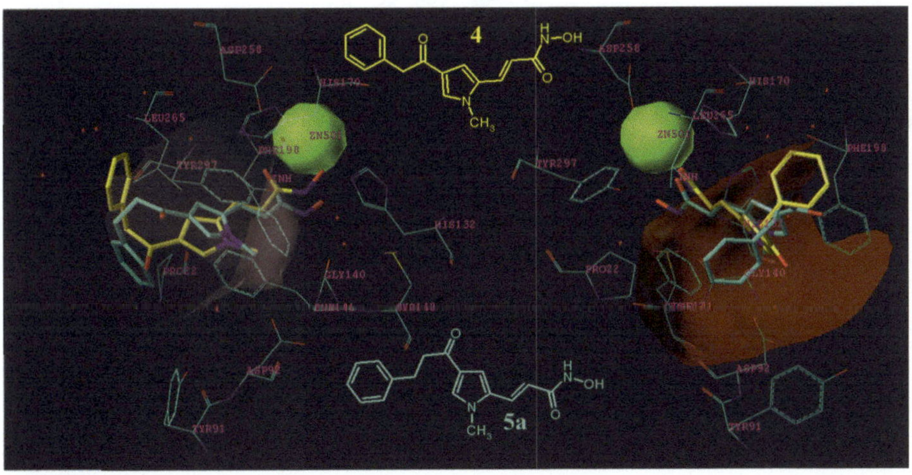

FIGURE 8

Orthographic view of **4** (yellow) and **5a** (cyan) binding modes into the HDAC1 catalytic core: (left) shaded in white, the N-methylpyrrole binding subpocket; (right) shaded in orange, the pyrrole C4-acyl subsite. The green sphere indicates the Zn ion position in the HDAC1.

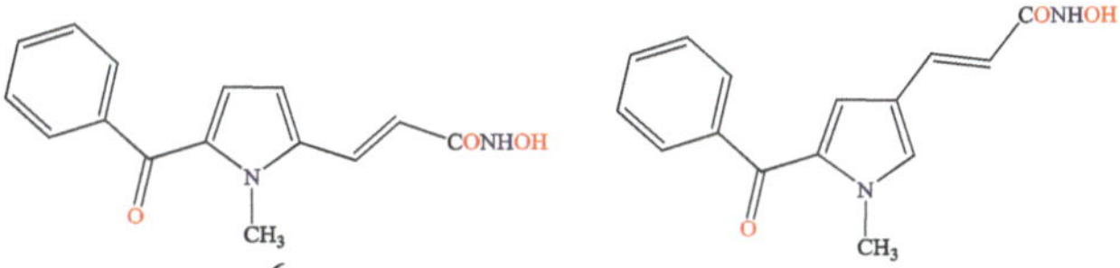

6

7

CHART 5

Structure of **3a** designed isomers.

Virtual rotation of N-methylpyrrole ring between benzoyl and N-hydroxy-2-propenamide moieties allowed us to design the two **3a** isomers **6** and **7** (Chart 1). Extensive binding mode analyses were undertaken to check whether HDAC1 enzyme inhibitory potency would be influenced by the earlier different benzoyl and N-hydroxypropenamide substituents rearrangement on pyrrole ring. Binding mode analyses of **6** and **7** into the modeled HDAC1 structure were investigated using the SAD procedure, DOCK, and AutoDock (Figure 9).

The results of the three dockings were energy rescored by a single-point minimization by the MACROMODEL molecular mechanics program, using AMBER all atom force field while using MOPAC2000 calculated point charges. Three bound conformations for both **6** and **7** were obtained and their associated HDAC inhibitory potencies predicted by means of the refined VALIDATE scoring function model (Table 6).

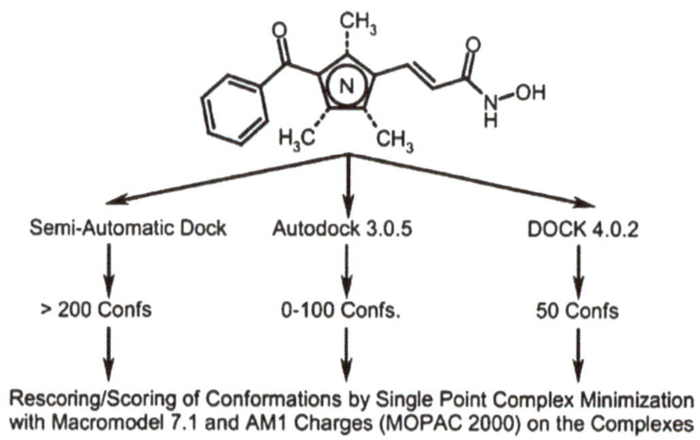

FIGURE 9
Docking procedure scheme.

Table 6 Minimized Complex Steric Energies (kJ/mol) (AMBER Force Field) Relative to APHA Isomers 6 and 7 Obtained by SAD, DOCK, and AutoDock

	VALIDATE Predicted pK_i			Experimental pIC_{50}	
Compd	**SAD**	**DOCK**	**AutoDock**	**HDAC1**	**HD2**
3a	5.76	4.76	7.31	5.31	5.42
6	5.87	7.31	7.76	6.64	6.60
7	7.13	7.55	7.47	6.11	7.30
TSA	8.61[a]	7.26	7.92	8.70	8.15
SAHA	6.69[a]	6.79	7.15	6.95	7.30

For comparison purposes, values for 3a, TSA, and SAHA are also reported.
[a]Predicted from X-ray experimental data.

In all cases predicted binding affinities were predicted from 1.3- to 354.8-times (**6**) and from 1.20- to 616.6-times (**7**) higher than those of **3a** against HDAC1 (Table 1). From these encouraging data, **6** and **7** were promptly synthesized and tested against both maize HD2 and mouse HDAC1 enzymes whose results fully assessed the computational studies (Table 6).

2.4 SB and LB Combined Approaches

After the disclosure of derivative **3** several other APHAs were synthesized and reported as HDACi (Mai et al., 2004a,b; 2006; Minucci et al., 2006; Ragno et al., 2004). In view of the constant increasing interest and to set up rational DD protocols for new HDACi these SBDD studies were coupled with LB approaches such as 3-D QSAR (Ragno et al., 2006). Guo et al. (2005) reported similar studies , nevertheless their 3-D QSAR were focused on a single chemical class. In our SB 3-D QSAR the application of the GRID/GOLPE (Cruciani and Watson, 1994) procedure was applied to a training set of 103 different HDACi all homogeneously tested against HD2 (Chart 6).

The derivation of a 3-D QSAR was twofold: gain insights into how steric, electrostatic, hydrophobic, and hydrogen-bonding interactions could influence the inhibitory enzyme activity; and to derive predictive models for designing and forecasting the activity of untested HDACis. The final 3-D QSAR model was developed using four different HDACi series (TAAs, SPACAs, APHA1, and APHA2) whose determined pIC_{50} spanned more than 5 log units. Training set molecules were SB aligned by manual docking procedure using either TSA-like or SAHA-like binding conformations, producing two different aligned training sets. Inactive compounds were included in the training set to develop 3-D QSAR models able to discriminate active versus inactive derivatives. Two external training sets (TsSet) were compiled to assess the model's predictive ability. A first TsSet (APHA_TS) comprised compounds belonging to APHA1 and APHA2

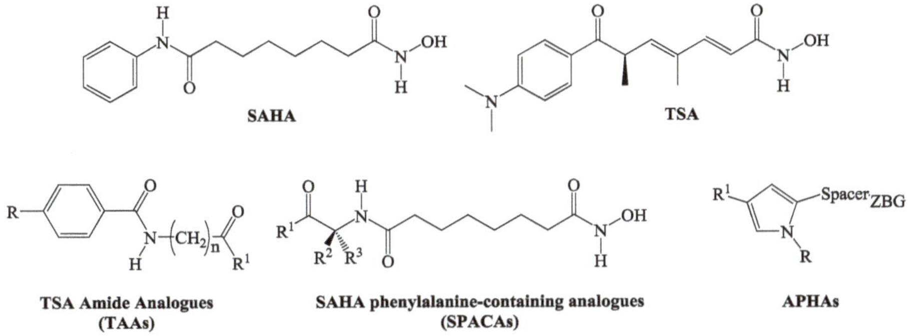

CHART 6
Structures of different series of HDAC inhibitors tested against Maize HD2 and included in the 3-D QSAR training set.

series (Mai et al., 2003a; 2005a) and was used to confirm the model's robustness, while the second contained a different chemical class of HDACi, uracil-based hydroxamic acids (UBHAs, UBHA_TS) (Mai et al., 2005c) and served as a real test in proving the model's predictability toward new chemical entities.

Initially, 3-D QSAR models developed for each series (TAAs, SPACAs, APHA1, and APHA2) were derived leading to highly statistical significant correlation coefficient (r^2), cross-validated correlation coefficient (q^2), and cross-validated Standard Deviation of Errors of Prediction ($SDEP_{CV}$) values in the ranges 0.95–0.99, 0.66–0.85, and 0.29–0.85, respectively. The obtained values were much higher of those of previously reported 3-D QSAR built on other HDACi congeneric series (Chen et al., 2003; Guo et al., 2005; Liu et al., 2005). The single chemical class self-consistency models supported the four series to be merged in a single training set. Still, high statistical values endowed models resulted in the display of r^2, q^2, and $SDEP_{CV}$ values in the ranges 0.83–0.90, 0.71–0.75, and 0.69–0.74, respectively (Table 7).

As listed in Table 7, 3-D QSAR models with and without inactive compounds were developed. On one hand the use of models containing information from inactive derivatives would certainly be of high utility in the design of new compounds, especially in the case of a virtual high throughput screening, allowing the medicinal chemist to directly filter only compounds that should have some level of activity and discarding those predicted in the same activity range of non-active derivatives. On the other hand, the inclusion of nonactive compounds, to which arbitrary values were assigned, is a source of spurious data making the model less precise in detecting the highly active ones. In any case, models A and B showed similar statistical profiles of those derived with only active HDACi (Table 7, models C and D). Table 6 also reports the SDEP valued for the external test sets APHA_TS and UBHA_TS, highlighting the good level of a model's predictive ability, in particular for model D, the one derived with SAHA-based

Table 7 3-D QSAR Models Derived on Different Training Set Composition (See Text and Cited Reference for Details)

Model	Series	Alignment	N[a]	Vars[b]	PC[c]	r^2	q^2	$SDEP_{CV}$	$SDEP_{testset}$ 1	$SDEP_{testset}$ 2
A	United	TSA-based	103	726	3	0.90	0.75	0.69	1.20	1.46
B	United	SAHA-based	103	829	2	0.83	0.71	0.74	1.00	1.39
C	United OA[d]	TSA-based	71	659	3	0.94	0.83	0.41	1.30	1.06
D	United OA[d]	SAHA-based	71	770	3	0.91	0.75	0.51	0.96	0.99

[a]*Number of compounds used in the model.*
[b]*Number of selected variables.*
[c]*Number of principal components that showed the maximum q^2 value.*
[d]*OA: Only active compounds were used in the training set.*

SBDD alignment and using the training set with only active compounds. One important feature of 3-D QSARs is the graphical representation of the model making its interpretation easier. To this partial least square (PLS) coefficients plot, the overall graphical representation of the 3-D QSAR model is used and compared to the compound present field plots for the 3-D QSAR interpretation. In Figure 10 yellow and cyan areas are close to the Zn-binding groups (ZBG) and although having opposite signs, both contribute positively to the activity.

The big cyan polyhedrons represent significant electrostatic and hydrogen bond interactions between the ZBG with zinc ion or active site residues. The ZBG position is also important: in fact, derivatives with an efficient ZBG have low activity because their chelation group is shifted away from zinc ion during the complex minimization. The big yellow polyhedron highlights furthermore a hydrogen bond between ZBG and the surrounding HDAC residues. Further big yellow polyhedrons run around the capping groups meaning that in this area hydrophobic interactions contribute positively to the activity. In fact, active compounds place their capping groups superimposed on that of TSA/SAHA (Figure 10(a)), while less active derivatives (Figure 10(b)) adopt a different binding conformation contributing negatively to the activity. Along PLS coefficient plots that give a global interpretation of the 3-D QSAR model, the activity contribution plot gives the possibility of displaying spatial regions that are individually important for the selected molecules. The activity contribution plot, different for every molecule within the training set, are obtained from the product of the coefficients by the actual field values for each molecule. In Figure 11 the activity contribution plots of selected highly active (Figure 11(a)–(c)) and poorly active compounds (Figure 11(d)–(f)) are reported as examples for each series. In the activity contribution plots cyan polyhedrons (negative)

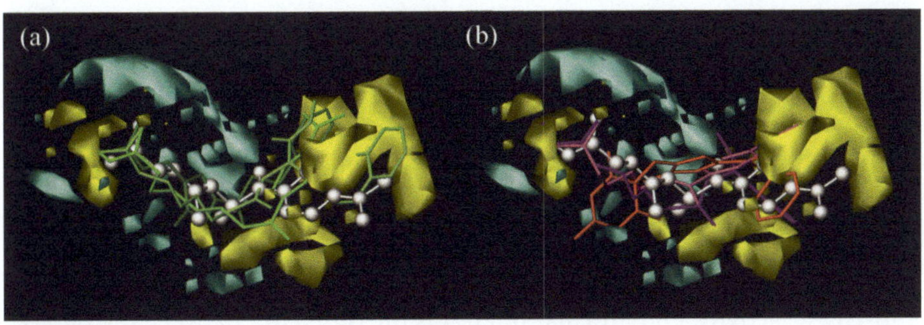

FIGURE 10
GRID/GOLPE PLS coefficient contour maps for the 3-D QSAR model D. To aid interpretation high active compounds (TAA_5a in green, SPACA_11o in light green, and APHA2_3a in dark green) are displayed in (a) and low active compounds (TAA_5c in pink, SPACA_11a in purple, and APHA1_25 in red) are in (b). For comparison purposes TSA is also displayed in white. Molecule names are those reported in Ragno et al., 2006.

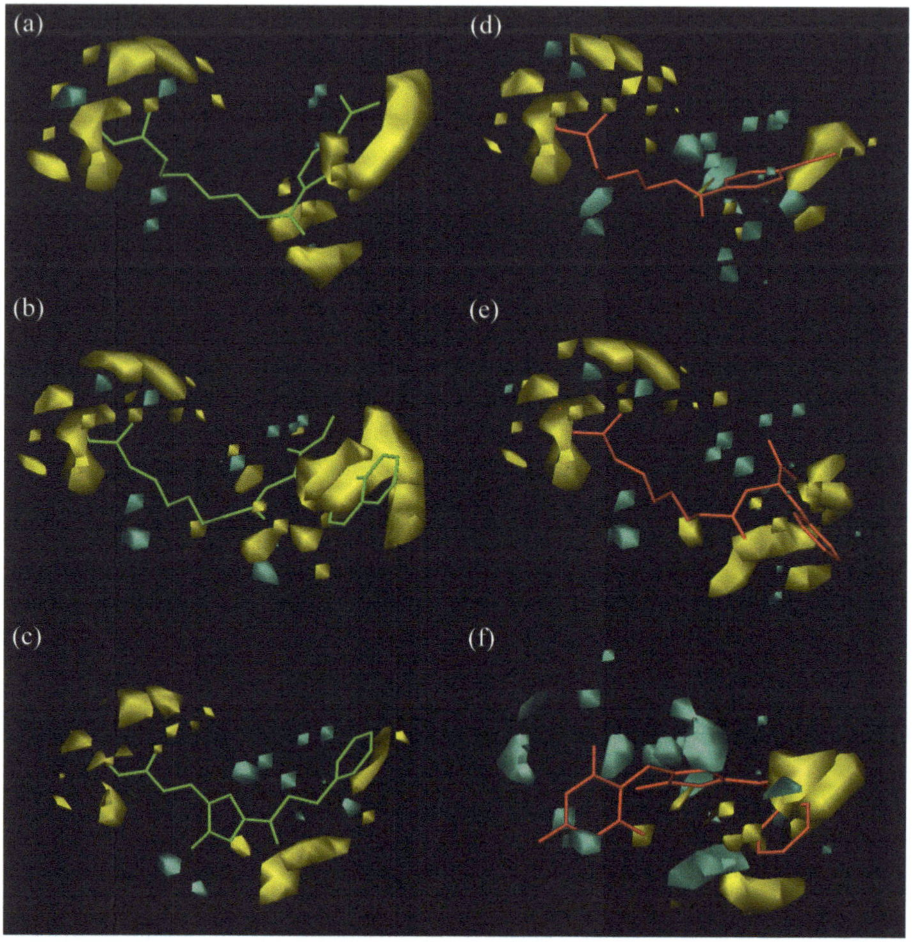

FIGURE 11

Activity contribution plots of selected representatives for each of the molecular series included in the training set. In green are reported the more active compounds and in red the lower active. (a) TAA_5a; (b) SPACA_11r; (c) APHA2_3a; (d) TAA_5c; (e) SPACA_11o; and (f) APHA1_25. Molecules name are those reported in Ragno et al., 2006.

show areas where a negative contribution to the inhibitory activity is associated, while yellow polyhedrons indicate areas of positive contribution to the anti-HDAC activity. Of particular interest is the difference between the plots of high active APHA derivative (Figure 11(c)) with that of Figure 11(f) (low active compound); it is possible to observe the conversion of some yellow colored polyhedrons into cyan colored ones.

As a rule of thumb, 3-D QSAR maps should not be used to extrapolate ligand/receptor interactions. Nevertheless the alignment of the training set was based

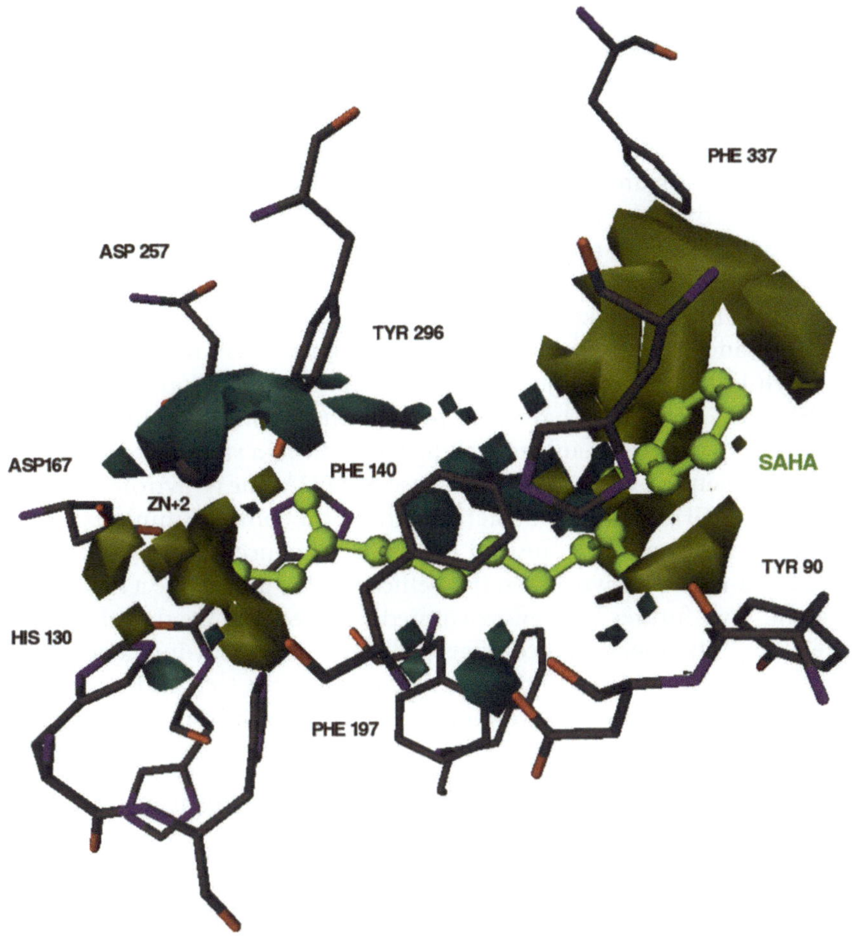

FIGURE 12

PLS coefficient maps superimposed to a 6 Å core of HDAC1. Bound conformation of SAHA is also displayed to aid interpretation.

on docking procedure and by minimizing the obtained complexes using a virtual HDAC1 model; furthermore, considering the previously cited HD2/HDAC1 IC_{50} relationship, the 3-D QSAR models represent the chance to check for matching between the HDAC1 receptor channel residues and the 3-D QSAR maps. In Figure 12, a 6 Å core of HDAC1 residues is overlapped with the 3-D PLS coefficients maps, and the close match between the HDAC1 residues and the PLS polyhedrons is evident.

The biggest, most important polyhedrons are close to crucial residues of the HDAC1 catalytic core: the big cyan polyhedron is situated near or over the Zn ion, Asp257, Asp167, and Tyr296 residues, likely identifying the region more involved in electrostatic interactions between ligand and enzyme. Analysis of the TSA/HDAC1 complex is quite straightforward; the cyan big contour expresses a favorable interaction, besides Zn and the ZBG group, between hydroxamate CO and the Tyr296 hydroxyl group. On the other hand, the yellow polyhedrons, overlapping the ZBG hydroxyl group, point toward His130, suggesting that a hydrogen bond could occur. This implies that to display an appreciable activity an inhibitor should bear both hydrogen bond acceptor and donor to support these interactions. Smaller yellow and cyan polyhedrons are spread along the receptor cavity underlying the importance of this region. This enzyme pocket, similar to a tube, has a wideness of 8 Å delimited by Phe197 and Phe140 as possible anchor points where the inhibitor could insert their linker portions (joining the ZBG and the cap groups). In this context, the APHA derivatives have their pyrrole rings (part of the linker portion) between the two phenylalanines making favorable π–π stacking interactions. If the linker moiety is branched or in the case of APHAs the N-pyrrole substituent is bigger than CH3, because of steric clash, upon complex minimization the ligand moves the entrance outward leading to a reduced metal chelation and lowered anti-HDAC activity, even if an efficient ZBG group like a hydroxamate is present. Two further big yellow-colored polyhedrons overlap the residues in the HDAC channel entrance, suggesting that important interactions occur between the enzyme rim and the ligand cap region. In fact all the most active compounds display their hydrophobic caps in this zone to make favorable interactions with either Phe337 or Tyr90.

This GRID/GOLPE SB 3-D QSAR study was the very first of an enlarged heterogeneous training set of 103 HDAC inhibitors.

2.5 LB and SB Studies of HDACi Class I/II Selectivity

In 2003, class II-selective HDAC inhibitors (Mai et al., 2003a) were discovered by chemical manipulation performed on the APHA **3a** template. To evaluate the HDAC class I/II selectivity of (aryloxopropenyl)pyrrolyl hydroxamides, the new derivatives were tested against two maize deacetylases, homologs enzymes of mammalian class I (HD1-B) and class II (HD1-A) HDACs (Brosch et al., 1992; 1996; Kolle et al., 1999; Lechner et al., 2000). As previously reported the maize system offered the advantage of available completely detached enzyme active forms, while human HDACs act as multiprotein and, in some cases, multi-HDAC containing complexes (Curtin and Glaser, 2003). Moreover, maize enzymes share acceptable sequence similarities with the human enzymes and are easier to purify. For these reasons, maize systems were used as a working horse to screen for newly synthesized compounds. Indeed properly substituted APHA-related derivatives appeared to be highly selective against the class II HDAC homologous HD1-A.

To investigate the structure-selectivity relationships, these GRID-independent descriptors (GRIND) of ALMOND software were used to derive 3-D QSAR models (Ragno et al., 2008). The training set 3-D structures were generated using the experimental bound conformation of TSA and SAHA as templates and then subjected to a simulated annealing (SA) procedure in continuum solvent simulation (GBSA). At first glance minima conformations divided in the training set molecules resulted in two clusters: straight-shaped conformations (Figure 13(c)) and bow-shaped conformations (Figure 13(b)). Further analysis of the dihedral angle Φ (Figure 13(a)) indicated the highest absolute values of the Φ dihedral angles were found associated with the most potent derivatives suggesting it as structural feature to discriminate HD1-B or HD1-A inhibitory activity (Figure 14).

The SA-derived conformations were then used to generate the GRIND descriptors and PLS models were generated and optimized by means of an iterative fractional factorial design (FFD) variable selection procedure while using five random groups for the cross-validations. Using as dependent variable the compounds HD1-B pIC_{50}, a series of sequential FFD selections led the definition of a final model with 223 descriptive variables and r^2 and q^2 values of 0.97 and 0.81, respectively (Figure 15). Analogously, a PLS model for HD1-A was derived with 181 descriptive variables and r^2 and q^2 values of 0.98 and 0.85, respectively.

For both models, PLS coefficient histograms were used to identify the most important pairs of nodes contributing negatively or positively to the anti-HD1-B or HD1-A activities. Finally to test the validity of these models, a test set was compiled with 14 unpublished (aryloxopropenyl)pyrrolyl hydroxyamides analogs and the HD1-B and HD1-A inhibitory activities were predicted. Interestingly HD1-B versus HD1-A selectivity of newly synthesized (aryloxopropenyl)pyrrolyl hydroxyamides were correctly predicted. Along the good predictions, further inspection revealed all the derivatives included in the test set

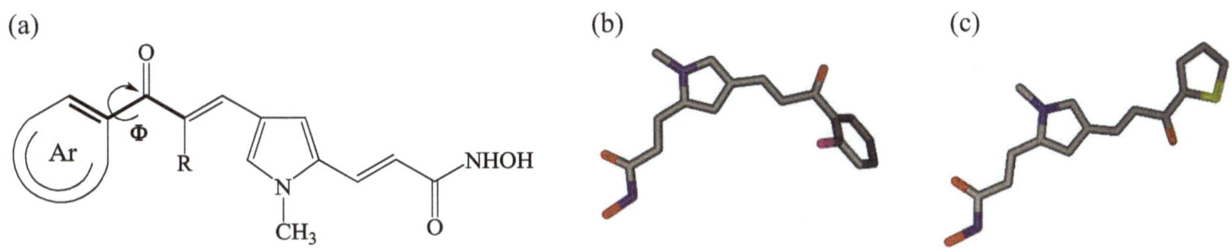

FIGURE 13
(a) General formula; (b) HD1-B most selective derivative as example of folded-shaped molecule; (c) HD1-A most selective compound as example of straight-shaped molecule.

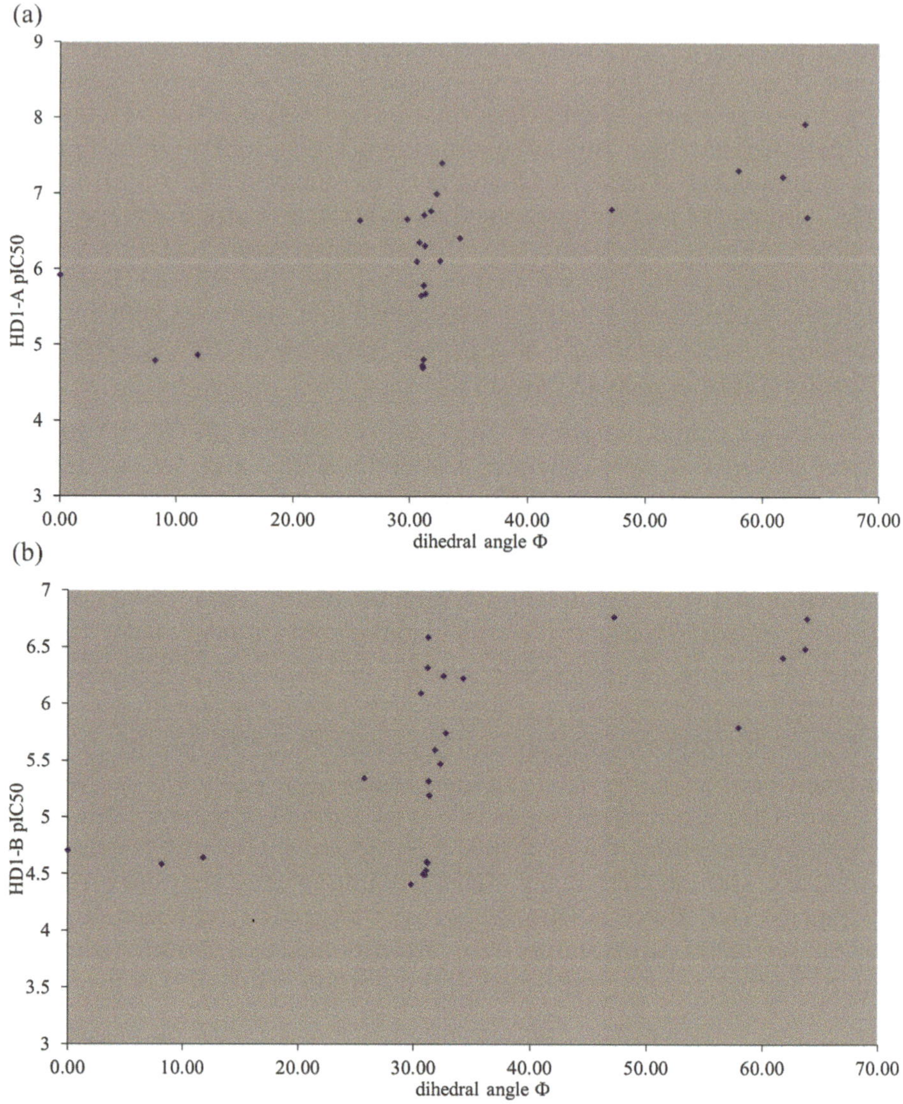

FIGURE 14

Graphical correlation between biological activities (pIC_{50}) and Φ dihedral angles (a) relative to HD1-B and (b) relative to HD1-A.

displayed an overall tendency to adopt straight SA conformations (Figure 16) and since they were all selective endowed for class II HDAC this was found in good agreement with the straighten/bowed hypothesis (Figure 17).

To further inspect structural features responsible for selectivity, molecular dynamics (MD) calculations were carried out on homology models for HD1-A and HD1-B.

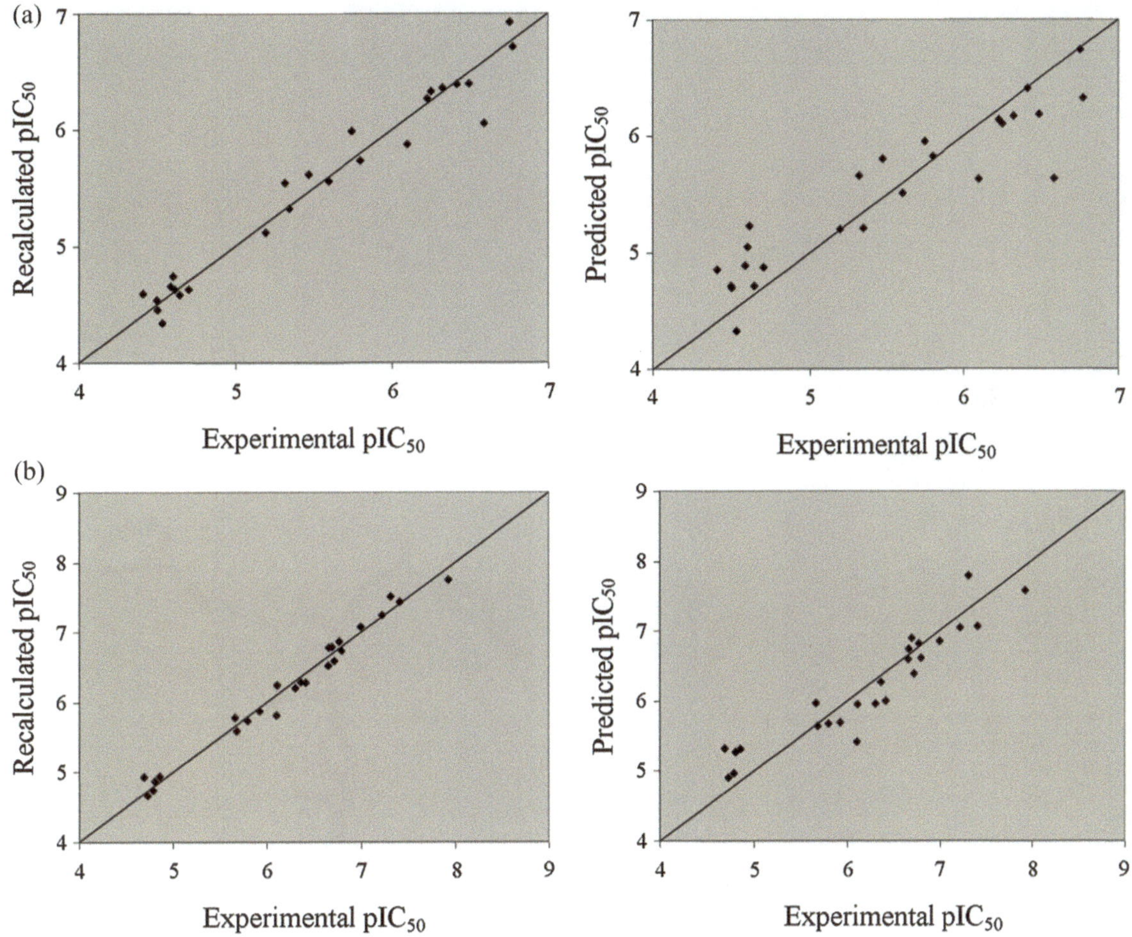

FIGURE 15

Fitting plot between the experimental and the calculated pIC_{50}. Right: cross-validation plot between the experimental and the predicted pIC_{50} (four principal components) for the HD1-B activity; (a) relative to HD1-B and (b) relative to HD1-A.

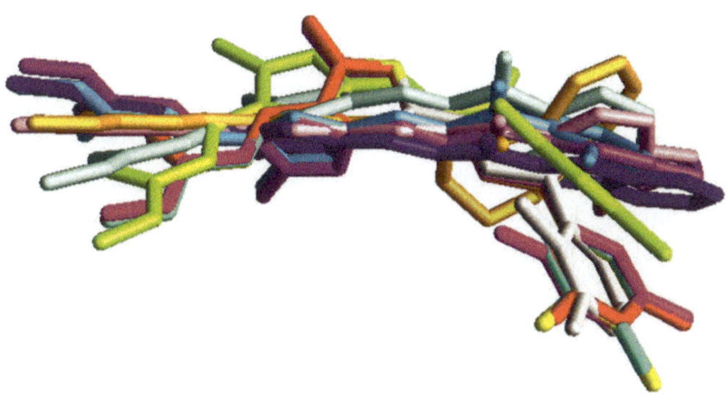

FIGURE 16

SA conformations of the test set compounds.

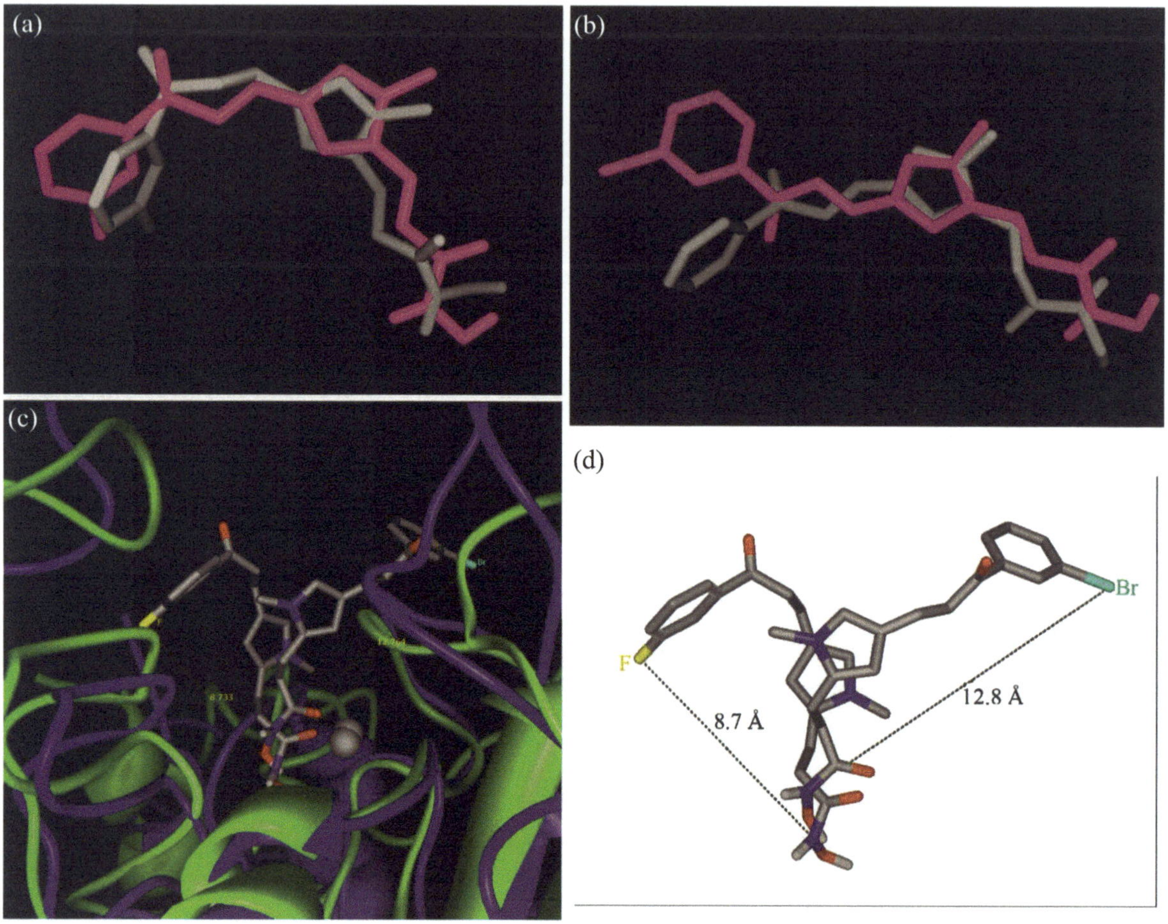

FIGURE 17

(a and b) Superimposition of the SA generated conformations (magenta) and the AutoDock proposed bound conformations (white) for HD1-B most selective derivative (left side) and HD1-A most selective compound (right side). (c) Docked conformations and the tube representation of HD1-B (green) and HD1-A (blue). (d) Overlapped docked conformations as obtained from SB alignment of the corresponding complexes.

By means of AutoDock 4, molecular docking was then carried out using minimized MD average structures with either HD1-B most selective derivative or HD1-A most selective compound. Direct comparison of the obtained binding modes revealed differences that reflected structural dissimilarities observed comparing HD1-A and HD1-B MD studies. Further inspections highlighted the likely ligand/protein interactions responsible for the experimental selectivities.

In conclusion, the 3-D QSAR models shed some light on the interpretation of the selectivity issue for the class I and II HDAC inhibitors, although for class I selectivity some uncertainty is still present due to the fact that highly HD1-B-selective compounds were not available.

2.6 A Comprehensive SBDD Study for All Zn-Based Human HDACs

The majority of HDACi reported so far inhibit all HDAC isoforms nonspecifically (so-called pan-inhibitors). SAHA and TSA are canonical pan-inhibitors, influencing all HDACs activity roughly equivalent potency (Khan et al., 2008). Selective HDAC inhibitors, which affect either a single HDAC isoform (isoform-selective HDACi) or several isoforms within a single class (class-selective HDACi), would be ideal chemical tools to elucidate the individual functions of each HDAC isoform. Definitely, selective HDAC inhibitor availability would be of great benefit to define molecular mechanisms connecting HDAC activity to cancer formation or other pathologies. Moreover, it is very likely that class-selective or isoform-selective HDAC inhibitors would provide more effective chemotherapy than pan-inhibitors. Although of clear usefulness and advantages, the design of selective HDAC inhibitors is highly challenging due to either high sequence similarity within the active sites of the isoforms or limited crystallographic data. Regardless of these challenges, class-selective and isoform-selective HDAC inhibitors have been reported within the past decade. Recently, at RCMD a pure structure-based 3-D QSAR protocol was set using as much as possible all HDAC available experimental 3-D structural data. The procedure was based on the Ortiz's pioneeristic work (Ortiz et al., 1995) called COMBINE, which relies on the assumption that the protein–receptor binding free energy, ΔG, can be approximated by a weighted sum of n terms, Δu, each describing the change in property u upon binding as described by the following equation.

$$\Delta G = \sum_{i=1}^{n} w_i \, \Delta u_i + C$$

Using this expression, biological activities can be predicted by assuming that these quantities are linear functions of ΔG. A revisited COMBINEr approach, previously derived and assessed (Ballante et al., 2012; Rotili et al., 2014), was used in place of the original methodology to develop a comprehensive model for virtually any HDAC inhibitor/isoform complex either experimental or derived by molecular modeling (homology modeling, docking, etc.). To build the model experimental 3-D structures of HDAC-2, -4, -7, and -8 cocrystallized with different ligands were retrieved from the Protein Data Bank, while HDAC-1, -3, -5, -6-1 (histone deacetylase domain), -6-2 (tubulin deacetylase domain), -9, -10, and -11 were obtained by homology modeling. Along with the protein structure, a set of 15 known HDAC inhibitors was used to compile a training set of 94 different complexes, covering as much as possible all the 11 isoforms and using biological data all determined in the same laboratory (Blackwell et al., 2008).

After training set complexes minimization and structure-based alignment the ligand/enzyme per-residues interaction were calculated by means of the AutoDock suite's autogrid module and then coupled with the experimental pK_i values for the subsequent statistical treatment. Partial least square calculation was used to derive the SB 3-D QSAR models, which were subjected to internal (cross-validation) and external (test-set) assessments. For internal validation either leave one out (LOO) or leave some out (LSO) techniques were used to measure models' robustness and eventual presence of chance correlation in Y-scrambling procedure. For the external validation three different test sets were used: a modeled test set (MTS), containing 76 complexes derived from a series of 10 HDACi docked into the 11 isoenzymes; a crystal test set (CTS) composed of nonhuman HDACs and available from PDB; and a test set composed of largazole bound to four HDAC isoforms (largazole test set).

Among the derived COMBINEr models (Table 8), the one developed using electrostatic and desolvation energies showed the best statistical profile, with conventional squared correlation coefficient (r^2) and lowest standard deviation error of calculation (SDEC) values: of 0.80 and 0.73, respectively (Figure 18). LOO and LSO cross-validation lead to the same statistical coefficient (q^2=0.76) with only two principal components (PCs) and suggesting a high level of robustness. Finally, the Y-scrambling procedure revealed an absolute absence of chance correlation.

An inspection of LSO cross-validation predictions, focusing on either inhibitor structures or HDAC isoform, emphasizes COMBINEr model ability to retain the correlation within various subgroups, either considering all the training-set

Table 8 Statistical Coefficients for Derived COMBINEr Models

#	Field	PC[a]	r^{2b}	SDEC[c]	q^2_{LSO}[d]	$SDEP_{LSO}$[e]	q^2_{LOO}[f]	$SDEP_{LOO}$[g]	Scrambled q^{2h}	
									% Positive Values	Max. Value
1	ELE	2	0.69	0.91	0.67	0.94	0.68	0.93	5	0.07
2	STE	2	0.27	1.40	0.14	1.52	0.15	1.51	n.d.	n.d.
3	DRY	2	0.46	1.21	0.34	1.33	0.36	1.32	n.d.	n.d.
4	ELE+STE	2	0.74	0.84	0.68	0.93	0.68	0.93	2	0.05
5	ELE+DRY	2	0.80	0.73	0.76	0.81	0.76	0.81	6	0.08
6	STE+DRY	3	0.54	1.11	0.33	1.34	0.35	1.33	n.d.	n.d.
7	ELE+DRY+STE	2	0.77	0.78	0.72	0.87	0.72	0.87	4	0.04

[a]*Number of principal components used.*
[b]*Conventional squared correlation coefficient.*
[c]*Standard deviation error of recalculation.*
[d]*Cross-validation coefficient for the LSO method.*
[e]*Standard deviation error of prediction for the LSO cross-validation with five groups.*
[f]*Cross-validation coefficient for the LOO method.*
[g]*Standard deviation error of prediction for the LOO cross-validation.*
[h]*Percentage of positive cross-validation coefficients for the scrambling method and its maximum values.*

inhibitors versus each HDAC (correlation of anti-HDAC inhibitors potency, Figure 19) or considering each inhibitor binding into different HDAC isoforms (correlation of selectivity, Figure 19). An interesting aspect of the obtained model is its deep inspections through istograms built using PLS coefficients (PLS coeffs), standard deviation errors (StDev), and PLS coeff × StDev values (Figure 20).

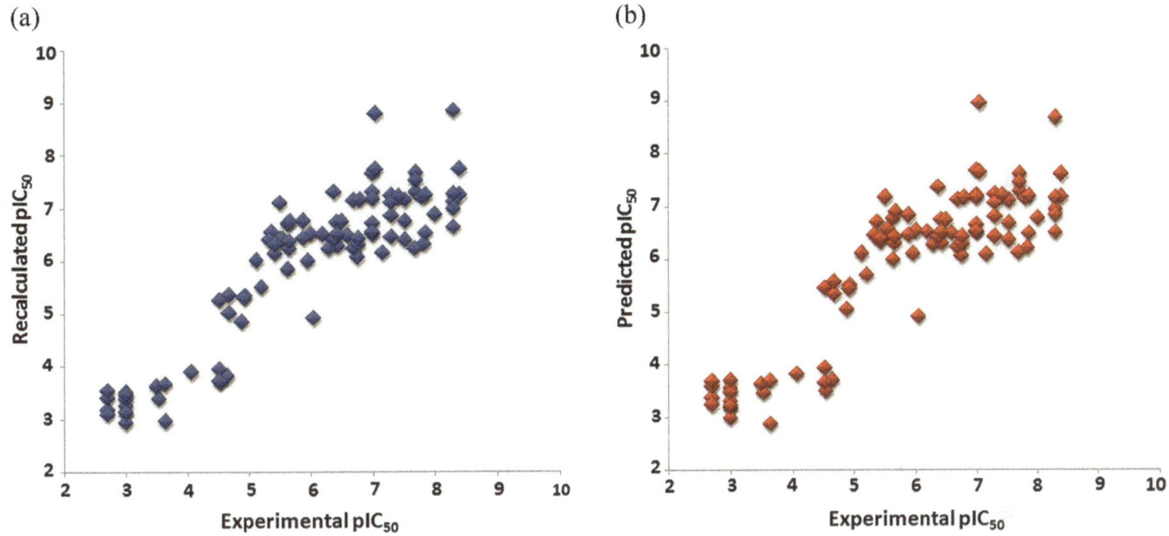

FIGURE 18

Fitting (a) and R5G-LSO cross-validation (b) plots, for the ELE + DRY model (Table 7).

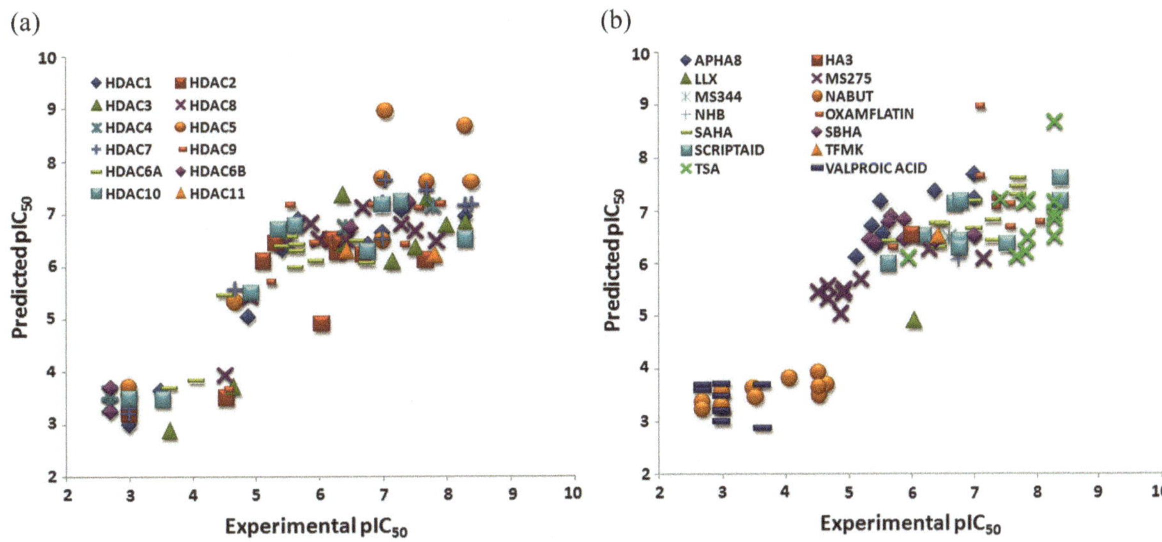

FIGURE 19

LSO Cross-validation predictions depicted by HDAC isoforms (a) and by inhibitor (b).

Interpretation of COMBINEr models can identify the residues relevant for differences in activity and quantify their relative importance. To this aim, the PLS-coefficient (Figure 20) and activity-contribution plots are useful. The former provides a global view and gives information on the entire training set. The sign and the magnitude of PLS coefficient of an energy term multiplied by the corresponding energy term (field) shows the influence of the corresponding residue on ligand binding. Through plots in Figure 20, isoform selectivity could be explained with only 34 residues and enzyme inhibiting power with only five residues (Figure 21); the graphical representation (Figure 21) overlapped exceptionally well with the generally accepted HDACi pharmacophore model. Fine-tuning of both potency and selectivity result from other contributions, and therefore, each isoform needs to be inspected individually.

Another aspect of the model is the potential to have a global view of the overall interactions by plotting the sums for either the ELE or DRY activity contributions for each training-set complex (Figure 22). From this results the DRY field contribution mostly modulating the activities (bigger red bars on bulkier compounds), while ELE contribution was shown to be more important in modulating the low activities of small-sized inhibitors (bigger blue bars on short fatty acid inhibitors), butirrate (NABUT), and valproic acid (VA), due to missing interactions with residue at the enzymes' rims.

Focusing on the ELE interactions, plots in Figure 20 indicated the ELE field to be highly important in the inner part (black-labeled residues) of the HDACs catalytic domains and relatively insignificant for residues forming the channel (blue-labeled residues in Figure 20) and those at the entrance rim (red-labeled residues in Figure 20). Regarding the channel-forming residues, 294 (at the edge between the channel and the bottom of the HDAC-binding sites) displayed the highest values in all three plots of Figure 20. Indeed, this residue (a conserved histidine for all HDACs) is primarily involved in modulating the potency between small inhibitors (NABUT and VA) and channel-filling inhibitors (i.e., SAHA and TSA). For NABUT and VA, diminished interactions with residue 294 accounted for 0.8 to 1.0 decrement in activity. To some extent, the fact that either NABUT or VA are carboxylic acids, which indicates that higher negative charge (NABUT and VA were modeled as carboxylates, thus bearing a discrete negative formal charge) in proximity to residue 294 is unproductive. Among the other channel-forming selected residues, 262 (always a Gly), 263 (mostly a Phe), and 264 (always a Cys), the most interesting is residue 263, which is involved in modulating the activity decrement for small compounds, in particular for VA.

Most of the ELE-selected residues (18 out of 27) are in the deep part of the channels around the catalytic Zn. Of particular interest are residues involved in the HDAC catalytic process conserved among the 12 isoforms, as follows: residues 253 (His), 254 (His), 292 (Asp), 392 (Asp), and 571 (Zn). In general the activity contribution associated with these five residues modulates the activity decrement

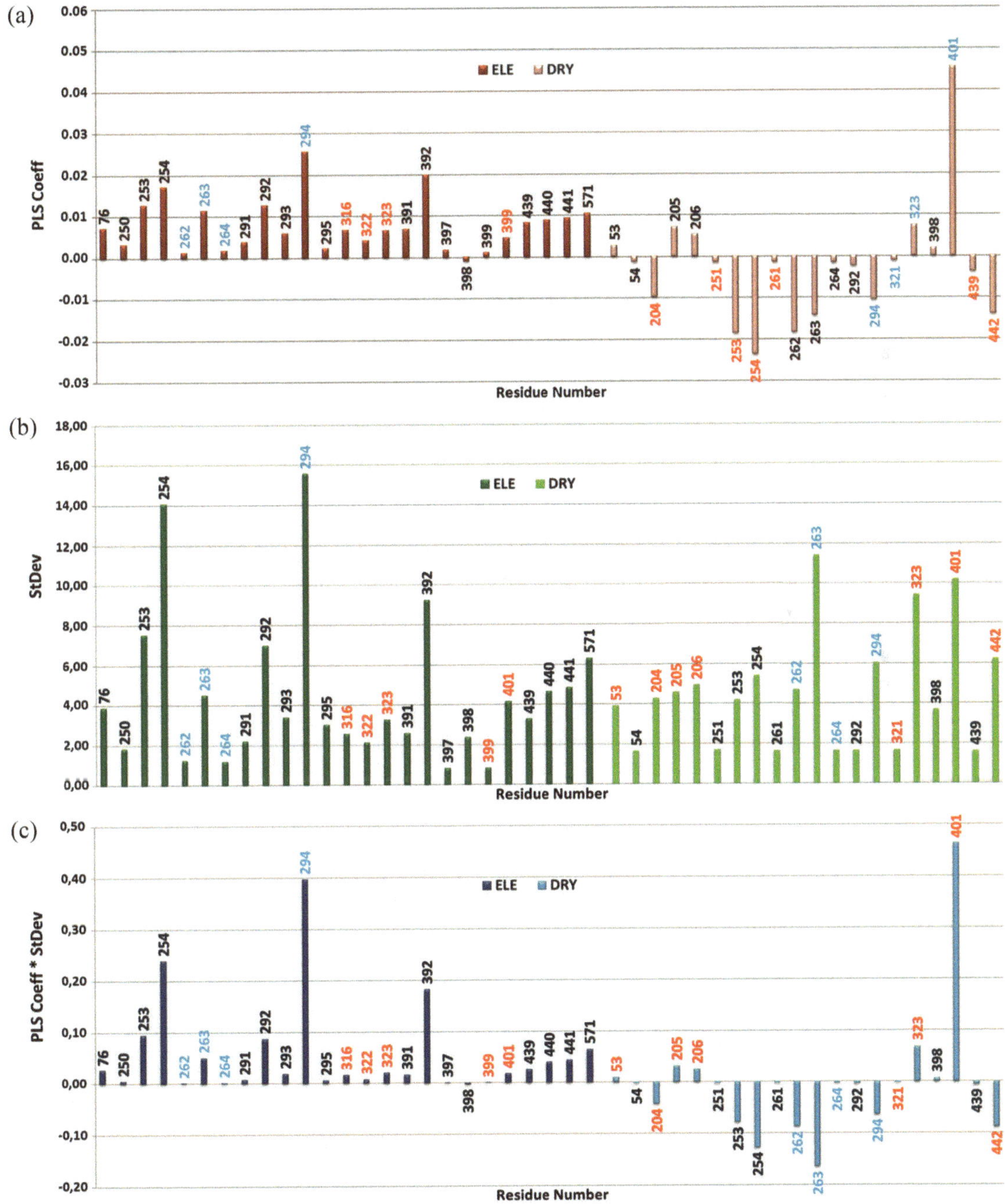

FIGURE 20

PLS coeff (a), StDev (b), and PLS coeff × StDev (c) plots for the ELE + DRY COMBINEr model. The residues were selected using a PLS coefficient threshold value of 0.001. For residue numbering refer to Silvestri et al. (2012). The residues located in the rim region are colored red, while the residues forming the central channel are blue, and those in proximity to the catalytic Zn ion are black.

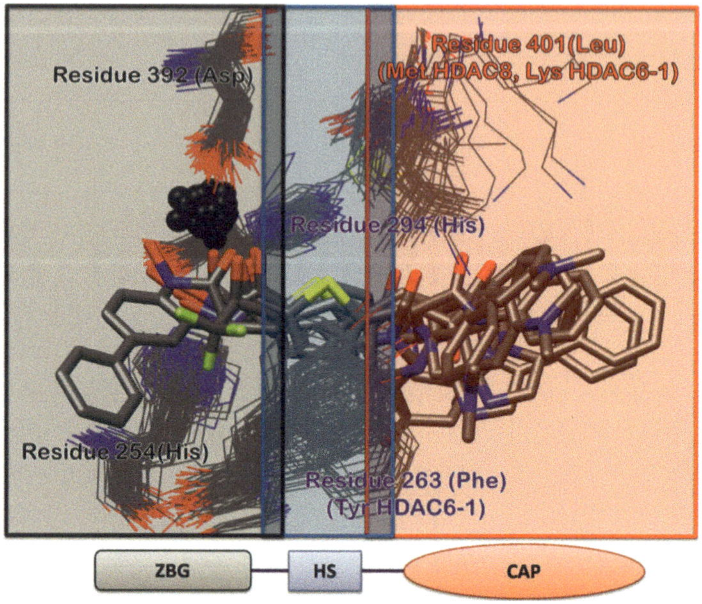

FIGURE 21

The four most important residues (MIRs, see text) from the COMBINEr model analysis. The labels and regions are color coded: the residues in the HDAC's rim region are shown in red, those forming the central tube channel are in blue, and those in the proximity of the catalytic Zn ion are in black. The Zn-binding region (black line box), the connection region (blue line box), and the CAP region (red line box) are also highlighted to recall the HDAC pharmacophore model depicted at the bottom. ZBG: Zn-binding group, HS: hydrophobic spacer, CAP: hydrophobic capping group.

for carboxylate-based ZBGs. As examples, residues 253 (SAHA in HDAC-1) and 254 (SAHA in HDAC-3, HDAC-4, and HDAC-6-2; and SBHA in HDAC-4 and HDAC-8) are associated with a positive activity contribution of about 0.1. On the other hand the DRY field gives a rough estimation of steric interactions and compared to ELE a totally different and more complicated scenario was observed on the relative importance of each residue. In general, the most important modulating interaction relates to 401Leu, replaced by Met in HDAC-8 or by Lys in HDAC-6-1. Upon deeper inspection (not considering the small-molecule complexes NABUT, VA, and NHB), only 27 of 94 activities were modulated by residue 401 with activity contributions ranging between 0.7 and 2.13. Apart from residue 401, the other residues play a major role in modulating the overall biological activities. Seven out of 10 residues (204, 253, 254, 262, 263, 294, and 442) are related to negative modulating values, while the other three (205, 206, and 323) are positive modulators. Residue 263 (Tyr for HDAC-6-1 and Phe for the others) located in the wall of the channel shows the largest range with larger negative values. No specific pattern is detected for residue 263 in modulating regarding the different enzyme classes or inhibitor structures. The small inhibitor NABUT is not influenced by residue 263, likely due to the fact that there are no

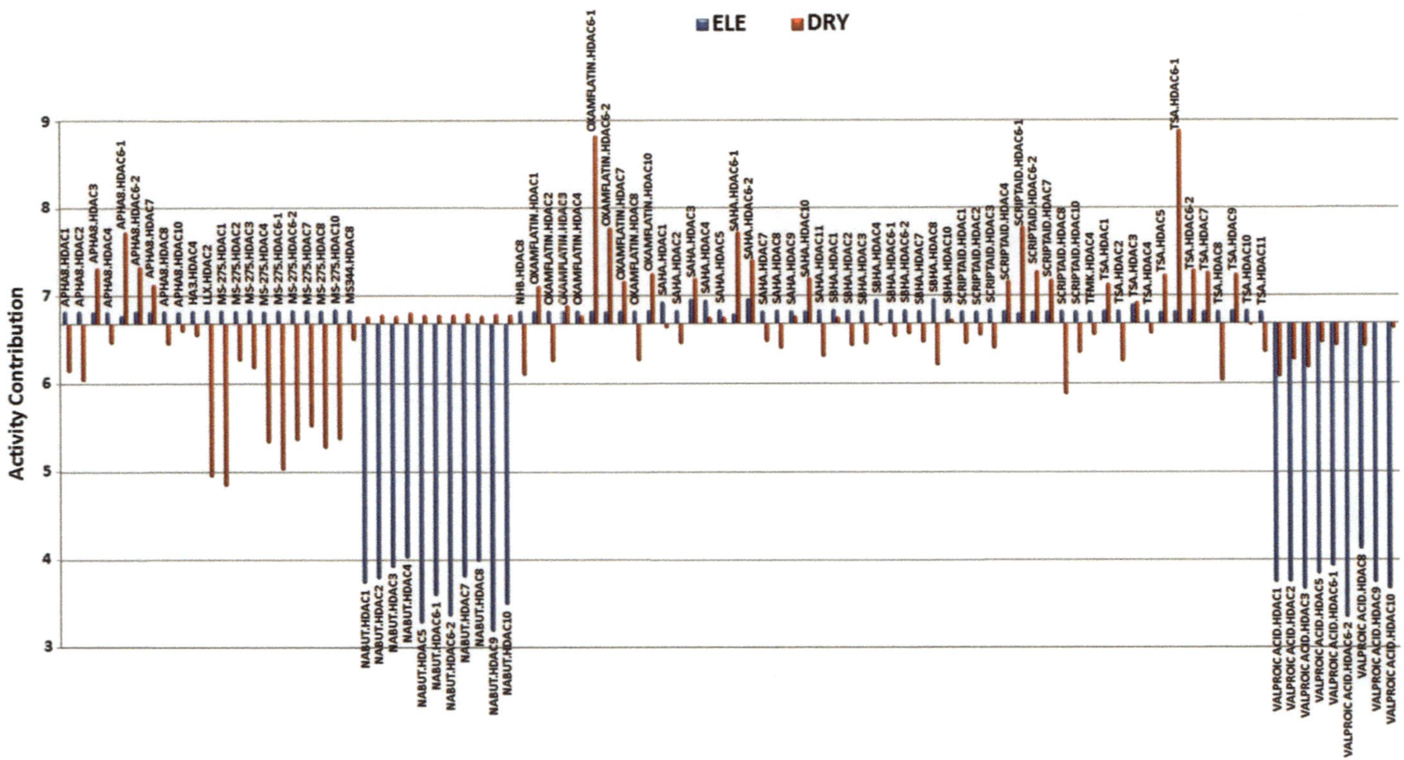

FIGURE 22

ELE and DRY total-activity contributions. The constant (PLS intercept) of the COMBINEr equation has a value of 6.68. The sum of ELE and DRY contribution is obtained by the algebraic sum of all per-residue contributions.

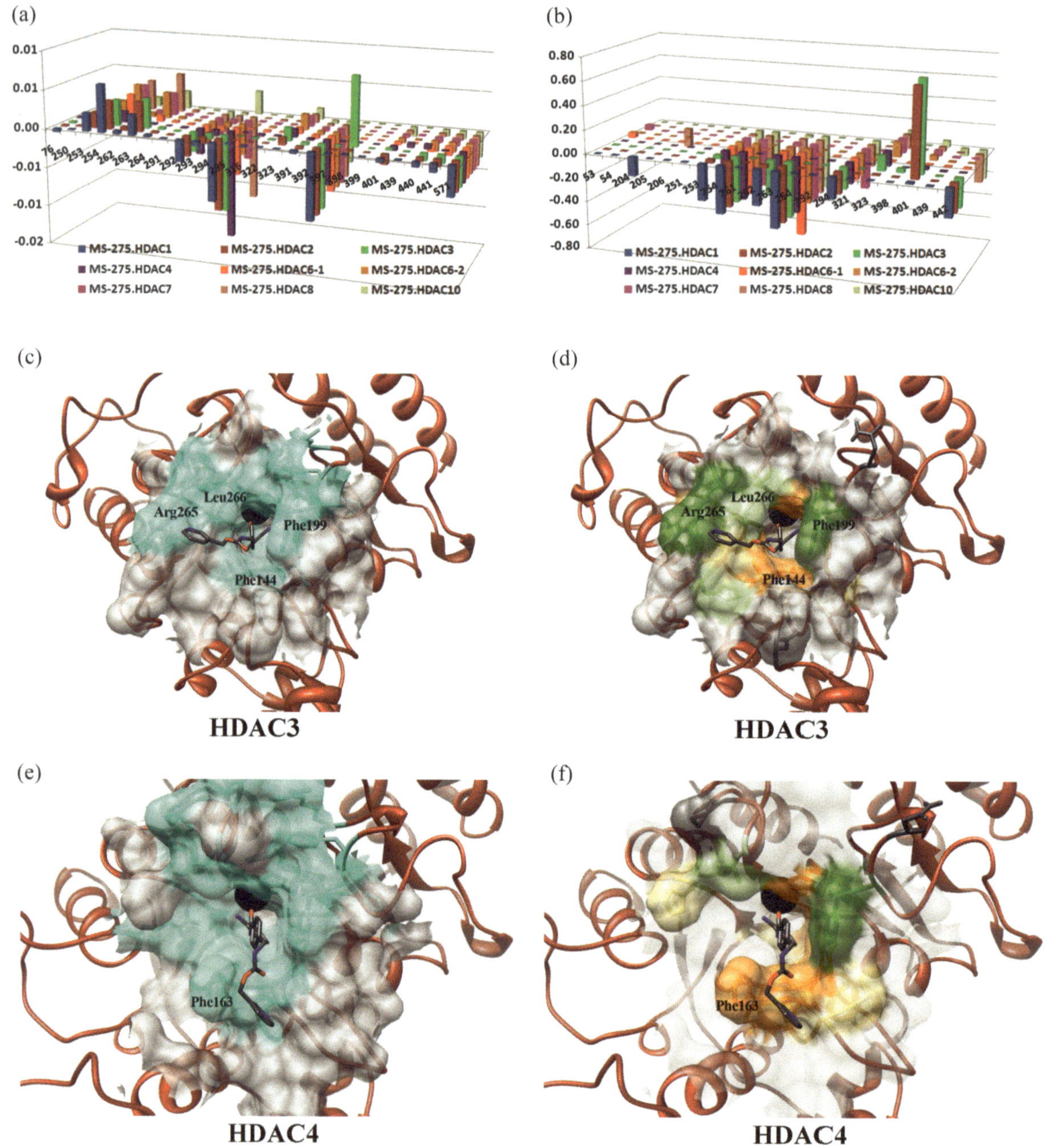

FIGURE 23

Activity contributions (a and b) for MS-275 and their graphical representations (c–e). (a), (c), and (e) in the left side account for the ELE field. The DRY is depicted on the right ((b), (d), and (f)). Residue surfaces are color coded: for ELE blue-based surfaces indicate a positive contribution (light blue if the contribution is less than 50% of maximum contribution for a given residue; dark blue indicate areas with higher contributions); red-based surfaces indicate negative contributions (light red for absolute contribution less than 50% of the corresponding residue; dark red for higher percentage of negative contribution). For the DRY field positive contributions are indicated in green (dark green: contribution higher than 50% of the maximum activity contribution; light green for less contribution); yellow colors are used to indicate negative DRY contributions (dark yellow: absolute contribution higher than 50% of the maximum activity contribution; light yellow for low negative contributions). Dark gray surfaces indicate zero contribution, while light gray are residues with PLS coefficients lower than 0.001 (see earlier in the model interpretation section). Only residues cited in the text are labeled.

direct contacts. Residue 442 (His for Class IIa and Tyr for the others) located in the bottom of the binding sites showed the largest range with larger negative values associated mainly with class I complexes, with particular allusion to HDAC-8 thus suggesting that interaction with this residue might be used to selectively avoid inhibition of HDAC-8. Residue 254 (His in the Zn-binding region) displayed the second-highest StDev value and negatively modulated mainly non-hydroxamate inhibitors making complexes (LLX, MS-275, and VA) in agreement with ELE interaction analysis. Residue 204 (of various nature present on the rim of six out of 12 HDACs) and 294 (His, a channel-forming residue) are also negative-modulating residues, but the associated low standard deviation indicates that no selectivity can be attributed to the DRY interactions; residue 204 seems to specifically modulate the inhibitory activity for HDAC-8 complexes.

Considering the high correlation between DRY and STE, interactions with residues 263 and 294 are of crucial importance for optimal fitting of inhibitors in the HDAC channels. Regarding the three DRY positive-modulating residues, 323, an aromatic side-chain-bearing residue missing in HDAC-1 and HDAC-11, shows the highest maximum-activity contribution and larger variability; maximum-activity contributions occur with APHA8 and TSA binding to either class I or class II enzymes. The other highly positively contributing residue 205 is peculiar for HDAC-6-1 (Tyr85) and thus uniquely modulates inhibition of this enzyme. Interaction and activity-contribution analyses suggested that useful insight into structural determinants exists for both HDAC isoforms and their inhibitors and therefore the COMBINEr model can be used to help optimize isoform-specific inhibitors. Derivation of rules to guide the structural basis for isoform selectivity required single analysis for each specific isoform model. A particular model's potential use could be highlighted by means of inhibitors seeking in term of maximum isoform selectivity, which led to select MS-275 and SCRIPTAID as template examples. MS-275 and SCRIPTAID displayed the largest activity variability, being MS-275 partially selective for class I HDACs and SCRIPTAID partially selective for class II.

Regarding MS-275, interactions with HDAC-3 (Class I HDAC) and class IIa HDAC-4 analyzed through the model can rationalize the higher activity of MS-275 for class I enzymes.

As shown in Figure 23, the residues responsible for this activity difference can be clearly highlighted. In fact, in view of electrostatic interactions, it is evident that low correlation with activity exists, and only gray or light blue surfaces are present in Figure 23(c) and (e). On the other hand, the DRY field seems very sensitive (Figure 23(d) and (f)) since high color variation clearly indicates those residues responsible for the higher activity of MS-275 against HDAC-3 (Phe199 and Arg265 are dark green colored). Other green-colored residues are also located around the rim, for example, Leu266. A few residues are colored yellow, such as residue 263 (Phe144 in Figure 23(d)), indicating that MS-275 anti-HDAC-3 activity could be improved by optimizing the interactions in the enzyme channel. Concerning MS-275/HDAC-4 complex, many DRY surfaces have turned from green to yellow, thus highlighting that residue 263 (HDAC4-Phe163) plays a major role in decreasing activity with many residues showing zero activity contribution.

In opposition to MS-275, SCRIPTAID was chosen as a selective class II inhibitor. Likewise for MS-275, the electrostatic interactions differentiated when comparing the activity contributions of HDAC-6 and HDAC-8 (Figure 24). Definitely, Figure 24(a) clearly indicates that the ELE contributions are below 0.02. So correspondingly to MS-275, DRY terms help rationalize the inhibitory activities of SCRIPTAID with HDAC-6 and HDAC-8. Most differences are located in the rim zone; specifically Lys267 in HDAC-6 is responsible for a strong positive contribution, while Met261, its counterpart in HDAC-8, displays a much smaller contribution.

Once the SB 3D QSAR models were assessed and interpreted, for future DD the first phase to be accomplished refers to how a new compound could be docked into the various HDACs with the minimum error. Cross-docking experiments were conducted by means of AutoDock and AutoDockVina programs to check which ones would have displayed the best docking accuracy (DA). AutoDockVina was found to be more accurate, displaying a docking accuracy of 75% for the best cluster poses (Table 9), therefore was selected as the preferred docking tool to study the binding modes of MTS test set. Docking experiments on MTS were done in binding pockets of all minimized HDAC structures and for each isoform all the poses were collected and clustered to select, as indicated by the docking assessment, the best cluster conformation.

A total of 76 HDAC-inhibitors complexes were compiled, and the ELE + DRY COMBINEr model was used to predict inhibitors activities. Figure 25 and Table 10 show the pIC_{50} predicted for the MTS external test set and statistical results (SDEPext and AAEP).

The model showed a good external predictivity with SDEP of 1.41 for the optimal principal components (2 PCs). Inspection of Figure 25 revealed that JMC-23 and MCL-4 were the worst predicted compounds. JMC-23 contains an oxime amide as a ZBG that can be interpreted as a modified version of the efficient hydroxamate moiety. This compound is a poor HDAC pan-inhibitor; the COMBINEr model fails to predict correctly 5 out of 11 activities. Regarding MCL-4, this is the hydroxamate version of MCL-3, while the latter is recognized as a very poor inhibitor with the correct trend, and MCL-4 was highly overpredicted in HDAC-4, HDAC-5, HDAC-7, and HDAC-9 complexes. Nevertheless the average pIC_{50} value for MCL-4 (exp = 5.18, pred = 6.31) was correctly calculated to be higher than that for MCL-3 (exp = 3.40, pred = 3.33).

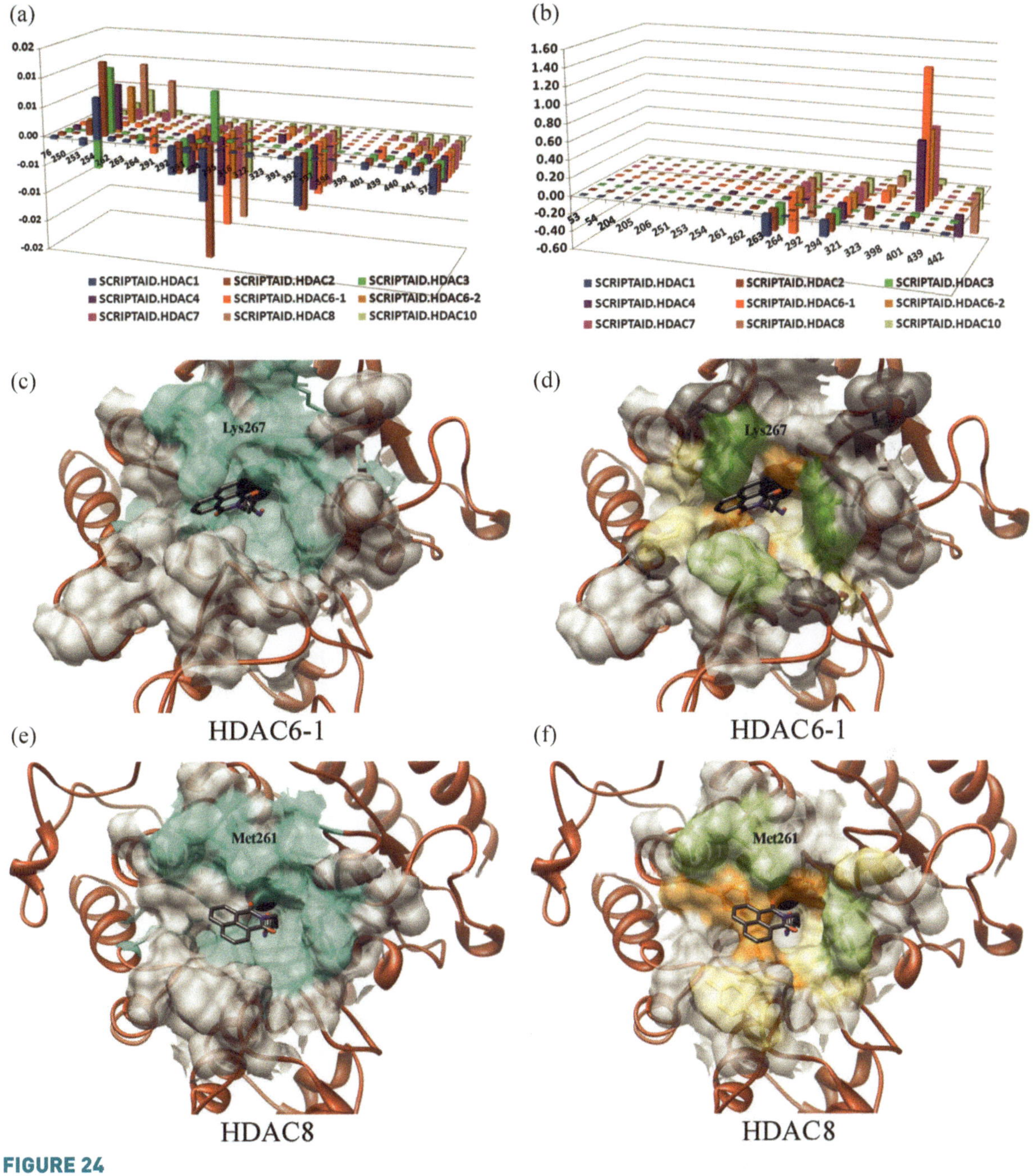

FIGURE 24

Activity contributions (a and b) for SCRIPTAID and their graphical representations (c–e). (a), (c), and (e) account for the ELE field. The DRY field is depicted on the right (b, d, and f). Residue surfaces are color coded: for the ELE blue-based surfaces indicate positive contributions (light blue if the contribution is less than 50% of maximum contribution for a given residue; dark blue indicates areas with higher contributions); red-based surfaces indicate negative contributions (light red for absolute contributions less than 50% of the corresponding residue; dark red for higher percentage of negative contributions). For the DRY field, positive contributions are indicated in green (dark green: contribution higher than 50% of the maximum activity contribution; light green for less contribution); yellow colors are used to indicate negative DRY contribution (dark yellow: absolute contribution higher than 50% of the maximum activity absolute contribution; light yellow for low negative contributions). Dark gray surfaces indicate zero contributions, while light gray are residues with PLS coefficients lower than 0.001 (see earlier in the model interpretation section).

Table 9 Redocking Results (RMSD)

Complex Name	AutoDock			AutoDockVina		
	Best Docked	Best Cluster	Best Fit	Best Docked	Best Cluster	Best Fit
LLX.HDAC2	0.48	0.48	0.48	0.24	0.24	0.24
HA3.HDAC4	5.25	4.76	4.4	3.87	2.34	1.93
TMFK.HDAC4	3.46	5.75	3.46	4.02	1.9	1.46
SAHA.HDAC7	10.36	10.36	2.18	2.45	2.45	1.88
TSA.HDAC7	6.06	6.06	1.4	2.19	2.19	1.21
APHA.HDAC8	5.4	2.26	2.26	1.43	1.43	1.43
SAHA.HDAC8	5.84	7.29	4.1	2.49	2.49	1.72
TSA.HDAC8	5.1	5.52	1.45	2.09	1.22	1.22
DA%	12.5	18.75	50	50	75	100

Values for best docked (the lowest energy docked conformation of the first cluster generated), best cluster (the lowest energy docked conformation of the most populated cluster), and best fit (the lowest energy conformation of the cluster showing the lowest RMSD value), obtained with the two docking programs.

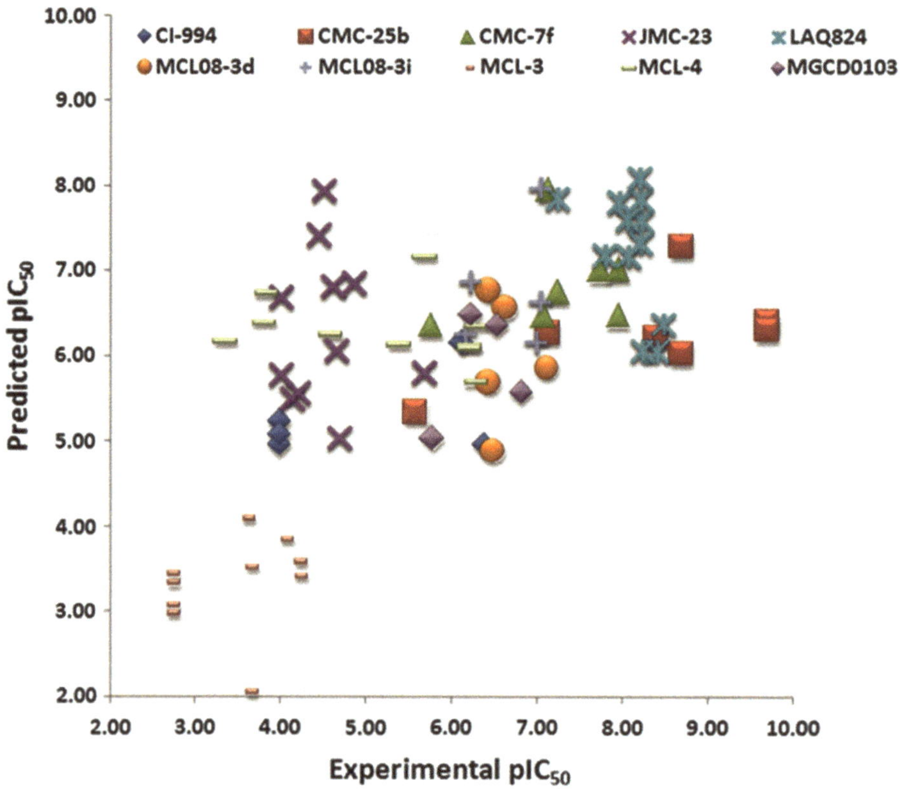

FIGURE 25

Experimental/predicted pIC_{50} for the MTS.

Table 10 Predicted pIC_{50} for the MTS

Principal Components	1	2	3	4	5
$SDEP_{ext}$	1.44	1.41	1.47	1.59	1.60
Average absolute error of prediction	1.13	1.10	1.16	1.25	1.25

The SDEP and the AAEP are reported for the first five PCs.

Comparisons of predictions for single HDAC isoforms (Figure 26) revealed that complexes of HDAC-2 and HDAC-3 were the best predicted with an AAEP of 0.53 and 0.65, respectively. Complexes related to HDAC-7, HDAC-9, HDAC-10, and HDAC-11 showed the highest AAEP values. For HDAC-9, HDAC-10, and HDAC-11, the worst predictions could be associated with the low number of complexes present in the training set. In general, the model was able to reproduce the activity of class I HDACs better than class II. Regarding HDAC-10 and HDAC-11, the smaller amount of experimental data in the training set was the probable cause for the failed activity-trend predictions (Figure 26, panels K and L). Notably the external SDEP value confirmed that the model at 2 PCs was indeed the most predictive as correctly indicated by the cross-validation runs.

The application of the COMBINEr model to the MTS proved the ability of the model in predicting the relative potency and the correct activity trend of a given series of inhibitors for 10 out of 12 HDAC isoforms even when the binding conformations of the test set inhibitors were obtained from docking.

Application of the COMBINEr model to the CTS test set further confirmed its good predictive ability of the reporting the correct trend and activity potencies with an AAEP values of only 0.71. This second test set proved that availability of correct bound conformations as those experimentally derived allow low error of prediction and furthermore the inclusion of bacterial HDACs (HDAH and HDLP) indicated the model useful to predict activities against nonhuman HDACs, potentially useful in the search for antiparasitic, antifungal, and antibacterial therapeutics. Finally, the third test set was used to see the model's ability to predict bioactivity of cyclotetrapeptide-like inhibitors like largazole. To some extent, the COMBINEr model was able to recognize the relative potency of largazole for HDAC-1, HDAC-2, and HDAC-6-1; while for HDAC-3, the predicted pIC_{50} was underestimated, indicating that further modeling of this class of inhibitors is necessary.

Concluding this section, this was the first example, and actually no other studies like this have been reported yet, of structure-based 3-D QSAR model using comparative binding-energy analysis that focused on the selectivity of the 11 human Zn-based histone deacetylase. The derived COMBINEr model showed

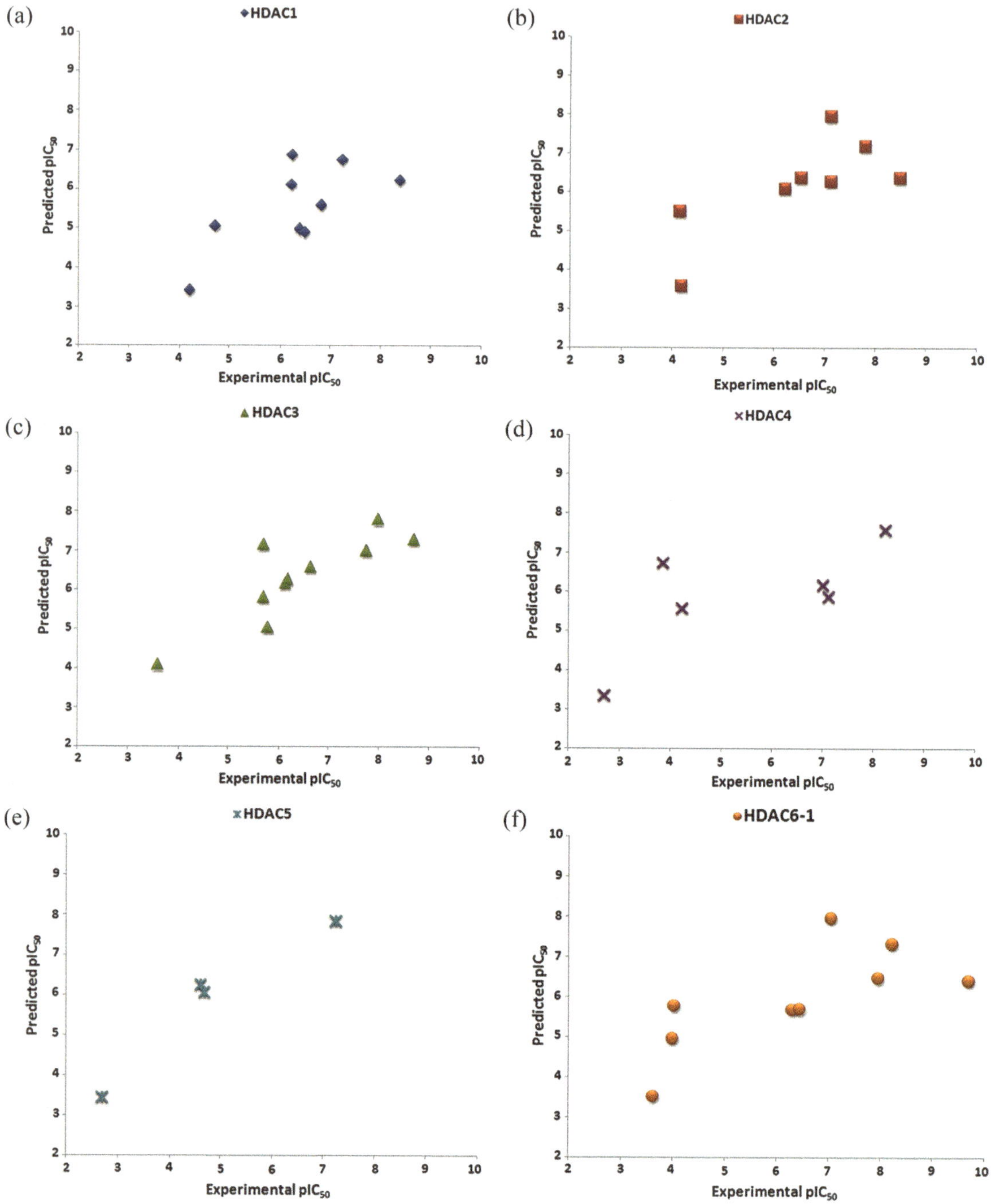

FIGURE 26
MTS predictions for single HDAC isoforms.

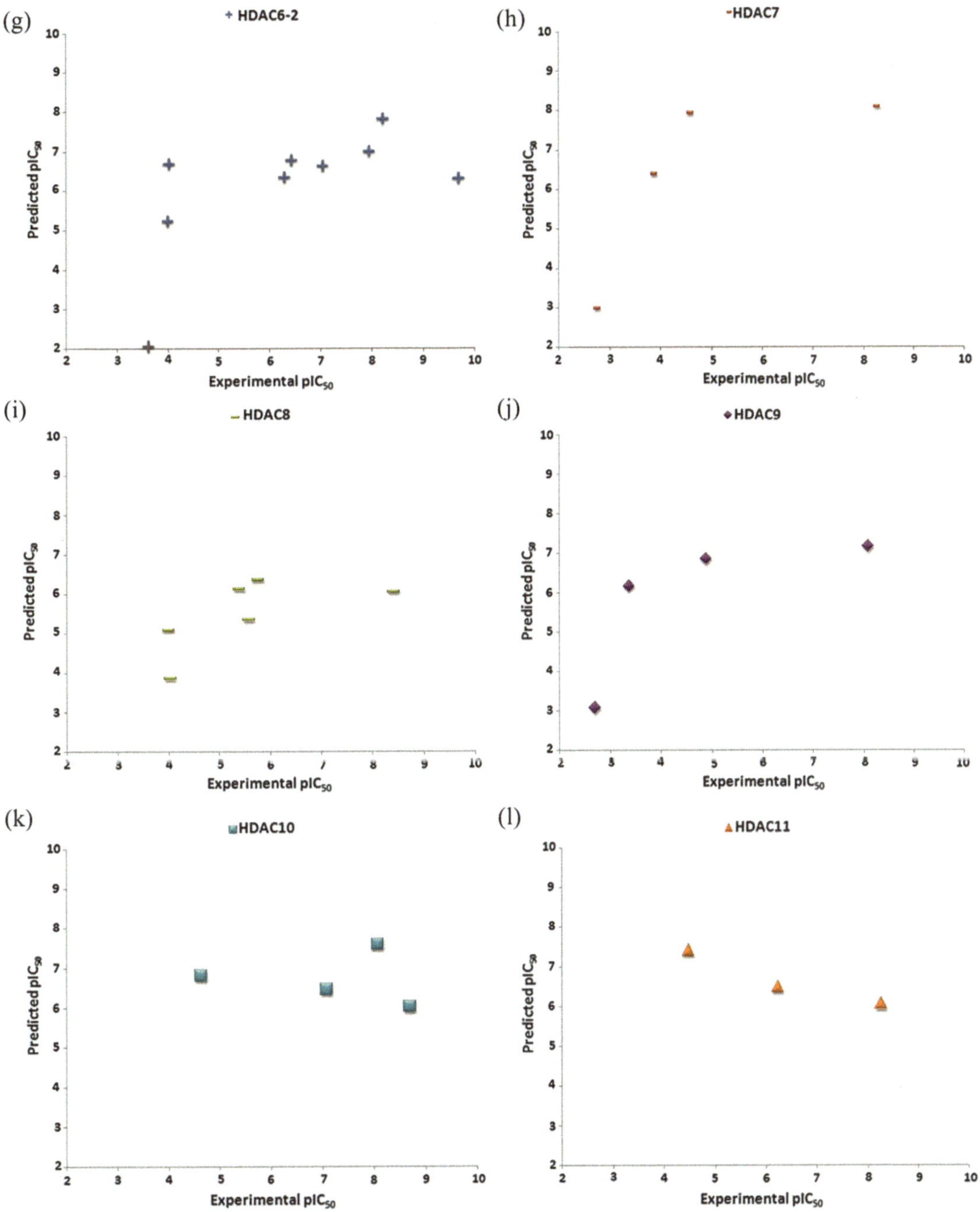

FIGURE 26—CONTINUED

good statistical coefficients, was predictive for the compounds in the test sets, and robust to cross-validation while omitting multiple data. Actually, the COMBINEr model is the only one available to rationalize the different activity profiles for almost any given HDAC inhibitor.

3. SB MODELING FOR NAD-HDAC (SIRTUINS)

3.1 Studies on Sirtinol Analogs

Along with the earlier investigations on Zn-based HDACi in 2004 some SBDDs were also conducted on sirtuin inhibitors (Mai et al., 2005a). Sirtuins are structurally and mechanistically distinct from Zn-based HDACs. They contain a conserved ~300 amino acid catalytic domain that catalyzes a two-step reaction that requires NAD^+ as a cosubstrate. The first series of reactions is cleavage of the nicotinamide–ribose glycosidic bond of NAD^+, release of free nicotinamide, and formation of a relatively long-lived peptidyl imidate intermediate. Second is the transfer of the acetyl group to ADP-ribose with production of O-acetyl-ADP-ribose (Figure 27).

The sirtuin family of enzymes are named after their founding member, the Sir2 (silent information regulator 2) protein of *Saccharomyces cerevisiae*. In yeast, Sir2 is critical for transcriptional silencing at three specific loci: the telomeres, ribosomal DNA, and the silent mating type loci. Sir2 and its homologs have gained considerable attention for their ability to mimic the diet known as caloric restriction, which extends lifespan in a variety of organisms, including yeast, *Caenorhabditis elegans*, rodents, and probably primates. Sir2-like proteins also deacetylate nonhistone proteins, including the tumor suppressor p53, the RNA polymerase I transcription factor TAFI68, bovine serum albumin, α-tubulin, the archaeal chromatin protein Alba, and acetyl-coenzyme A synthetase. The human Sir2 ortholog, SIRT1, has been implicated in a variety of important disease-related processes including inflammatory responses, cell defense and survival, and fatty acid metabolism, and finding molecules that modulate these enzymes is considered a possible route for disease treatment.

Molecules that inhibit the class I/II HDACs are ineffective in inhibiting sirtuins and vice versa. That said, class I/II HDAC inhibitors do induce differential changes in gene expression of sirtuin mRNAs in cultured cells, specifically up-regulation of SIRT2, -4, and -7 and down-regulation of SIRT1, -5, and -6 mRNAs. At the time this study was ongoing only a few specific sirtuin inhibitors were available, including sirtinol, splitomicin, and nicotinamide (Chart 7).

Sirtinol and its analogs drew some attention to its 2-hydroxy-1-naphthaldehyde moiety linked to a 2-amino-N-(1-phenylethyl)benzamide portion through an aldimine linkage, and provides an interesting chemical template from which new analogs could be developed. Therefore, some Sirtinol analogs were

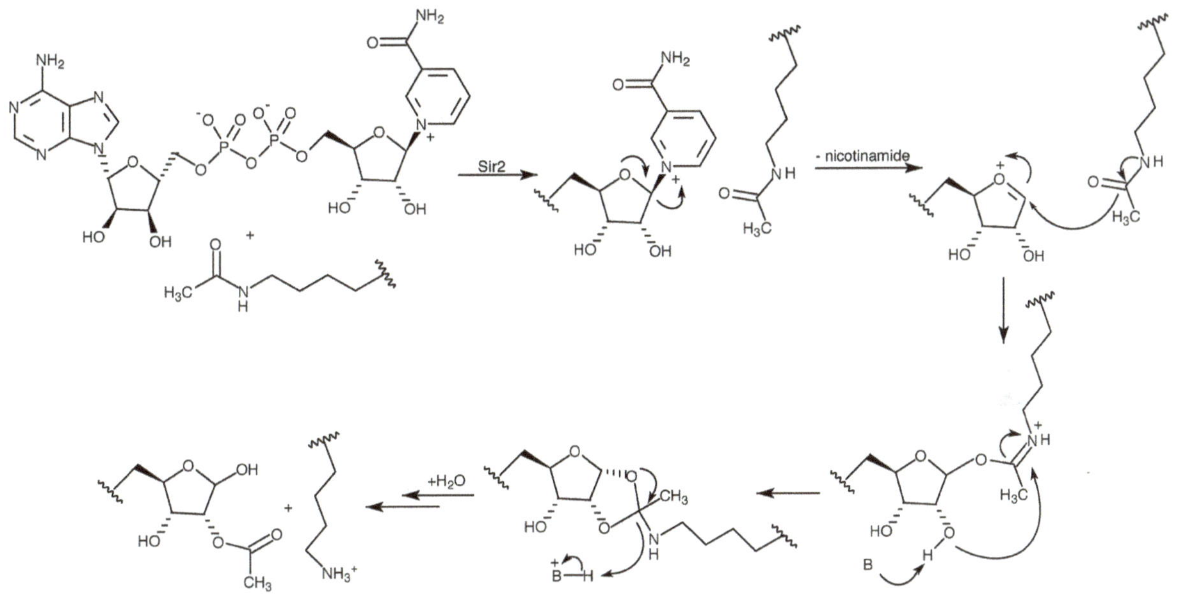

FIGURE 27
Deacetylation mechanism of sirtuins.

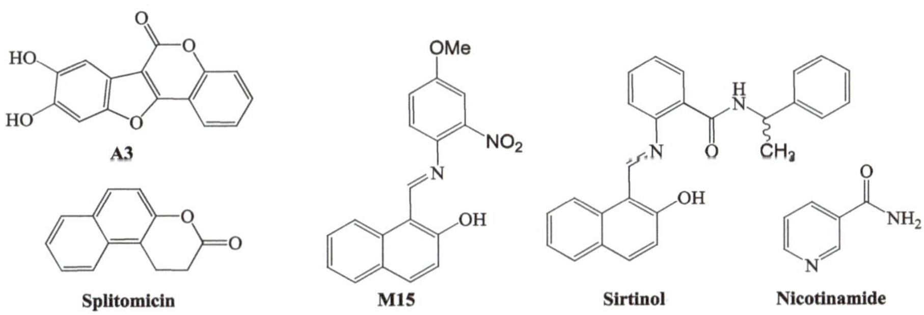

CHART 7
Known NAD⁺-dependent (class III) HDAC inhibitors.

prepared and tested as inhibitors of yeast Sir2, human SIRT1, and human SIRT2 in vitro, with sirtinol as the reference compound (Table 11).

To inspect binding modes of small molecule inhibitors A3, M15, (*R*)-sirtinol **8h**, (*S*)-sirtinol **8i**, and splitomicin, docking studies were conducted by means of AutoDock using Sir2-Af2/NAD⁺/p53 modeled ternary complex. To ascertain whether these compounds could act as competitive or noncompetitive inhibitors, docking was conducted using either the free Sir2-Af2 enzyme or the complex with one of the substrates. Although it was reported that NAD⁺ would likely bind only to the preformed Sir2/acetylated peptide binary complex,

Table 11 Yeast Sir2 (ySir2) and Human SIRT1 (hSIRT1) Inhibitory Activity of Sirtinol Analogues **8a-i**

		% Inhibtn (100 μM)		IC_{50} (μM)	
Compd	**Structure**	**ySir2**	**hSIRT1**	**ySir2**	**hSIRT1**
Sirtinol	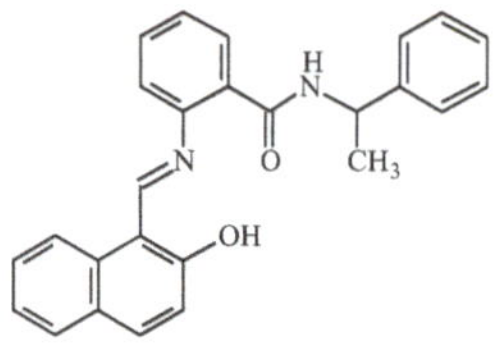	65.6	44.5	48 ± 4	131 ± 11
8a	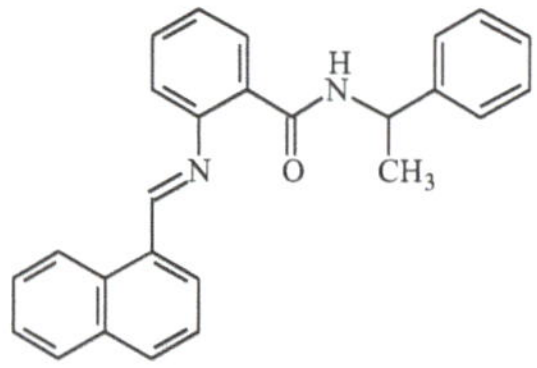	10.0	33.0	ND	ND
8b	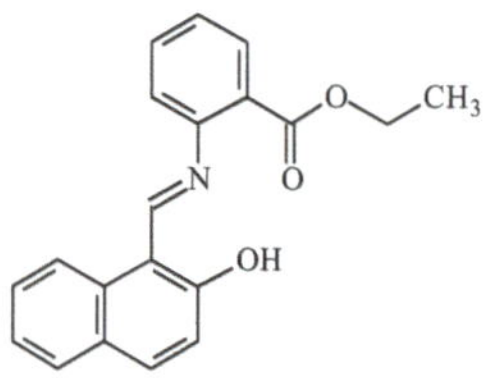	20.0	9.0	ND	ND
8c	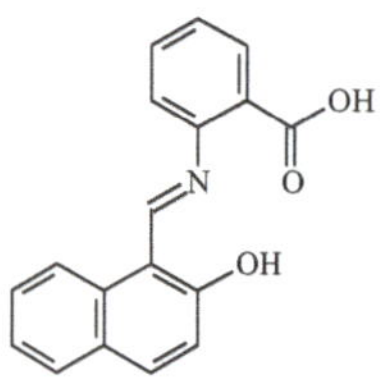	5.0	26.0	ND	ND
8d	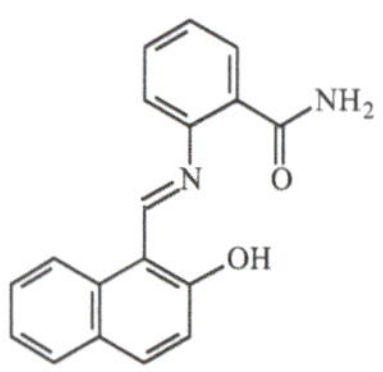	No Inhibition	No Inhibition		
8e	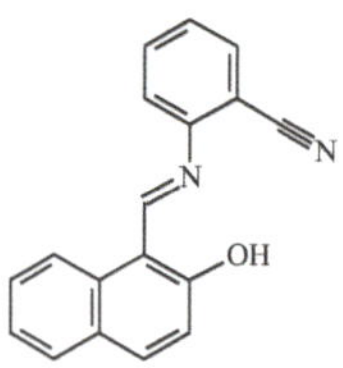	15.0	20.0	ND	ND

Table 11 Yeast Sir2 (ySir2) and Human SIRT1 (hSIRT1) Inhibitory Activity of Sirtinol Analogues **8a-i** *Continued*

		% Inhibtn (100 μM)		IC_{50} (μM)	
Compd	**Structure**	**ySir2**	**hSIRT1**	**ySir2**	**hSIRT1**
8f (*meta*-Sirtinol)	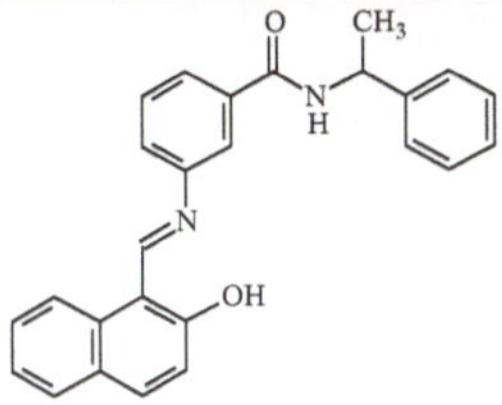	53.9	63.0	72±3	59±2
8g (*para*-Sirtinol)	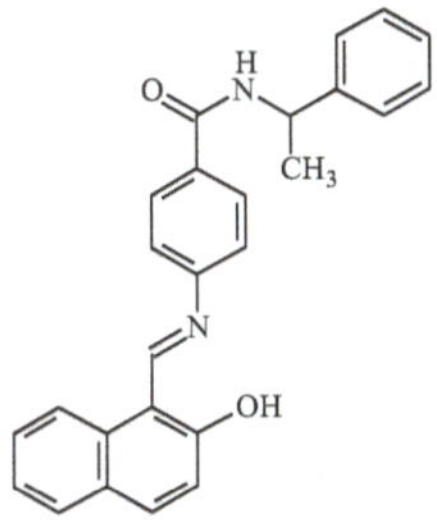	59.8	82.6	33±1	13±2
8h ((*R*)-Sirtinol)	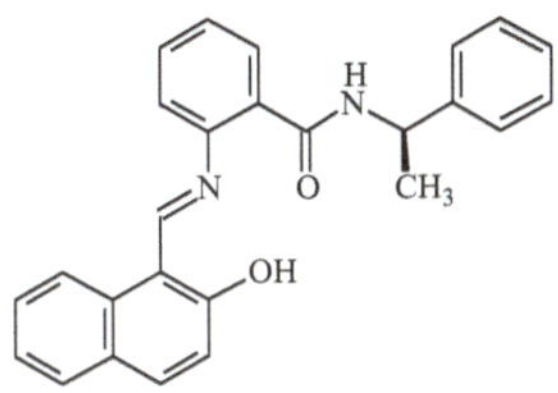	55.5	61.8	62±5	55±5
8i ((*S*)-Sirtinol)	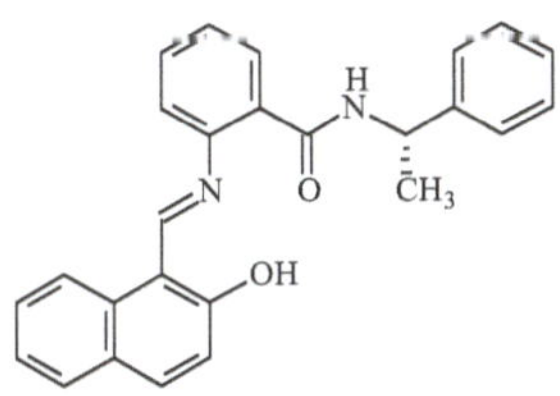	57.0	80.2	68±4	67±4

docking experiments were also undertaken using the less likely Sir2-Af2/NAD+ complex to ascertain the possible competitive inhibitory action against the Ac-K site.

3.2 Docking of A3

Against the free Sir2-Af2 enzyme and the Sir2-Af2/p53 complex, although in a reverse orientation, A3 was found preferably in the NAD+ binding site (Figure 28) acting as a competitive inhibitor for this substrate. On the other

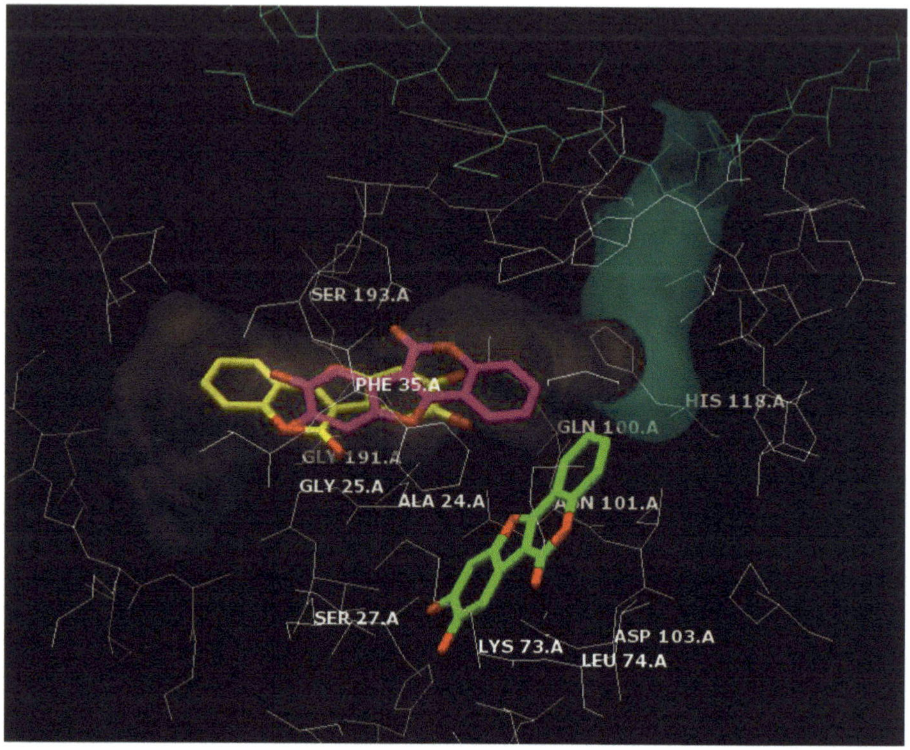

FIGURE 28

A3 docked conformations. The A3 binding mode obtained in the absence of both the NAD+ and Ac-K substrates is displayed in yellow-colored carbon atoms. In green-colored carbon atoms the A3 docked conformation is reported in the presence of only the NAD+ substrate. In magenta-colored carbon atoms the A3 docked conformation is reported in the presence of only the p53 substrate. For interpretation purposes the NAD+ and Ac-K volumes are displayed in white and cyan, respectively. The p53 (cyan wire) and 5Å core of the Sir2-Af2 (white wire) are also displayed. For the sake of clarity hydrogen atoms are omitted.

hand, in the presence of NAD+, A3 seemed to act primarily as noncompetitive inhibitor, occupying a separate site, close but distinct from those of NAD+ and Ac-K binding sites. In the NAD+ competitive bound conformation the docked A3 molecule fully superimposed the NAD+ scaffold, making a favorable interaction with both hydrophobic and hydrophilic residues (Ala24, Gly25, Phe35, Arg36, Gln100, Asn101, His118, Gly191, Ser193). Thus A3 could act as a noncompetitive or NAD+ competitive inhibitor without affecting the acetylated peptide binding. In the noncompetitive bound conformation the molecule of A3 seated in the nicotinamide escape tunnel makes favorable contacts with the side chains and backbone atoms of hydrophilic residues (Ser27, Lys73, Asn101, and Asp103).

3.3 Docking of M15

In the free Sir2-Af2 enzyme and the Sir2-Af2/NAD$^+$ complex, AutoDock found the same binding conformation that seems to act as a competitive inhibitor for the Ac-K site. Differently, against the Sir2-Af2/p53 complex the docked M15 molecule only partially superimposes the NAD$^+$ binding site. In the p53 competitive bound conformations the nitromethoxyphenyl moiety of M15 partially occupies the Ac-K binding site, making favorable interactions with hydrophobic residues (Val162, Val163, Leu164, and Phe165), while the naphthol group interacts with the residue of the nicotinamide escape tunnel, making favorable contacts with the side chains and backbone atoms of Lys73, Asn101, Ile102, and Asp103. In the NAD$^+$ competitive bound conformation the docked M15 only partially superimposes the NAD$^+$ binding site, suggesting that this inhibitor would likely not bind in the presence of the p53 substrate (Figure 29). Thus it seems that M15 could act as an anti-Sir2 agent by inhibiting the binding of the acetylated substrate.

3.4 Docking of (*R*)-Sirtinol (8h)

As seen for the previous docking of M15, AutoDock suggested a similar binding conformation also for (*R*)-sirtinol (**8h**) either in the free Sir2-Af2 enzyme or the Sir2-Af2/NAD$^+$ complex (Figure 8). The molecule seemed to act as a competitive inhibitor for the Ac-K site, but differently from what was observed for M15. (*R*)-sirtinol (**8h**) occupies the Ac-K entrance channel in a particular fashion, overlapping part of the experimental acetylated p53 substrate. The two bound conformations differed only in the position of the 2-naphthol moiety; in particular the free enzyme bound conformation makes a π-stacking interaction with the Tyr197, while in the **2** Sir2-Af2/NAD$^+$ AutoDock proposed binding mode, the 2-naphthol group is flipped about 180°, so that it lacks the π-stacking interaction but displays a strong hydrogen bond interaction with the carbonyl group of Tyr197 (distance $OH_{naphthol}$---$O{=}C_{Tyr197} = 1.91$ Å). The other side of the two **8h** conformations is perfectly superimposed, displaying two further hydrogen bonds, a strong one between the **2** carboxamide oxygen and the Gly166 NH (distance $C{=}O_2$---$HN_{Gly166} = 2.03$ Å) and a weak one between the carboxamide NH and the Tyr197 CO (distance $N\text{-}H_2$---$O{=}C_{Tyr197} = 2.41$ Å). Furthermore, the superimposition of the two **8h** bound conformations with the experimental acetylated p53 shows how the inhibitor can mime the substrate. In particular, the naphthol moiety partially occupies the spatial region of the p53 Leu12-Phe14; the central *ortho*-disubstituted benzene ring overlaps the hydrophobic parts of the Ac-K side chain; the carbonyl group plays the role of the Ac-K CO; the terminal phenyl ring overlaps the p53-Lys10.

The docking of **8h** against the Sir2-Af2/p53 binary complex places the molecule in a different space that partially overlaps NAD$^+$ and fails to make strong

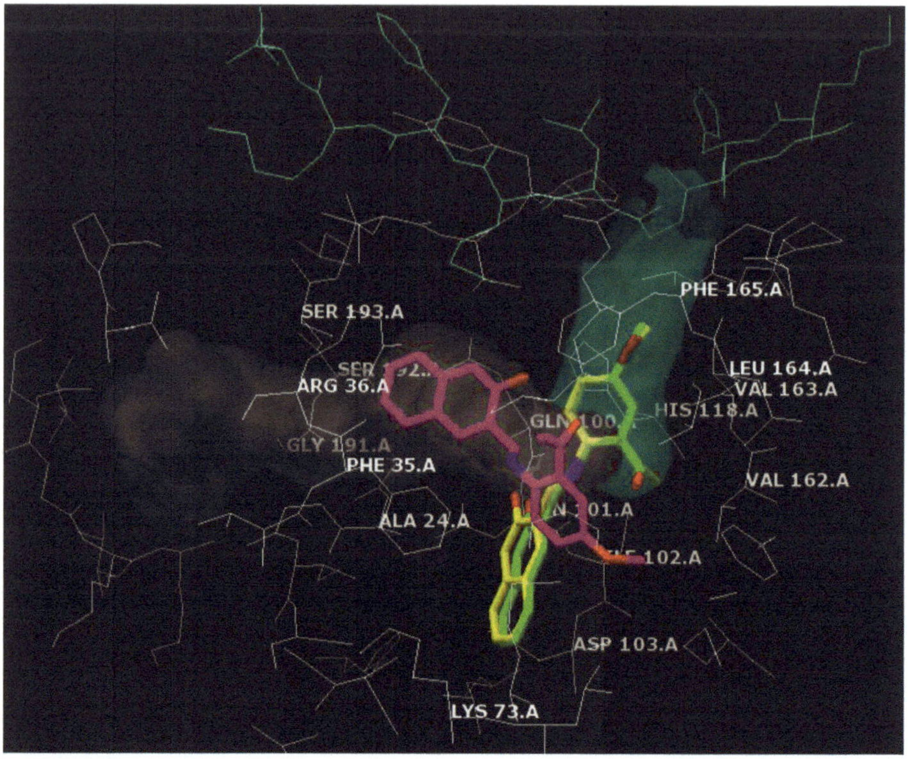

FIGURE 29

M15 docked conformations. In yellow-colored carbon atoms the M15 binding mode obtained in the absence of both the NAD+ and Ac-K substrates is displayed. In green-colored carbon atoms the M15 docked conformation is reported in the presence of only the NAD+ substrate. In magenta-colored carbon atoms the M15 docked conformation is reported in the presence of only the p53 substrate. For interpretation purposes the NAD+ and Ac-K volumes are displayed in white and cyan, respectively. The p53 (cyan wire) and 5 Å core of the Sir2-Af2 (white wire) are also displayed. For the sake of clarity hydrogen atoms are omitted.

interactions with the complex, suggesting that this NAD+ competitive inhibition mechanism is highly unlikely (Figure 30).

3.5 Docking of (*S*)-Sirtinol (8i)

Similar to our results with the (*R*) enantiomer, our analysis indicated that (*S*)-sirtinol (**8i**) likely competed for the Ac-K binding site. Again, similar to **8h**, the docked conformations obtained with the free and NAD+ bound enzymes are almost coincident, but they do not display 2-naphthol flipping (compare Figure 30 with Figure 31). Superimposition of the two enantiomers showed that in the model the 2-naphthol of **8i** is not coplanar with that of **2**, and would not allow the (*S*) enantiomer to form either the $OH_{naphthol}$---$O{=}C_{Tyr197}$

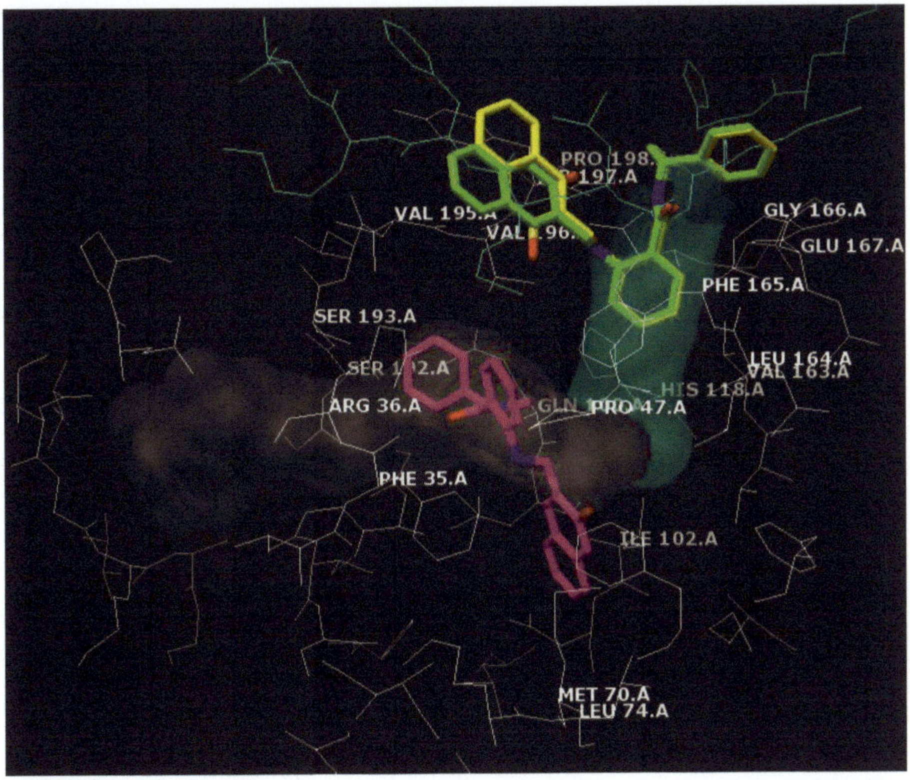

FIGURE 30

(*R*)-Sirtinol (**8h**) docked conformations. In yellow-colored carbon atoms the **2** binding mode obtained in the absence of both the NAD+ and Ac-K substrates is displayed. In green-colored carbon atoms the **8h** docked conformation is reported in the presence of only the NAD+ substrate. In magenta-colored carbon atoms the **8h** docked conformation is reported in the presence of only the p53 substrate. For interpretation purposes the NAD+ and Ac-K volumes are displayed in white and cyan, respectively. The p53 (cyan wire) and 5 Å core of the Sir2-Af2 (white wire) are also displayed. For the sake of clarity hydrogen atoms are omitted.

hydrogen bond or the π-stacking interaction. Moreover, the inversion of the chiral center makes the **8i** carboxamide to be shifted, preventing the formation of the $C{=}O_{carboxamide}$---HN_{Gly166} hydrogen bond. In general, the two enantiomers **8h** and **8i** share a common binding mode, and thus they should show similar anti-Sir2 (Sir2-Af2) activities.

The docking of **8i** against the Sir2-Af2/p53 binary complex, similar to what was observed for **8h**, predicts that **8i** partially occupies the NAD+ binding site, suggesting that this binding mode is the only one possible in the presence of bound acetylated substrate.

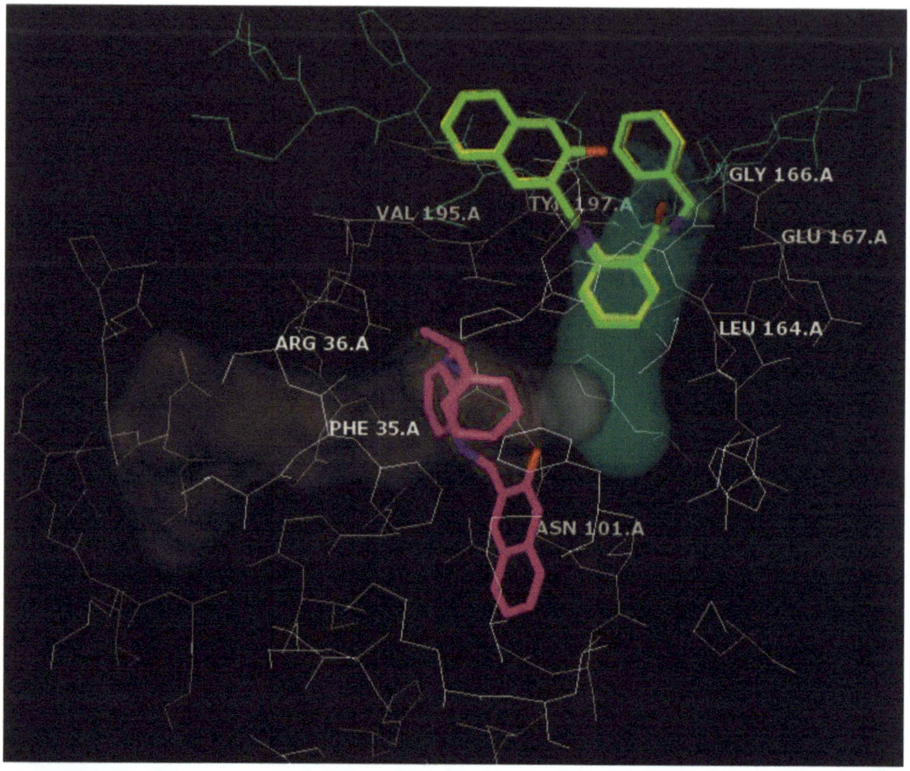

FIGURE 31

(*S*)-Sirtinol (**8i**) docked conformations. In yellow-colored carbon atoms the **3** binding mode obtained in the absence of both the NAD+ and Ac-K substrates is displayed. In green-colored carbon atoms the **8i** docked conformation is reported in the presence of only the NAD+ substrate. In magenta-colored carbon atoms the **8i** docked conformation is reported in the presence of only the p53 substrate. For interpretation purposes the NAD+ and Ac-K volumes are displayed in white and cyan, respectively. The p53 (cyan wire) and 5 Å core of the Sir2-Af2 (white wire) are also displayed. For the sake of clarity hydrogen atoms are omitted.

3.6 Docking of Splitomicin

By comparing the docking of splitomicin against the free Sir2-Af2 or the NAD+ and p53 binary complexes, it became clear that splitomicin might not compete for the Ac-K binding site but rather it could inhibit NAD+ binding. In docking experiments conducted against either the free enzyme or the p53 bound form, splitomicin overlaps the NAD+ structure, placing the lacton moiety in proximity to the phosphodiester bridge, while the naphthalene ring overlaps the nicotinamide ribose (not shown). In the Sir2-Af2/NAD+ complex (the less likely arrangement) our modeling indicates that splitomicin might bind in the nicotinamide escape channel making contacts with surrounding residues Phe35, Gln100, Asn101, Ile102, Asp103, and His118 (Figure 32).

FIGURE 32
Splitomicin docked conformations. In yellow-colored carbon atoms the splitomicin binding mode obtained in the absence of both the NAD^+ and Ac-K substrates is displayed. In green-colored carbon atoms the splitomicin docked conformation is reported in the presence of only the NAD^+ substrate. In magenta-colored carbon atoms the splitomicin docked conformation is reported in the presence of only the p53 substrate. For interpretation purposes the NAD^+ and Ac-K volumes are displayed in white and cyan, respectively. The p53 (cyan wire) and 5 Å core of the Sir2-Af2 (white wire) are also displayed. For the sake of clarity hydrogen atoms are omitted.

In conclusion, the docking of five synthetic inhibitors both in the absence of Sir2 substrates or in the presence of p53 or NAD^+ indicated that the five inhibitors might inhibit the Sir2 enzyme via different mechanisms. A3 and splitomicin seem to bind preferably in the NAD^+ site or in a noncompetitive side pocket blocking the nicotinamide escape, whereas (*R*)- and (*S*)-sirtinol, by mimicking the backbone of the p53, might act as Ac-K binding-site competitors by blocking the entrance channel of the acetylated lysine. Lastly, M15 seems to bind in the bottom of the Ac-K binding site likely interfering with the transfer of the acetyl group, thereby acting as a noncompetitive inhibitor (Figure 7). The docking of these small molecules against the Sir2-Af2/NAD^+ complex agree with a report (Borra et al., 2004) that is very unlikely that the free enzyme and the Sir2-Af2/p53 complex are the only forms that can be inhibited.

3.7 Multidocking Experiments (Unpublished Studies)

Although the previous docking studies were in agreement with some experimental data, some uncertainties remained on the inhibitors binding modes, likely due to the high reliability of AutoDock scoring function. To overcome this ambiguity a multidocking procedure was set up to study the binding mode of an AGK-2a reported SIRT2 inhibitor (Outeiro et al., 2007).

It has been extensively reported for sirtuins that zinc domain during cosubstrate modifications undergoes a rotation toward the Rossmann fold domain (Moniot et al., 2013; North and Verdin, 2004; Sakkiah et al., 2013) (Figure 33(a)). A deep study reported that this domain movement is related to $NAD^+ \rightarrow$adenosine diphosphate ribose (ADPr) transformation in which the leaving nicotinamide moiety, when still bound, seem to maintain an open conformation (Sanders et al., 2010). This large protein movement results in a switch between open (SIRT/NAD^+ binary complex$\rightarrow SIRT_{open}$) and closed (SIRT/ADPr binary complex$\rightarrow SIRT_{closed}$) SIRT enzyme conformations. SIRT3 inhibitors have been reported to bind to either SIRT3/NAD^+ or SIRT3/ADPr binary complexes (Gertz et al., 2013). Such structural information is not yet available for SIRT2. Understanding of the structural basis of small molecule-SIRT2 interactions is crucial for in silico design of novel and potent SIRT2 inhibitors. Due to lack of any details on the possible AGK2 and MC2494 inhibition mechanisms, a structure-based (SB) approach on both the open and closed conformation has been undertaken by means of structural information available from PDB. Among the possible binding sites that would be reasonable that the compounds could

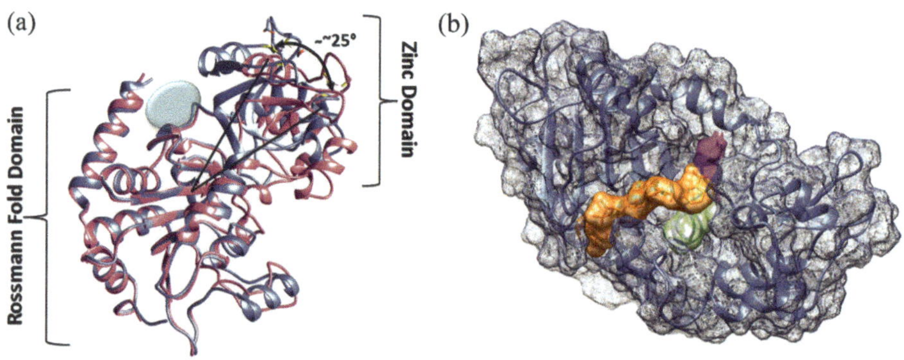

FIGURE 33

(a) Superimposition of the open (pdb entry code: 3ZGO, $SIRT_{open}$, colored in magenta) and the closed (pdb entry code: c-3ZGV, $SIRT_{close}$, colored in cyan) SIRT2 conformations. The zinc domain shift is highlighted and the transparent blue ovale indicate the opened NAD^+ binding site. (b) SIRT2 (in cartoon) binding sites. The SIRT2 surface is shown in mesh, while the NAD^+ (orange), the acetyl lysine (violet), and the leaving nicotinamide (green) binding sites are displayed in transparent solid surfaces.

bind to are the NAD^+, acetyl lysine (Aly), or the escaping nicotinamide pockets (Figure 33(b)) (Sanders et al., 2010).

In the applied SB protocol either the $SIRT2_{open}$ or $SIRT2_{close}$ were modeled in three different forms: without any ligand (APO-SIRT2), complexed with the processed NAD^+ cofactor adenosine diphosphate ribose (ADPr) (SIRT2-ADPr), and complexed with the unmodified NAD^+ (SIRT2-NAD) (Figure 34).

To explore the possible inhibition mechanism of AGK2, molecular docking studies with seven different program/score combinations were run in the prepared six SIRT2 closed and opened apo-proteins or binary complexes to check whether the molecules could bind in one of the cited binding sites (Sanders et al., 2010). The entire SB procedure is schematized in Figures 35 and 36; similar to virtual screening application the final goal is the identification of the putative binding mode of the two inhibitors using a convergence protocol.

Inspection of the docked poses in all the closed SIRT2 conformations revealed that only a few poses of AGK2 can reach the expected binding sites. In fact upon the zinc domain rotation a reduced space results in the leaving nicotinamide SIRT2 area where several sirtuin inhibitors were found to bind (Gertz et al., 2013; Sanders et al., 2010); moreover since no poses were found in the Aly binding site, it can be speculated that the molecules' sizes prevent any conformation to fit in. All the proposed lowest energy/score conformations

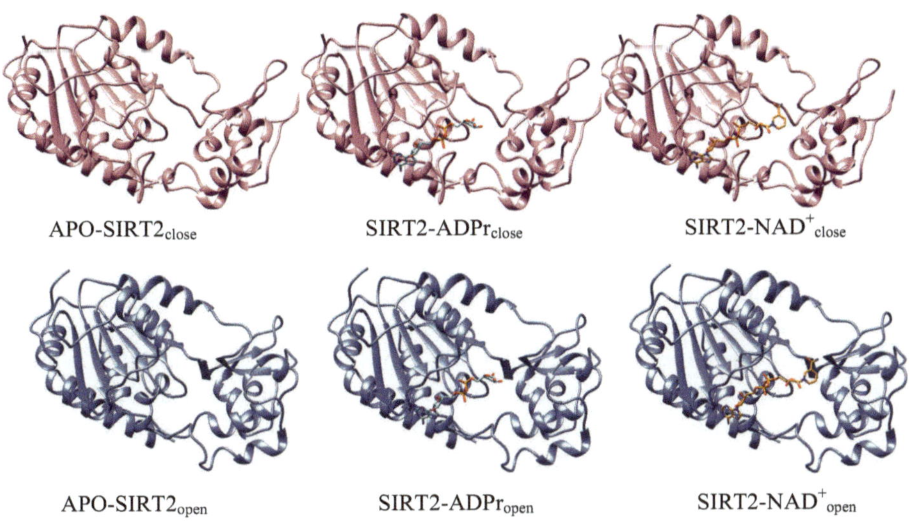

FIGURE 34

Crystal structures of the apo-enzymes and the binary complex with both NAD^+ and ADPr .The structures in pink indicate the close form, while the blue structures indicate the open form.

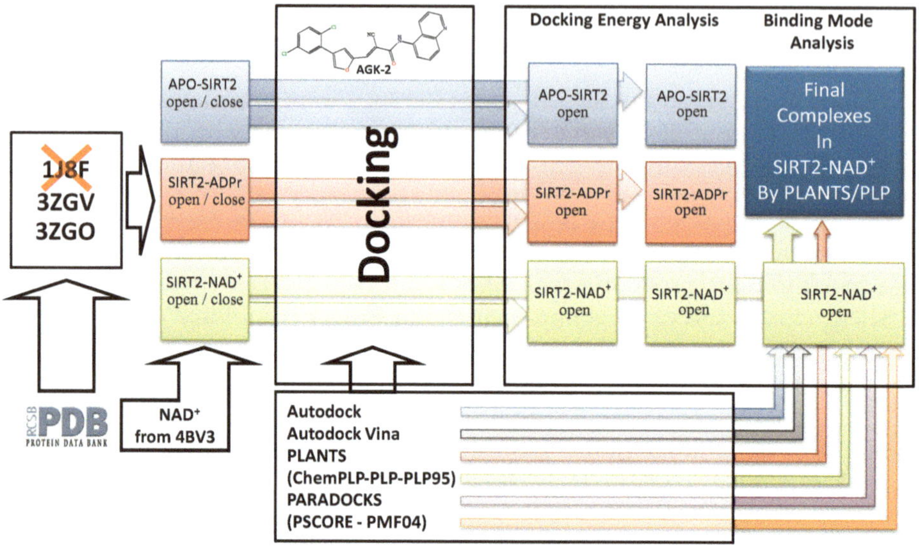

FIGURE 35

Procedure flow to study the AGK2 binding mode.

were found to bind in a disordered pattern (Figure 37(a)). On the contrary, the docked poses in the open SIRT2 conformation were mainly found to bind in the NAD^+, Aly, or nicotinamide escaping binding site proximities (Figure 37).

Furthermore a direct comparison among the docking energies (compare values in Tables 12 and 13) indicated that, on average, more stable complexes could be obtained for the $SIRT2_{open}$ conformations.

Once established that the compounds prefer to bind to the $SIRT2_{open}$ conformations, the docking scores listed in Table 2 were further compared and clearly indicated that the most favorable interactions are among the $SIRT2-NAD^+{}_{open}$-AGK2 and $SIRT2-NAD^+{}_{open}$-MC2494 ternary complexes. To a slightly lesser extent also the SIRT2-$ADPr_{open}$-AGK2 and SIRT2-$ADPr_{open}$-MC2494 are predicted to have high probability to form. Definitively APO-$SIRT2_{open}$-AGK2 binary complex had the least scores and therefore it could be excluded that the compound could act as NAD^+ displacer. The latter statement is somehow in contrast with the previous report on AGK2 binding mode exploration, where the inhibitors were found to bind in the NAD^+ site, but only an APO-SIRT2 homolog structure was used (Outeiro et al., 2007). AGK2 docking suggested binding conformations in either $SIRT2-NAD^+{}_{open}$ or SIRT2-$ADPr_{open}$ binary complexes are almost superimposable, therefore to avoid redundancy only the inhibitor binding mode in $SIRT2-NAD^+{}_{open}$ was described in detail. The seven docking combinations and AGK2 binding conformations, although in good agreement and sharing the same binding space, appear not fully superimposable with average RMSD values ranging from 8.21 to 11.28.

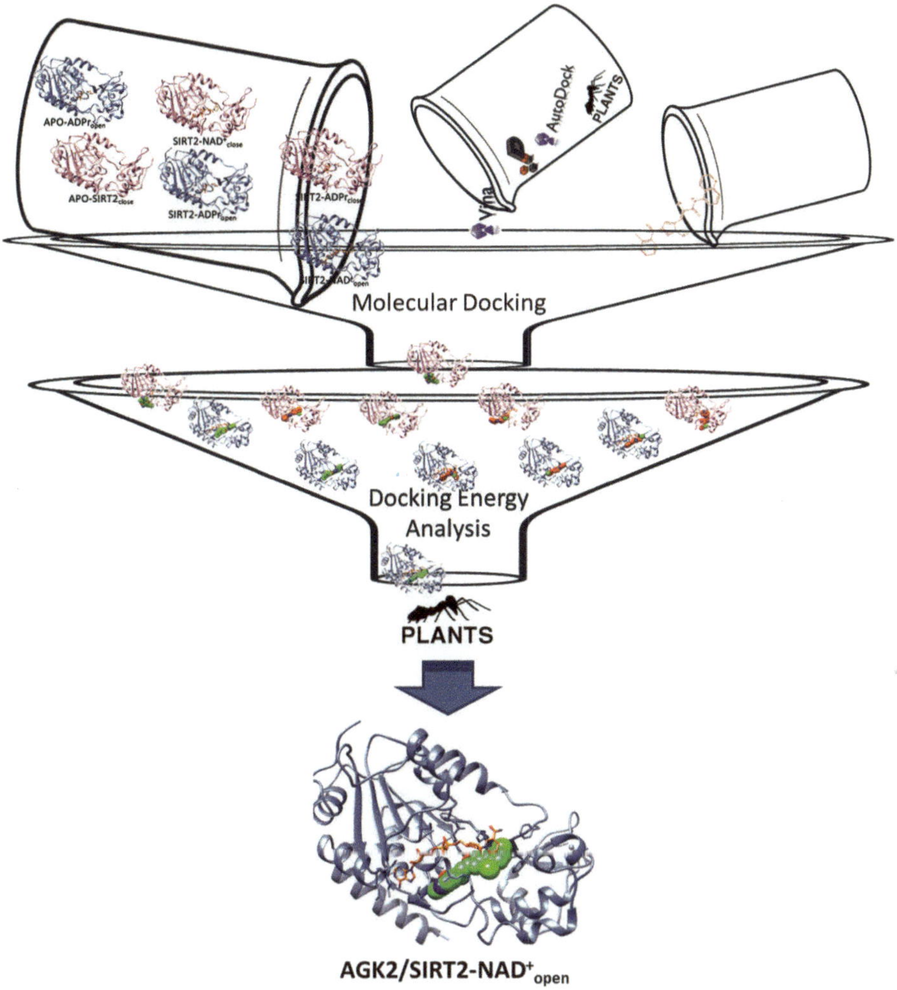

FIGURE 36

Convergence structure-based procedure to study the AGK2 and MC2494 binding modes.

In order to compare the scores of Table 13, the normalized values, obtained by dividing each program score for the value obtained in the APO-SIRT2$_{open}$ conformation, was considered (Table 15). This transformation highlights that AGK2 could be described by PARADOCKS/PMF04. Nevertheless, the docked conformations proposed by the program PARADOCKS display the larger disagreement when compared with those obtained from all the other docking programs (see Table 14). For this reason, the binding modes proposed by PLANTS/PLP were selected. Further application of this multidocking protocol on newly synthesized AGK-2 analogs are ongoing and will be reported elsewhere.

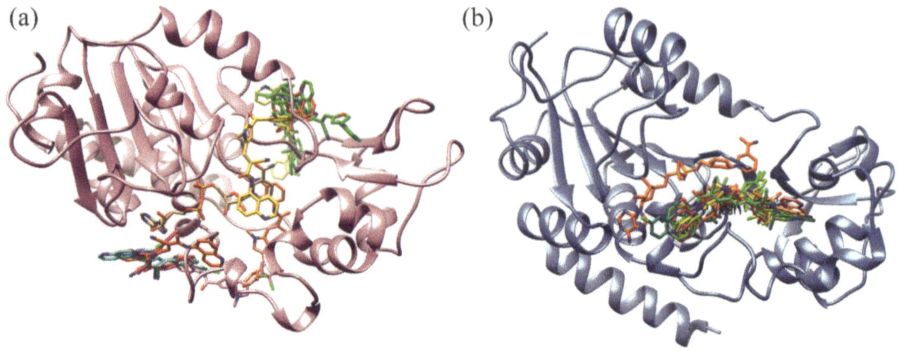

FIGURE 37

AGK2 docking poses. (a) The disordered pattern in the APO-SIRT2$_{close}$. (b) A convergence binding site in the SIRT2-NAD$^{+}{}_{open}$.

Table 12 Docking Scores for the Lowest Energy Poses of AGK2 Obtained from the Different Docking Programs in the Closed SIRT2 Conformations

Docking Program	APO-SIRT2$_{close}$	SIRT2-ADPr$_{close}$	SIRT2-NAD$^{+}{}_{close}$
AutoDock	−9.77	−9.68	−8.12
Vina	−9.90	−9.00	−8.90
PLANTS/CHEMPLP	−111.90	−84.55	−83.12
PLANTS/PLP	−102.15	−77.83	−77.28
PLANTS/PLP95	−166.31	−110.24	−117.44
PARADOCKS/Pscore	−875.36	−774.63	−559.87
PARADOCKS/PMF04	−711.80	−507.38	−536.24

Table 13 Docking Scores for the Lowest Energy Poses of AGK2 Obtained from the Different Docking Programs in the Opened SIRT2 Conformations

Docking Program	APO-SIRT2$_{open}$	SIRT2-ADPr$_{open}$	SIRT2-NAD$^{+}{}_{open}$
AutoDock	−9.77	−10.89	−10.76
Vina	−9.70	−9.90	−10.30
PLANTS/CHEMPLP	−87.05	−99.12	−102.64
PLANTS/PLP	−85.20	−95.48	−97.82
PLANTS/PLP95	−134.44	−142.79	−141.65
PARADOCKS/Pscore	−582.25	−657.63	−680.78
PARADOCKS/PMF04	−621.25	−605.51	−749.51

Table 14 RMSD Values Calculated for the AGK2 and MC2494 Poses

Inh	Docking Program Combination		AutoDock	Vina	PLANTS			PARADOCKS		Average
					CHEMPL	PLP	PLP95	PSCORE	PMF04	
AGK2	AUTODOCK		–	13.51	5.93	4.26	10.78	9.88	10.68	9.17
	VINA		13.51	–	13.35	13.55	14.63	5.52	4.36	10.82
	PLANTS	CHEMPL	5.93	13.35	–	3.59	9.55	10.51	11.16	9.01
		PLP	4.26	13.55	3.59	–	9.91	10.19	10.96	8.74
		PLP95	10.78	14.63	9.55	9.91	–	10.93	11.88	11.28
	PARADOCKS	PSCORE	9.88	5.52	10.51	10.19	10.93	–	2.22	8.21
		PMF04	10.68	4.36	11.16	10.96	11.88	2.22	–	8.54

The values were calculated considering each program docking pose as the reference structure; average values are reported in the last column.

Table 15 Normalized Docking Scores from Table 13

Docking Program	APO-SIRT2$_{open}$	SIRT2-ADPr$_{open}$	SIRT2-NAD$^{+}_{open}$
AutoDock	1	1.11	1.10
Vina	1	1.02	1.06
PLANTS/CHEMPLP	1	1.14	1.18
PLANTS/PLP	1	1.12	1.15
PLANTS/PLP95	1	1.06	1.05
PARADOCKS/PSCORE	1	1.13	1.17
PARADOCKS/PMF04	1	0.97	1.21

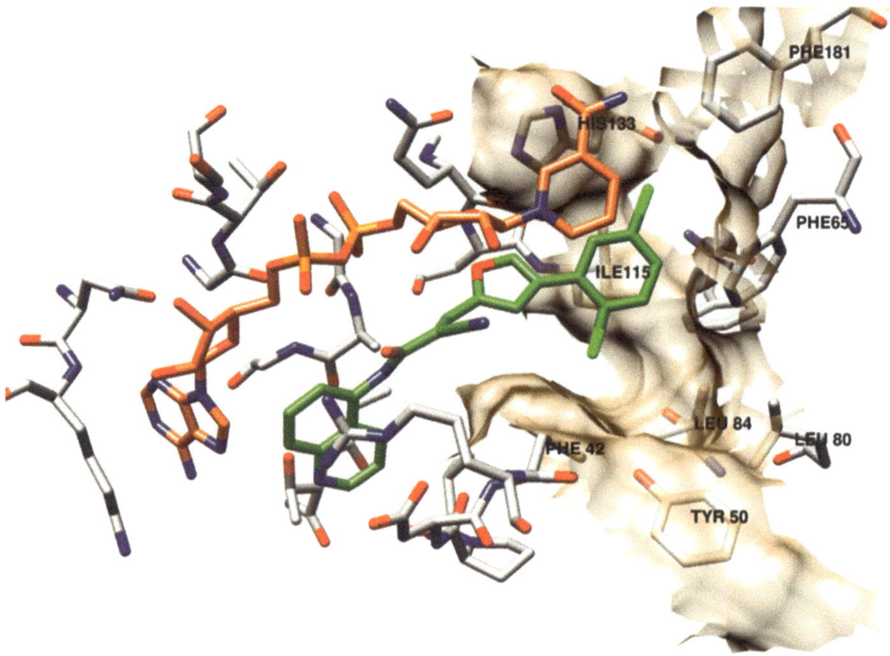

FIGURE 38

Details on the AGK2 (green) binding mode. The SIRT2 surface interacting with the benzene rings is represented in gold.

In conclusion, the multidocking approach led to define in a convergent way the very likely AGK-2 binding mode (Figure 38) in which the compounds bind close to NAD and the AGK2 2,5-dichlorophenyl group and are mainly placed in the vicinity of the NAD^{+} nicotinamide, filling a small pocket formed by residues Phe65, Ile115, His133, and Phe181, likely avoiding the Ac-K substrate interaction with the NAD cofactor.

4. CONCLUSIONS

In this chapter, application of various structure-based modeling techniques and protocols to HDAC enzymes have been reviewed and it is definitively proved that SBDD is a valuable tool to design HDACi. A lot of work still remains to improve both the reliability and prediction ability of molecular docking and 3D QSAR models (classical CoMFA like or COMBINEr) as specialized external scoring functions. Nevertheless the reported studies fully ratify that it is just a matter of time and it is the author's conviction that early new improved protocols will enable the design of potent selective HDACi (against Zn- or NAD-based enzymes) to be used as essential chemical tools to aid the decryption of each isoform biological role.

REFERENCES

Ballante, F., Musmuca, I., Marshall, G.R., Ragno, R., 2012. Comprehensive model of wild-type and mutant HIV-1 reverse transciptases. J. Comput. Aided Mol. Des. 26 (8), 907–919. http://dx.doi.org/10.1007/s10822-012-9586-6.

Blackwell, L., Norris, J., Suto, C.M., Janzen, W.P., 2008. The use of diversity profiling to characterize chemical modulators of the histone deacetylases. Life Sci. 82 (21–22), 1050–1058.

Borra, M.T., Langer, M.R., Slama, J.T., Denu, J.M., 2004. Substrate specificity and kinetic mechanism of the Sir2 family of NAD$^+$-dependent histone/protein deacetylases. Biochemistry 43 (30), 9877–9887.

Brosch, G., Georgieva, E.I., Lopez-Rodas, G., Lindner, H., Loidl, P., 1992. Specificity of Zea mays histone deacetylase is regulated by phosphorylation. J. Biol. Chem. 267 (29), 20561–20564.

Brosch, G., Goralik-Schramel, M., Loidl, P., 1996. Purification of histone deacetylase HD1-A of germinating maize embryos. FEBS Lett. 393 (2–3), 287–291.

Chen, H.-F., Kang, J.-H., Li, Q., Zeng, B.-S., Yao, X.-J., Fan, B.-T., Yuan, S.-G., Panay, A., Doucet, J.P., 2003. 3D-QSAR study on apicidin inhibit histone deacetylase. Chin. J. Chem. 21 (12), 1596–1607.

Corelli, F., Massa, S., Stefancich, G., Mai, A., Artico, M., Panico, S., Simonetti, N., 1987. Antibacterial and antifungal compounds. VIII. Synthesis and antifungal activity of pyrrol derivatives similar to trichostatin A. Farmaco Sci. 42 (12), 893–903.

Cruciani, G., Watson, K.A., 1994. Comparative molecular field analysis using GRID force-field and GOLPE variable selection methods in a study of inhibitors of glycogen phosphorylase b. J. Med. Chem. 37 (16), 2589–2601.

Curtin, M., Glaser, K., 2003. Histone deacetylase inhibitors: the Abbott experience. Curr. Med. Chem. 10 (22), 2373–2392.

Doweyko, A.M., 2006. Three-Dimensional Quantitative Structure-Activity Relationship: The State of the Art.

Gertz, M., Fischer, F., Nguyen, G.T., Lakshminarasimhan, M., Schutkowski, M., Weyand, M., Steegborn, C., 2013. Ex-527 inhibits Sirtuins by exploiting their unique NAD$^+$-dependent deacetylation mechanism. Proc. Natl. Acad. Sci. U.S.A. 110 (30), E2772–E2781.

Guo, Y., Xiao, J., Guo, Z., Chu, F., Cheng, Y., Wu, S., 2005. Exploration of a binding mode of indole amide analogues as potent histone deacetylase inhibitors and 3D-QSAR analyses. Bioorg. Med. Chem. 13 (18), 5424–5434.

Head, R.D., Smythe, M.L., Oprea, T.I., Waller, C.L., Green, S.M., Marshall, G.R., 1996. VALIDATE: a new method for the receptor-based prediction of binding affinities of novel ligands. J. Am. Chem. Soc. 118 (16), 3959–3969.

Khan, N., Jeffers, M., Kumar, S., Hackett, C., Boldog, F., Khramtsov, N., Qian, X., Mills, E., Berghs, S.C., Carey, N., Finn, P.W., Collins, L.S., Tumber, A., Ritchie, J.W., Jensen, P.B., Lichenstein, H.S., Sehested, M., 2008. Determination of the class and isoform selectivity of small-molecule histone deacetylase inhibitors. Biochem. J. 409 (2), 581–589.

Kolle, D., Brosch, G., Lechner, T., Pipal, A., Helliger, W., Taplick, J., Loidl, P., 1999. Different types of maize histone deacetylases are distinguished by a highly complex substrate and site specificity. Biochemistry 38 (21), 6769–6773.

Kuntz, I.D., Blaney, J.M., Oatley, S.J., Langridge, R., Ferrin, T.E., 1982. A geometric approach to macromolecule-ligand interactions. J. Mol. Biol. 161 (2), 269–288.

Langer, T., 2010. Pharmacophores in drug research. Mol. Inf. 29 (6–7), 470–475. http://dx.doi.org/10.1002/minf.201000022.

Lechner, T., Lusser, A., Pipal, A., Brosch, G., Loidl, A., Goralik-Schramel, M., Sendra, R., Wegener, S., Walton, J.D., Loidl, P., 2000. RPD3-type histone deacetylases in maize embryos. Biochemistry 39 (7), 1683–1692.

Liu, B., Lu, Ai-J., Liao, C.-Z., Liu, H.-B., Zhou, J.-J., 2005. 3D-QSAR of sulfonamide hydroxamic acid HDAC inhibitors. Wuli Huaxue Xuebao 21 (3), 337.

Lu, Q., Wang, D.-S., Chen, C.-S., Hu, Y.-D., Chen, C.-S., 2005. Structure-based optimization of phenylbutyrate-derived histone deacetylase inhibitors. J. Med. Chem. 48 (17), 5530–5535. http://dx.doi.org/10.1021/jm0503749.

Mai, A., Cerbara, I., Valente, S., Massa, S., Walker, L.A., Tekwani, B.L., 2004a. Antimalarial and antileishmanial activities of aroyl-pyrrolyl-hydroxyamides, a new class of histone deacetylase inhibitors. Comments Antimicrob. Agents Chemother. 48 (4), 1435–1436. http://dx.doi.org/10.1128/AAC.48.4.1435-1436.2004.

Mai, A., Massa, S., Cerbara, I., Valente, S., Ragno, R., Bottoni, P., Scatena, R., Loidl, P., Brosch, G., 2004b. 3-(4-Aroyl-1-methyl-1H-2-pyrrolyl)-N-hydroxy-2-propenamides as a new class of synthetic histone deacetylase inhibitors. 2. Effect of pyrrole-C2 and/or -C4 substitutions on biological activity. J. Med. Chem. 47 (5), 1098–1109. http://dx.doi.org/10.1021/jm030990+.

Mai, A., Massa, S., Lavu, S., Pezzi, R., Simeoni, S., Ragno, R., Mariotti, F.R., Chiani, F., Camilloni, G., Sinclair, D.A., 2005a. Design, synthesis, and biological evaluation of sirtinol analogues as class III histone/protein deacetylase (Sirtuin) inhibitors. J. Med. Chem. 48 (24), 7789–7795. http://dx.doi.org/10.1021/jm050100l.

Mai, A., Massa, S., Pezzi, R., Rotili, D., Loidl, P., Brosch, G., 2003a. Discovery of (Aryloxopropenyl)pyrrolyl hydroxyamides as selective inhibitors of class IIa histone deacetylase homologue HD1-a. J. Med. Chem. 46 (23), 4826–4829. http://dx.doi.org/10.1021/jm034167p.

Mai, A., Massa, S., Pezzi, R., Simeoni, S., Rotili, D., Nebbioso, A., Scognamiglio, A., Altucci, L., Loidl, P., Brosch, G., 2005b. Class II (IIa)-selective histone deacetylase inhibitors. 1. Synthesis and biological evaluation of novel (aryloxopropenyl)pyrrolyl hydroxyamides. J. Med. Chem. 48 (9), 3344–3353. http://dx.doi.org/10.1021/jm049002a.

Mai, A., Massa, S., Ragno, R., Cerbara, I., Jesacher, F., Loidl, P., Brosch, G., 2003b. 3-(4-Aroyl-1-methyl-1H-2-pyrrolyl)-N-hydroxy-2-alkylamides as a new class of synthetic histone deacetylase inhibitors. 1. Design, synthesis, biological evaluation, and binding mode studies performed through three different docking procedures. J. Med. Chem. 46 (4), 512–524. http://dx.doi.org/10.1021/jm021070e.

Mai, A., Massa, S., Ragno, R., Esposito, M., Sbardella, G., Nocca, G., Scatena, R., Jesacher, F., Loidl, P., Brosch, G., 2002. Binding mode analysis of 3-(4-benzoyl-1-methyl-1H-2-pyrrolyl)-N-hydroxy-2-propenamide: a new synthetic histone deacetylase inhibitor inducing histone hyperacetylation, growth inhibition, and terminal cell differentiation. J. Med. Chem. 45 (9), 1778–1784. http://dx.doi.org/10.1021/jm011088+.

Mai, A., Massa, S., Rotili, D., Pezzi, R., Bottoni, P., Scatena, R., Meraner, J., Brosch, G., 2005c. Exploring the connection unit in the HDAC inhibitor pharmacophore model: novel uracil-based hydroxamates. Bioorg. Med. Chem. Lett. 15 (21), 4656–4661. http://dx.doi.org/10.1016/j.bmcl.2005.07.081.

Mai, A., Massa, S., Valente, S., Simeoni, S., Ragno, R., Bottoni, P., Scatena, R., Brosch, G., 2006. Aroyl-pyrrolyl hydroxyamides: influence of pyrrole C4-phenylacetyl substitution on histone deacetylase inhibition. ChemMedChem 1 (2), 225–237. http://dx.doi.org/10.1002/cmdc.200500015.

Marshall, G.R., Arimoto, R., Ragno, R., Head, R.D., 2000a. Predicting Affinity: the Sina Qua Non of Activity. Abstr. Pap. Am. Chem. Soc. 219, 056–COMP.

Marshall, G.R., Head, R.D., Ragno, R., 2000b. Affinity Prediction: the Sine Qua Non. J. Med. Chem. 43, 1320–1328.

Massa, S., Artico, M., Corelli, F., Mai, A., Di Santo, R., Cortes, S., Marongiu, M.E., Pani, A., La Colla, P., 1990. Synthesis and antimicrobial and cytotoxic activities of pyrrole-containing analogues of trichostatin A. J. Med. Chem. 33 (10), 2845–2849.

Massa, S., Mai, A., Sbardella, G., Esposito, M., Ragno, R., Loidl, P., Brosch, G., 2001. 3-(4-aroyl-1H-pyrrol-2-yl)-N-hydroxy-2-propenamides, a new class of synthetic histone deacetylase inhibitors. J. Med. Chem. 44 (13), 2069–2072.

Matter, H., Baringhaus, K.-H., Naumann, T., Klabunde, T., Pirard, B., 2001. Computational approaches towards the rational design of drug-like compound libraries. Comb. Chem. High. Throughput Screen. 4 (6), 453–475. http://dx.doi.org/10.2174/1386207013330896.

Minucci, S., Pelicci, P.G., Mai, A., Ballarini, M., Gargiulo, G., & Massa, S., 2006. WO2006037761A1.

Mjalli, A.M.M., Harris, A.L., 1996. Lead generation using combinatorial chemistry in drug discovery. J. Biomol. Screen. 1 (1), 17–18. http://dx.doi.org/10.1177/108705719600100108.

Mohamadi, F., Richards, N.G.J., Guida, W.C., Liskamp, R., Lipton, M., Caufield, C., Chang, G., Hendrickson, T., Still, W.C., 1990. Macromodel—an integrated software system for modeling organic and bioorganic molecules using molecular mechanics. J. Comput. Chem. 11 (4), 440–467.

Moniot, S., Schutkowski, M., Steegborn, C., 2013. Crystal structure analysis of human Sirt2 and its ADP-ribose complex. J. Struct. Biol. 182 (2), 136–143.

Morris, G.M., Huey, R., Lindstrom, W., Sanner, M.F., Belew, R.K., Goodsell, D.S., Olson, A.J., 2009. AutoDock4 and AutoDockTools4: Automated docking with selective receptor flexibility. J. Comput. Chem. 30 (16), 2785–2791.

North, B.J., Verdin, E., 2004. Sirtuins: Sir2-related NAD-dependent protein deacetylases. Genome Biol. 5 (5), 224.

Ortiz, A.R., Pisabarro, M.T., Gago, F., Wade, R.C., 1995. Prediction of drug binding affinities by comparative binding energy analysis. J. Med. Chem. 38 (14), 2681–2691.

Outeiro, T.F., Kontopoulos, E., Altmann, S.M., Kufareva, I., Strathearn, K.E., Amore, A.M., Volk, C.B., Maxwell, M.M., Rochet, J.C., McLean, P.J., Young, A.B., Abagyan, R., Feany, M.B., Hyman, B.T., Kazantsev, A.G., 2007. Sirtuin 2 inhibitors rescue alpha-synuclein-mediated toxicity in models of Parkinson's disease. Science 317 (5837), 516–519.

Ragno, R., Mai, A., Massa, S., Cerbara, I., Valente, S., Bottoni, P., Scatena, R., Jesacher, F., Loidl, P., Brosch, G., 2004. 3-(4-Aroyl-1-methyl-1H-pyrrol-2-yl)-N-hydroxy-2-propenamides as a new class of synthetic histone deacetylase inhibitors. 3. Discovery of novel Lead compounds through structure-based drug design and docking studies. J. Med. Chem. 47 (6), 1351–1359. http://dx.doi.org/10.1021/jm031036f.

Ragno, R., Simeoni, S., Rotili, D., Caroli, A., Botta, G., Brosch, G., Massa, S., Mai, A., 2008. Class II-selective histone deacetylase inhibitors. Alignment-independent GRIND 3-D QSAR, homology and docking studies. Eur. J. Med. Chem. 43 (3), 621–632. http://dx.doi.org/10.1016/j.ejmech.2007.05.004.

Ragno, R., Simeoni, S., Valente, S., Massa, S., Mai, A., 2006. 3-D QSAR studies on histone deacetylase inhibitors. A GOLPE/GRID approach on different series of compounds. J. Chem. Inf. Model. 46 (3), 1420–1430. http://dx.doi.org/10.1021/ci050556b.

Rotili, D., Tarantino, D., Nawrozkij, M.B., Babushkin, A.S., Botta, G., Marrocco, B., Cirilli, R., Menta, S., Badia, R., Crespan, E., Ballante, F., Ragno, R., Esté, J.A., Maga, G., Mai, A., 2014. Exploring the role of 2-chloro-6-fluoro substitution in 2-alkylthio-6-benzyl-5-alkylpyrimidin-4(3H)-ones: effects in HIV-1-infected cells and in HIV-1 reverse transcriptase enzymes. J. Med. Chem. 57 (12), 5212–5225. http://dx.doi.org/10.1021/jm500284x.

Sakkiah, S., Arooj, M., Cao, G.P., Lee, K.W., 2013. Insight the C-site pocket conformational changes responsible for sirtuin 2 activity using molecular dynamics simulations. PLoS One 8 (3), e59278.

Sanders, B.D., Jackson, B., Marmorstein, R., 2010. Structural basis for sirtuin function: what we know and what we don't. Biochim. Biophys. Acta 1804 (8), 1604–1616.

Silvestri, L., Ballante, F., Mai, A., Marshall, G.R., Ragno, R., 2012. Histone deacetylase inhibitors: structure-based modeling and isoform-selectivity prediction. J. Chem. Inf. Model. 52 (8), 2215–2235. http://dx.doi.org/10.1021/ci300160y.

Wang, D.-F., Wiest, O.G., & Helquist, P., 2005. Homology modeling, docking and molecular dynamics simulation of class I HDACs, Poster 229. In: The 230th ACS National Meeting, in Washington, DC, August 28–September 1, 2005.

Wang, D.-F., Wiest, O., Helquist, P., Lan-Hargest, H.-Y., Wiech, N.L., 2004. On the function of the 14 Å long internal cavity of histone deacetylase-like protein: implications for the design of histone deacetylase inhibitors. J. Med. Chem. 47 (13), 3409–3417. http://dx.doi.org/10.1021/jm0498497.

Xie, A.-H., Li, B.-Y., Liao, C.-Z., Li, Z.-B., Lu, X.-P., Shi, L.-M., Zhou, J.-J., 2004. Docking study of HDAC implication for benzamide inhibitors binding mode. Wuli Huaxue Xuebao 20 (6), 569–572.

CHAPTER 7

Searching Histone Deacetylase Inhibitors under Computational Procedures

O.J. Zacarías-Lara, D.E. González-Juárez, J. Correa-Basurto
Laboratorio de Modelado Molecular, Bioinformática y Diseño de Fármacos de la Escuela Superior de Medicina, Instituto Politécnico Nacional, Mexico City, Mexico

CONTENTS

1. INTRODUCTION

In recent years, genomic and genetic variations have been associated with different human diseases. These variations have allowed us to understand several disease etiologies. Furthermore, this information is a tool that is now being used for designing new therapeutic agents supported by strong molecular target evidence (Berdasco and Esteller, 2013) and in the future to contribute to a medical personal treatment for a great number of diseases including cancer. Also, it is known that epigenetic and genetic alterations contribute to the onset and progression of cancer, due to the treatment fail by certain biomolecules, whose function is to regulate some processes that preserve genetic information to prevent the expression of antioncogene or protein markers that can lead to the development of any pathology without early identification. For this reason, the study of all epigenetic modifications that are reversible or heritable genetic changes without changes in DNA sequence are the starting point for trying to explain the mechanisms involved with cancer, where there is an overexpression of epigenetic regulators such as histone deacetylases enzymes (HDAC) (Baylin and Jones, 2011; Dawson and Kouzarides, 2012). There is a classification of these epigenetic regulators based on their functions: the epigenetic writers, which establish marks in DNA or histones; others, like the epigenetic erasers like HDAC, which remove these marks. Together, epigenetic regulators play an important role regulating various DNA-dependent processes including DNA transcription, DNA replication, and DNA repair.

Epigenetic writers and erasers have an enzymatic nature and relationship with the progression of cancer. Recently, medical chemistry investigators have been focused on developing pharmacological inhibitors for many of these enzymes, particularly HDAC again, as anticancer agents (Dawson et al., 2012). So, taking advantage of modern bioinformatics tools as docking and molecular dynamics (MD) simulations, it is possible to rationally design new drug candidates that show potent anticancer efficacy while decreasing

Epi-Informatics. http://dx.doi.org/10.1016/B978-0-12-802808-7.00007-1

the side effects. Recently experimental evidence increased the knowledge that cancer is not only produced by a fail in a unique cell receptor. Instead, cells require a great number of internal processes where multiple biological receptors are involved, which contributes significantly to the deregulated cell proliferation as is observed in cancer (Geromichalos, 2007); some targets that share an apoptotic role with HDAC are the antiapoptotic protein Bcl-2 (B-cell lymphoma) and the G protein-coupled estrogen receptor 1 (GPER1) (Lindemann et al., 2007).

Due to this, it is clearly necessary to develop chemical compounds, changing the old approach in drug discovery based on one molecule for one target. The new strategy is designing multitarget agents by targeting simultaneously various bioreceptors implicated in multiple pathological processes such as the case of cancer. The initiation and progression of cancer is often caused by the interference in processes related to many genes, proteins, and signaling pathways involved in damaged cells (Borisy et al., 2003).

2. HISTONES

There are several human diseases associated with gene information. The genes are located inside the DNA that regulates the protein expression that participates in cell homeostasis. However, several environmental factors (physical, chemical, biological, etc.) are capable of inducing human diseases by affecting the gene expression. These environmental factors could induce epigenetic changes, in which there are neither mutations nor deletion of DNA. Epigenetic changes include methylation, acetylation, and ubiquitination, among other chemical modifications of DNA or protein close to DNA such as histones, which are catalyzed by several related enzymes. The most studied epigenetic changes have been associated with HDAC and histone acetyltransferase (HAT) activities due to inhibitory or activating properties by several endogen and/or exogenous factors including synthetic drugs.

In eukaryotic cells, genes are complexes with core histones and other chromosomal proteins in the form of chromatin. The basic repeating unit of chromatin, the nucleosome, includes two copies of each of the four core histones H2A, H2B, H3, and H4 wrapped by 146 bp of DNA. With the aid of additional proteins, including histone H1, the nucleosomes are further packaged into 30-nm fibers with six nucleosomes per turn in a spiral or solenoid arrangement. The 30-nm fiber unfolds to generate a template for transcription, an 11-nm fiber or beads on a string, by a mechanism that is not entirely clear. However, it is thought that unfolding involves posttranslational modifications, particularly acetylation of the core histone amino-terminal tails. The 11-nm fiber is also repressive to processes requiring access of proteins (transcription factors) to DNA. As an important component of the nucleosome, each core histone is composed of a structured, three-helix domain called the histone fold and two unstructured 15–30 residues at the amino termini commonly referred to as tails. Although the histone tails are necessary for the formation of the nucleosome, they are required for nucleosome–nucleosome interaction and for establishing transcriptionally repressive chromatin, referred to as heterochromatin. Transcriptionally active chromatin within the nucleus is referred as euchromatin. The core histone tails are susceptible to a great variety of covalent modifications, including acetylation, phosphorylation, methylation, ubiquitination, among others. Patterns of lysine acetylation along regions of nucleosomes have distinct biological meanings, through recruitment of specific proteins as well as structural changes to the chromatin fiber. Thus, the histone tails are exposed; that is a crucial feature for the role of the nucleosome in transcriptional regulation. There are different types of protein complexes capable of altering the chromatin, and these may act in a physiological context to modulate DNA accessibility. One family includes multiprotein complexes that utilize the energy derived from ATP hydrolysis to mobilize or alter the structure of nucleosomes. The other family includes proteins that modify principally the histone tails by adding covalently small molecules that affect the protein charges (Zhang and Reinberg, 2001).

As previously mentioned, the reversible acetylation and deacetylation of the ε-amino groups of lysine residues at the N-terminal domain of histone are catalyzed by HATs and HDACs, respectively, which have been found to support a substantial degree of epigenetic regulation associated to gene expression (Figure 1).

HDACs, as part of multiprotein complexes, catalyze the removal of acetyl groups from lysine residues on histones and nonhistone proteins, cytoskeletal

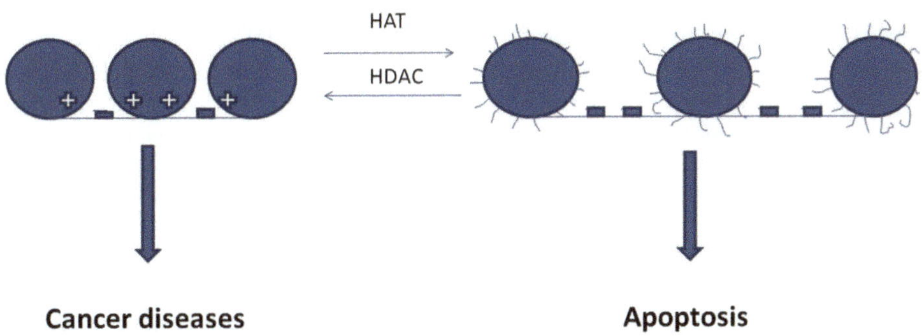

FIGURE 1

Histones are deacetylate and acetylate by HDAC and HAT, respectively, which have relationships with cancer diseases.

proteins, nuclear import factors, and molecular chaperones. This affects gene expression, differentiation, growth arrest, and/or apoptosis in transformed cell cultures (Glozak et al., 2005; Witter et al., 2007).

The hyperacetylation of histone proteins leads to a relaxed chromatin structure that results in the recruitment of various transcription factors (Figure 1). These modifications influence how tightly or loosely the chromatin compacts, and thereby plays a regulatory role in gene expression. Structural changes into chromatin are capable of inducing carcinogenesis and cancer progression. Silencing the tumor suppressor genes is associated with promoter DNA hypermethylation, chromatin hypoacetylation, and decreased histone lysine acetylation. The level of histone acetylation is modulated by HATs and HDACs; the latter are zinc-dependent metalloenzymes. The catalytic functions of these two groups of enzymes are well balanced in normal cells, but aberrant deacetylation of histone as well as other nonhistone proteins will cause many diseases, of which cancer is the most dreadful. Thus, HDAC has become an important target enzyme for anticancer therapies (Huang et al., 2012; Young et al., 2008).

HDACs have been widely studied since these proteins represent emerging targets for the functional modification of the pathogenesis of cancer (Minucci and Pelicci, 2006), HIV, and other diseases (Wightman et al., 2012; Wiech et al., 2009). HDACs are a family of 18 proteins. These enzymes are divided into four groups based on their homology domain, cofactors, and ability to shuttle between the nucleus and the cytoplasm. HDAC isoforms 1, 2, 3, and 8 belong to class I, and isoforms 4, 5, 6, 7, 9, and 10 belong to class II. Class I isoforms are usually located in the nucleus and require zinc (Zn) as a cofactor for enzyme activity, while class II isoforms can be located in either the nucleus or the cytoplasm, depending on the cell's metabolic requirements. Class III isoforms, commonly known as sirtuins, require NAD^+ as a cofactor (Bolden et al., 2006). Class IV contains only one isoform, HDAC11, which, compared to other HDAC isoforms, significantly differs in the composition of its domain. It is believed that HDAC11 diverged from the other isoforms first in evolutionary terms. Most HDAC inhibitors (HDACIs) are active against both class I and II HDACs (Minucci and Pelicci, 2006; Bolden et al., 2006) due to their great structural similarities at the catalytic site.

3. TARGETING HISTONES

It is known that several human diseases have been associated with certain genetic problems and defects that alter the regulation of epigenetic proteins capable of regulating transcriptional mechanisms, which are identified as a key response for the progress and development of cancer, neurological and immunological disorders, among other diseases (Berdasco and Esteller, 2013).

One of the most studied proteins implicated in epigenetic alterations are HDACs. This is because of their enzymatic activity, acting together with other factors like transcriptional regulators to maintain equilibrated changes related with gene activation or repression by certain modifications in chromatin structure (Bannister and Kouzarides, 2011). Specifically, some of these genetic changes have been described as a result of the lysine acetylation of histones. This process neutralizes positive charges of the histone lysine residues and allows relaxation of the chromatin, enabling access of all transcription machinery for gene activation. When the transcription event has finished, in contrast with acetylation, it is necessary to remove those acetyl groups from histones to induce newly chromatin condensation, stopping the transcription, and repressing the expression of genes that could have serious consequences in cellular processes (Haberland et al., 2009). According to the histone families and based on their homology, they are involved in several processes such as proliferation, cell cycle regulation, and apoptosis; this results in DNA changes without sequence alterations, often observed in human diseases, particularly in cancer (Segre and Chiocca, 2011). For this, HDACs are considered important therapeutic targets for treatment of many human cancers. Therefore, HDACs are approached by a great number of investigators around the world for the design, synthesis, and experimentation of new drug candidates capable of inhibiting the activity of HDACs overexpressed in a broad range of cancers types (Gryder et al., 2012).

The great challenge in drug design targeting HDAC is because they have a very similar binding site (catalytic site), which could explain the side effects of the compounds called pan-inhibitors, which are characterized by having a hydroxamic acid or apolar chain (Benedetti et al., 2014). For that reason, rational drug design involves the investigation of new molecules that complement the molecular target in shape and charge, aiming to reach the active site and/or additional binding sites of the enzyme (Nussinov and Tsai, 2012). This could be approached by in silico methods such as molecular docking and MD simulations accompanied with other bioinformatics tools and prediction of ADMET properties. Docking and MD simulations are capable of attempting to predict noncovalent bonds among macromolecules or between a macromolecule and small molecules. Therefore, molecular docking has been used widely to increase the effectiveness to obtain potential drugs. However, when docking is performed, the protein structure is generally modeled without or with limited conformational mobility associated with residue side chain motions. Furthermore, using a native protein conformation ignoring important dynamic aspects of protein–ligand binding (Knegtel et al., 1997) could omit several biological properties required to achieve the ligand–protein complex. Taking into account that proteins have certain physiological properties necessary for coupling with drugs or other molecules, MD simulations can simulate and acquire several protein conformations (Figure 2), which could be

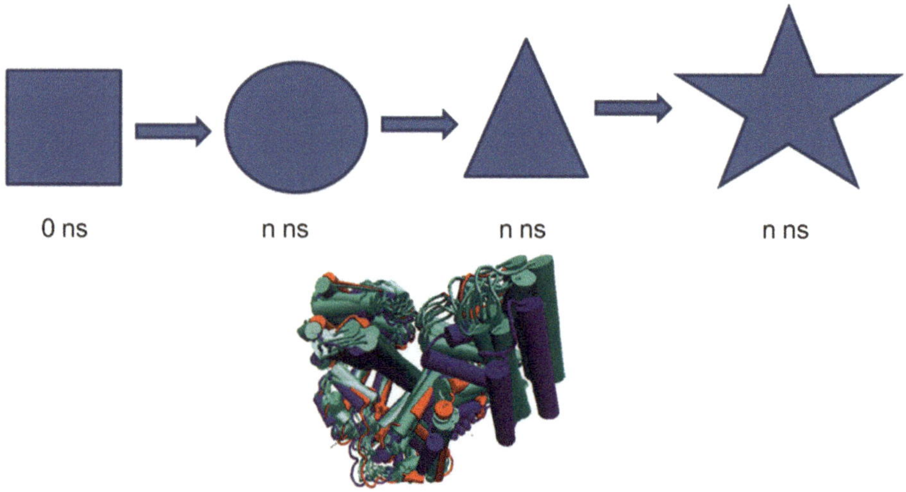

FIGURE 2

Different structural conformation that can suffer a protein study under MD simulations to depict different binding sites. Below is shown the human serum albumin study under MD simulations during 5 ns (Deeb et al., 2010).

used to establish a mode of binding between protein (using several snapshot) and different chemical compounds. In addition, one could perform MD simulations of the ligand–protein complex.

The MD simulation of the ligand–protein complex could help to obtain the free energy values using more robust calculations that include entropy properties. However, the combination of docking and MD simulations may provide valuable information regarding shape-related interactions, non-bonded interactions, and electrostatic properties of both the ligand and the protein (Bello et al., 2013). The combined approach (docking and MD simulations) can help determine the possible factors that affect the affinity and binding mode of a drug–receptor complex. This information is very important for drug design due to the complexity of the action site in every biological target. Recently, some structural details about the HDAC8 binding site were described, combining docking and MD simulations (Bermúdez-Lugo et al., 2012) and experimentally with crystal structures (Micelli and Rastelli, 2015). The structural data provides descriptions of additional binding sites that could help to design new selective compounds for HDAC8. This data suggest that the traditional pharmacophores are limited to some chemical cores and consequently not capable of discriminating HDAC enzymes.

Current inhibitors of HDACs share a common pharmacophore that include (1) a zinc binding group (ZBG) or warhead to chelate the catalytic zinc ion,

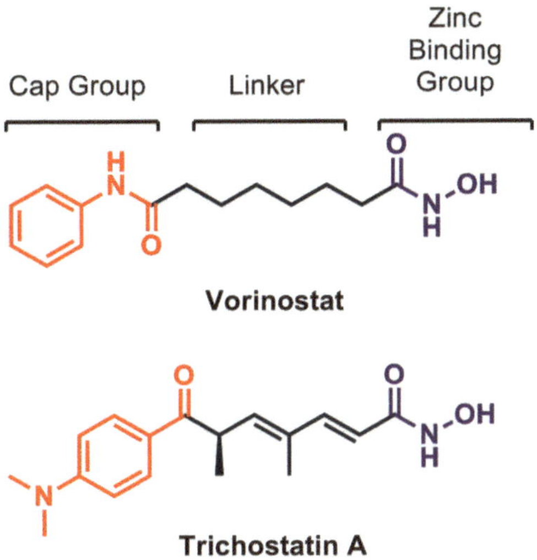

SCHEME 1

Chemical structures of Vorinostat and Trichostatin A.

(2) a capping group binding on the surface of the active site pocket, and (3) a linker between the ZBG and the cap (Bieliauskas et al., 2007).

Vorinostat (SAHA) and Trichostatin (TSA), shown in Scheme 1, are some of the most studied compounds in this way. Both are HDAC inhibitors that chelate the zinc ion in the catalytic pocket with their hydroxamic acid group, which has been described as the main mechanism of HDAC inhibition (Weerasinghe et al., 2008). However, there are small molecules such as butyric and valproic acid that inhibit the HDAC activity but lack the hydroxamic acid moiety. Our hypothesis is that those small molecules reach additional binding cavities with hydrophobic side chains, as previously described by our work group (Bermúdez-Lugo et al., 2012; Sixto-López et al., 2014). This hypothesis suggests that molecules that have these additional pharmacophore groups will have selective activity for HDAC8.

Although the effects of the cap group and the linker have been explored by different groups for developing new drugs able to inhibit HDAC (Pontiki and Hadjipavlou-Litina, 2012), it is generally thought that the ZBG, to a significant extent, is responsible for the potency and sometimes the selectivity of HDAC inhibitors (Methot et al., 2008). Hydroxamate is the most effective zinc binding group known in HDAC inhibitors (Fennell et al., 2008). However, hydroxamates have poor pharmacokinetics due to rapid clearance (Bradner et al., 2010) and are associated with severe toxicity due to nonspecific metal binding (Summers et al., 1987). However, the suberoyl-anilide-hydroxamic acid (SAHA) (Vorinostat) was approved as the first inhibitor of

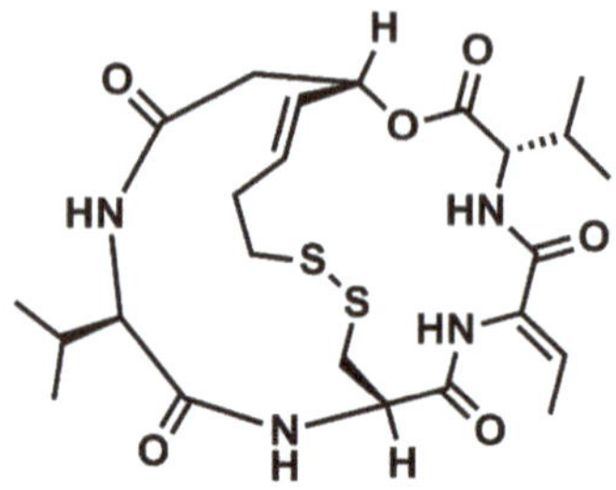

SCHEME 2
Chemical structure of Romidepsin.

histone deacetylase in 2006 by the US Food and Drug Administration (FDA) for the treatment of cutaneous T cell lymphoma CTCL (Marks and Breslow, 2007). Another HDAC inhibitor also approved by the FDA in 2009 for the treatment of CTCL and other types of cancer is the depsipeptide Romidepsin (Scheme 2), a natural product obtained from the bacteria *Chromobacterium violaceum*, considered a stable prodrug to inhibit class I enzymes and activated by reduction after uptake into the cells (Furumai et al., 2002).

Due to all pharmacokinetics and toxicity problems associated with hydroxamic acid inhibitors of HDAC, a new kind of molecules called second-generation HDAC inhibitors have been prepared and used in clinical assays (Suzuki et al., 1999). Examples are benzamide compounds such as MS-275 (Entinostat) and MGCD0103 (Mocetinostat) shown in Scheme 3, that are able to inhibit HDACs in micromolar concentrations, associating their mechanism of action with the capability to reach the pocket foot in the catalytic site of enzyme. However, this is not clearly understood yet due to the geometry and molecular properties of benzamides in contrast with hydroxamic acid derivatives (Bressi et al., 2010). The advantage of this compound is that it has better physicochemical properties than TSA; this goal is approached by investigators to provide the basis for the development of ideal HDAC inhibitors free of the problems associated with hydroxamates.

In this context other classes of chemical compounds such as short-chain fatty acids, butanoic acid, valproic acid (VPA), and 4-phenylbutanoic acid (Scheme 4) are agents that exhibit certain potency for HDACs with IC_{50} values for inhibition in the micromolar range (Lea and Tulsyan, 1995; Newmark and Young, 1995; Phiel et al., 2001; Gore and Carducci, 2000). Furthermore, new valproic acids have been developed, however, their potency is very close to VPA, increasing the water solubility (Luna-Palencia et al., 2014). Furthermore, we are developing new compounds with aryl moieties with better potency (Silva-Trujillo et al., 2015), which have been tested on HeLa cells, increasing the potency in

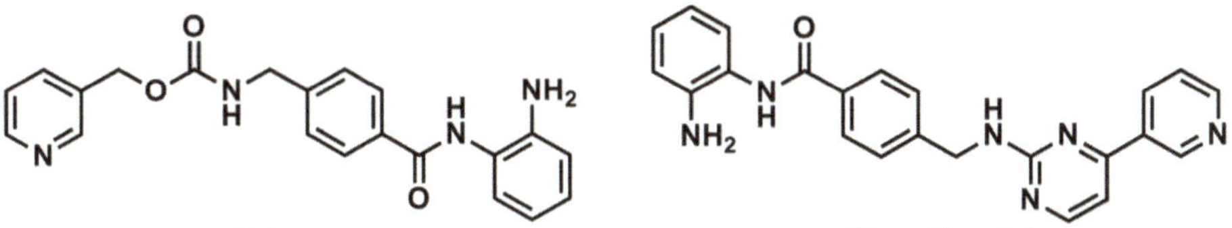

Etinostat Mocetinostat

SCHEME 3

Chemical structures of Entinostat and Mocetinostat.

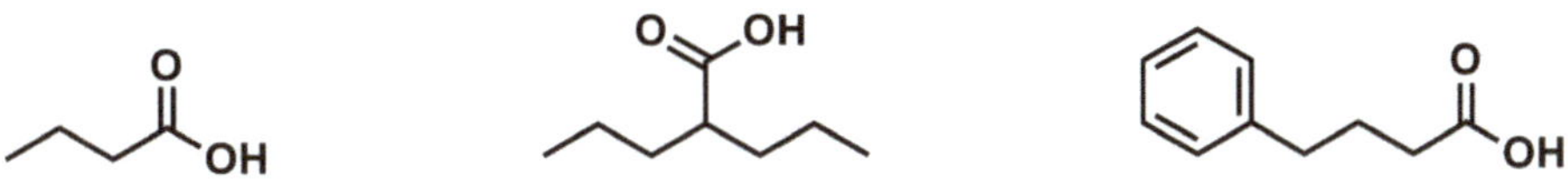

Butanoic acid Valproic acid Phenylbutanoic acid

SCHEME 4

Chemical structures of butanoic, valproic, and phenylbutanoic acids.

a micromolar range. These compounds were developed using computational strategies considering the pharmacophore groups of SAHA and VPA (patent request MX/a/2013/002766).

The potency of short-chain fatty acids is generally low due to the short side chains that limit the contact with the catalytic pocket of HDACs (Acharya et al., 2005). However there are hypotheses that short-chain fatty acids mediate HDAC inhibition through nonspecific hydrophobic interactions with surface residues located at the enzyme pocket entrance and/or the hydrophobic region inside the tube-like pocket. According to this, it is rationalized that the weak potency of short-chain fatty acids in HDAC inhibition is attributed to their inability to access the Zn^{2+} cation in the active-site pocket, which plays a pivotal role in the deacetylation catalysis (Lu et al., 2004).

For this, in the design of new HDAC inhibitors, it is necessary for all chemical compounds to include the pharmacophore structure responsible for inhibition, consisting of the three-component working model (cap, linker, and ZBG) in order to develop more potent and selective inhibitors. However, since it has been possible to obtain crystal structures of HDAC, specifically HDAC8 in complex with some inhibitors (Somoza et al., 2004; Vannini et al., 2004) and in silico methods were developed and used for drug design, it has been possible to reveal new information about different interactions and details about the mechanism of inhibition of HDAC activity (Vannini et al., 2007; Thangapandian et al., 2010). The synergy of crystallography and in silico methods is a promising strategy to understand various aspects of the deacetylation mechanism and guide the design of potential cancer chemotherapeutic agents. As mentioned before, docking and MD simulations are bioinformatics tools useful for this purpose. For example, our work group has identified that HDAC8 has several additional sites that are much conserved or that emerge during MD simulations (Sixto-Lopéz et al., 2014).

4. GENERAL STEPS FOR DEVELOPMENT OF NEW DRUGS

Several years ago the identification of drugs were in some cases accidental. However, now by using bioinformatics, chemoinformatics, and molecular modeling tools it is possible to design drugs using the protein structures considering also the ADMET (absorption, distribution, metabolism, excretion, and toxicity) properties (Rosales-Hernández and Correa-Basurto, 2015). Taking into account ADMET and binding properties one could start to select the best candidates by designing the pharmacophore cores according to structural retrieval from the protein by either protein–ligand complexes from experimental assays or by docking studies. Then, one can build a global scoring that includes ADMET and binding properties retrieved from docking or MD simulations. The ADMET properties could include the Lipinski's rules, referring to important parameters that determinate parameters of the chemical compounds for their in vitro or in vivo experimental evaluation. These parameters could give an idea of the potential application for considering a ligand such as a good drug candidate (Castillo-Garit et al., 2008). However, when there are several chemical values they need to be submitted to Qualitative Structure Activity Relationship (QSAR) analyses and principal component analyses (PCA) or screening workflow systems (Figure 3) to accelerate and automate drug searching (Rosales-Hernández and Correa-Basurto, 2015).

It is important to mention that the first strategy to begin a drug development it is necessary to identify a target disease and obtain all information about the pathology mechanisms, the drug alternatives that are employed in the treatment for cure or stop the malignance, evaluate positive and negative aspects in relation with drugs used in the therapeutic. These include pharmacokinetics properties of the molecules (ADMET), side effects, and cost of drug production (Choi and Yu, 2014). The second step or strategy in drug development is to identify on a cellular or molecular level the physiological factors involved in the pathogenesis; in other words, this refers to knowing which biological macromolecules in an organ or cell trigger a disorder for the disease progress. Thereby, one could focus on target-to-drug design, principally, because there are different biological targets like membrane receptors, enzymes, intracellular receptors, ionic channels, nuclear receptors, nucleic acids, and so on, which are structurally different (Landry and Gies, 2008). This gives a clear idea of

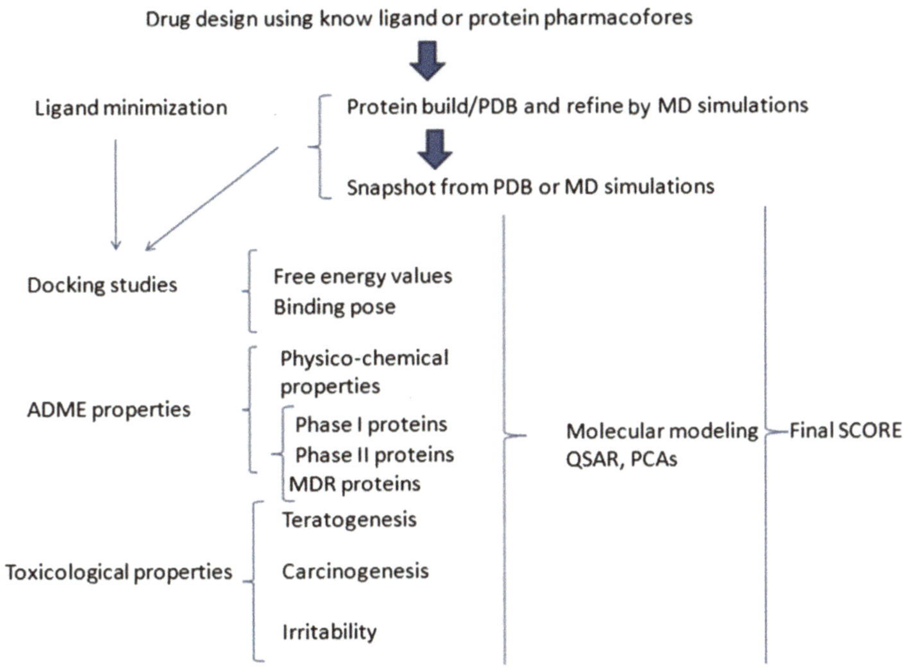

FIGURE 3
Parameter used for drug development apply to HDAC inhibitors to get a final score to select the best candidates. *Modified from Rosales-Hernández and Correa-Basurto (2015).*

the chemical characteristics that a drug or molecule needs to make possible its arrival at one of these biological target sites.

The next step is to establish or propose bioassays (in vitro, in vivo, ex vivo) for future evaluation of the pharmacology of drug candidates. But one can ask the question, how does one select a chemical compound for its application as a drug? To begin answering this question one needs to consider that a drug candidate could have natural or synthetic origin. If one knows the mechanism of action of any drug used for the treatment of any disease and the pharmacophore structure responsible of this action, one could construct or design new derivatives for better selectivity and potency than the native structure, with the purpose of decreasing or avoiding the side and toxic effects (Yang, 2010). After drug design, it is possible to use bioinformatics tools (docking, MD simulations) for establishing whether a molecule is capable of binding on a biological target (Tintori, 2008). Finally, with this virtual screening one could decide the best and efficient strategy for its preparation; this includes the synthetic route or the chemical modifications of the pharmacophore structure necessary to obtain a drug candidate employed in the biological experimentation to determinate its application and explain its target recognition related to disease progress (Estrada et al., 2000).

5. DEVELOPING NEW STRATEGIES TO GET SELECTIVE COMPOUNDS ON HISTONES

To identify the best drug candidate that could reach the target selectively, one needs to explore additional protein cavities named hot-spots. These can be explored testing several snapshots from a reported structure deposited at the Protein Data Bank or retrieve the snapshot from MD simulations (Ramírez-Durán et al., 2013). There are several ligands that are structurally different and they are capable of reaching a target, because these targets are flexible and they are capable of showing several sites that recognize different structural ligands (Méndez-Luna et al., 2015). These strategies allow one to identify potential additional binding sites that help to incorporate new pharmacophore moieties as happens with a set of compounds (not published data) that were designed after the structural studies of HDAC8 by our work group (Bermúdez-Lugo et al., 2012 and Sixto-López et al., 2014). There are other reports that employ pharmacophore-based virtual screening coupled with docking and MD simulations to describe the binding poses (Hou et al., 2015). Thus, the evidence that computational resources are needed as a new strategy to get new HDAC inhibitors, which could act selectively by adding pharmacophore moieties capable of reaching sites located in the proteins surface being their hallmarks.

6. DEVELOPING NEW STRATEGIES TO MULTITARGET COMPOUNDS INCLUDING HISTONES

Due to the great advantages of targeting multiple intracellular ways there are several recent work groups focusing on taking several pharmacophores to design multitarget compounds. These strategies could be guided by using in silico tools, particularly docking and MD simulations. The advantages are to avoid the multitherapy drugs and the drug resistance, and the effectiveness could be increased (Figure 4). HDACs are inhibited by small molecules, which are easily chemically modified. For example, in HDAC8 one could accept compounds with hydroxamic acid and hydrophobic properties, two potential binding sites, as was observed in our contributions (Bermúdez-Lugo et al., 2012 and Sixto-López et al., 2014). By adding other pharmacophores capable of reaching other proteins such as Bcl-2 protein and GPR30 one can target a way to relate to the induction of death in cancer cells by apoptosis.

Targeting several biological targets has the advantage that when there is drug resistance of one target, the multitarget ligand recognizes other biological targets, making the chemotherapy effective. In our work group we are designing new multitarget compounds (unpublished data) using the structural data study and reported for HDAC8 (Bermúdez-Lugo et al., 2012 and Sixto-López et al., 2014), for Bcl-2 (Ilizaliturri-Flores et al., 2014) and GPR30 (Méndez-Luna et al., 2015).

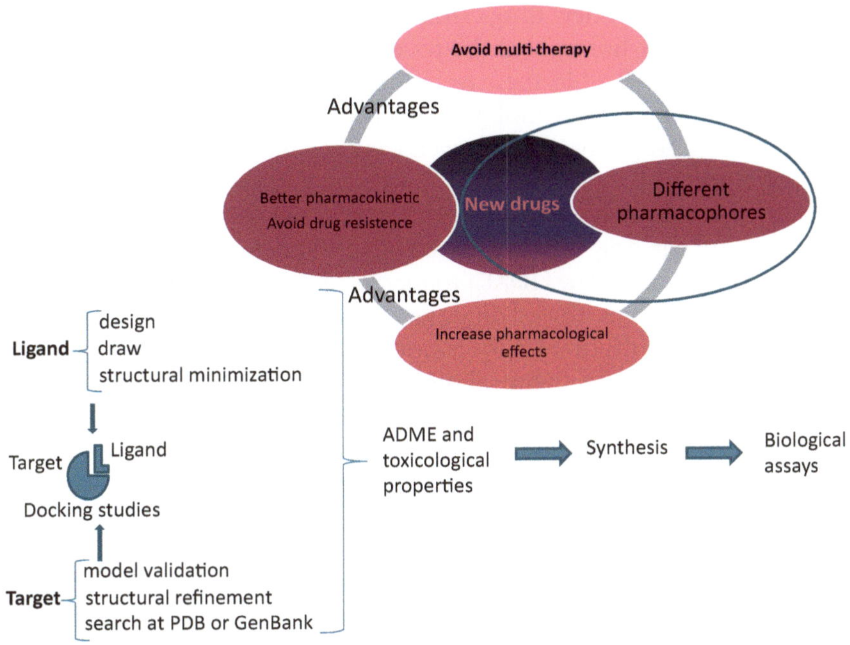

FIGURE 4

With computational resources (mixing docking and MD simulations) it is possible to describe the pharmacophore groups of different targets related in the same diseases getting compounds with several pharmacological advantages to calculate their DMET properties to select the best candidates to be synthesized and biologically tested.

Our experiences in the multitarget compounds have been applied to design compounds against Alzheimer's disease with several advantages (Hernández-Rodríguez et al., 2014). However, it is not new that there are several published works focused on the development of new multitarget compounds. Several of these studies use computational resources (Fang et al., 2015).

In conclusion, one can use computational resources to get small molecules that act selectively on one target, or reach different targets, with their corresponding advantages and disadvantages, pharmacologically as well as the complications in the chemical synthesis.

7. CONCLUSIONS

It is evident that issues associated with diseases such as cancer require a quick and easy solution, especially all those related to the treatment employed: poor pharmacokinetic and toxicity by chemotherapeutics, the development of drug resistance by cancer cells, and the expensive cost of anticancer drugs and their multitherapy use. Through MD and molecular docking it is possible to carry out a rational design of new drug candidates, especially for expanding the use of a great molecular library with better advantages in treatment, control, or prevention of cancer. By this, these techniques could help to conduct a rational drug design, respecting structural parameters of ligand or prototype molecules before particular properties in the target. For example, for HDAC, it is possible to find significant HDAC8 structural details that rule the ligand recognition process such as additional cavities with chemical details, which can be taken into account for the drug design. These HDAC8 structural properties could be important in designing new HDACI to increase HDAC8 selectivity, a particular issue present in the majority of drugs listed above as inhibitors. Furthermore, the use of these bioinformatics tools not only can contribute to the research with target details. If the chemical and pharmacological requirements for a ligand to interact correctly with its macromolecular target receptor are known, it could be possible to introduce different pharmacophoric groups providing the ligand with a larger ability to bind selectively to a set of desired bioreceptors. Thus the result of this design will be multitarget compounds, which minimize toxicity problems, solve pharmacokinetics issues, reduce costs and time in synthetic preparation, use low doses, avoid multitherapy, and remove chemo-resistance by cancer cells. With this purpose, is now a great opportunity to contribute to drug discovery and excite the pharmaceutical industry.

REFERENCES

Acharya, M.R., Sparreboom, A., Venitz, J., Figg, W.D., 2005. Rational development of histone deacetylase inhibitors as anticancer agents: a review. Mol. Pharmacol. 68 (4), 917–932.

Bannister, A.J., Kouzarides, T., 2011. Regulation of chromatin by histone modifications. Cell Res. 21, 381–395.

Baylin, S.B., Jones, P.A., 2011. A decade of exploring the cancer epigenome-biological and translational implications. Nat. Rev. Cancer 11, 726–734.

Bello, M., Martínez-Archundia, M., Correa-Basurto, J., 2013. Automated docking for novel drug discovery. Expert Opin. Drug Discov. 8 (7), 821–834.

Benedetti, R., Conte, M., Altucci, L., 2014. Targeting histone deacetylases in diseases: where are we? Antioxid. Redox Signal 23 (1), 99–126.

Berdasco, M., Esteller, M., 2013. Genetic syndromes caused by mutations in epigenetic genes. Hum. Genet. 132, 359–383.

Bermúdez-Lugo, J.A., Perez-González, O., Rosales-Hernández, M.C., Ilizaliturri-Flores, I., Trujillo-Ferrara, J., Correa-Basurto, J., 2012. Exploration of the valproic acid binding site on histone deacetylase 8 using docking and molecular dynamic simulations. J. Mol. Model 18 (6), 2301–2310.

Bieliauskas, A., Weerasinghe, S.V.W., Pflum, M.K.H., 2007. Structural requirements of HDAC inhibitors: SAHA analogs functionalized adjacent to the hydroxamic acid. Bioorg. Med. Chem. Lett. 17, 2216–2219.

Bolden, J.E., Peart, M.J., Johnstone, R.W., 2006. Anticancer activities of histone deacetylase inhibitors. Nat. Rev. Drug Discov. 5 (9), 769–784.

Borisy, A.A., Elliott, P.J., Hurst, N.W., Lee, M.S., Lehár, J., Price, E.R., et al., 2003. Systematic discovery of multicomponent therapeutics. Proc. Natl. Acad. Sci. U.S.A. 100, 7977–7982.

Bradner, J.E., West, N., Grachan, M.L., Greenberg, E.F., Haggarty, S.J., Warnow, T., Mazitschek, R., 2010. Chemical phylogenetics of histone deacetylases. Nat. Chem. Biol. 6 (3), 238–243.

Bressi, J.C., Jennings, A.J., Skene, R., Wu, Y., Melkus, R., De Jong, R., O'Connell, S., Grimshaw, C.E., Navre, M., Gangloff, A.R., 2010. Exploration of the HDAC2 foot pocket: synthesis and SAR of substituted N-(2-aminophenyl) benzamides. Bioorg. Med. Chem. Lett. 20, 3142–3145.

Castillo-Garit, J.A., Marrero-Ponce, Y., Torrens, F., Garcia-Domenech, R., 2008. Estimation of ADME properties in drug discovery: predicting Caco-2 cell permeability using atom-based stochastic and non-stochastic linear indices. J. Pharm. Sci. 97, 1946–1976.

Choi, Y.H., Yu, A.M., 2014. ABC transporters in multidrug resistance and pharmacokinetics, and strategies for drug development. Curr. Pharm. Des. 20 (5), 793–807.

Dawson, M.A., Kouzarides, T., 2012. Cancer epigenetics: from mechanism to therapy. Cell 150, 12–27.

Dawson, M.A., Kouzarides, T., Huntly, B.J., 2012. Targeting epigenetic readers in cancer. N. Engl. J. Med. 367, 647–657.

Deeb, O., Rosales-Hernández, M.C., Gómez-Castro, C., Garduño-Juárez, R., Correa-Basurto, J., 2010. Exploration of human serum albumin binding sites by docking and molecular dynamics flexible ligand-protein interactions. Biopolymers 93 (2), 161–170.

Estrada, E., Uriarte, E., Montero, A., Teijeira, M., Santana, L., De Clercq, E., 2000. A novel approach for the virtual screening and rational design of anticancer compounds. J. Med. Chem. 43, 1975–1985.

Fang, J., Li, Y., Liu, R., Pang, X., Li, C., Yang, R., He, Y., Lian, W., Liu, A.L., Du, G.H., 2015. Discovery of multitarget-directed ligands against Alzheimer's disease through systematic prediction of chemical-protein interactions. J. Chem. Inf. Model 55 (1), 149–164.

Fennell, K.A., Moellmann, U., Miller, M.J., 2008. Syntheses and biological activity of amamistatin B and analogs. J. Org. Chem. 73, 1018–1024.

Furumai, R., Matsuyama, A., Kobashi, N., Lee, K.H., Nishiyama, M., Nakajima, H., Tanaka, A., Komatsu, Y., Nishino, N., Yoshida, M., Horinouchi, S., 2002. FK228 (depsipeptide) as a natural prodrug that inhibits class I histone deacetylases. Cancer Res. 62, 4916–4921.

Geromichalos, G.D., 2007. Importance of molecular computer modeling in anticancer drug development. J. BUON 1, 101–118.

Glozak, M.A., Sengupta, N., Zhang, X., Seto, E., 2005. Acetylation and deacetylation of non-histone proteins. Gene 363, 15–23.

Gore, S.D., Carducci, M.A., 2000. Modifying histones to tame cancer: clinical development of sodium phenylbutyrate and other histone deacetylase inhibitors. Expert Opin. Invest Drugs 9, 2923–2934.

Gryder, B.E., Sodji, Q.H., Oyelere, A.K., 2012. Targeted cancer therapy: giving histone deacetylase inhibitors all they need to succeed. Future Med. Chem. 4 (4), 505–524.

Haberland, M., Montgomery, R.L., Olson, E.N., 2009. The many roles of histone deacetylases in development and physiology: implications for disease and therapy. Nat. Rev. Genet. 10, 32–42.

Huang, D., Li, X., Wei, Y., Xiu, Z., 2012. A novel series of L-2-benzyloxycarbonylamino-8-(2-pyridyl)-disulfidyloctanoic acid derivatives as histone deacetylase inhibitors: design, synthesis and molecular modeling study. Eur. J. Med. Chem. 52, 111–122.

Hernández-Rodríguez, M., Correa-Basurto, J., Martínez-Ramos, F., Padilla-Martínez, L.L., Benítez-Cardoza, C.G., Mera-Jiménez, E., Rosales-Hernández, M.C., 2014. Design of multi-target compounds as AChE, BACE1, and amyloid-β(1-42) oligomerization inhibitors: in silico and in vitro studies. J. Alzheimers Dis. 41 (4), 1073–1085.

Hou, X., Du, J., Liu, R., Zhou, Y., Li, M., Xu, W., Fang, H., 2015. Enhancing the sensitivity of pharmacophore-based virtual screening by incorporating customized ZBG features: a case study using histone deacetylase 8. J. Chem. Inf. Model 55 (4), 861–871.

Ilizaliturri-Flores, I., Correa-Basurto, J., Benítez-Cardoza, C.G., Zamorano-Carrillo, A., 2014. A study of the structural properties and thermal stability of human Bcl-2 by molecular dynamics simulations. J. Biomol. Struct. Dyn. 32, 1707–1719.

Knegtel, R.M.A., Kuntz, I.D., Oshiro, C.M., 1997. Molecular docking to ensembles of protein structures. J. Mol. Biol. 266, 424–440.

Landry, Y., Gies, J.P., 2008. Drugs and their molecular targets: an updated overview. Fund. Clin. Pharmacol. 22 (1), 1–18.

Lea, M.A., Tulsyan, N., 1995. Discordant effects of butyrate analogues on erythroleukemia cell proliferation, differentiation and histone deacetylase. Anticancer Res. 15, 879–883.

Lindemann, R.K., Newbold, A., Whitecross, K.F., Cluse, L.A., Frew, A.J., Ellis, L., Williams, S., Wiegmans, A.P., Dear, A.E., Scott, C.L., Pellegrini, M., Wei, A., Richon, V.M., Marks, P.A., Lowe, S.W., Smyth, M.J., Johnstone, R.W., 2007. Analysis of the apoptotic and therapeutic activities of histone deacetylase inhibitors by using a mouse model of B cell lymphoma. Proc. Natl. Acad. Sci. 104, 8071–8076.

Lu, Q., Yang, Y.T., Chen, C.S., Davis, M., Byrd, J.C., Etherton, M.R., Umar, A., Chen, C.S., 2004. Zn^{2+}-chelating motif-tethered short-chain fatty acids as a novel class of histone deacetylase inhibitors. J. Med. Chem. 47, 467–474.

Luna-Palencia, G.R., Martinez-Ramos, F., Vasquez-Moctezuma, I., Fragoso-Vazquez, M.J., Mendieta-Wejebe, J.E., Padilla-Martínez, I.I., Sixto-Lopez, Y., Mendez-Luna, D., Trujillo-Ferrara, J., Meraz-Rios, M.A., Fonseca-Sabater, Y., Correa-Basurto, J., 2014. Three amino acid derivatives of valproic acid: design, synthesis, theoretical and experimental evaluation as anticancer agents. Anti-Cancer Agent Med Chem 14 (7), 984–993.

Marks, P.A., Breslow, R., 2007. Dimethyl sulfoxide to vorinostat: development of this histone deacetylase inhibitor as an anticancer drug. Nat. Biotechnol. 25 (1), 84–90.

Méndez-Luna, D., Martínez-Archundia, M., Maroun, R.C., Ceballos-Reyes, G., Fragoso-Vázquez, M.J., González-Juárez, D.E., Correa-Basurto, J., 2015. Deciphering the GPER/GPR30-agonist and antagonists interactions using molecular modeling studies, molecular dynamics, and docking simulations. J. Biomol. Struct. Dyn. 33 (10), 2161–2172. http://dx.doi.org/10.1080/07391102.2014.994102.

Methot, J.L., Chakravarty, P.K., Chenard, M., Close, J., Cruz, J.C., Dahlberg, W.K., Fleming, J., Hamblett, C.L., Hamill, J.E., Harrington, P., Harsch, A., Heidebrecht, R., Hughes, B., Jung, J., Kenific, C.M., Kral, A.M., Meinke, P.T., Middleton, R.E., Ozerova, N., Sloman, D.L., Stanton, M.G., Szewczak, A.A., Tyagarajan, S., Witter, D.J., Secrist, J.P., Miller, T.A., 2008. Exploration of the internal cavity of histone deacetylase (HDAC) with selective HDAC1/HDAC2 inhibitors (SHI-1:2). Bioorg. Med. Chem. Lett. 18 (3), 973–978.

Minucci, S., Pelicci, P.G., 2006. Histone deacetylase inhibitors and the promise of epigenetic (and more) treatments for cancer. Nat. Rev. Cancer 6 (1), 38–51.

Micelli, C., Rastelli, G., 2015. Histone deacetylases: structural determinants of inhibitor selectivity. Drug Discov. Today 20 (6), 718–735. http://dx.doi.org/10.1016/j.drudis.2015.01.007.

Newmark, H.L., Young, C.W., 1995. Butyrate and phenylacetate as differentiating agents: practical problems and opportunities. J. Cell Biochem. 22, 247–253.

Nussinov, R., Tsai, C.J., 2012. The different ways through which specificity works in orthosteric and allosteric drugs. Curr. Pharm. Des. 18 (9), 1311–1316.

Pontiki, E., Hadjipavlou-Litina, D., 2012. Histone deacetylase inhibitors (HDACIs). Structure-activity relationships: history and new QSAR perspectives. Med. Res. Rev. 32 (1), 1–165.

Phiel, C.J., Zhang, F., Huang, E.Y., et al., 2001. Histone deacetylase is a direct target of valproic acid, a potent anticonvulsant, mood stabilizer, and teratogen. J. Biol. Chem. 276, 36734–36741.

Rosales-Hernández, M.C., Correa-Basurto, J., 2015. The importance of employing computational resources for the automation of drug discovery. Expert Opin. Drug Discov. 10 (3), 213–219.

Ramírez-Durán, L.A., Rosales-Hernández, M.C., Hernández-Rodríguez, M., Mendieta-Wejebe, J.E., Trujillo-Ferrara, J., Correa-Basurto, J., 2013. Mapping myeloperoxidase to identify its promiscuity properties using docking and molecular dynamics simulations. Curr. Pharm. Des. 19 (12), 2204–2215.

Segre, C.V., Chiocca, S., 2011. Regulating the regulators: the posttranslational code of class I HDAC1 and HDAC2. J. Biomed. Biotechnol. 690–848.

Summers, J.B., Gunn, B.P., Mazdiyasni, H., Goetze, A.M., Young, P.R., Bouska, J.B., Dyer, R.D., Brooks, D.W., Carter, G.W., 1987. In vivo characterization of hydroxamic acid inhibitors of 5-lipoxygenase. J. Med. Chem. 30, 2121–2126.

Suzuki, T., Ando, T., Tsuchiya, K., Fukazawa, N., Saito, A., Mariko, Y., Yamashita, T., Nakanishi, O., 1999. Synthesis and histone deacetylase inhibitory activity of new benzamide derivatives. J. Med. Chem. 42, 3001–3003.

Silva-Trujillo, A., Correa-Basurto, J., Romero-Castro, A., Albores, A., Mendieta-Wejebe, J.E., 2015. A simple validated RP-HPLC bioanalytical method for the quantitative determination of a novel valproic acid arylamide derivative in rat hepatic microsomes. Biomed. Chromatogr. 29 (4), 523–528. http://dx.doi.org/10.1002/bmc.3307.

Sixto-López, Y., Gómez-Vidal, J.A., Correa-Basurto, J., 2014. Exploring the potential binding sites of some known HDAC inhibitors on some HDAC8 conformers by docking studies. Appl. Biochem. Biotechnol. 173, 1907–1926.

Somoza, J.R., Skene, R.J., Katz, B.A., Mol, C., Ho, J.D., Jennings, A.J., Luong, C., Arvai, A., Buggy, J.J., Chi, E., Tang, J., Sang, B.C., Verner, E., Wynands, R., Leahy, E.M., Dougan, D.R., Snell, G., Navre, M., Knuth, M.W., Swanson, R.V., McRee, D.E., Tari, L.W., 2004. Structural snapshots of human HDAC8 provide insights into the class I histone deacetylases. Structure 12, 1325–1334.

Tintori, C., 2008. Targets looking for drugs: a multistep computational protocol for the development of structure-based pharmacophores and their applications for hit discovery. J. Chem. Inf. Model 48 (11), 2166–2179.

Thangapandian, S., John, S., Sakkiah, S., Lee, K.W., 2010. Ligand and structure based pharmacophore modeling to facilitate novel histone deacetylase 8 inhibitor design. Eur. J. Med. Chem. 45 (10), 4409–4417.

Vannini, A., Volpari, C., Filocamo, G., Casavola, E.C., Brunetti, M., Renzoni, D., Chakravarty, P., Paolini, C., De Francesco, R., Gallinari, P., Steinkühler, C., Di Marco, S., 2004. Crystal structure of a eukaryotic zinc-dependent histone deacetylase, human HDAC8, complexed with a hydroxamic acid inhibitor. Proc. Natl. Acad. Sci. 101 (42), 15064–15069.

Vannini, A., Volpari, C., Gallinari, P., Jones, P., Mattu, M., Carfi, A., De Francesco, R., Steinkühler, C., Di Marco, S., 2007. Substrate binding to histone deacetylases as shown by the crystal structure of the HDAC8-substrate complex. EMBO Rep. 8 (9), 879–884.

Witter, D.J., Belvedere, S., Chen, L., Secrist, J.P., Mosleyd, R.T., Miller, T.A., 2007. Benzo[b]thiophene-based histone deacetylase inhibitors. Bioorg. Med. Chem. Lett. 17 (16), 4562–4567.

Wiech, N.L., Fischer, J.F., Helquist, P., Wiest, O., 2009. Inhibition of histone deacetylases: a pharmacological approach to the treatment of non-cancer disorders. Curr. Top. Med. Chem. 9 (3), 257–271.

Weerasinghe, S.V., Estiu, G., Wiest, O., Pflum, M.K., 2008. Residues in the 11 A channel of histone deacetylase 1 promote catalytic activity: implications for designing isoform-selective histone deacetylase inhibitors. J. Med. Chem. 51 (18), 5543.

Wightman, F., Ellenberg, P., Churchill, M., Lewin, S.R., 2012. HDAC inhibitors in HIV. Immunol. Cell Biol. 90 (1), 47–54.

Yang, S.Y., 2010. Pharmacophore modeling and applications in drug discovery: challenges and recent advances. Drug Discov. Today 15 (11–12), 444–450.

Young, A.M., Jung, J.H., Jin, N.Y., Sik, K.H., 2008. A natural histone deacetylase inhibitor, Psammaplin A, induces cell cycle arrest and apoptosis in human endometrial cancer cells. Gynecol. Oncol. 108, 27–33.

Zhang, Y., Reinberg, D., 2001. Transcription regulation by histone methylation: interplay between different covalent modifications of the core histone tails. Genes Dev. 15, 2343–2360.

CHAPTER 8

Current Development of Protein Arginine Methyltransferase Inhibitors

Kun Qian, Y. George Zheng

Department of Pharmaceutical and Biochemical Sciences, College of Pharmacy, University of Georgia, Athens, GA, USA

CONTENTS

1. INTRODUCTION

Epigenetics is a prevailing event in living cells that records the developmental and environmental cues by modifying genetic material without altering the nucleotide sequences (Riddihough and Zahn, 2010). Epigenetic regulations determine the normal production of proteins and cellular differentiation, and are associated with many pathogenic pathways in human diseases. Typically, chromatin modification, such as methylation, leaves a chemical mark on DNA, histone, or other related nuclear proteins, which are responsible for switching the conformational transition of chromatin between transcriptionally active and inactive states (Finley and Copeland, 2014). Enzymes that deposit the marks on the histones are considered histone writers, such as protein methyltransferases (PMTs) (Wang et al., 2007). PMTs are associated with specific genetic alterations in human pathological conditions, such as proliferative disorders, metabolic disorders, blood disorders, and cancers (Copeland et al., 2013). Therefore, PMTs are considered potential therapeutic targets (Copeland, 2012).

2. PROTEIN ARGININE METHYLTRANSFERASES

PMTs include two families: PKMTs (protein lysine methyltransferases) and PRMTs (protein arginine methyltransferases). The methylation of arginine residues is catalyzed by the PRMTs. To date, nine PRMT members (PRMT1–PRMT9) have been identified in human cells (Yang and Bedford, 2013). PRMTs are considered epigenetic regulators because histone tails are one of their primary targets (Di Lorenzo and Bedford, 2011). Other than functioning on the histones, PRMTs also methylate other various cellular proteins (Table 1). In general, most of the PRMTs prefer to recognize methylate glycine- and arginine-rich (GAR) motifs in their substrates, except for CARM1(PRMT4), which

Epi-Informatics. http://dx.doi.org/10.1016/B978-0-12-802808-7.00008-3

Table 1 Selected Substrates of PRMTs and Their Cancer Relevance (Bedford and Richard, 2005; Yost et al., 2011; Yang and Bedford, 2013)

PRMTs	Substrates	Cancer Relevance
PRMT1	H4, NAB2p, NPL3p, ERαR260, FOXO1	Overexpressed/aberrant in breast, prostate, lung, colon, bladder cancer, and leukemia
PRMT2	H4	Overexpressed in breast cancer
PRMT3	rpS2, PABPN1	High level activity in breast tumors
CARM1	H3, PABP1, p300/CBP	Overexpressed in breast, prostate, and colorectal cancer
PRMT5	H3, H4, H2A, FEN1, MBP, Sm D1	Overexpression/increased activity in gastric, colorectal, lung cancer, lymphoma, and leukemia
PRMT6	H3, H4, H2A, PRMT6, HIV tat	Overexpressed in bladder and lung cancer
PRMT7	H4, H2A, Fibrillarin	Involved in breast cancer metastasis
PRMT8	PRMT8, EWS (Pahlich et al., 2008)	Somatic mutations were found in ovarian, skin, and large intestine cancer
PRMT9	Unknown	Unknown

instead has an affinity toward proline, glycine, methionine, and arginine-rich (PGM) motifs. PRMT5 also has shown an affinity for PGM motifs in some cases (Bedford et al., 1998; Cheng et al., 2007). The reported crystal structures of PRMTs are shown in Figure 1. The nine human PRMTs are proposed to share a conserved catalytic site (Figure 2), which is organized around a Rossman fold for cofactor binding and a β-barrel for substrate binding (Cheng et al., 2005). Some PRMTs may contain domains for protein–protein interaction at N-terminus, for example, the SH_3domain in PRMT2, zinc finger of PRMT3, PH domain in CARM1, and a TIM barrel in PRMT5. Evidences show that these unique domains are responsible for recruitment of other proteins, formation of homo-oligomers and may also take part in substrate binding (Schapira and Ferreira de Freitas, 2014).

3. CATALYTIC ACTIVITY OF PRMTs

The catalytic reaction mediated by PRMTs are shown in Figure 3. The terminal guanidino group of an arginine residue on the protein substrates of PRMTs can be methylated in three different ways: monomethylation, asymmetrical dimethylation, and symmetrical dimethylation. The product arginines have been named as MMA, aDMA, and sDMA, respectively, and each of them has different functional consequences (Bedford, 2007). The nine mammalian PRMTs are classified into three types: type I (PRMT1, PRMT2, PRMT3, PRMT4, PRMT6, and PRMT8), type II (PRMT5, PRMT9), or type III (PRMT7)

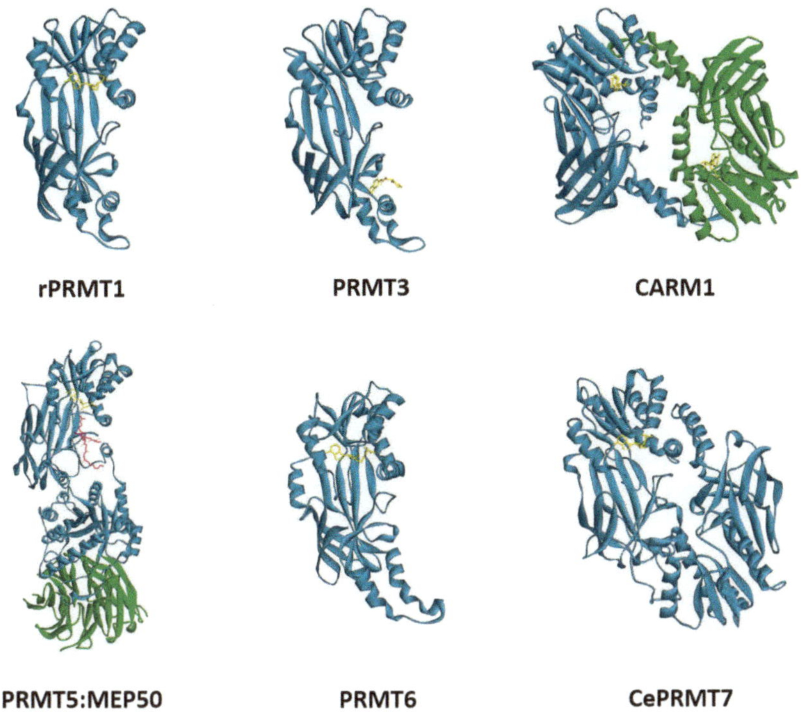

FIGURE 1

Crystal structures of PRMTs. Enzymes are indicated in blue or green, ligands are indicated in yellow. PDB codes: PRMT1, 1OR8 (rat, with SAH); PRMT3, 3SMQ (human, with an allosteric inhibitor); CARM1, 3B3F (rat, dimer, with SAH); PRMT5-MEP50 complex, 4GQB (human, MEP50 is indicated in green, with a SAM analog); PRMT6, 4HC4 (human, with SAH).

(Bedford and Richard, 2005; Di Lorenzo and Bedford, 2011; Yang et al., 2015; Zurita-Lopez et al., 2012). Both type I and type II PRMTs catalyze dimethylation via an MMA (mono-methylated arginine) intermediate, then type I PRMTs further turn MMA into aDMA while type II PRMTs allow formation of sDMA. Type III PRMTs can only monomethylate the substrate. PRMT7 has been assigned into type III, but some studies show its ability to catalyze sDMA (Lee et al., 2005; Zurita-Lopez et al., 2012). PRMT9 is a newly discovered PRMT, and has been assigned to type II (Cook et al., 2006; Yang et al., 2015). The global arginine level in cells is 1500:3:2:1 for Arg:ADMA:MMA:SDMA (Bedford, 2007). Among all, PRMT1 as the primary type I enzyme has been found to be responsible for more than 85% of the total methylation in cells; PRMT5 is the primary type II enzyme (Herrmann et al., 2005). Studies have been shown that the PRMT1 knockout or PRMT5 knockdown cells resulted in substrate scavenging by other PRMTs, which indicated the dynamic interplay between different types of PRMTs (Dhar et al., 2013).

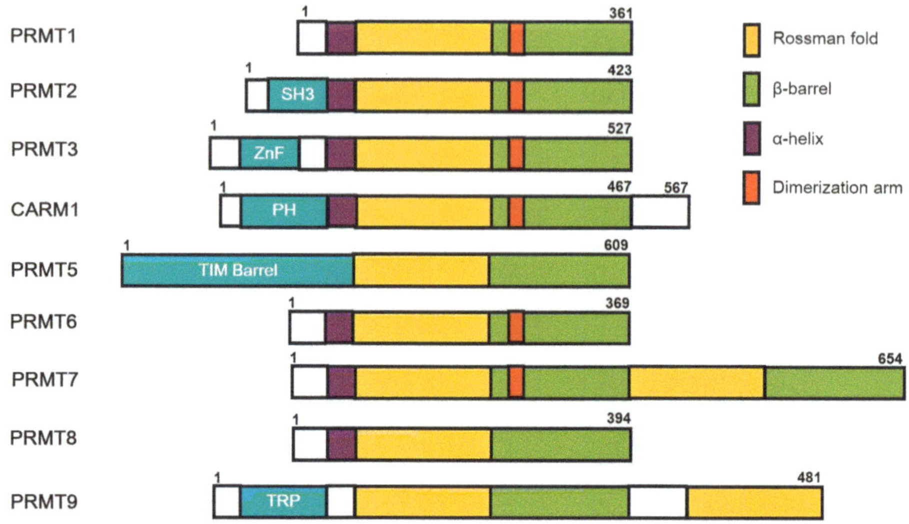

FIGURE 2

Domain architecture of human PRMTs.

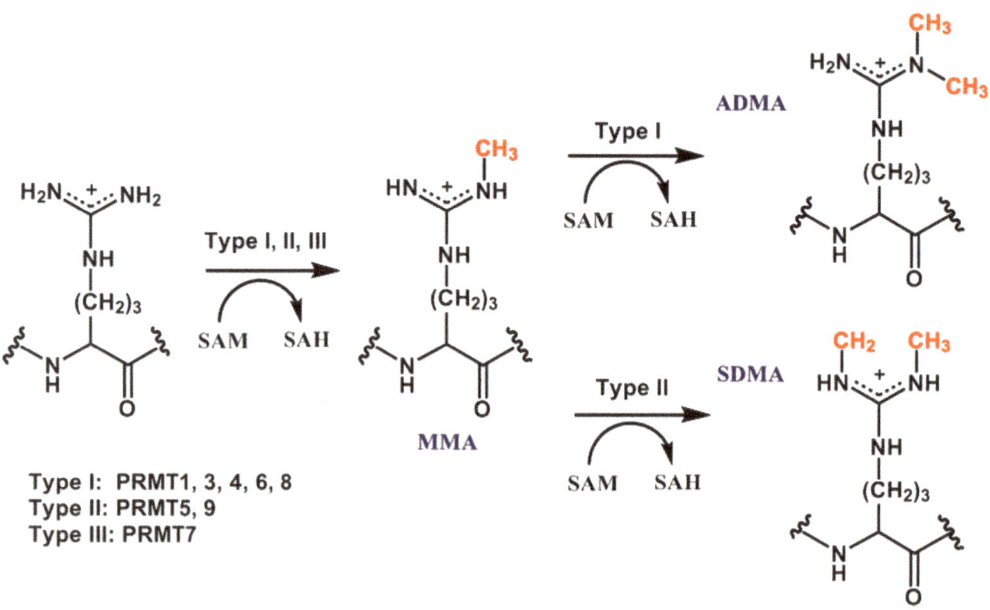

FIGURE 3

Arginine methylation reaction and types of PRMTs. Only side chain of arginine is shown. Methyl group transferred from SAM is indicated in red.

A S_N2-like mechanism was proposed for the methylation process. PRMTs transfer a methyl group from *S*-adenosyl methionine (AdoMet or SAM) to a guanidine nitrogen of arginine, generating *S*-adenosyl homocysteine (AdoHcy or SAH) and the methyl arginine (Bedford and Richard, 2005). The two conserved glutamate residues (i.e., E144 and E153 of PRMT1) interacting with the substrate arginine were found to be crucial in charge redistribution during the catalysis (Zhang et al., 2013; Zhang and Cheng, 2003; Zhang et al., 2000). Evidence also suggested that a conserved methionine residue was essential for discriminating between sDMA and aDMA formation (Antonysamy et al., 2012; Gui et al., 2014, 2013). Each addition of a methyl group to an arginine residue removes a potential hydrogen bond donor and changes its shape (Bedford and Clarke, 2009). Bulkiness and hydrophobicity is introduced by the additional mono- or dimethyl group(s) on the arginine residue of protein substrates and affect protein–protein interactions. However, methylation does not neutralize the cationic charge of an arginine residue (Tripsianes et al., 2011).

4. PHARMACOLOGICAL SIGNIFICANCE

Protein arginine methylation is an abundant modification that has been implicated in, but not limited to, signal transduction, gene transcription, DNA repair, and mRNA splicing (Yang and Bedford, 2013). An increasing amount of evidence shows that abnormality of PRMTs is associated with many diseases, including cancer (Yang and Bedford, 2013). Table 1 shows a brief summarization of the known relevance between PRMTs and cancer. For example, PRMT1 is an essential component of MLL (mixed-lineage leukemia) oncogenic transcriptional complex. One of the major target sites, H4R3, is a suggestive marker for the diagnosis of prostate cancer (Cheung et al., 2007). Aberrant expression of PRMT1 and CARM1 has been observed in breast cancer (Frietze et al., 2008; Scorilas et al., 2000). PRMT5 acts as strong repressor of numerous genes, and the overexpression of PRMT5 has been observed in a variety of lymphoma, leukemia, gastric carcinoma, and immortalized fibroblast cells (Kim et al., 2005; Pal and Sif, 2007; Wang et al., 2008). PRMT6 is a transcriptional repressor and PRMT7 can down-regulate sensitized cancer cells (Di Lorenzo and Bedford, 2011). Overall, since dysregulation of PRMTs has been observed in diverse types of cancers, and the modulation of their levels affects cancer cell growth, these enzymes are considered potential therapeutic targets for cancer treatment (Yost et al., 2011).

5. PRMT INHIBITORS

With mounting evidence suggesting that PRMTs play crucial roles both physiologically and pathologically, understanding their functions in various pathways become significantly in demand. Currently, many of the functions and mechanisms of PRMTs still remain unclear. Together with the rapid development of molecular biotechnologies, organic inhibitors have been widely used as chemical tools to interrogate enzyme functions. PRMT inhibitors also have become important probes for basic biology and oncology studies. Since overexpression of PRMTs are found to be associated with many human diseases, they are considered ideal prognostic biomarkers and therapeutic targets. Therefore, PRMT inhibitors are potential therapeutic agents against diseases and cancers. Below we categorized the reported PRMT inhibitors by their structure features and their targeting enzymes. Also, based on the current state of PRMT inhibitor development, we provided our perspectives on the future direction and challenges in this field.

5.1 Cofactor Mimics

S-Adenosyl methionine (SAM/AdoMet) (Figure 4, **1**) is a small organic cofactor participating in enzyme catalysis of PRMTs and many other methyltransferases. The structural analogs of SAM can inhibit the activity of SAM-dependent methyltransferases by competing with SAM binding. One example is *S*-adenosyl-homocysteine (SAH, AdoHcy) (Figure 4, **2**), the reaction product from SAM after the methyl group is removed. However, it is a substrate of 5′-AdoHcy hydrolase, which means that it has a limited stability in vivo. Another SAM analog is sinefungin (Figure 4, **3**), which is a natural streptomycine antibiotic. SAH and sinefungin were originally used to inhibit mRNA methyltransferases; later they were found to have inhibition for PMTs (Pugh et al., 1977, 1978). Many cofactor mimics have been widely used in PRMT protein crystallization, as indicated in Figure 1.

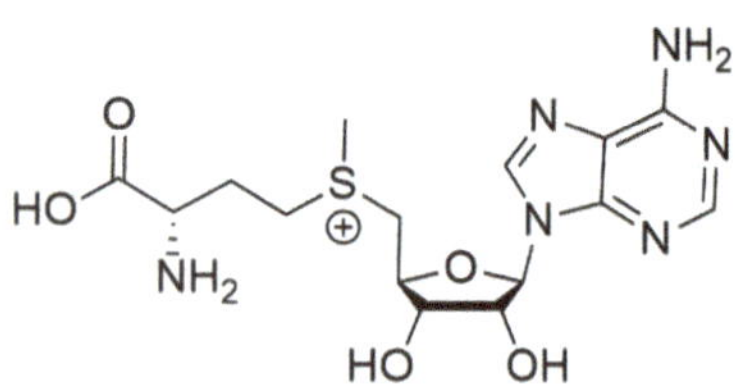
1, AdoMet, SAM

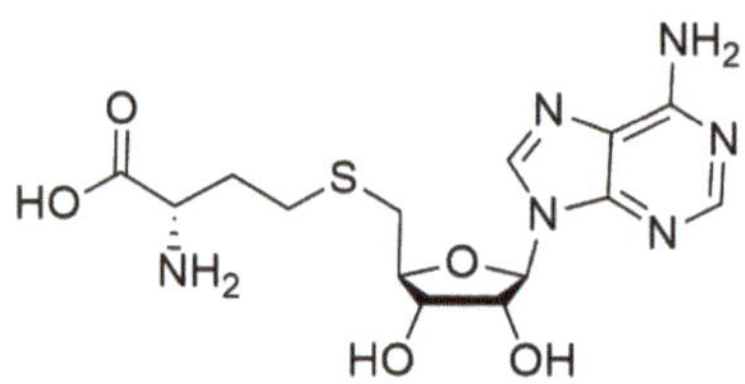
2, AdoHcy, SAH

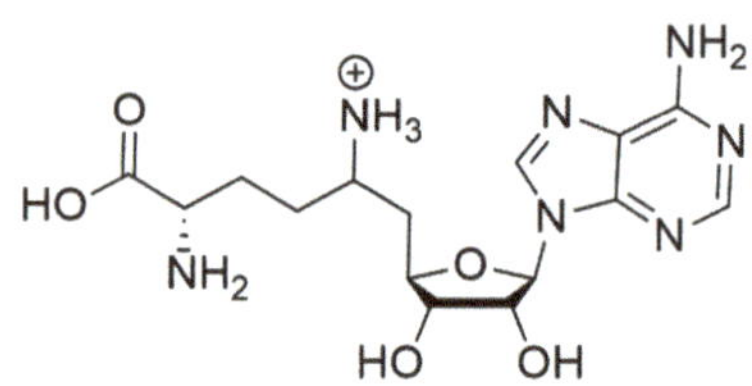
3, Sinefungin

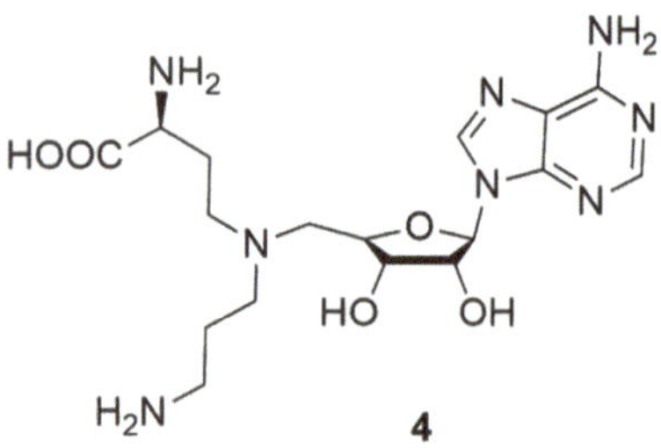
4

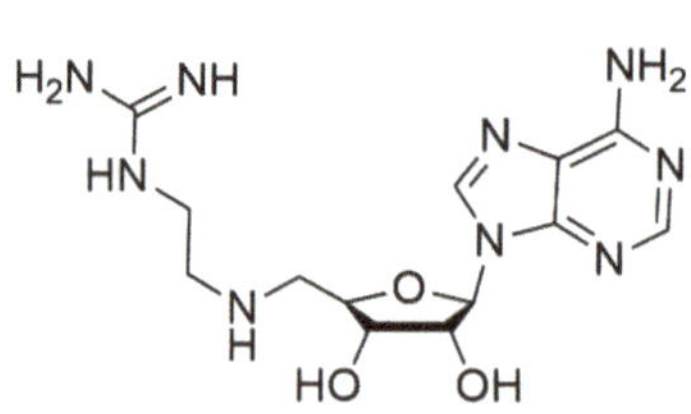
5

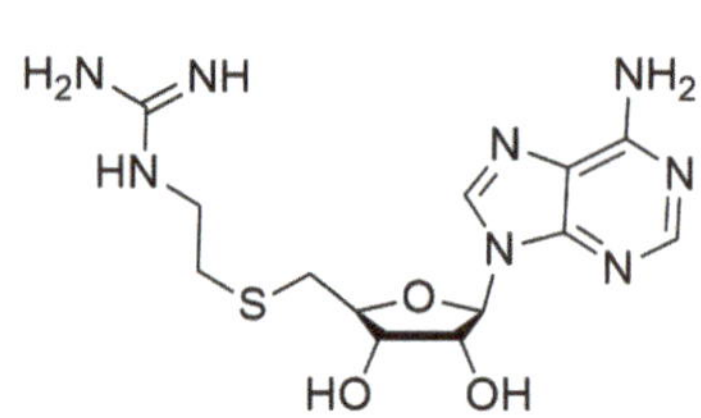
6

FIGURE 4
Cofactor mimics as PRMT inhibitors.

Extensive work has been done to develop SAM analogs as PRMT inhibitors. Dowden et al. (2011) synthesized several SAM derivatives, and found that compound **4** (Figure 4) inhibited PRMT1 with an IC_{50}value of 3.9 μM, and was inactive against CARM1 and SETD7. As described in their studies, the structure–activity relationship (SAR) analysis of the synthesized SAM derivatives has shown a way to achieve selective inhibition through various PRMT binding site sequences. Interestingly, the docking results of compound **4** suggested a bisubstrate binding mode, although this finding lacked binding and kinetic proof. The idea of designing a small molecule bisubstrate inhibitor has been applied by van Haren et al. (2015). Among the six SAM analogs synthesized and evaluated, compound **5** (Figure 4) had an IC_{50} value of 3.2 μM for PRMT6, and was mostly inactive for PRMT1, CARM1, and G9a. Compound **6** (Figure 4) was surprisingly selective for G9a with an IC_{50} value of 3.2 μM. Although the docking poses suggested that these compounds occupied both the substrate and cofactor binding site, there is no following study to support this hypothesis. Besides, the activity of the inhibitors mentioned above has not been characterized in cells.

The idea of designing structure mimics of SAM certainly leads to PRMT inhibitors with promising potency. However, whether it is possible to achieve selectivity among PTMs remains to be challenged (Copeland et al., 2009). Recent study of DOT1L selective inhibitors showed that the enzyme-active site may bear a certain degree of flexibility due to the structural nature of activation loop and substrate binding loop (Daigle et al., 2011; Yu et al., 2012). The SAM analog inhibitors of DOT1L occupied the SAM binding site with a better affinity than SAM. DOT1L is the only PKMT that doesn't contain SET domain but is much similar to the structure of PRMTs, and since the PRMTs also contain flexible α-X and α-Y helices in the catalytic pocket, it will be intriguing to see whether this mechanism of inhibition is plausible for PRMTs.

5.2 AMI Compounds and Their Analogs

In 2004, the first small molecule inhibitors of PRMTs, named as AMIs (arginine methyltransferase inhibitors) (Figure 5, **7–9**), were identified by Cheng et al. via a high-throughput screening of a diverse 9000-compound library (Cheng et al., 2004). The ELISA-based assay was established to detect methylation of Np13p, an RNA binding protein, by Hmt1p (yeast arginine methyltransferase) or PRMT1. Nine hits were identified to indiscriminately inhibit all tested type IPRMTs (PRMT1, PRMT3, PRMT4, PRMT6), but did not have activity on PKMTs. IC_{50} values of the nine inhibitors ranged from 0.19 to 16.3 μM for PRMT1. The leading compound **7** (Figure 5, AMI-1) inhibited PRMT1 with an IC_{50} value of 8.8 μM and did not compete for SAM binding. It was confirmed to be cell permeable, nontoxic, and inhibited in vitro methylation of GFP-Npl3 in Hela cells in a concentration-dependent manner.

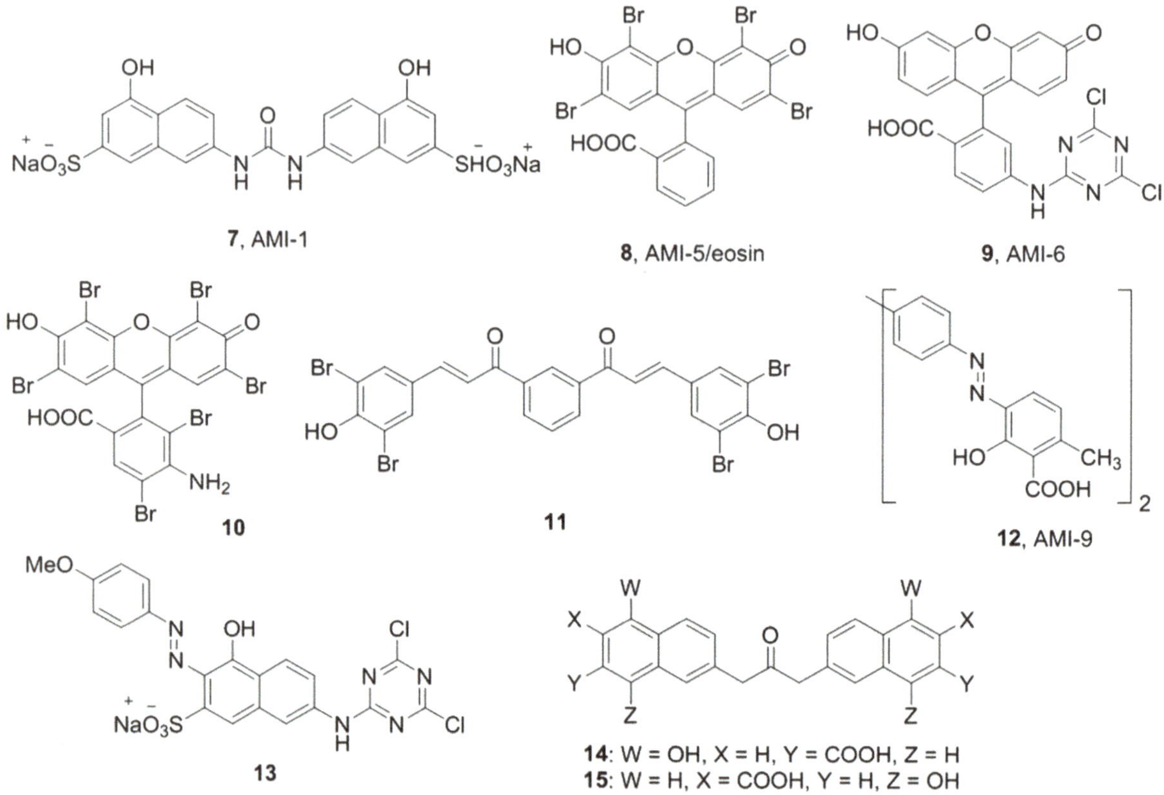

FIGURE 5
Selected structures of AMIs.

Subsequent studies showed that **7** was inactive for type II PRMT5 (Bonham et al., 2010), and it inhibited sirtuin (a histone deacetylase) with an IC_{50}value of 32 μM (Trapp et al., 2007).

Following the discovery of AMIs, researchers applied computational approaches to help investigate more small-molecule inhibitors for PRMTs. Ragno et al. (2007) published the molecular modeling studies based on the dye-like scaffold of compound **8** (AMI-5/eosin), compound **9** (AMI-6), and their analogs. They generated the homology model of PRMT1 from the crystal structure of rat PRMT1 (PDB: 1OR8), yeast homolog RMT1/Hmt1 (PDB: 1G6Q), and rat PRMT3 (PDB: 1F3L). Guided by the docking poses, compound **10** (Figure 5) was synthesized and identified to have an IC_{50}value of 4.8 μM for PRMT1. The consistency of the modeling studies with the biological results gave confidence that it would be worth further utilizing this approach for inhibitor identification.

On the other hand, inhibitor design was pursued through structure modifications. Mai et al. (2008) synthesized simplified analogs based on the pharmacophore of compound **8** to screen against a panel of PRMTs and PKMTs. Among all, compound **11** (Figure 5), with an IC_{50} value of 10 μM for RmtA, could inhibit CARM1 but not PRMT1 or SET7/9 at 100 μM. Many of the analogs induced apoptosis of human leukemia cells. Additionally, Bonham et al. (2010) designed compound **13**, which combined the structural features of **7**, **9**, and **12** (Figure 5). It was less polar than **7** and retained its potency (PRMT1 IC_{50} = 4.2 μM; CARM1 IC_{50} = 2.6 μM). Further characterizations of compound **13** (Figure 5) have shown that it inhibited both type I and type II PRMTs (PRMT5, PRMT6, PRMT8), but was mostly inactive against SET7/9. The cellular activity of compound **13** indicated an enhancement of T helper cell proliferation without affecting viability. In the same year, Castellano et al. (2010) synthesized carboxy analogs of **7**. The PRMT1 IC_{50} values of **7**, **14**, and **15** (Figure 5) are 92.1, 298.0, and 111.7 μM, respectively. Compound **7** presented some minor inhibition while both **14** and **15** were inactive against SET7/9. Molecular modeling provided an explanation that **14** and **15** established new interactions with the exposed residues in an active site, which positively contributed in binding process.

AMIs have made a significant impact in advancing the PRMT inhibitor development. However, there are certain limitations for this type of compound. First of all, researchers have not found an inhibitor with potency below micromolar value. Second, according to Feng et al. (2010), compound **7** probably interacts with the substrate H4 instead of the PRMTs. This proposed mechanism also explained the fact that these compounds were not selective for the enzymes using H4 or H4 peptides as substrates in the assays, and they were mostly inactive to the same enzymes when using other substrates such as H3 or H3 peptides. It is also necessary to point out that the potency parameter (IC_{50} or Ki) derived under different assays or conditions may not be consistent; for example, the potency of **7** was reported to be 8.8 μM on PRMT1 upon discovery, however, later studies showed the IC_{50} value was 92.1 and 76.9 μM (Castellano et al., 2010; Cheng et al., 2004; Wang et al., 2012). Besides, since the binding mode of the AMIs is not determined, there is an emerging need of the inhibitor-enzyme cocrystallized X-ray structures.

5.3 Dapsone Derivatives, Diamidine Compounds, and Other PRMT1 Inhibitors

Dapsone derivatives and diamidine compounds were identified as PRMT inhibitors by Spannhoff et al. (2007a) via a target-based screening approach. The author generated homology models of human PRMT1 and RmtA based on the rat PRMT3 X-ray structure (PDB: 1F3L). The Rita homology model containing SAH was then used to dock 1630 drug-like compounds derived from the

NCI diversity set (140,000 compounds). Compound **17** (Figure 6, allantodapsone) and compound **19** (Figure 6, stilbamidine) were two of the best leads that had IC_{50} values of 57 and 1.7 μM for PRMT1, respectively. The docking results on the human PRMT1 homology model suggested that these two compounds contain basic or polar groups interacting with acidic residue Glu152 in the active site. In cellular assays, these two compounds led to hypomethylation and blocked estrogen receptor activation. Subsequently, Bissinger et al. (2011) reported the SAR of allantodapsone (Figure 6, **17**) analogs and identified an improved inhibitor bearing abis-chloroacetyl amide (Figure 6, **18**). The IC_{50} value was 1.5 μM for PRMT1 and it was inactive against CARM1 and SET7/9.

The symmetric phenylamidine structure in stilbamidine (Figure 6, **19**) has been identified as an important pharmacophore. Yan et al. (2014) explored the use of this pharmacophore and identified a series of diamidine compounds as potent and selective PRMT1 inhibitors. Compound **20** (Figure 6, DB75) inhibited PRMT1 activity with an IC_{50} of 9.4 μM and was selective for PRMT1 over CARM1 (>42-fold), PRMT5 (>18-fold), and PRMT6 (>30-fold). SAR analysis and molecular modeling results suggested the interactions between the diamidine functionality and the acidic residues (Glu129, Glu144, and Glu153) in the catalytic site are essential for the overall binding. Compound **20** was primarily competitive with the substrate and noncompetitive with the cofactor. It was also cell permeable and blocked the proliferation of several leukemia cell lines with different lesions.

During the development of PRMT1 inhibitors, researchers have been utilizing and improving various virtual screening methodology in the drug discovery process. Spannhoff et al. (2007b) reported compound **22** (Figure 6, RM65, IC_{50} = 55 μM), which was identified via a fragment-based virtual screening approach. They started with screening of the HKI database (about 9000 compounds) for novel fragment-like leads, and ended up using 900 compounds for RmtA homology model docking followed by primary biological screening. Interestingly, the docking result of **22** on human PRMT1 suggested a bisubstrate inhibition mode in the active site. This compound was further characterized as cell permeable and could cause hypomethylation in cancer cells. Later, Heinke et al. (2009) developed an approach that combined ligand- and target-based virtual screenings. Approximately 328,000 compounds from the Chembridge compound collection were screened by docking into a homology model of human PRMT with SAH. Top-ranking compounds then filtered through a pharmacophore model, which was generated from the known PRMT1 inhibitor structures. Nine compounds were identified to have micromolar potency for human PRMT1; the best among nine was compound **21** (Figure 6) with an IC_{50} value of 12.75 μM. More recently, Xie et al. (2014) identified compounds **26** (Figure 6, DCLX069) and **27** (Figure 6, DCX078) through structure-based virtual screening with IC_{50} values of 17.9 and 26.2 μM for PRMT1. Both compounds were found to block cell proliferation in breast cancer, liver cancer, and acute myeloid leukemia cell lines.

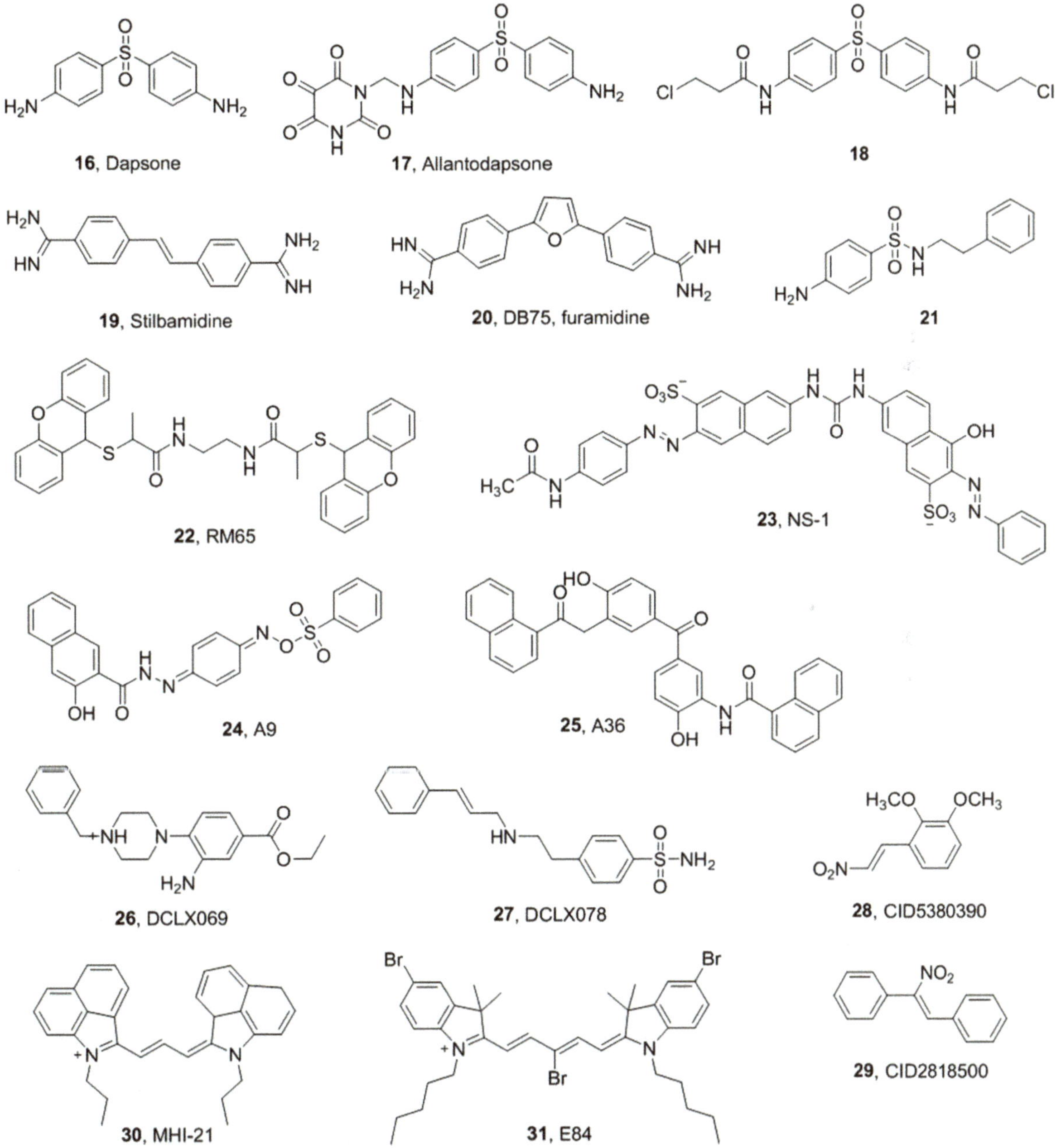

FIGURE 6

Dapsone (compound **16**) and dapsone derivatives (compounds **17** and **18**), diamidine analogs (compounds **19** and **20**), and other PRMT1 inhibitors.

Some of the inhibitors identified by virtual screening strategy have a different mechanism of inhibition. Instead of directly binding to the enzyme, they may target the substrate instead. Feng et al. (2010) reported the identification of compound **23** (Figure 6, NS-1, naphthalene-sulfo derivative one) via structure-based virtual screening of the ChemBridge small molecule compound collection (400,000 compounds), based on rat PRMT1 structure (PDB: 1OR8). Top 50 hits from the screening were evaluated by a radiometric biochemical assay. Compound **23** (Figure 6) was then identified with an IC_{50}value of 12.7 μM for PRMT1, while the IC_{50} value of compound **19** was only 105.7 μM in this assay. Kinetic studies suggested a substrate competitive, cofactor noncompetitive inhibition mechanism. Interestingly, further studies showed that **23** directly interacted with substrate H4 but not PRMT1. Wang et al. (2012) reported the discovery of PRMT1 inhibitors **24** (Figure 6, A9) and **25** (Figure 6, A36) through pharmacophore-based virtual screening. Pharmacophore models were generated from known PRMT inhibitors and used to filter the SPECS database (>300,000 compounds). The top 102 compounds from the screening were evaluated, and compounds **24** and **25** were identified to have PRMT1 IC_{50} values of 41.7 μM and 12.0 μM, respectively. IC_{50} value of compound **7** was tested to be 76.9 μM in this assay. Kinetic data has shown that **24** was a substrate-competitive and cofactor noncompetitive inhibitor, while **25** was noncompetitive for both substrate and cofactor. As with compound **22**, binding studies suggested that they were targeting substrate H4. Compound **24** significantly inhibited the proliferation of castrate-resistant prostate cancer cells.

In addition to the virtual screening method, Dillon et al. (2012) developed a high-throughput screening assay based on a fluorescence polarization change of the probe labeled on SAM-binding cysteine residues of the PRMT1 active site. They successfully identified two mechanism-based inhibitors, compounds **28** (Figure 6, CID5380390) and **29** (Figure 6, CID2818500), with IC_{50} values of 23 μM and 11 μM, respectively. These compounds also inhibited PRMT8 activity, due to the conserved cysteine residues of PRMT1 and PRMT8, but they were inactive against CARM1 and SETD7/9.

Moreover, some PRMT1 inhibitors have strong fluorescence properties that are explored as imaging probes. Sinha et al. (2012) synthesized and evaluated carbocyanine dyes as PRMT1 inhibitors and imaging agents. Compound **30** (Figure 6) presented low micromolar inhibition on PRMT1 (IC_{50} = 4.1 μM), and was noncompetitive by binding to an allosteric site. This compound was used in optical and fluorescent microscopy to show the chromatin interfering property of cells. However, this compound had very limited selectivity among other PRMTs. Hu et al. (2015) further studied the SAR of cyanine compounds and was able to identify compound **31** (Figure 6, E84), with IC_{50} value of 3.38 μM for PRMT1, and have 6- to 25-fold selectivity over CARM1, PRMT5, and PRMT8. It also inhibited cellular PRMT1 activity and blocked leukemia cell proliferation.

To conclude, the development of PRMT1 inhibitors has been the most vigorous in the field. Compared to AMIs, these inhibitors have much better selectivity. The discovery of PRMT1 inhibitors greatly rely on the advancement of computational tools. The above examples included some of the frequently used strategies, which are the structure-/target-based, fragment-based, and ligand-/pharmacophore-based virtual screening. In many cases, the homology model of human PRMT1 was generated because the existing crystal structure of rat PRMT1 (PDB: 1OR8) was crystallized in a nonphysiological condition (pH = 4.7) (Zhang and Cheng, 2003). The molecular modeling method was used to help analyze the ligand–protein interactions and guide the chemical modifications. It is also intriguing to see that many inhibitors discovered through this methodology are known drugs that have been applied in other medical practices. For example, compound **19** (Figure 6, stilbaminine) is a treatment for fungal infections and compound **20** (Figure 6, furamidine) is for the treatment of parasitic infections such as malaria (Bocobo et al., 1953; Purfield et al., 2008). Given the fact that computational methods are efficient ways to identify drug-like small molecule inhibitors, there are certain limitations. First, theoretical models can often result in artifacts. This may explain that only very few inhibitors from the virtual screening hits were confirmed to have actual activity in biochemical screening assays. On the other hand, since docking simulation needs to define a certain grid or pocket, which is often the enzyme active site, it is impossible to find out whether the inhibitors actually bind to the other sites on the enzyme. Besides, this class of inhibitors has not been successfully cocrystallized with PRMT1, thus the binding analysis results generated from molecular docking are not conclusive. The remaining challenge for this class of inhibitors is to further develop selective inhibitors with nanomolar or picomolar potency. Computational approaches might be extremely helpful in the lead optimization process.

5.4 Allosteric PRMT3 Inhibitors

Siarheyeva et al. (2012) reported the discovery of the first allosteric PRMT3 inhibitor (Figure 7, compound **32**) proved by the inhibitor-enzyme cocrystal structure. By screening a library containing 16,000 diverse compounds, the authors identified compound **32** to have an IC_{50} value of 1.6 μM for the full-length human PRMT3. It was selective over other PKMTs and PRMTs, including G9a, GLP, SUV39H2, SETD7, SETD8, PRMT1, CARM1, PRMT5, and PRMT8. This inhibitor bound to a novel allosteric pocket located at the interface of the PRMT3 dimerization arm, but did not prevent homodimerization of PRMT3. It had a Kd value of 9.5 μM, and was noncompetitive with both SAM and the H4 peptide substrate. Interestingly, the cyclohexenyl moiety of the inhibitor interacted with the α-Y segment, which induced the conformational change that disordered the folding of the α-X helix on

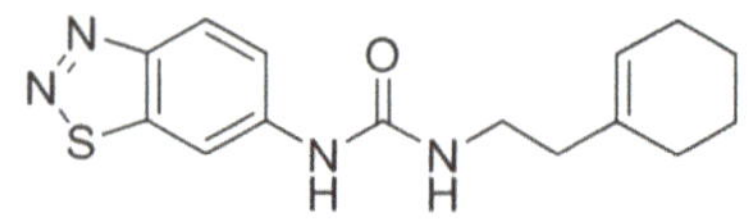

32

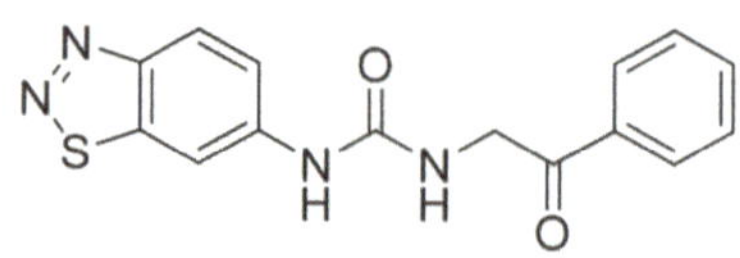

33

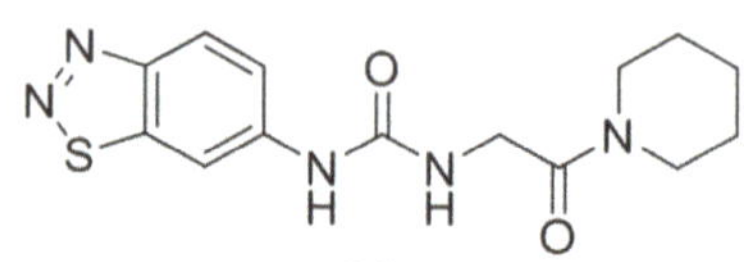

34

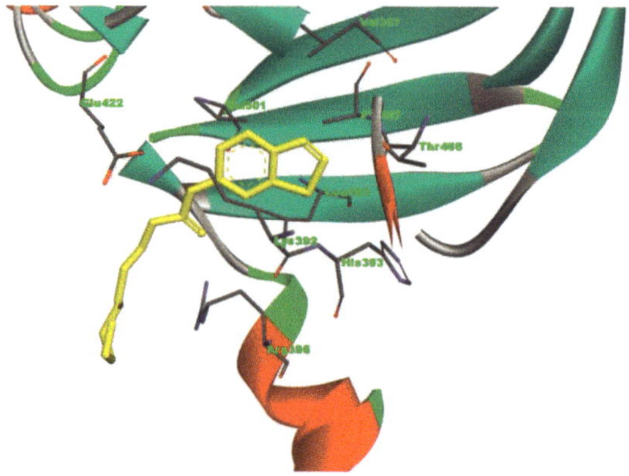

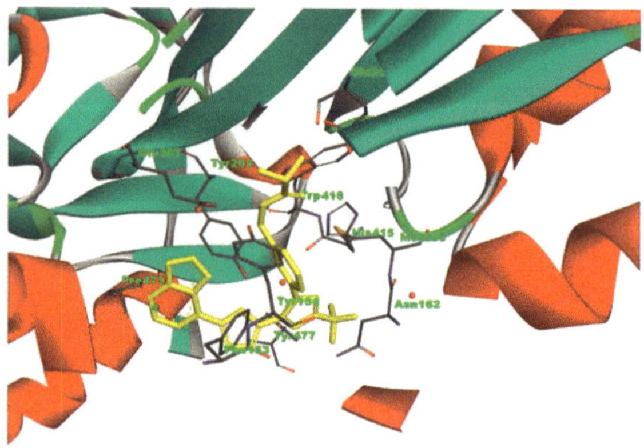

FIGURE 7

Selected PRMT3 allosteric inhibitors and the cocrystal structure of PRMT3 and CARM1 with their inhibitors. Inhibitors are indicated in yellow. Interacting amino acid residues are indicated in sticks. Left, PRMT3 (PDB: 3SMQ); right, CARM1 (2Y1W).

the cofactor binding site and resulting in an inactive α-helix structure. This disruption was proposed to have a direct influence on enzymatic activity of PRMT3. Furthermore, Liu et al. (2013) improved the inhibitor potency via SAR studies. Among all the analogs that have been synthesized, compound **34** (Figure 7) had the best potency(IC_{50} = 230 nM). Resolved crystal structure of PRMT3 with **33** (Figure 7, IC_{50} = 2.5 μM) showed that it bound to the same site as **32**. Compound **33** showed excellent selectivity for PRMT3 over other protein methyltransferases, including G9a, GLP, SUV39H2, PRMT5, SETD7, PRC2, SETD8, SETDB1, SUV420H1, SUV420H2, MLL1, SMYD3, SMYD2, DOT1L, and DNMT1. However, cellular activities of **33** and **34** remained unknown.

The above results suggest that the allosteric binding site of PRMT3 can be exploited to yield potent and selective inhibitors. According to the multiple sequence alignment of type I PRMTs based on PRMT3 residues interacting with the cocrystallized inhibitors (**32** and **33**, Figure 7), there is significant sequence variability by the comparison (Liu et al., 2013; Siarheyeva et al., 2012). This finding indicates that although this pocket is not shown in the other PRMT crystal structures, it is still quite likely that allosteric selective inhibitors can be exploited by targeting this site in other PRMTs. Also, due to the dynamic nature of α-X and α-Y in PRMTs, it is exciting to see whether other PRMT allosteric binding sites can be explored.

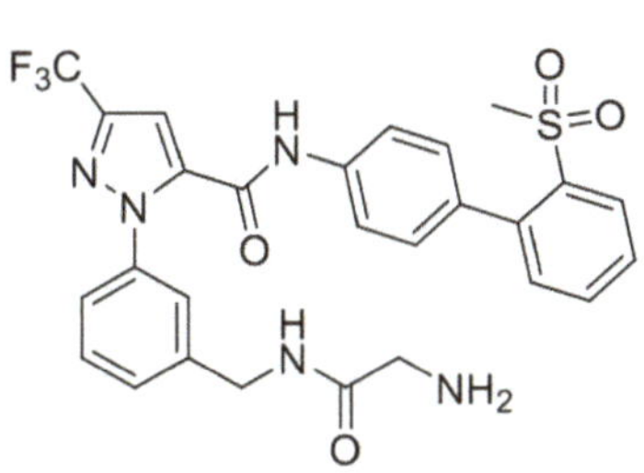

35

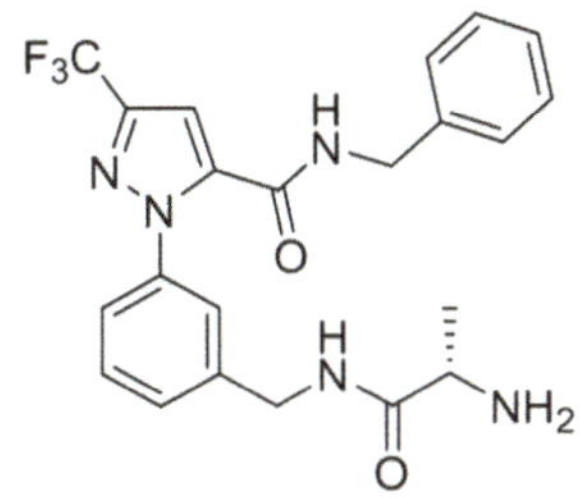

36

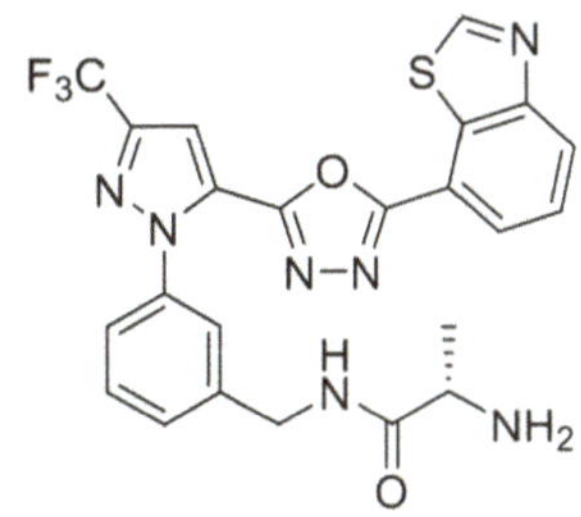

37

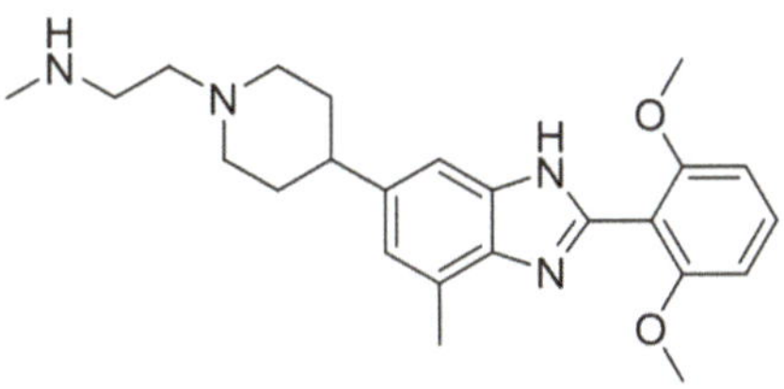

38

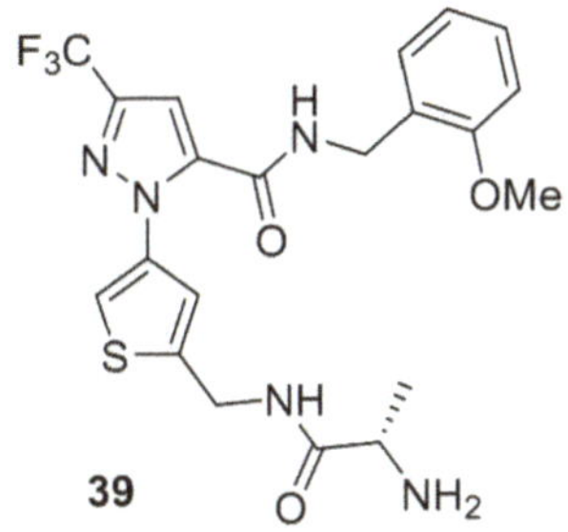

39

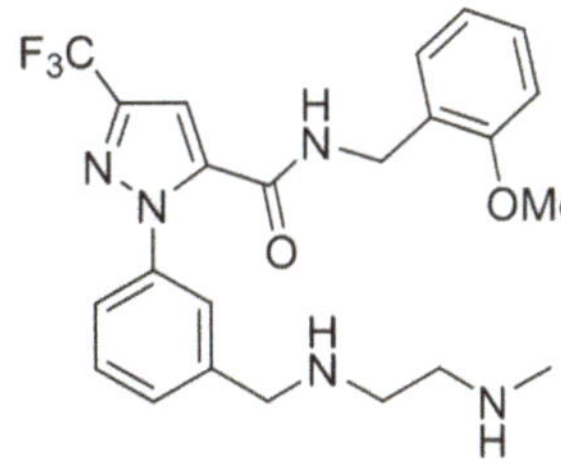

40

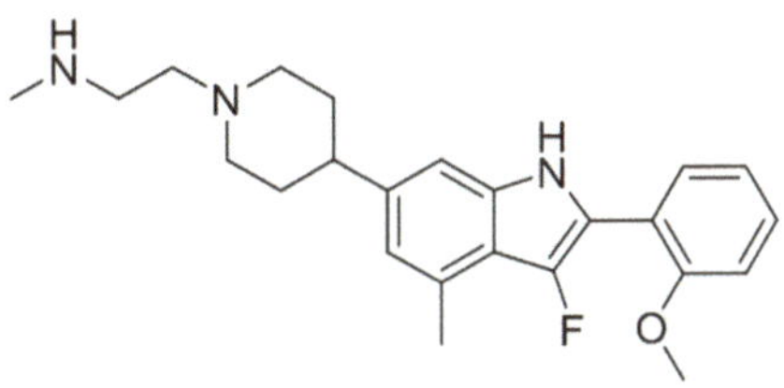

41

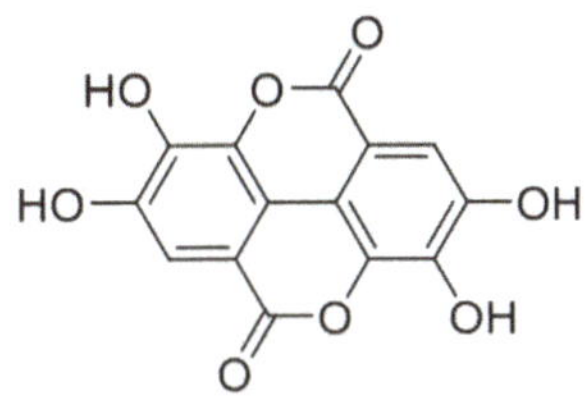

42, TBBD

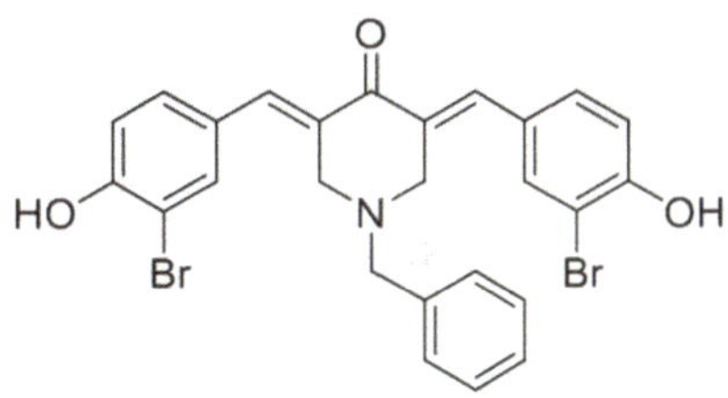

43

FIGURE 8
Selected CARM1 inhibitors.

5.5 Pyrazoles, Imidazoles, Indoles, and Other CARM1(PRMT4) Inhibitors

Purandare et al. (2008) identified pyrazole amide as an initial hit (Figure 8, **35**, IC_{50} = 1.8 μM) by a high-throughput screening, which identified the compound **36** as a potent and selective inhibitor of CARM1 (IC_{50} = 80 nM), albeit poor permeability shown by PAMPA and poor PK properties. Further hit-to-lead optimization of this class of inhibitors led to compound **37** (Figure 8, IC_{50} = 40 nM) (Huynh et al., 2009). Compounds **36** and **37** were found to be significantly less potent against PRMT1 and PRMT3 (IC_{50} > 25 μM). Through SAR analysis, the replacement of the amide functionality with the 1,3,4-oxadiazole moiety in **37** improved membrane permeability. Concurrently, Wan et al. (2009) identified benzo[d]imidazole-related analogs as CARM1 inhibitors. Modification made from SAR analysis led to the most potent analog, compound **38** (Figure 8,

IC_{50} = 70 nM). It was significantly less active against PRMT1 and 3 (IC_{50} > 25 μM). Allan et al. (2009) developed compound **39** (Figure 8) as a CARM1 inhibitor with IC_{50} value of 60 nM, which was also inactive against PRMT1 and SET7/9 (IC_{50} > 100 μM). However, it did not show cellular activity. Therrien et al. (2009) further synthesized 1,2-diamine compounds as CARM1 inhibitors, represented by compound **40** (Figure 7). Although it only had an IC_{50} value of 0.2 μM and did not show cellular activity, the PK profile was improved. Sack et al. (2011) described the crystal structures of an improved pyrazole/indole-containing inhibitor with CARM1 and SAH/sinefungin. Compound **41** (Figure 8) is the analog of **38** that has an IC_{50} value of 30 nM for CARM1, while IC_{50} > 10 μM for PRMT1 and PRMT3. The inhibitor was shown to occupy the substrate binding site and the surrounding pocket located in the interface between the N- and C-terminal domains in the presence of SAH (Figure 7, right). Overall, the pyrazole/imidazole/indole CARM1 inhibitors have nanomolar potency, however there were no further in vivo studies reported, possibly due to the low cell permeability or cytotoxic side effects. The other limitations are that the selectivity profile of these compounds are not complete and there is lack of inhibition mechanism elucidation.

Other than the pyrazole/imidazole/indole compounds, Selvi et al. (2010) identified compound **42** (Figure 8, TBBD), which is a natural product from pomegranate extract, as a CARM1 inhibitor. Compound **41** inhibited CARM1 activity in a dose-dependent manner, while the IC_{50} value was not determined. It did not inhibit G9a and histone acetyltransferase CBP/p300. This compound was noncompetitive with bothH3 and SAM. ITC experiments showed minimal interaction between **42** and CARM1 alone, while the $K_D$value of CARM1-histone 3 complex was 4 μM. This data suggested a partial inhibition mechanism mediated via its interaction with the enzyme–substrate complex. Also possible is that compound **42** may bind to the histone H3 substrate, in a similar way to NS-1. Additionally, **42** did not reduce the p21 expression in HeLa cells as observed in H1299 and HEK293T cells. The other set of CARM1 inhibitors was reported by Cheng et al. (2011). Compound **43** had IC_{50} values of 8.6 μM for CARM1, and had low or no inhibition on a panel of PRMTs (PRMT1, PRMT3, PRMT5, and PRMT6) and PKMTs (SET7, DOTL1, Suv39H1, and G9a). In human prostate cancer cells, compound **43** showed a significant dose-dependent reduction of the PSA promoter activity. Although this set of CARM inhibitors had significantly increased cellular activity and better selectivity profiling, the potency was not as satisfying as the pyrazoles.

5.6 PRMT5 Inhibitors

Despite the fact that PRMT5 is the major type II enzyme and also an important therapeutic target, very few inhibitors are known for PRMT5. In 2015, Alinari et al. (2015) reported compound **44** (Figure 9, CMP5) as a first-in-class,

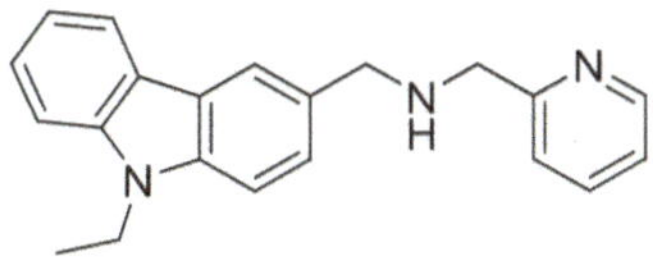

44, CMP5

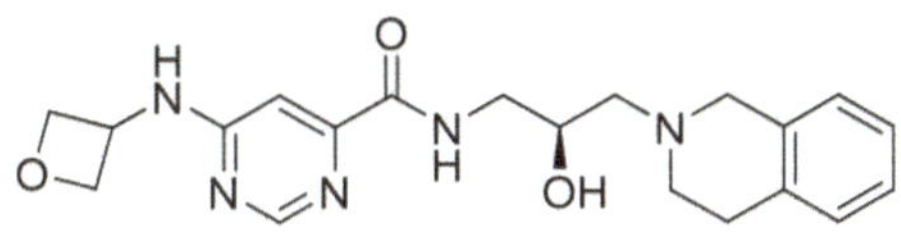

45, EPZ015666

FIGURE 9
Selected PRMT5 inhibitors.

small-molecule PRMT5 inhibitor that blocked initiation and maintenance of B lymphocyte transformation. It was identified via structure-based screening based on a human PRMT5 catalytic site model, which is generated from rat PRMT1 crystal structure (PDB: 1OR8). After screening against the ChemBridge CNS-Set™ library (10,000 small-molecule compounds), they further docked the hits and selected the top eight compounds that had the lowest binding energy for further characterization. Cell-based immunofluorescence assay showed that **44** selectively blocked symmetric dimethylation on H4R3 by inhibiting PRMT5 activity, while others had no effect on H4R3 methylation. Compound **44** did not affect asymmetric methylation of H4R3, and was found to be inactive against PRMT1, CARM1, and PRMT7. Cytotoxicity studies suggested that **44** was selectively toxic to lymphoma cells but not as toxic to normal resting B lymphocytes. The other PRMT5 inhibitor, compound **45** (Figure 9, EPZ015666), was disclosed by Epizyme, Inc., although there is no published article available (Duncan et al., 2014). It was claimed to have a cell biochemical K_i of 8 nM, and >20,000-fold K_i of other PMTs. Compound **45** was tested in both in vitro and in vivo models of mantle cell lymphoma, showing methyl mark inhibition and tumor suppression.

5.7 Peptide Inhibitors

There are more than 50 peptides or peptide-based therapeutics and diagnostic agents that have been approved by the FDA (Hamzeh-Mivehroud et al., 2013). There has been an increasing amount of studies conducted on peptide inhibitors both for probe designing purposes or therapeutic uses. Osborne et al. (2008) reported an in situ synthesis strategy of a bisubstrate peptide inhibitor for PRMT1 (Figure 10). The peptide inhibitor **47** was generated through a PRMT1-mediated transfer using a SAM analog compound **46** (Figure 10, AAI), which was found to have an IC_{50} value of 18.5 μM. This study indicated that a chemoenzymatic generation of bisubstrate inhibitors was plausible for PRMTs, although the selectivity and cellular activity data was not available. The same group has reported the other PRMT1 peptide inhibitors based on the N terminus of histone H4 incorporating fluoro- and chloroacetamidine warheads (Figure 10, **48**: C21 and **49**: F21) (Obianyo et al., 2010). Peptide **48** had an IC_{50} value of 1.8 μM for PRMT1, while **49** was 94 μM. Peptide **48** was 4.9-fold

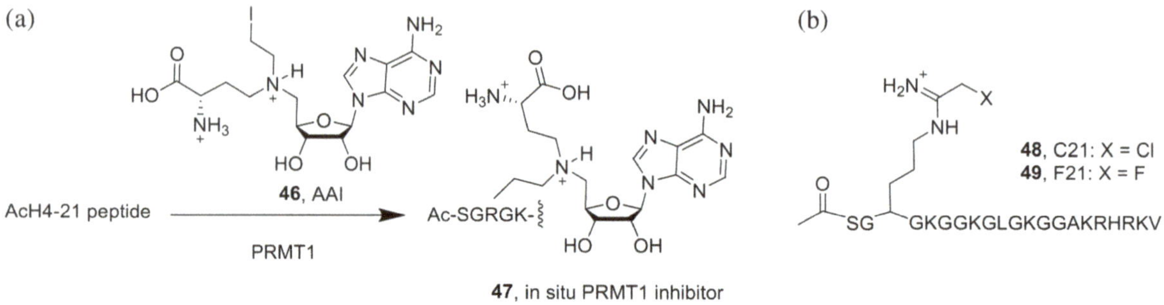

FIGURE 10

(a) In situ generation of a bisubstrate peptide inhibitor. (b) Selected bisubstrate peptide inhibitors.

more selective for PRMT1 compared to PRMT6, and was mostly inactive for PRMT3 and CARM1. These peptides were irreversible inhibitors that could form covalent bonds to PRMT1 through an S_N2 attack, and were also proven to selectively inhibit PRMT1 activity in cells.

Frankel et al. have done a series of studies on the development of peptide inhibitors for PRMTs. The first ones reported were N^{η}-substituted arginyl peptides **50** to **53** (Figure 11, R1-1 to R1-4) (Lakowski et al., 2010). Their IC_{50} values ranged from 4.82 to 14.2 μM for PRMT6, and from 27.5 to 56.5 μM for PRMT1. The IC_{50} values of these peptide inhibitors were 2.5- to 5-fold lower than the methylation product R1-aDMA. The same group synthesized a series of R1 peptides bearing various N^{η}-substitutions, and described the preferences of PRMTs for those modified peptide substrate-inhibitors (Thomas et al., 2014). Peptides **54** and **55** (Figure 11, **54**: R1-can and **55**: R1–NO_2) could not be methylated and had micromolar level potency for PRMT1, PRMT4, and PRMT6 (for PRMT1, **54**: IC_{50} = 30.4 μM, **55**: IC_{50} = 26.0 μM). Subsequently, Hart et al. (2011) reported partial bisubstrate R1 peptide inhibitor **56** (Figure 11, R1-Orn, PRMT6 IC_{50} = 36.7 μM, not active for PRMT1 and PRMT4) and **57** (Figure 11, R1-Lys, PRMT1 IC_{50} = 13.9 μM, PRMT4 IC_{50} = 35.7 μM, PRMT6 IC_{50} = 29.0 μM), suggesting a six-carbon linker might better mimic the transition state model proposed for PRMT catalytic mechanism. They also pursued a similar approach and synthesized HIV-$Tat_{48\text{-}60}$ peptide analogs (Figure 11, **58**) as potent PRMT substrate-inhibitors, which have 1.31–11.9 μM K_i values for PRMT1 and 19.9–77.3 μM K_i values for PRMT6 (Hart et al., 2012).

Therapeutic peptides have great potential as anticancer agents, because they can easily achieve target specificity through rational design. Currently, the modification of PRMT peptide inhibitors are mainly focused on the substrate arginine residues. Due to the fact that many PRMTs share the same substrates, specific inhibition may rely on the differences in a PRMT substrate binding site environment. The limitation of peptide inhibitors is their low stability in vivo and poor membrane penetration.

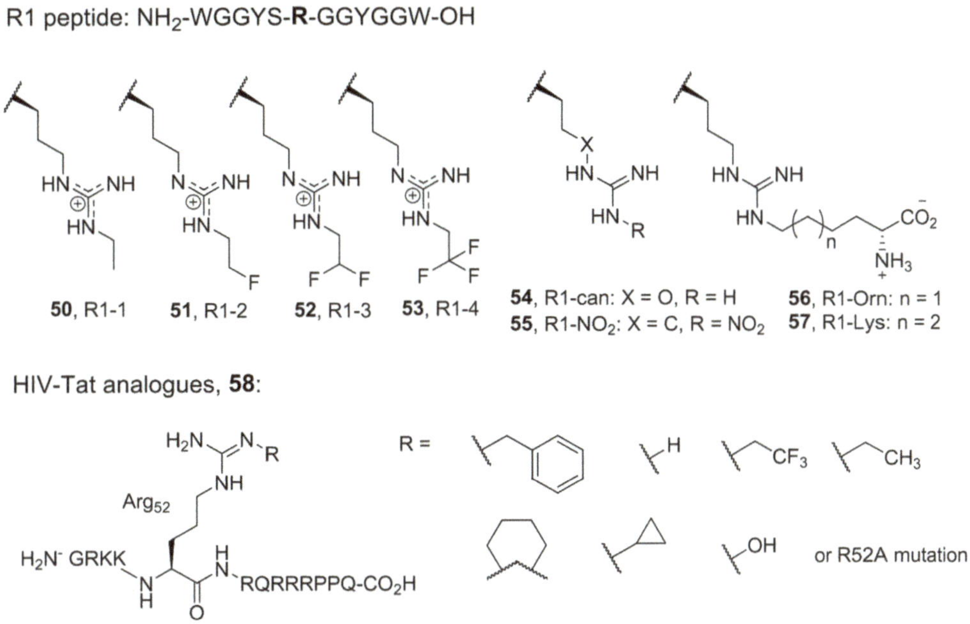

FIGURE 11
Selected peptide inhibitors.

6. CONCLUSION

In this chapter, we summarized the current development of PRMT inhibitors to advocate their importance in advancing the field of epigenetics. PRMTs are important enzymes that have been found to participate in abundant biological processes. Accumulating evidence has suggested the inevitable relationship between abnormal PRMTs and various diseases, including cancer. Therefore, inhibitors of PRMTs have drawn plenty of attention by researchers in the field of drug discovery and epigenetic research. Tremendous progress has been made in the development of PRMT inhibitors. The computation-based method has been widely used in hit identification, including structure-/target-based, ligand-/pharmacophore-based, and fragment-based virtual screening. Homology modeling is used to compensate the missing structure of certain enzyme targets. For lead optimization, in many cases it was done via the traditional medicinal chemistry strategy. The obvious shortage of this method is that a large amount of synthesis is required to gather sufficient information, which usually costs a lot of time and labor. With the help of cocrystal structures and molecular docking, the structure-guided design becomes a popular and much more efficient way to conduct chemical modification on lead compounds.

Based on the current state of the field of developing PRMT inhibitors, there are certain challenges and opportunities. First of all, current selective inhibitors do not cover all the PRMT members. Great efforts have been devoted to PRMT1 inhibitors, although none are for PRMT7, PRMT8, and PRMT9. This is partially due to the very limited understanding of these enzymes compared to other PRMTs. Second, many reported inhibitors are not well characterized. Some of the common pitfalls are incomplete selectivity profile, lack of mechanistic studies and binding mode analysis, or no cellular activity data. Investigation of these inhibitors in animal models is greatly needed. Such missing information can be critical for the future application of the inhibitors. Also, due to the inconsistency of IC_{50} or Ki values from different assays, it is necessary to include reported PRMT inhibitors as a control, in order to compare with newly discovered inhibitors. Last but not least, PRMT inhibitors with nano- or picomolar potency are needed for preclinical and clinical studies. It is a pleasure to see that a few patents were newly filed for PRMT inhibitors, including PRMT1, CARM1, and PRMT5 inhibitors, which are all proposed to be drug candidates (Chesworth et al., 2014a,b; Duncan et al., 2014; Mitchell et al., 2014). We are quite optimistic, and look forward to see more PRMT inhibitors become therapeutic agents.

LIST OF ACRONYMS AND ABBREVIATIONS

aDMA Asymmetrical mono methylated arginine
AMIs Arginine methyltransferase inhibitors
CARM1 Co activator-associated arginine methyltransferase1
DNMT1 DNA methyltransferase 1
DOT1L Disruptor of telomeric silencing 1-like
ELISA Enzyme-linked immunosorbent assay
EZH1 Enhancer of zeste homolog 1
EZH2 Enhancer of zeste 2 polycomb repressive complex 2 subunit
FDA Food and Drug Administration
G9a EHMT2, Euchromatic histone-lysine *N*-methyltransferase 2
GAR Glycine and arginine rich
GFP Green fluorescent protein
H3 Histone 3
H4 Histone 4
H4R3 Histone 4 arginine 3
HTS High-throughput screening
MLL Mixed lineage leukemia
MMA Monomethylation of arginine
MS Mass spectrometry
NCI National Cancer Institute
PAMPA Parallel artificial membrane permeability assay
PDB Protein Database
PGM Proline, glycine, and methionine rich
PK Pharmacokinetic
PKMT Protein lysine methyltransferase
PMT Protein methyltransferase
PRMT Protein arginine methyltransferase

SAH *S*-5′-Adenosyl-L-homocysteine
SAHH SAH hydrolase
SAM *S*-5′-Adenosyl-L-methionine
sDMA Symmetrical mono methylation of arginine
SETD7 SET domain containing (lysine methyltransferase) 7
SETD8 SET domain containing (lysine methyltransferase) 8
SETDB1 SET domain, bifurcated 1
SMYD SET and MYND domain containing
SPA Scintillation proximity assay
SUV39H1 Suppressor of variegation 3–9 homolog 1

ACKNOWLEDGMENTS

This work was supported by funds provided by NIH grant R01GM086717.

REFERENCES

Alinari, L., Mahasenan, K.V., Yan, F., Karkhanis, V., Chung, J.H., Smith, E.M., Quinion, C., Smith, P.L., Kim, L., Patton, J.T., Lapalombella, R., Yu, B., Wu, Y., Roy, S., De Leo, A., Pileri, S., Agostinelli, C., Ayers, L., Bradner, J.E., Chen-Kiang, S., Elemento, O., Motiwala, T., Majumder, S., Byrd, J.C., Jacob, S., Sif, S., Li, C., Baiocchi, R.A., 2015. Selective inhibition of protein arginine methyltransferase 5 blocks initiation and maintenance of b-cell transformation. Blood 125 (16), 2530–2543. http://dx.doi.org/10.1182/blood-2014-12-619783.

Allan, M., Manku, S., Therrien, E., Nguyen, N., Styhler, S., Robert, M.F., Goulet, A.C., Petschner, A.J., Rahil, G., Robert Macleod, A., Deziel, R., Besterman, J.M., Nguyen, H., Wahhab, A., 2009. *N*-Benzyl-1-heteroaryl-3-(trifluoromethyl)-1*H*-pyrazole-5-carboxamides as inhibitors of co-activator associated arginine methyltransferase 1 (CARM1). Bioorg. Med. Chem. Lett. 19 (4), 1218–1223.

Antonysamy, S., Bonday, Z., Campbell, R.M., Doyle, B., Druzina, Z., Gheyi, T., Han, B., Jungheim, L.N., Qian, Y., Rauch, C., Russell, M., Sauder, J.M., Wasserman, S.R., Weichert, K., Willard, F.S., Zhang, A., Emtage, S., 2012. Crystal structure of the human PRMT5:MEP50 complex. Proc. Natl. Acad. Sci. U.S.A. 109 (44), 17960–17965.

Bedford, M.T., 2007. Arginine methylation at a glance. J. Cell Sci. 120 (Pt 24), 4243–4246.

Bedford, M.T., Clarke, S.G., 2009. Protein arginine methylation in mammals: who, what, and why. Mol. Cell 33 (1), 1–13.

Bedford, M.T., Reed, R., Leder, P., 1998. WW domain-mediated interactions reveal a spliceosome-associated protein that binds a third class of proline-rich motif: the proline glycine and methionine-rich motif. Proc. Natl. Acad. Sci. U.S.A. 95 (18), 10602–10607.

Bedford, M.T., Richard, S., 2005. Arginine methylation an emerging regulator of protein function. Mol. Cell 18 (3), 263–272.

Bissinger, E.M., Heinke, R., Spannhoff, A., Eberlin, A., Metzger, E., Cura, V., Hassenboehler, P., Cavarelli, J., Schule, R., Bedford, M.T., Sippl, W., Jung, M., 2011. Acyl derivatives of *p*-aminosulfonamides and dapsone as new inhibitors of the arginine methyltransferase hPRMT1. Bioorg. Med. Chem. 19 (12), 3717–3731.

Bocobo, F.C., Curtis, A.C., Harrell, E.R., Conner, V., 1953. In vitro fungistatic activity of stilbamidine, propamidine, pentamidine and diethylstilbestrol. J. Invest. Dermatol. 21 (3), 149–156.

Bonham, K., Hemmers, S., Lim, Y.H., Hill, D.M., Finn, M.G., Mowen, K.A., 2010. Effects of a novel arginine methyltransferase inhibitor on t-helper cell cytokine production. FEBS J. 277 (9), 2096–2108.

Castellano, S., Milite, C., Ragno, R., Simeoni, S., Mai, A., Limongelli, V., Novellino, E., Bauer, I., Brosch, G., Spannhoff, A., Cheng, D., Bedford, M.T., Sbardella, G., 2010. Design, synthesis and biological evaluation of carboxy analogues of arginine methyltransferase inhibitor 1 (AMI-1). ChemMedChem 5 (3), 398–414.

Cheng, D., Cote, J., Shaaban, S., Bedford, M.T., 2007. The arginine methyltransferase CARM1 regulates the coupling of transcription and mRNA processing. Mol. Cell 25 (1), 71–83.

Cheng, D., Valente, S., Castellano, S., Sbardella, G., Di Santo, R., Costi, R., Bedford, M.T., Mai, A., 2011. Novel 3,5-bis(bromohydroxybenzylidene)piperidin-4-ones as coactivator-associated arginine methyltransferase 1 inhibitors: enzyme selectivity and cellular activity. J. Med. Chem. 54 (13), 4928–4932.

Cheng, D., Yadav, N., King, R.W., Swanson, M.S., Weinstein, E.J., Bedford, M.T., 2004. Small molecule regulators of protein arginine methyltransferases. J. Biol. Chem. 279 (23), 23892–23899.

Cheng, X., Collins, R.E., Zhang, X., 2005. Structural and sequence motifs of protein (histone) methylation enzymes. Annu. Rev. Biophys. Biomol. Struct. 34, 267–294.

Chesworth, R., Mitchell, L.H., Shapiro, G., Kuntz, K.W., 2014a. Arginine Methyltransferase Inhibitors and Uses Thereof. Google Patents.

Chesworth, R., Moradei, O.M., Shapiro, G., Jin, L., Babine, R.E., 2014b. CARM1 Inhibitors and Uses Thereof. Google Patents.

Cheung, N., Chan, L.C., Thompson, A., Cleary, M.L., So, C.W., 2007. Protein arginine-methyltransferase-dependent oncogenesis. Nat. Cell Biol. 9 (10), 1208–1215.

Cook, J.R., Lee, J.H., Yang, Z.H., Krause, C.D., Herth, N., Hoffmann, R., Pestka, S., 2006. FBXO11/PRMT9, a new protein arginine methyltransferase, symmetrically dimethylates arginine residues. Biochem. Biophys. Res. Commun. 342 (2), 472–481.

Copeland, R.A., 2012. Protein methyltransferase inhibitors as personalized cancer therapeutics. Drug Discov. Today Ther. Strateg. 9 (2–3), e83–e90.

Copeland, R.A., Moyer, M.P., Richon, V.M., 2013. Targeting genetic alterations in protein methyltransferases for personalized cancer therapeutics. Oncogene 32 (8), 939–946.

Copeland, R.A., Solomon, M.E., Richon, V.M., 2009. Protein methyltransferases as a target class for drug discovery. Nat. Rev. Drug Discov. 8 (9), 724–732.

Daigle, S.R., Olhava, E.J., Therkelsen, C.A., Majer, C.R., Sneeringer, C.J., Song, J., Johnston, L.D., Scott, M.P., Smith, J.J., Xiao, Y., Jin, L., Kuntz, K.W., Chesworth, R., Moyer, M.P., Bernt, K.M., Tseng, J.C., Kung, A.L., Armstrong, S.A., Copeland, R.A., Richon, V.M., Pollock, R.M., 2011. Selective killing of mixed lineage leukemia cells by a potent small-molecule DOT1L inhibitor. Cancer Cell 20 (1), 53–65.

Dhar, S., Vemulapalli, V., Patananan, A.N., Huang, G.L., Di Lorenzo, A., Richard, S., Comb, M.J., Guo, A., Clarke, S.G., Bedford, M.T., 2013. Loss of the major type I arginine methyltransferase PRMT1 causes substrate scavenging by other PRMTs. Sci. Rep. 3, 1311.

Dillon, M.B., Bachovchin, D.A., Brown, S.J., Finn, M., Rosen, H., Cravatt, B.F., Mowen, K.A., 2012. Novel inhibitors for PRMT1 discovered by high-throughput screening using activity-based fluorescence polarization. ACS Chem. Biol. 7 (7), 1198–1204.

Dowden, J., Pike, R.A., Parry, R.V., Hong, W., Muhsen, U.A., Ward, S.G., 2011. Small molecule inhibitors that discriminate between protein arginine *N*-methyltransferases PRMT1 and CARM1. Org. Biomol. Chem. 9 (22), 7814–7821.

Duncan, K.W., Chesworth, R., Munchhof, M.J., 2014. PRMT5 Inhibitors and Uses Thereof. Google Patents.

Feng, Y., Li, M., Wang, B., Zheng, Y.G., 2010. Discovery and mechanistic study of a class of protein arginine methylation inhibitors. J. Med. Chem. 53 (16), 6028–6039.

Finley, A., Copeland, R.A., 2014. Small molecule control of chromatin remodeling. Chem. Biol. 21 (9), 1196–1210.

Frietze, S., Lupien, M., Silver, P.A., Brown, M., 2008. CARM1 regulates estrogen-stimulated breast cancer growth through up-regulation of E2F1. Cancer Res. 68 (1), 301–306.

Gui, S., Gathiaka, S., Li, J., Qu, J., Acevedo, O., Hevel, J.M., 2014. A remodeled protein arginine methyltransferase 1 (PRMT1) generates symmetric dimethylarginine. J. Biol. Chem. 289 (13), 9320–9327.

Gui, S., Wooderchak-Donahue, W.L., Zang, T., Chen, D., Daly, M.P., Zhou, Z.S., Hevel, J.M., 2013. Substrate-induced control of product formation by protein arginine methyltransferase 1. Biochemistry 52 (1), 199–209.

Hamzeh-Mivehroud, M., Alizadeh, A.A., Morris, M.B., Church, W.B., Dastmalchi, S., 2013. Phage display as a technology delivering on the promise of peptide drug discovery. Drug Discov. Today 18 (23–24), 1144–1157.

Heinke, R., Spannhoff, A., Meier, R., Trojer, P., Bauer, I., Jung, M., Sippl, W., 2009. Virtual screening and biological characterization of novel histone arginine methyltransferase PRMT1 inhibitors. ChemMedChem 4 (1), 69–77.

Herrmann, F., Lee, J., Bedford, M.T., Fackelmayer, F.O., 2005. Dynamics of human protein arginine methyltransferase 1(PRMT1) in vivo. J. Biol. Chem. 280 (45), 38005–38010.

Hu, H., Owens, E.A., Su, H., Yan, L., Levitz, A., Zhao, X., Henary, M., Zheng, Y.G., 2015. Exploration of cyanine compounds as selective inhibitors of protein arginine methyltransferases: synthesis and biological evaluation. J. Med. Chem. 58 (3), 1228–1243.

Huynh, T., Chen, Z., Pang, S., Geng, J., Bandiera, T., Bindi, S., Vianello, P., Roletto, F., Thieffine, S., Galvani, A., 2009. Optimization of pyrazole inhibitors of coactivator associated arginine methyltransferase 1 (CARM1). Bioorg. Med. Chem. Lett. 19 (11), 2924–2927.

't Hart, P., Lakowski, T.M., Thomas, D., Frankel, A., Martin, N.I., 2011. Peptidic partial bisubstrates as inhibitors of the protein arginine *N*-methyltransferases. ChemBioChem 12 (9), 1427–1432.

t Hart, P., Thomas, D., van Ommeren, R., Lakowski, T.M., Frankel, A., Martin, N.I., 2012. Analogues of the HIV-Tat peptide containing N^{η}-modified arginines as potent inhibitors of protein arginine *N*-methyltransferases. MedChemComm 3 (10), 1235–1244.

van Haren, M., van Ufford, L.Q., Moret, E.E., Martin, N.I., 2015. Synthesis and evaluation of protein arginine *N*-methyltransferase inhibitors designed to simultaneously occupy both substrate binding sites. Org. Biomol. Chem. 13 (2), 549–560.

Kim, J.M., Sohn, H.Y., Yoon, S.Y., Oh, J.H., Yang, J.O., Kim, J.H., Song, K.S., Rho, S.M., Yoo, H.S., Kim, Y.S., Kim, J.G., Kim, N.S., 2005. Identification of gastric cancer-related genes using a cDNA microarray containing novel expressed sequence tags expressed in gastric cancer cells. Clin. Cancer Res. 11, 473–482.

Di Lorenzo, A., Bedford, M.T., 2011. Histone arginine methylation. FEBS Lett. 585 (13), 2024–2031.

Lakowski, T.M., 't Hart, P., Ahern, C.A., et al., 2010. N^{η}-Substituted arginyl peptide inhibitors of protein arginine *N*-methyltransferases. ACS Chem. Biol. 11, 1053–1063.

Lee, J.H., Cook, J.R., Yang, Z.H., Mirochnitchenko, O., Gunderson, S.I., Felix, A.M., Herth, N., Hoffmann, R., Pestka, S., 2005. PRMT7, a new protein arginine methyltransferase that synthesizes symmetric dimethylarginine. J. Biol. Chem. 280 (5), 3656–3664.

Liu, F., Li, F., Ma, A., Dobrovetsky, E., Dong, A., Gao, C., Korboukh, I., Liu, J., Smil, D., Brown, P.J., Frye, S.V., Arrowsmith, C.H., Schapira, M., Vedadi, M., Jin, J., 2013. Exploiting an allosteric binding site of PRMT3 yields potent and selective inhibitors. J. Med. Chem. 56 (5), 2110–2124.

Mai, A., Cheng, D., Bedford, M.T., Valente, S., Nebbioso, A., Perrone, A., Brosch, G., Sbardella, G., De Bellis, F., Miceli, M., 2008. Epigenetic multiple ligands: mixed histone/protein methyltransferase, acetyltransferase, and class iii deacetylase (sirtuin) inhibitors. J. Med. Chem. 51 (7), 2279–2290.

Mitchell, L.H., Shapiro, G., Chesworth, R., Boriack-Sjodin, P.A., Moradei, O.M., Kuntz, K.W., 2014. PRMT1 Inhibitors and Uses Thereof. Google Patents.

Obianyo, O., Causey, C.P., Osborne, T.C., Jones, J.E., Lee, Y.H., Stallcup, M.R., Thompson, P.R., 2010. A chloroacetamidine-based inactivator of protein arginine methyltransferase 1: design, synthesis, and in vitro and in vivo evaluation. ChemBioChem 11 (9), 1219–1223.

Osborne, T., Weller Roska, R.L., Rajski, S.R., Thompson, P.R., 2008. In situ generation of a bisubstrate analogue for protein arginine methyltransferase 1. J. Am. Chem. Soc. 130 (14), 4574–4575.

Pahlich, S., Zakaryan, R.P., Gehring, H., 2008. Identification of proteins interacting with protein arginine methyltransferase 8: the Ewing sarcoma (EWS) protein binds independent of its methylation state. Proteins 72 (4), 1125–1137.

Pal, S., Sif, S., 2007. Interplay between chromatin remodelers and protein arginine methyltransferases. J. Cell Physiol. 213, 306–315.

Pugh, C.S.G., Borchardt, R.T., Stone, H., 1977. Inhibition of Newcastle disease virion messenger RNA (guanine-7-)-methyltransferase by analogues of *S*-adenosylhomocysteine. Biochemistry 16 (17), 3928–3932.

Pugh, C.S.G., Borchardt, R.T., Stone, H., 1978. Sinefungin, a potent inhibitor of virion mRNA (guanine-7-)-methyltransferase, mRNA(nucleoside-2′-)-methyltransferase, and viral multiplication. J. Biol. Chem. 253 (12), 4075–4077.

Purandare, A.V., Chen, Z., Huynh, T., Pang, S., Geng, J., Vaccaro, W., Poss, M.A., Oconnell, J., Nowak, K., Jayaraman, L., 2008. Pyrazole inhibitors of coactivator associated arginine methyltransferase 1 (CARM1). Bioorg. Med. Chem. Lett. 18 (15), 4438–4441.

Purfield, A.E., Tidwell, R.R., Meshnick, S.R., 2008. Interactions of DB75, a novel antimalarial agent, with other antimalarial drugs in vitro. Antimicrob. Agents Chemother. 52 (6), 2253–2255.

Ragno, R., Simeoni, S., Castellano, S., Vicidomini, C., Mai, A., Caroli, A., Tramontano, A., Bonaccini, C., Trojer, P., Bauer, I., 2007. Small molecule inhibitors of histone arginine methyltransferases: homology modeling, molecular docking, binding mode analysis, and biological evaluations. J. Med. Chem. 50 (6), 1241–1253.

Riddihough, G., Zahn, L.M., 2010. What is epigenetics? Science 330 (6004), 611.

Sack, J.S., Thieffine, S., Bandiera, T., Fasolini, M., Duke, G.J., Jayaraman, L., Kish, K.F., Klei, H.E., Purandare, A.V., Rosettani, P., Troiani, S., Xie, D., Bertrand, J.A., 2011. Structural basis for carm1 inhibition by indole and pyrazole inhibitors. Biochem. J. 436 (2), 331–339.

Schapira, M., Ferreira de Freitas, R., 2014. Structural biology and chemistry of protein arginine methyltransferases. Med. Chem. Commun. 5 (12), 1779–1788.

Scorilas, A., Black, M.H., Talieri, M., Diamandis, E.P., 2000. Genomic organization, physical mapping, and expression analysis of the human protein arginine methyltransferase 1 gene. Biochem. Biophys. Res. Commun. 278 (2), 349–359.

Selvi, B.R., Batta, K., Kishore, A.H., Mantelingu, K., Varier, R.A., Balasubramanyam, K., Pradhan, S.K., Dasgupta, D., Sriram, S., Agrawal, S., Kundu, T.K., 2010. Identification of a novel inhibitor of coactivator-associated arginine methyltransferase 1 (CARM1)-mediated methylation of histone H3 arg-17. J. Biol. Chem. 285 (10), 7143–7152.

Siarheyeva, A., Senisterra, G., Allali-Hassani, A., Dong, A., Dobrovetsky, E., Wasney, G.A., Chau, I., Marcellus, R., Hajian, T., Liu, F., Korboukh, I., Smil, D., Bolshan, Y., Min, J., Wu, H., Zeng, H., Loppnau, P., Poda, G., Griffin, C., Aman, A., Brown, P.J., Jin, J., Al-Awar, R., Arrowsmith, C.H., Schapira, M., Vedadi, M., 2012. An allosteric inhibitor of protein arginine methyltransferase 3. Structure 20 (8), 1425–1435.

Sinha, S.H., Owens, E.A., Feng, Y., Yang, Y., Xie, Y., Tu, Y., Henary, M., Zheng, Y.G., 2012. Synthesis and evaluation of carbocyanine dyes as PRMT inhibitors and imaging agents. Eur. J. Med. Chem. 54, 647–659.

Spannhoff, A., Heinke, R., Bauer, I., Trojer, P., Metzger, E., Gust, R., Schüle, R., Brosch, G., Sippl, W., Jung, M., 2007a. Target-based approach to inhibitors of histone arginine methyltransferases. J. Med. Chem. 50 (10), 2319–2325.

Spannhoff, A., Machmur, R., Heinke, R., Trojer, P., Bauer, I., Brosch, G., Schule, R., Hanefeld, W., Sippl, W., Jung, M., 2007b. A novel arginine methyltransferase inhibitor with cellular activity. Bioorg. Med. Chem. Lett. 17 (15), 4150–4153.

Therrien, E., Larouche, G., Manku, S., Allan, M., Nguyen, N., Styhler, S., Robert, M.-F., Goulet, A.-C., Besterman, J.M., Nguyen, H., 2009. 1,2-Diamines as inhibitors of co-activator associated arginine methyltransferase 1 (CARM1). Bioorg. Med. Chem. Lett. 19 (23), 6725–6732.

Thomas, D., Koopmans, T., Lakowski, T.M., Kreinin, H., Vhuiyan, M.I., Sedlock, S.A., Bui, J.M., Martin, N.I., Frankel, A., 2014. Protein arginine *N*-methyltransferase substrate preferences for different *N*η-substituted arginyl peptides. ChemBioChem 15 (11), 1607–1613.

Trapp, J., Meier, R., Hongwiset, D., Kassack, M.U., Sippl, W., Jung, M., 2007. Structure-activity studies on suramin analogues as inhibitors of NAD+-dependent histone deacetylases (sirtuins). ChemMedChem 2 (10), 1419–1431.

Tripsianes, K., Madl, T., Machyna, M., Fessas, D., Englbrecht, C., Fischer, U., Neugebauer, K.M., Sattler, M., 2011. Structural basis for dimethylarginine recognition by the tudor domains of human SMN and SPF30 proteins. Nat. Struct. Mol. Biol. 18 (12), 1414–1420.

Wan, H., Huynh, T., Pang, S., Geng, J., Vaccaro, W., Poss, M.A., Trainor, G.L., Lorenzi, M.V., Gottardis, M., Jayaraman, L., 2009. Benzo[d]imidazole inhibitors of coactivator associated arginine methyltransferase 1 (CARM1)—hit to lead studies. Bioorg. Med. Chem. Lett. 19 (17), 5063–5066.

Wang, G.G., Allis, C.D., Chi, P., 2007. Chromatin remodeling and cancer, part I: covalent histone modifications. Trends Mol. Med. 13 (9), 363–372.

Wang, J., Chen, L., Sinha, S.H., Liang, Z., Chai, H., Muniyan, S., Chou, Y.W., Yang, C., Yan, L., Feng, Y., Li, K.K., Lin, M.F., Jiang, H., Zheng, Y.G., Luo, C., 2012. Pharmacophore-based virtual screening and biological evaluation of small molecule inhibitors for protein arginine methylation. J. Med. Chem. 55 (18), 7978–7987.

Wang, L., Pal, S., Sif, S., 2008. Protein arginine methyltransferase 5 suppresses the transcription of the RB family of tumor suppressors in leukemia and lymphoma cells. Mol. Cell Biol. 28 (20), 6262–6277.

Xie, Y., Zhou, R., Lian, F., Liu, Y., Chen, L., Shi, Z., Zhang, N., Zheng, M., Shen, B., Jiang, H., Liang, Z., Luo, C., 2014. Virtual screening and biological evaluation of novel small molecular inhibitors against protein arginine methyltransferase 1 (PRMT1). Org. Biomol. Chem. 12 (47), 9665–9673.

Yost, J.M., Korboukh, I., Liu, F., Gao, C., Jin, J., 2011. Targets in epigenetics: inhibiting the methyl writers of the histone code. Curr. Chem. Genomics 5, 72–84.

Yan, L., Yan, C., Qian, K., Su, H., Kofsky-Wofford, S.A., Lee, W.C., Zhao, X., Ho, M.C., Ivanov, I., Zheng, Y.G., 2014. Diamidine compounds for selective inhibition of protein arginine methyltransferase 1. J. Med. Chem. 57 (6), 2611–2622.

Yang, Y., Bedford, M.T., 2013. Protein arginine methyltransferases and cancer. Nat. Rev. Cancer 13 (1), 37–50.

Yang, Y., Hadjikyriacou, A., Xia, Z., Gayatri, S., Kim, D., Zurita-Lopez, C., Kelly, R., Guo, A., Li, W., Clarke, S.G., Bedford, M.T., 2015. PRMT9 is a type II methyltransferase that methylates the splicing factor SAP145. Nat. Commun. 6, 6428.

Yu, W., Chory, E.J., Wernimont, A.K., Tempel, W., Scopton, A., Federation, A., Marineau, J.J., Qi, J., Barsyte-Lovejoy, D., Yi, J., Marcellus, R., Iacob, R.E., Engen, J.R., Griffin, C., Aman, A., Wienholds, E., Li, F., Pineda, J., Estiu, G., Shatseva, T., Hajian, T., Al-Awar, R., Dick, J.E., Vedadi, M., Brown, P.J., Arrowsmith, C.H., Bradner, J.E., Schapira, M., 2012. Catalytic site remodelling of the DOT1L methyltransferase by selective inhibitors. Nat. Commun. 3, 1288.

Zhang, R., Li, X., Liang, Z., Zhu, K., Lu, J., Kong, X., Ouyang, S., Li, L., Zheng, Y.G., Luo, C., 2013. Theoretical insights into catalytic mechanism of protein arginine methyltransferase 1. PLoS One 8 (8), e72424.

Zhang, X., Cheng, X., 2003. Structure of the predominant protein arginine methyltransferase PRMT1 and analysis of its binding to substrate peptides. Structure 11 (5), 509–520.

Zhang, X., Zhou, L., Cheng, X., 2000. Crystal structure of the conserved core of protein arginine methyltransferase PRMT3. EMBO J. 19 (14), 3509–3519.

Zurita-Lopez, C.I., Sandberg, T., Kelly, R., Clarke, S.G., 2012. Human protein arginine methyltransferase 7 (PRMT7) is a type III enzyme forming ω-NG-monomethylated arginine residues. J. Biol. Chem. 287 (11), 7859–7870.

CHAPTER 9

Molecular Design of Compounds Targeting Histone Methyltransferases

Alberto Del Rio, Greta Varchi
Institute of Organic Synthesis and Photoreactivity (ISOF), National Research Council (CNR), Bologna, Italy

CONTENTS

1. INTRODUCTION

Epigenetic alterations in chromatin involve methylation of DNA patterns, several forms of histone modifications, and noncoding RNAs like microRNAs. All these processes are known to modulate the structure of chromatin leading to the activation or silencing of gene expression (Herceg and Ushijima, 2010; Baylin, 2008; Jones and Baylin, 2007; Kouzarides, 2007; Kelly et al., 2010). In particular, histone modification is one of the main mechanisms of chromatin remodeling that occurs through a variety of posttranslational modifications (PTMs) at the N-terminal tails of histone proteins. These PTMs include acetylation, methylation, phosphorylation, ubiquitylation, sumoylation, glycosylation, ADP-ribosylation, carbonylation, citrullination, and biotinylation, among others that are being uncovered (Gardner et al., 2011; Sidoli et al., 2012; Andreoli and Del Rio, 2014, 2015). For example, histone tails can have their lysine residues acetylated, methylated, or ubiquitilated and arginine residues can be methylated or serine and threonine residues can be phosphorylated (Andreoli and Del Rio, 2015; Cosgrove et al., 2004; Cosgrove and Wolberger, 2005; Cruickshank et al., 2010; Bannister and Kouzarides, 2011; Imhof, 2006; Mellor et al., 2008; Weake and Workman, 2008; Loyola and Almouzni, 2007; De Koning et al., 2007). The fine regulation of histone PTMs and DNA methylation is controlled and catalyzed by different classes of enzymes whose existence and functions have been elucidated with an extraordinary progression in the last decade (Bannister and Kouzarides, 2011; Santos-Rosa and Caldas, 2005; Golbabapour et al., 2011; Rius and Lyko, 2012; Chi et al., 2010; Copeland et al., 2010; Dhanak, 2012). From the chemical point of view, epigenetic modifications occurring at histone tails are reversible nuclear chemical reactions that are due to enzymes able to exercise opposing catalytic effects (Kouzarides, 2007; Andreoli and Del Rio, 2015; Chi et al., 2010; Andreoli et al., 2013). Most histone modifying enzymes simply act by producing specific covalent modifications that, taken together, elicit a complex and concerted processing.

Epi-Informatics. http://dx.doi.org/10.1016/B978-0-12-802808-7.00009-5

Histone methylation has emerged as one of the most important PTMs in recent years. Consequently, protein methyltransferases (PMT), and in particular histone methyltransferases (HMT), have been ascribed a focal role as new targets for pharmaceutical development since they have been found to be overexpressed or repressed in several types of pathological conditions, especially cancer. In particular, HMT have been found to be correlated with genesis and progression of breast cancer, leukemia, myeloma, and ovarian, prostate, and kidney cancers, among others. Several insightful reviews were written recently describing catalytic mechanisms and specific biological roles of these attractive enzymes (Copeland, 2012; Martinet et al., 2011; Copeland et al., 2009; Nimura et al., 2010; Daniel et al., 2011; Spannhoff et al., 2009; Ferrari and Pasini, 2013; Piunti and Pasini, 2011; Fandy, 2009; Yoo and Medina-Franco, 2011; Albert and Helin, 2010; Burgers et al., 2002). As anticipated, due to their importance in different pathological conditions, several drug discovery programs have been launched in order to design specific compounds able to modulate them (Copeland et al., 2009). Therefore, an important effort is currently made for the development of new methods and technologies able to help the discovery of small-molecule inhibitors with therapeutic potential (Heightman, 2011; Kooistra and Helin, 2012). In this chapter we present and discuss HMTs, their biological role, and the perspectives for considering these enzymes as new targets for drug discovery initiatives that take advantage of computer-aided molecular design techniques for the identification and the de novo design of new specific inhibitors.

2. PROTEIN METHYLTRANSFERASES

2.1 Protein Methylation

Histone tails are rich in lysines (K) and arginines (R) that constitute positive residues available to interact electrostatically with the negatively charged phosphate backbone of DNA. Interestingly, these residues are positioned in a regular way on the histone tails where two to three uncharged residues typically space out K and R positions. Of note, lysine and arginine methylation is limited to few sites located in the histones H3 and H4 (Andreoli and Del Rio, 2014). Some examples of the positions targeted by HMTs are H3K4, H3K9, H3K27, H3K36, H3K79, and H4K20. One peculiarity of the histone methylation is the sequential nature of this reaction that can occur in multiple steps where a site can be mono-, di-, or trimethylated (Figure 1).

Because of this multistep catalytic mechanism, histone methylation appears to be one of the modifications that is most finely tuned since, upon the first methylation step (Figure 1), the additions of methyl groups to histone sites produce relevant changes in the physicochemical properties like polarity, volume, and basicity of the modified residue. In terms of these properties

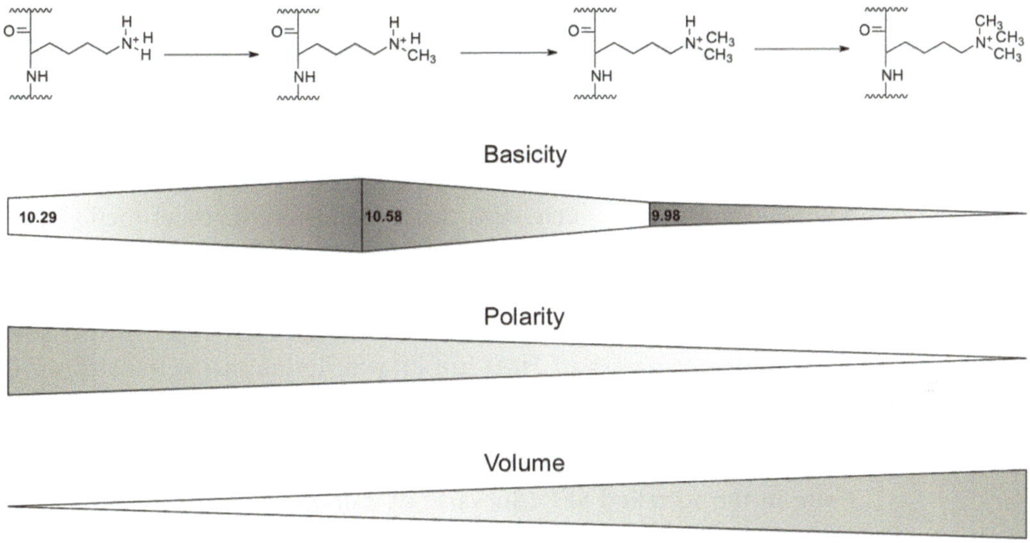

FIGURE 1
Peculiarities of histone lysine methylation.

the monomethylation of lysine lowers the polarity while increasing basicity and volume properties. The dimethylation further increases steric hindrance and lipophilicity but lowers the basicity of the ε-nitrogen. Interestingly, the most relevant physicochemical modification is obtained by the trimethylation, where a complete abrogation of basic property occurs. This fact gives to the trimethylated lysines ($LysMe_3$) the unique feature of being charged quaternary amines, stable, and not subjected to acid/base equilibriums like primary, secondary, and ternary amines. Moreover, the interaction of quaternary amines can span from polar to nonpolar behavior exemplifying this posttranslational modification as one of the most versatile for the global rearrangement of the histone core.

Arginine is known to undergo only two steps of methylation, mono- and dimethylation. Monomethylation of arginine can occur in one of the two primary nitrogens of the guanidine portion increasing volume, polarity, and basicity of the residue. The second methylation step can be symmetrical or asymmetrical, where the symmetrical happens if the substitution occurs on the non-previously methylated nitrogen while the asymmetrical occurs on the already methylated nitrogen (Andreoli and Del Rio, 2015). Even for this second step, and irrespectively to the symmetry of this methylation, basicity, steric hindrance, and lipophilicity increase on the targeted residue.

The action of methyltransferases is opposed to enzymes that remove methyl groups from histone, which are the histone demethylases. The removal of methyl groups takes place via an oxidation process mediated by FAD, for lysine-specific demethylases (LSD), or by Fe^{2+}, for demethylases of the Jumonji C family (JmjC), in which the metal ion is coordinated to α-ketoglutarate (α-KG) and three conserved residues, to form formaldehyde as a by-product. Lysine demethylation also occurs as a multistep process where these demethylases can perform mono-, di-, or tridemethylation, resulting in a complex kinetic equilibrium where each step depends on the other methylation steps, still emphasizing the fine-tuning mechanism of histone methylation.

2.2 Structure and Function of Protein Methyltransferases

Two different classes of HMT are known; lysine methyltransferases (PKMT) and arginine methyltransferases (PRMT), which are encoded in 51 and 45 genes (Copeland et al., 2009; Nimura et al., 2010; Copeland and Richon, 2011; Di Lorenzo and Bedford, 2011), respectively. PKMTs share a conserved active site in the so-called SET (Su(Var)3–9, Enhancer of zeste, Trithirax) domain. The only known exception is the DOT1L, which has PKMT activity without having the SET domain in its structure and shares a higher homology toward PRMTs. For this reason it is often reported in the PRMT family tree diagrams (Copeland, 2012; Copeland et al., 2009; Nimura et al., 2010; Copeland and Richon, 2011). In order to methylate certain lysine or arginine residues, HMTs use a reactive S-adenosyl methionine (SAM), which leaves a methyl group to the respective Lys or Arg residue, becoming S-adenosyl homocysteine (SAH) (Figure 2). The structural characteristic that differentiates HMTs to other SAM-binding proteins like DNA methyltransferases is their elongated active site geometry. In HMTs the SAM binding pocket entrance is in the opposite position of the hydrophobic and narrow histone (Lys or Arg) binding pocket and these two tunnels have a contact area where methylation of histones occurs (Copeland, 2012; Martinet et al., 2011; Copeland et al., 2009; Nimura et al., 2010; Daniel et al., 2011; Spannhoff et al., 2009; Ferrari and Pasini, 2013; Piunti and Pasini, 2011; Fandy, 2009; Yoo and Medina-Franco, 2011; Albert and Helin, 2010; Burgers et al., 2002; Chang et al., 2010).

Several crystallographic structures of PKMTs are available in the protein data bank (Andreoli et al., 2013). As anticipated, with the exception of DOT1L, all PKMTs share a canonical SET domain (Copeland, 2012; Copeland et al., 2009; Nimura et al., 2010; Copeland and Richon, 2011) and have SAM, SAH, or inhibitors cocrystallized or are bound with peptide protein partners. PKMTs with resolved SET domain include MLL1, EHMT1, EHMT2, SUV39H2, NSD1, SETD3, SETD6, SETD7, SETD8, SETD2, SETMAR, ASH1L, SUV420H1, SUV420H2, SMYD1, SMYD2, SMYD3, PRDM1, PRDM4, PRDM10, PRDM11, and PRDM12. A recent list of available crystallographic structures of PKMTs was recently reviewed by our group (Andreoli et al., 2013). Similar to PKMTs, several PRMTs have also been solved and in the PDB one could find structures

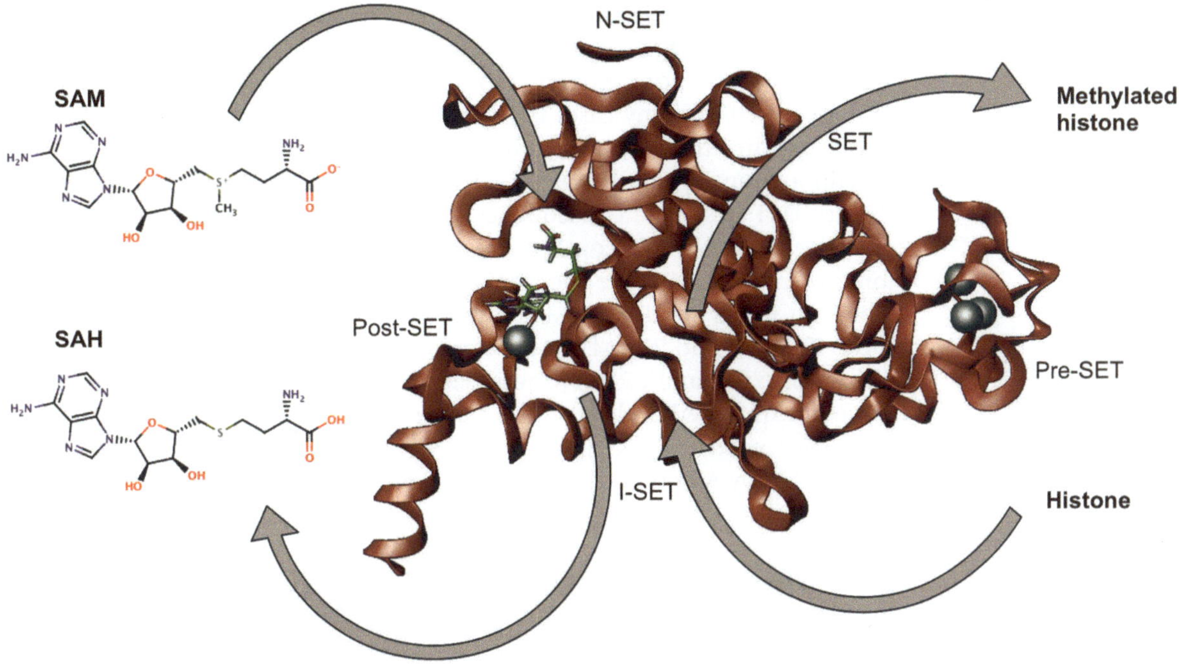

FIGURE 2
Structure of histone methyltransferases and representation of the catalytic mechanism (based on the g9A PDB code: 208J). Marvin was used for drawing and displaying chemical structures, Marvin 6.0.0, ChemAxon (http://www.chemaxon.com).

of PRMT1, PRMT2, PRMT3, ECE2, METTL11A, among others. Even in this case, interestingly, some structures are bound with recently developed compounds, providing new insightful information for the development of specific arginine methyltransferase inhibitors (Andreoli et al., 2013).

3. CLINICOPATHOLOGICAL ROLE OF PKMTs

In the last decade several cancer pathologies have been associated with specific epigenetic changes but the way in which epigenetic modifications are regulated is still largely unknown. In this context PKMTs and PRMTs are playing a prominent role as many of them have been linked in humans or in mouse models to the genesis, progression, aggressiveness, and increased invasion of several forms of cancer (Albert and Helin, 2010), as well as other pathologies or conditions like inflammatory-related and neurodegenerative diseases. The polycomb group transcriptional repressor EZH2, for instance, is overexpressed in many different cancer types and has been proposed as a molecular marker of prostate cancer progression and metastasis. It has also been linked to breast cancer aggressiveness and progression of bladder tumors.

Consequently, potent inhibitors of EZH2 are currently being tested in clinical trials for different types of cancers. Similarly, NSD2 is associated with tumor aggressiveness and poor prognosis in several types of cancer. It is important to emphasize that in many cases the specific molecular reasons for which HMT contributes directly to disease development are still to be elucidated. Indeed, it is still a daunting challenge to demonstrate whether the disregulation of a specific HMT is a mere consequence of the disease state itself or whether it constitutes a veritable oncogenic driver. Nevertheless, it should be noted that, while for many HMTs the mechanisms of disregulation are still too distant to be well understood, intragenic mutations have been described for MLL3, NSD1, PRDM1, and PRDM2 in different diseases, and other HMTs like NSD3, EHMT1, SMYD2, and PRDM12 reside in chromosomal regions that are either frequently amplified or deleted in cancer. An insightful review regarding some of the best validated examples of HMTs contributing to tumorigenesis has been written by Albert and Helin (Albert and Helin, 2010).

4. DRUG DISCOVERY OF NEW EPI-DRUGS FOR HISTONE METHYLTRANSFERASES

As mentioned above, epigenetics became a thriving field of research and major efforts are currently directed toward the discovery of new small-molecules able to modulate these target proteins (Sippl and Jung, 2009, 2010). Recent success stories document the potential to successfully interfere with the epigenetic code with small organic molecules (Dhanak, 2012). Beside DNA methylation and noncoding RNAs, histone deacetylases of class I, II, and IV have been, de facto, the first family of enzymes that opened the way to histone epigenetic therapies. The first preclinical and clinical results have been obtained in the last years and led to the perception that many other 'epi-drugs' might result in new single or combination therapies to control the process of genesis and progression of several forms of cancer. Because a great number of methyltransferases governs the mechanisms underlying the epigenetic regulation occurring through histone (Andreoli and Del Rio, 2014, 2015; Andreoli et al., 2013) and, because the alterations of the dynamic balance between heterochromatin and euchromatin states has critical consequence in silencing and transcription processes, it became essential to understand how the action of HMTs and their combinations modify epigenomics and how the biochemical functions of these enzymes can be dissected in order to envision new epigenetic therapies. Indeed, the validation of new HMTs for pharmaceutical discovery is becoming increasingly important to insure major success of drug development pipelines. In this direction, the research has been in part fuelled by the increasing availability of structural information (Andreoli et al., 2013; Barth and Imhof, 2010). Nevertheless, daunting questions still need to be fully assessed to guide the discovery of new chemicals; for instance, it is still to be clarified whether selective or promiscuous compounds would be the most appropriate to generate effective and safe new drugs. Several reviews discussing these aspects have been written recently (Kelly et al., 2010; Andreoli et al., 2013; Vellore and Baron, 2014; Hamm and Costa, 2011; Mai, 2014; Drewes, 2012).

4.1 Computer-Aided Molecular Design Approaches

Computer-aided drug design (CADD) techniques are used for the rapid assessment of chemical libraries in order to guide and speed up the early-stage development of new active compounds. CADD entails a vast number of computational methodologies like virtual screening, virtual library design, lead optimization, de novo design, and so forth. Because of the proven ability of computational techniques to guide the selection of new hit compounds, computational chemistry and chemoinformatics are still scientific disciplines in full bloom and, in particular, the term virtual screening (VS), introduced in the late 1990s, describes the usage of computational algorithms and models for the identification of novel bioactive molecules through the analysis of large collections of chemical databases. Indeed, while in the last 20 years or so the expansion of combinatorial chemistry and high-throughput screening (HTS) techniques represented an important step in the discovery of new compounds, the importance of in silico techniques has been associated with the reduced costs and the increased time-efficacy to identify new potential active compounds to propose for the subsequent steps of the drug discovery pipeline. Different methodologies can be applied to perform virtual screenings like molecular docking, pharmacophore searching, or with simple techniques such as similarity and/or fingerprint searching. While HTS still constitute a valuable method to test systematically libraries of compounds, especially in pharmaceutical industry, VS became critically valuable for small and academic organizations in which hit discovery cannot be usually conducted through expensive technologies.

In principle, VS technique can be divided into ligand- and structure-based drug design techniques (LBDD and SBDD): the first category usually takes advantage of information from known bioactive compounds (ligand) and are essential tools when structural information of a biological target is missing or when the molecular design is not directed toward a target-centric approach but point to modulate cellular pathways or phenotypic traits without a precise knowledge of the mechanism of action. On the other hand, SBDD usually exploit three-dimensional (3D) structure of the biological target (protein) in order to identify putative modulators of the protein activity. In the last years, the growing availability of protein structures, resolved by structural biologists, progressively raised the possibility to deploy SBDD. Thus, the popularity of

VS techniques has been unquestionably fuelled by the availability of the 3D structure of biological targets. A clear example in the context of epigenetics is the increasing number of available high-quality crystallographic structures of histone modifying enzymes (Andreoli et al., 2013).

Among computational approaches that can help drive the discovery of new bioactive compounds, a prominent workflow is the screening of large databases of readily available molecules; therefore it is important to define the source of chemical data that are available for this purpose. Different chemical databases have been developed in the last years and have been reviewed by our group (Moura Barbosa and Del Rio, 2012). Of note, there is a rising interest in databases of natural products and dietary components that are known to represent a vast chemical diversity with underlying scaffold complexities and architectures. Indeed, most natural products are assorted by chemical groups reflecting novel molecular properties/features as compared to synthetic compounds, and available drugs and chemoinformatic analyses have demonstrated the adherence to drug- and lead-like rules, purporting the idea that several classes can be considered as pharmaceutically-relevant entities (Medina-Franco et al., 2012; Del Rio et al., 2014). A particularly appealing feature for industrial settings in the context of developing natural compounds is to insure the appropriate intellectual property protection requested for the pharmaceutical development and, in this direction, it should be noted also that natural products are known to provide line principles for novel library design in combinatorial chemistry and targeted compound libraries inspired by nature. A particular class of natural compounds that can be considered as dietary components are flavoring substances like menthol, camphor, and anethol, which are discrete chemical entities that usually are considered "generally recognized as safe" (GRAS) compounds (Medina-Franco et al., 2012). Interestingly, the comparison of collections of compounds including GRAS, natural products, approved drugs, and datasets from commercial molecules by means of chemoinformatic analysis demonstrated that GRAS products are an important source of bioactive compounds that possess all the characteristics for drug discovery and nutraceutical purposes (Moura Barbosa and Del Rio, 2012; Medina-Franco et al., 2012; Del Rio et al., 2014).

4.2 CADD Approaches on Protein Methyltransferases

The common small molecule SAM is required as a cofactor for the catalytic mechanism of HMTs. Because SAM or SAH have been structurally characterized with a variety of HMTs—showing the possibility to accommodate organic molecules in the cosubstrate binding site—HMTs have been found to be generally druggable targets. In addition, other small molecules have been shown to bind the substrate binding site, offering a second way to target these family of enzymes with small organic compounds (Figure 3). A comprehensive study by

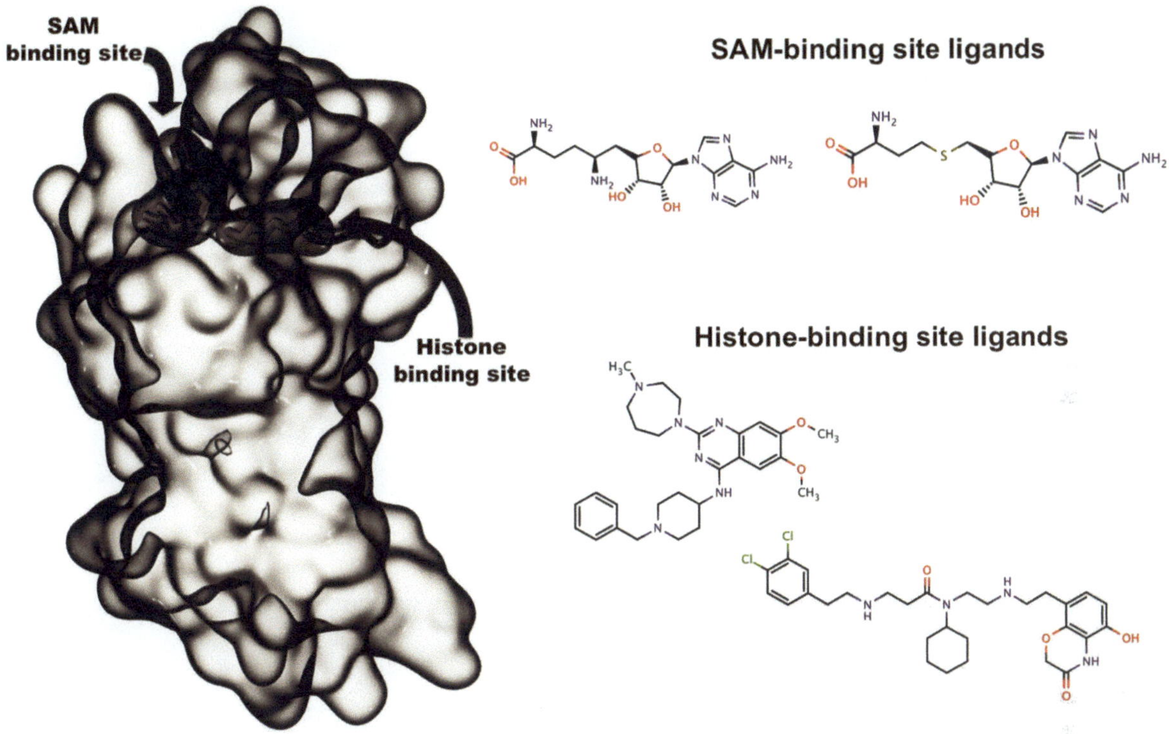

FIGURE 3

Depiction of SAM and histone binding sites of SETD8 methyltransferase and examples of known ligands. Assessing the mode of action implies a full understanding of the more appropriate mechanism of inhibition operated by small molecules.

Richon *et al.* (Richon et al., 2011) defined the number of PMT-related proteins that exist in humans and helped to understand their relatedness with respect to catalytic active domain structure. Authors performed a systematic survey of the human genome for proteins that display structural alignment with SET-domain PKMTs, and likewise with PRMTs, for which biochemical evidence of enzymatic activity has been demonstrated. Through multiple, iterative database searches and multiple sequence alignments, they defined the family trees for human PKMTs and PRMTs reporting useful information for chemical biologists and drug discovery scientists aiming at identifying selective ligands for these classes of enzymes. On the basis of these studies, HMTs acquired a rising interest in the world of drug discovery and computational techniques have been used in some cases (described below) to identify novel interacting compounds (Copeland et al., 2009).

A still challenging question concerning HMTs is whether drug design approaches should be addressed to target the SAM binding site or the substrate binding site, where Lys or Arg residues of the histones are methylated (Figure 4)

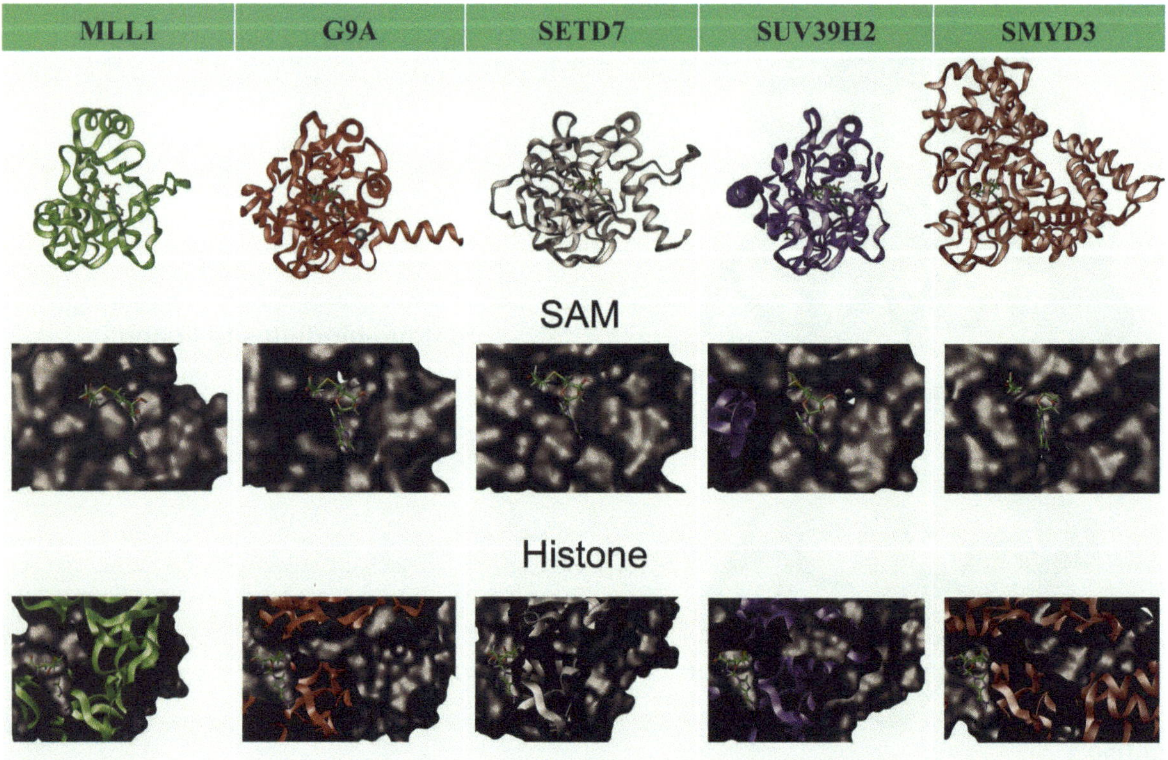

FIGURE 4
Depiction of the site mapping for five different methyltransferases. SAM and histone binding sites can have different shapes and plasticity degrees to accommodate small-molecule modulators.

and what the advantages of pursuing one or both strategies are. Structural studies have demonstrated that the SAM binding site is more buried with respect to the substrate binding site; this feature should be expected to translate in a more convenient possibility to identify potent inhibitors competing with the SAM. However, targeting the substrate binding site might have the advantage that, since specific structural motifs are required to select the proper histone substrate, inhibitors competing with the histone might result in more selective compounds. In fact, it should be emphasized that the SAM is used as a substrate by many other proteins like DNA methyltransferases and also non-epigenetic enzymes. This raises the question whether the pharmacological modulation of the SAM binding site may guarantee an adequate selectivity against other SAM-binding proteins.

The first examples of virtual screening studies on PMT were carried out on the histone arginine methyltransferase PRMT1. Ragno *et al.* (Ragno et al., 2007) described a homology model, a docking study, and a 3D-QSAR application in search of novel dye-like derivatives as inhibitors of PRMT1, while Heinke *et al.* screened the Chembridge database consisting of some 328,000 compounds against a homology model of human PRMT1 and identified, by hierarchical VS, novel drug-like compounds with activity below 35 μm (Heinke et al., 2009). In addition, the methyltransferase G9a was one of the most studied PMT because physiological and pathological functions indicated a clear involvement in disease. Consequently, potent inhibitors have been described by means of computational techniques. Ahrens *et al.* described an efficient hit-finding approach using G9a as a model. In their approach a knowledge-based selection of a compound library of 900,000 compounds was used by means of different computational techniques demonstrating that the use of a combination of virtual screening methods can efficiently enrich the hit rate even in a small screening set (Ahrens et al., 2012).

There is still a number of PKMTs or PRMTs that are currently not validated and have not yet been targeted by small molecules. Recently, in our group we have been interested in components of the KMT that are involved in colorectal tumorigenesis, namely EZH2, SUV39H1, and SMYD3. While compounds targeting EZH2 are currently in clinical trials for the treatment of human cancers, no SMYD3-specific compound has been identified so far. In order to preferentially target cancer cells, in a recent study we envisioned to specifically target SMYD3 catalytic function by means of small-molecule inhibitors that we could identify with in silico techniques. To this purpose, based on the crystallography structure of SMYD3, we targeted the histone binding site and set up a structure-based high-throughput docking screening workflow (Figure 5) by using the Asinex subset of the CoCoCo database (Del Rio et al., 2010). We could identify the BCI-121 compound that was able to inhibit SMYD3 catalytic activity and its specific methylation on histones (Figure 5) (Peserico et al., 2015). Further investigation with molecular modeling techniques revealed that BCI-121 binds in the inner part of the lysine channel connecting the cofactor site and the histone binding site. Interestingly, the opening of the substrate binding pocket revealed several acidic residues with no direct interaction with the compound, suggesting that modeling results can be of use for further optimization of the chemical structures in order to improve potency and selectivity of new compounds.

Interestingly, quantum mechanical/molecular mechanical (QM/MM) and free energy simulations have also been used to better understand the different methylation-transfer steps from the SAM to the target lysine/methyl-lysine, pointing out how methylation steps and point mutation on PMT can affect the efficiency of the methylation (Yao et al., 2012). QM/MM and free energy

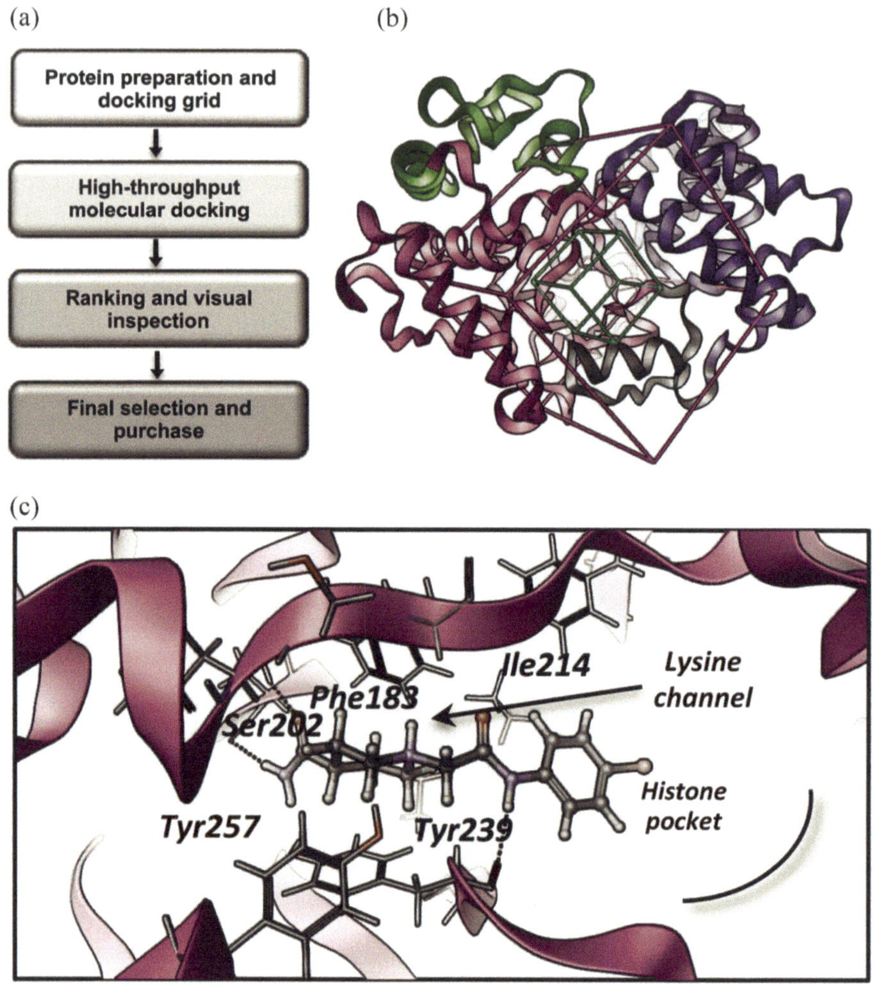

FIGURE 5

Computer-aided drug design approaches applied for the identification of compound targeting the histone lysine methyltransferase SMYD3 (Figure reuse license number: 3638781382978). (a) Schematic representation of the virtual screening procedure. (b) Structure of full-length SMYD3 highlighting the N-terminal SET domain (purple), the MYND domain (yellow), the post-SET domain (grey), and the C-terminal region (blue). The boxes define the area of the protein—centered on the histone binding site—that was considered to perform the docking screening calculation. The green box identifies the area containing at least one atom of the putative ligand, while the purple box identifies the area where all atoms of the ligands should lie. (c) Predicted binding mode of BCI-121 in the histone binding site.

simulation have been also used to study PRMT5 and, in particular, the symmetric methylation catalyzed by this protein arginine methyltransferase (Yue et al., 2015).

Because the substrate of HMTs are histones, which is the protein part of the nucleosome, increasing attention has been dedicated to the discovery of competing peptides that are able to inhibit HMTs. Smadbeck *et al.* have conceived a three-stage computational de novo peptide design method to design inhibitory peptide of EZH2 finding out new compounds with IC_{50} of 13.5 μM. (Smadbeck et al., 2014)

Finally, it is worth noting that a number of multiprotein complexes govern catalytic mechanisms associated with histone methyltransferases. Knowing how these molecular machineries are constituted, how their dynamic behavior influences the catalytic activities, and how this behavior ultimately influences the biological processes (for instance, of tumorigenesis and cancer progression) is of extreme importance. For example, the methyltransferase EZH2, which is part of the core of the polycomb repressive complex (PRC2), was recently subject of crystallographic studies indicating protein–protein interaction patterns within the PRC2 (Han et al., 2007). Similarly, in many cases, a low amount of information is available, at a molecular level, about the flexibility and the conformational ability of epigenetic enzymes to recognize and interact with histone substrates. It becomes an even more complicated task when additional components, like corepressors, are required (e.g., in LSD1). In this direction molecular modeling techniques promise to be a helpful tool to explore these molecular mechanisms.

5. CONCLUSIONS

It is unquestionable that epigenetics research will play a major role in the near future for the development of new targeted therapies. Indeed, while histone methyltransferases are important enzymes for normal cell functions, like gene expression, transcriptional regulation, heterochromatin formation, and X-chromosome inactivation, a significant amount of data suggest that overexpression of these enzymes can result in the aberrant silencing and play a role in cancer, inflammation, and neurodegenerative diseases. Biological advances implicate that there is a vast interest for developing new methods and technologies for the identification of specific compounds for the pharmaceutical industry. In this context computational techniques are receiving increasing attention because of the availability of structural data. Successful pipelines have been carried out for the identification of specific compounds for some HMTs like PRMT1, EZH2, and SMYD3. It is expected that the usage and combination of various in silico approaches could exceedingly constitute an effective step toward the identification, development, and pharmacological definition of new compounds for these families of enzymes and we hope that this chapter will stimulate new and original initiatives in this direction.

ACKNOWLEDGMENTS

This work was supported by the Associazione Italiana per la Ricerca sul Cancro (AIRC)—Emilia Romagna Start-Up grant 6266.

REFERENCES

Andreoli, F., Del Rio, A., 2014. Physicochemical modifications of histones and their impact on epigenomics. Drug Discov. Today 19, 1372–1379.

Andreoli, F., Del Rio, A., 2015. Computer-aided molecular design of compounds targeting histone modifying enzymes. Comput. Struct. Biotechnol. J. 13, 358–365.

Andreoli, F., Barbosa, A.J.M., Parenti, M.D., Del Rio, A., 2013. Modulation of epigenetic targets for anticancer therapy: clinicopathological relevance, structural data and drug discovery perspectives. Curr. Pharm. Des. 19, 578–613.

Albert, M., Helin, K., 2010. Histone methyltransferases in Cancer. Semin. Cell Dev. Biol. 21, 209–220.

Ahrens, T., Bergner, A., Sheppard, D., Hafenbradl, D., 2012. Efficient hit-finding approaches for histone methyltransferases: the key parameters. J. Biomol. Screen. 17, 85–98.

Baylin, S.B., 2008. The Molecular Basis of Cancer, vol. 2. Elsevier.

Bannister, A.J., Kouzarides, T., 2011. Regulation of chromatin by histone modifications. Cell Res. 21, 381–395.

Burgers, W.A., Fuks, F., Kouzarides, T., 2002. DNA methyltransferases get connected to chromatin. Trends Genet. 18, 275–277.

Barth, T.K., Imhof, A., 2010. Fast signals and slow marks: the dynamics of histone modifications. Trends Biochem. Sci. 35, 618–626.

Cosgrove, M.S., Boeke, J.D., Wolberger, C., 2004. Regulated nucleosome mobility and the histone code. Nat. Struct. Mol. Biol. 11, 1037–1043.

Cosgrove, M.S., Wolberger, C., 2005. How does the histone code work? Biochem. Cell Biol. 83, 468–476.

Cruickshank, M.N., Besant, P., Ulgiati, D., 2010. The impact of histone post-translational modifications on developmental gene regulation. Amino Acids 39, 1087–1105.

Chi, P., Allis, C.D., Wang, G.G., 2010. Covalent histone modifications—miswritten, misinterpreted and mis-erased in human cancers. Nat. Rev. Cancer 10, 457–469.

Copeland, R.A., Olhava, E.J., Scott, M.P., 2010. Targeting epigenetic enzymes for drug discovery. Curr. Opin. Chem. Biol. 14, 505–510.

Copeland, R.A., 2012. Protein methyltransferase inhibitors as personalized cancer therapeutics. Drug Discov. Today Ther. Strateg. 9, e83–e90.

Copeland, R.A., Solomon, M.E., Richon, V.M., 2009. Protein methyltransferases as a target class for drug discovery. Nat. Rev. Drug Discov. 8, 724–732.

Copeland, R.A., Richon, V.M., 2011. The human protein methyltransferases. Chem. Biol. Drug des. 78, 199–210.

Chang, Y., Ganesh, T., Horton, J.R., Spannhoff, A., Liu, J., Sun, A., Zhang, X., Bedford, M.T., Shinkai, Y., Snyder, J.P., Cheng, X., 2010. Adding a lysine mimic in the design of potent inhibitors of histone lysine methyltransferases. J. Mol. Biol. 400, 1–7.

De Koning, L., Corpet, A., Haber, J.E., Almouzni, G., 2007. Histone chaperones: an escort network regulating histone traffic. Nat. Struct. Mol. Biol. 14, 997–1007.

Di Lorenzo, A., Bedford, M.T., 2011. Histone arginine methylation. FEBS Lett. 585, 2024–2031.

Del Rio, A., Barbosa, A.J.M., Caporuscio, F., Mangiatordi, G.F., 2010. CoCoCo: a free suite of multiconformational chemical databases for high-throughput virtual screening purposes. Mol. Biosyst. 6, 2122–2128.

Del Rio, A., Da Costa, F.B., 2014. Molecular approaches to explore natural and food-compound modulators in cancer epigenetics and metabolism. In: Martinez-Mayorga, K., Medina-Franco, J.L. (Eds.), Foodinformatics: Applications of Chemical Information to Food Chemistry. Springer International Publishing, Cham, pp. 131–149.

Dhanak, D., 2012. Cracking the code: the promise of epigenetics. ACS Med. Chem. Lett. 3, 521–523.

Daniel, F.I., Cherubini, K., Yurgel, L.S., de Figueiredo, M.A.Z., Salum, F.G., 2011. The role of epigenetic transcription repression and DNA methyltransferases in cancer. Cancer 117, 677–687.

Drewes, G., 2012. Future strategies in epigenetic drug discovery. Drug Discov. Today Ther. Strateg. 9, e121–e127.

Ferrari, K.J., Pasini, D., 2013. Regulation and function of DNA and histone methylations. Curr. Pharm. Des. 19(4), 719–733.

Fandy, T.E., 2009. Development of DNA methyltransferase inhibitors for the treatment of neoplastic diseases. Curr. Med. Chem. 16, 2075–2085.

Gardner, K.E., Allis, C.D., Strahl, B.D., 2011. Operating on chromatin, a colorful language where context matters. J. Mol. Biol. 409, 36–46.

Golbabapour, S., Abdulla, M.A., Hajrezaei, M., 2011. A concise review on epigenetic regulation: insight into molecular mechanisms. Int. J. Mol. Sci. 12, 8661–8694.

Herceg, Z., Ushijima, T., 2010. Introduction: Epigenetics and Cancer, first ed. vol. 70. Elsevier Inc.

Heightman, T.D., 2011. Chemical biology of lysine demethylases. Curr. Chem. Genomics 5, 62–71.

Hamm, C.A., Costa, F.F., 2011. The impact of epigenomics on future drug design and new therapies. Drug Discov. Today 16, 626–635.

Heinke, R., Spannhoff, A., Meier, R., Trojer, P., Bauer, I., Jung, M., Sippl, W., 2009. Virtual screening and biological characterization of novel histone arginine methyltransferase PRMT1 inhibitors. Chem. Med. Chem. 4, 69–77.

Han, Z., Xing, X., Hu, M., Zhang, Y., Liu, P., Chai, J., 2007. Structural basis of EZH2 recognition by EED. Structure 15, 1306–1315.

Imhof, A., 2006. Epigenetic regulators and histone modification. Brief. Funct. Genomic Proteomic. 5, 222–227.

Jones, P.A., Baylin, S.B., 2007. The epigenomics of cancer. Cell 128, 683–692.

Kouzarides, T., 2007. Chromatin modifications and their function. Cell 128, 693–705.

Kelly, T.K., De Carvalho, D.D., Jones, P.A., 2010. Epigenetic modifications as therapeutic targets. Nat. Biotechnol. 28, 1069–1078.

Kooistra, S.M., Helin, K., 2012. Molecular mechanisms and potential functions of histone demethylases. Nat. Rev. Mol. Cell Biol. 13, 297–311.

Loyola, A., Almouzni, G., 2007. Marking histone H3 variants: how, when and why? Trends Biochem. Sci. 32, 425–433.

Mellor, J., Dudek, P., Clynes, D., 2008. A glimpse into the epigenetic landscape of gene regulation. Curr. Opin. Genet. Dev. 18, 116–122.

Martinet, N., Michel, B.Y., Bertrand, P., Benhida, R., 2011. Small molecules DNA methyltransferases inhibitors. Med. Chem. Comm. 3, 263.

Mai, A., 2014. Targeting epigenetics in drug discovery. Chem. Med. Chem. 9, 415–417.

Moura Barbosa, A.J., Del Rio, A., 2012. Freely accessible databases of commercial compounds for high-throughput virtual screenings. Curr. Top. Med. Chem. 12, 866–877.

Medina-Franco, J.L., Martínez-Mayorga, K., Peppard, T.L., Del Rio, A., 2012. Chemoinformatic analysis of GRAS (generally recognized as safe) flavor chemicals and natural products. PLoS One 7, e50798.

Nimura, K., Ura, K., Kaneda, Y., 2010. Histone methyltransferases: regulation of transcription and contribution to human disease. J. Mol. Med. Berl. 88, 1213–1220.

Piunti, A., Pasini, D., 2011. Epigenetic factors in cancer development: polycomb group proteins. Future Oncol. 7, 57–75.

Peserico, A., Germani, A., Sanese, P., Barbosa, A.J., Di Virgilio, V., Fittipaldi, R., Fabini, E., Bertucci, C., Varchi, G., Moyer, M.P., Caretti, G., Del Rio, A., Simone, C., 2015. A SMYD3 small-molecule inhibitor impairing cancer cell growth. J. Cell. Physiol. 230, 2447–2460.

Rius, M., Lyko, F., 2012. Epigenetic cancer therapy: rationales, targets and drugs. Oncogene 31, 4257–4265.

Richon, V.M., Johnston, D., Sneeringer, C.J., Jin, L., Majer, C.R., Elliston, K., Jerva, L.F., Scott, M.P., Copeland, R.A., 2011. Chemogenetic analysis of human protein methyltransferases. Chem. Biol. Drug Des. 78, 199–210.

Ragno, R., Simeoni, S., Castellano, S., Vicidomini, C., Mai, A., Caroli, A., Tramontano, A., Bonaccini, C., Trojer, P., Bauer, I., Brosch, G., Sbardella, G., 2007. Small molecule inhibitors of histone arginine methyltransferases: homology modeling, molecular docking, binding mode analysis, and biological evaluations. J. Med. Chem. 50, 1241–1253.

Sidoli, S., Cheng, L., Jensen, O.N., 2012. Proteomics in chromatin biology and epigenetics: elucidation of post-translational modifications of histone proteins by mass spectrometry. J. Proteomics 75, 3419–3433.

Santos-Rosa, H., Caldas, C., 2005. Chromatin modifier enzymes, the histone code and cancer. Eur. J. Cancer 41, 2381–2402.

Spannhoff, A., Sippl, W., Jung, M., 2009. Cancer treatment of the future: inhibitors of histone methyltransferases. Int. J. Biochem. Cell Biol. 41, 4–11.

Sippl, W., Jung, M., 2009. Epigenetic drug discovery special issue. Bioorg. Med. Chem. 19, 3603–3604.

Sippl, W., Jung, M., 2010. Epigenetic Targets in Drug Discovery, vol. 48. http://dx.doi.org/10.1002/9783527627073.

Smadbeck, J., Peterson, M.B., Zee, B.M., Garapaty, S., Mago, A., Lee, C., Giannis, A., Trojer, P., Garcia, B.A., Floudas, C.A., 2014. De novo peptide design and experimental validation of histone methyltransferase inhibitors. PLoS One 9. http://dx.doi.org/10.1371/journal.pone.0090095.

Vellore, N.A., Baron, R., 2014. Epigenetic molecular recognition: a biomolecular modeling perspective. Chem. Med. Chem. 9, 484–494.

Weake, V.M., Workman, J.L., 2008. Histone ubiquitination: triggering gene activity. Mol. Cell 29, 653–663.

Yoo, J., Medina-Franco, J.L., 2011. Discovery and Optimization of Inhibitors of DNA Methyltransferase as Novel Drugs for Cancer Therapy. 1(1). http://dx.doi.org/10.4236/cmb.2011.11002.

Yao, J., Chu, Y., An, R., Guo, H., 2012. Understanding product specificity of protein lysine methyltransferases from QM/MM molecular dynamics and free energy simulations: the effects of mutation on SET7/9 beyond the Tyr/Phe switch. J. Chem. Inf. Model. 52, 449–456.

Yue, Y., Chu, Y., Guo, H., 2015. Computational study of symmetric methylation on histone arginine catalyzed by protein arginine methyltransferase PRMT5 through QM/MM MD and free energy simulations. Molecules 20, 10032–10046.

CHAPTER 10

Computational Chemical Biology of Methyllysine Histone Effectors

Dmitri Kireev

Center for Integrative Chemical Biology and Drug Discovery, UNC Eshelman School of Pharmacy, University of North Carolina at Chapel Hill, Chapel Hill, NC, USA

1. INTRODUCTION

There is growing evidence that many diseases are caused by miswritten, miserased, or misinterpreted epigenetic marks on histone proteins (Chi et al., 2010; Portela and Esteller, 2010) that are responsible for the cell differentiation and phenotypic changes under the influence of the environment (Goldberg et al., 2007). The molecular basis of the epigenetic regulation is chromatin, a high-order assembly consisting of nucleosomes containing 147 base-pair DNA fragments wound around histone octamers (Cooper and Hausman, 2013). The disordered N-terminal histone tails undergo numerous posttranslational modifications (PTMs) (Kouzarides, 2007), collectively called the histone code (Strahl and Allis, 2000), which represent the key regulatory factor. The translation of the epigenetic information is mediated by specific proteins called histone effectors (readers), which can bind to the histone PTMs and consequently regulate gene expression by impacting the shape and function of chromatin (Yun et al., 2011). The histone effectors play multiple functional roles. They can recruit various writer enzymes or transcription factors, serve as linkers, or act as transcription factors themselves (Taverna et al., 2007). Many effector proteins contain several reader modules and can bind more than one epigenetic mark (either simultaneously or sequentially). In recent years, a number of histone readers have been implicated in the pathogenesis of various diseases and hence are being seriously considered as potential drug targets.

The current review focuses on one particular family of histone effectors whose members read methyllysine (KMe) PTMs. We briefly describe the role of lysine methylation marks in cell development and regulation. Furthermore, KMe effector proteins, their structure, function, and potential as therapeutic targets will be introduced. We then discuss how computational approaches can help the epigenetic community to validate KMe effectors as therapeutic targets and initiate the drug discovery effort. Initially, computational studies focused on the physical basis of how the effector pockets recognize specific methylation states. More recently, much computationally supported effort has been made to

CONTENTS

Epi-Informatics. http://dx.doi.org/10.1016/B978-0-12-802808-7.00010-1

develop chemical probes that can readily switch on and off proteins of interest. We conclude by exemplifying the latest work in which molecular simulations were used to understand the structural mechanics of multidomain effectors.

2. LYSINE METHYLATION: CHEMISTRY AND BIOLOGY

Lysine methylation is the most prolific letter of the histone code. Three states of lysine methylation (mono-, di-, and tri-) have been observed on histones. Methylated lysine residues include H3K4, H3K9, H3K27, H3K36, H3K79, H4K20, H1K26, H2BK5, and H2AK36 (Figure 1) (Musselman, Lalonde et al., 2012), which are written by lysine methyltransferases (KMTs) and erased by lysine demethylases (KDMs) (Black et al., 2012). Some of these marks determine specific outcomes (e.g., H3K9Me3 is clearly linked to gene repression), while others may contribute to more than one, context-dependent, effect (Kouzarides, 2007). Deregulation of the KMe histone code has been implicated in cancer and other diseases due to abnormal gene expression and chromatin instability (Kondo et al., 2003; Rivenbark et al., 2009; Portela and Esteller, 2010). There is growing evidence that both genetic and chromatin alterations contribute to the cancerogenesis (Wang et al., 2007a). It has also been shown that aberrant modification or significant loss of epigenetic marks at loci of cardiac genes may account for the development of congenital heart disease (Vallaster et al., 2012).

The spatial arrangements of methylation marks act through the recruitment of KMe effectors that mediate transcriptional activation or repression (Taverna et al., 2007). In general, the localization of chromatin effectors to

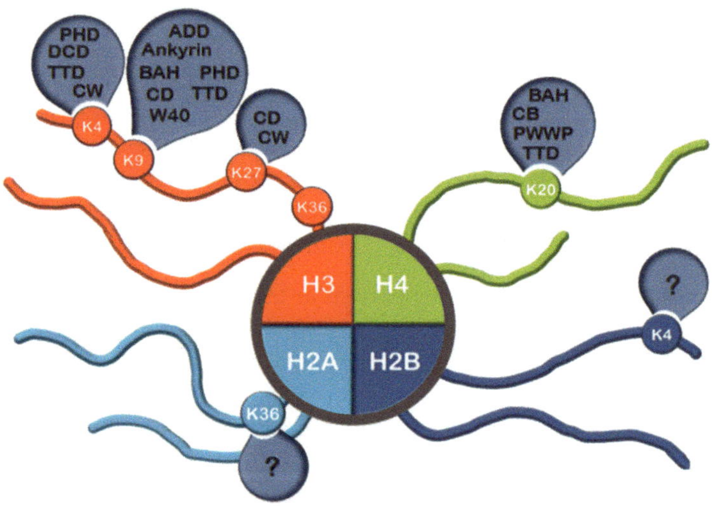

FIGURE 1
Currently known histone methyllysine sites and their respective effector domains.

particular chromatin loci determines the way they interpret the chromatin states (Suganuma and Workman, 2011). Effector recruitment often combines interaction with histone marks and specific DNA regions (Moazed, 2011). Also, one mark is often able to recruit more than one effector protein (Ruthenburg et al., 2007). Overall, this intricacy supports the notion of a code, in which combinatorial PTM patterns form words that determine the specificity of the effector recruitment (Lee et al., 2010). In this code, histone methylation represents a major complex system in which the interplay and global dynamics of readers and writers of the histone code may be subtly altered (Gardner et al., 2011).

3. METHYLLYSINE EFFECTORS

The known functions of the epigenetic effectors in general, and KMe effectors in particular, are extremely diverse and most of them still await of detailed characterization. The fact that KMe reader proteins not only recognize the methylation marks but also distinguish between various methylation states (mono-, di-, or trimethylation) adds to the complexity of the histone code. Multiple enzymes and other chromatin-binding proteins contain one or more KMe effector modules, some of which can simultaneously bind to more than one mark (Wigle and Copeland, 2013). The abundance of KMe marks combined with diverse effector functions is consistent with a great variety of effector proteins and the constituent KMe-binding modules (Arrowsmith et al., 2012). The typical sequence length is ~800 residues, with a few outliers as small as 251 residues (CBX-7) or as large as 1972 residues (53BP1). Their intrinsically disordered sections may reach ~200 residues in length and confer substantial rotational and translational freedom to each structured domain.

3.1 Families and Structure

The currently known structural KMe effector families include ATRX-DNMT3-DNMT3L (ADD), Ankyrin, Bromo-Adjacent Homology (BAH), Chromodomain, Double Chromodomain (DCD), Chromo-barrel, Malignant Brain Tumor (MBT), Plant Homeodomain (PHD), Pro-Trp-Trp-Pro (PWWP), Tandem-Tudor domain (TTD), Tudor, WD40, and CW (Figure 2) (Musselman et al., 2012). Their sequence length rarely exceeds 100 residues. Recent progress in high throughput crystallography allowed obtaining a large number of high-resolution structures for representatives of all of the above families (Adams-Cioaba and Min, 2009). All structures determined to date share a similar KMe-binding pocket composed of two to four aromatic side chains and an aspartic acid (Figure 2), whose specific composition and geometry determines selectivity with respect to a particular methylation state (Adams-Cioaba and Min, 2009). Two histone-effector recognition modes have currently been observed (Taverna et al., 2007): (1) cavity insertion (Figure 2(a),(d), and (h)), in which the interaction only involves the KMe residue and the effector's KMe-binding pocket, and (2) surface groove

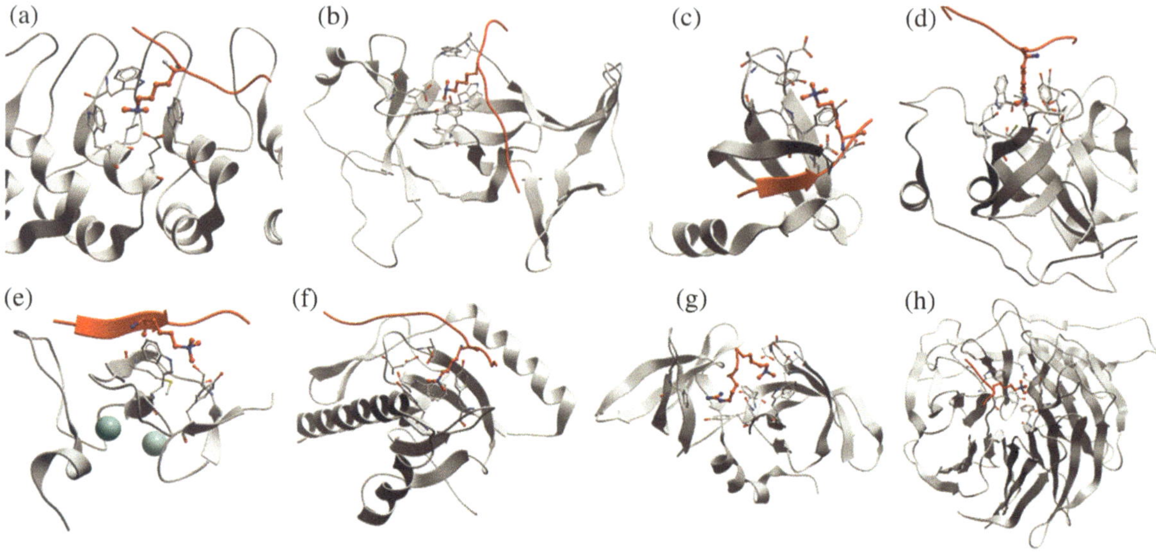

FIGURE 2

Structural folds of methyllysine binding interfaces in representative methyllysine effector domains including Ankyrin (a), BAH (b), Chromo (c), MBT (d), PHD (e), PWWP (f), Tandem Tudor (g), and WD40 (h).

(Figure 2(b),(c),(e),(f)), in which the reader module features an extended groove on its surface that interacts with four to six histone residues in a sequence-specific manner.

Six of the KMe-effector families (Chromo, Chromo-barrel, MBT, PWWP, Tudor, and TTD) are considered as structurally related, due to a distinctive β-barrel architecture, and form the Royal superfamily. Chromodomains are the smallest Royal members, composed of four β-strands and an α-helix, which are specific for KMe3 marks. Trimethylated histones adopt surface-groove binding mode to chromodomains through hydrogen bonds and electrostatic contacts involving up to nine histone residues (Kaustov et al., 2011). Other Royal domains use the cavity-insertion mode to recognize their cognate histone substrates. For instance, the Tudor domains of PHF1 or PHF19 bind to H3K36Me3, while the Tandem Tudor of 53BP1 recognizes H4K20Me2 in a sequence-unspecific manner (Botuyan et al., 2006; Ballaré et al., 2012; Musselman et al., 2012). MBT binds lower methylation states, KMe and KMe2 (Li et al., 2007; Min et al., 2007; Guo et al., 2009). The known MBT effector proteins contain up to four MBT domains, of which only one is competent in histone substrate binding with no sequence specificity observed (Bonasio et al., 2009). The remaining Royal domain, PWWP, selectively although very weakly bind H3K36Me2, H3K36Me3, and H4K20Me (Wu et al., 2011).

PHD fingers, well-studied H3K4Me3 effectors make extensive contacts with the histone substrate, imparting a high degree of specificity (Li, 2006; Pena, 2006; Musselman and Kutateladze, 2011). The histone tail makes an antiparallel β-strand bound to a β-sheet of the protein. Besides K4Me3 binding to the aromatic cage, R2 occupies an adjacent binding pocket and the N-terminal amino group binds to the effector backbone. Similar interactions have also been observed for the zinc finger CW bound to the N-terminal histone tetramer (He et al., 2010) and in the DCD of the CHD1 effector protein (Flanagan et al., 2005).

WD40 domain utilizes its aromatic cage situated on the extremity of the channel formed by seven β-sheets, such as the EED WD40 domain binding to trimethyllysine residues associated with repressive chromatin marks including H3K27me3, H3K9me3, H4K20me3, and H1K26me3 (Margueron et al., 2009). Finally, the Ankyrin repeats in G9a (Collins et al., 2008) and GLP and the BAH module in ORC1 (Kuo et al., 2012) make multiple polar and van der Waals interactions with histone peptides showing high specificity for H3K9Me2, H3K9Me3, and H4K20me2.

3.2 Biological Function

Functionally, binding of KMe effectors to their cognate histone marks produces a vast diversity of responses, many of which are yet to be characterized and understood. Perhaps the most studied example involves the formation and spreading of heterochromatin, triggered and maintained by binding of Polycomb and HP1 proteins to the H3K9 and H3K27 methylation marks (Lachner, O'Carroll et al., 2001; Fischle, 2003). Some other KMe effectors act by recruiting secondary chromatin modifiers that can either write new marks or erase the existing ones. For instance, PWWP domain of Pdp1 binds to H4K20Me and recruits Set9 that converts H4K20Me into H4K20Me3 (Wang et al., 2009). Ultimately, KMe readers can also recruit factors directly involved in DNA transcription, replication, or repair. For example, the 53BP1 Tudor domain binds to H4K20Me and localizes the rest of the repair machinery to a DNA double-strand break (Jungmichel and Stucki, 2010). In many signaling pathways involving KMe effectors, they are not self-sufficient chromatin binders. For example, HP1 needs the presence of ORC to bind to H3K9Me3-rich heterochromatin (Prasanth et al., 2010) or CBX7, a subunit of the PRC1 complex, in addition to its chromodomain reading K27me2, also recognizes an antisense noncoding RNA transcribed from the INK4 locus (Yap et al., 2010). Remarkably, in spite of significant amounts of matter displaced and large-scale downstream consequences caused by the recruitment of the KMe effectors to the chromatin, the interaction between individual reader domains and their cognate histone marks is relatively weak, with affinities ranging from 30 to 100 μM (Bonasio et al., 2009; Hoppmann et al., 2011; Kaustov et al., 2011; Sanchez and Zhou, 2011; Wu et al., 2011).

3.3 Role in Disease

A number of KMe effectors display genetic alterations causing severe conditions, including cancer. At the molecular level, such genetic alterations may cause either diminished expression of tumor suppressor genes or increased expression of oncogenes. Also, in some cases, genetic alterations result in an addiction of cancer cells on the altered effector protein for proliferation. To cite but a few examples, it has been shown that UHRF1 overexpression underlies DNA hypomethylation in cancer cells and that senescence would be a primary way to restrict tumorigenesis due to epigenetic disruption (Mudbhary et al., 2014). Also, 53BP1 has been identified as an essential factor for sustaining the growth arrest induced by *Brca1* deletion. Depletion of 53BP1 abrogates the ATM-dependent checkpoint response and G2 cell-cycle arrest triggered by the accumulation of DNA breaks in *Brca1*-deleted cells (Bouwman et al., 2010). Another compelling example is a *NUP98–PHF23* fusion, associated with acute myeloid leukemia (AML) in humans, leading to myeloid, erythroid, T-cell, and B-cell leukemia in mice (Gough et al., 2014).

3.4 Methyllysine Effectors as Potential Drug Targets

As it has been described in the previous section, there is a growing body of evidence that certain KMe effectors are strongly implicated in disease and hence can be considered as potential therapeutic targets, in particular for cancer treatment. Therefore, small molecule modulators of these effectors are expected to inhibit proliferation of cancer cells and to improve the survival of patients bearing the respective genetic alterations. However, the details of cellular function for most KMe effectors, including those implicated in cancer, are yet to be understood. Such understanding might be improved by using small molecules (called chemical probes) that can specifically bind to and modulate the function of a protein. In recent years, high-quality, well-characterized chemical probes emerged as useful tools to validate novel therapeutic targets (Bunnage et al., 2013). Developing chemical probes for the functional analysis of KMe code keepers in vivo should support new discoveries in this field and help to elucidate their function and biological activities, in addition to providing initial hits for subsequent drug development. In particular, chemical probes enable (1) an improved temporal resolution when observing biological effects for an effector of interest (rapid exposure and elimination of effects are possible), (2) mechanistic selectivity (a probe can potentially target separate functions of a protein), and (3) ease of delivery due to cell permeability (Weiss et al., 2007). However, despite an impressive pool of available X-ray structures and bioassays, very few low-weight KMe modulators have been made available so far. The reason behind this shortage is probably that pharmaceutical firms, the major actors in small-molecule discovery, are reluctant to work on targets whose therapeutic potential is still uncertain. Therefore, the academic research community is expected to deliver a critical mass of chemical compounds to probe histone methylation biology.

Currently, a compelling example of academic probe discovery has been demonstrated by the Bradner group at Harvard Medical School, who developed a cell-permeable, small-molecule JQ1 binding to bromodomains, acetyl-lysine binding proteins (Filippakopoulos et al., 2010). JQ1 was shown to displace the BRD4 fusion oncoprotein from chromatin, causing antiproliferative effects in BRD4-dependent cell lines and patient-derived xenograft models (Filippakopoulos et al., 2010). Several more potent small-molecule probes for bromodomain effectors have been reported since then (Ferguson et al., 2013; Hay et al., 2014; Owen and Trzupek, 2014). Less significant progress has been achieved in developing chemical probes for KMe effectors despite their great potential for drug development. It is still an open question whether these proteins are druggable at all (i.e., whether it is possible to develop high-affinity small-molecule binders for them). At present, only one potent KMe-effector antagonist, UNC1215 (James et al., 2013), has been reported as a probe to investigate the biology of L3MBTL3, a putative KMe effector having a role in hematopoiesis (Northcott et al., 2009) and implicated in meduloblastomas (Arai and Miyazaki, 2005). UNC1215 was shown to increase the cellular mobility of L3MBTL3 and revealed a KMe-dependent interaction of L3MBTL3 with BCLAF1, a protein implicated in DNA damage repair and apoptosis (James et al., 2013). The L3MBTL3 example suggests that KMe effectors are probably difficult, but tractable targets and continued effort needs to be applied to develop potent chemical probes for other KMe effectors. Such probes may then become valuable starting points for lead discovery programs when the respective effectors are validated as drug targets.

4. WHAT ROLE FOR COMPUTATIONAL APPROACHES?

Given the abundance of structural information on KMe effectors from virtually all structural families, an important role in the quest for novel small-molecule inhibitors of these proteins should be given to techniques and approaches that are able to efficiently utilize this wealth of structural data. Currently, a broad range of computational techniques exist to support the structure-based molecular design. These techniques differ in the ways they model the structure and intermolecular interactions and, accordingly, differ in accuracy and computational resources required. Hence, each computational tool is expected to match a specific application with its associated accuracy requirements, schedule, computational resources, and budget. In the next sections we will review recent studies focused on KMe effector proteins and involving a significant computational contribution. The computational methods used, their relevance, and impact will be discussed.

The studies discussed in this chapter can be grouped into three broad categories. The first (and earliest) category includes studies focused on understanding the physical principles of histone-effector recognition. Such understanding would, on the one hand, reveal the inner workings of the gene regulation machine and, on the other, help to design small molecules capable of efficiently dislodging the KMe effectors from their cognate histone substrates. The second category involves later work aimed at small-molecule hit finding and hit-to-lead progression. Since KMe effectors are a totally new target class, the medicinal chemists have not yet developed intuitive knowledge about privileged chemical motifs that could be used for hit generation. Hence, computational approaches, such as virtual screening, are expected to be a major source of initial hits. Finally, in the third category we find studies in which large-scale molecular simulations are used to help the chromatin biologists better understand the structural mechanics underlying the KMe effectors' function. Epigenetic proteins are not an easy subject for canonical structural biology tools, such as X-ray crystallography or NMR. Due to their multidomain architecture and long-disordered regions they almost never appear in solved structures in full length, while structures for a large number of isolated structural domains were made available in recent years. Computational simulations then play a role of a high-resolution optical microscope to visualize the dynamics and assess the energetics of KMe effector proteins.

5. ELUCIDATION OF EFFECTOR-SUBSTRATE RECOGNITION

While the functional role of various histone methylation states has been subject to intense investigations since the early 2000s (Gardner et al., 2011), the atomic-scale details of KMe recognition by respective effectors remained elusive until very recently. In particular, it was unclear how the lysine binding pockets of these effectors are tuned by nature for selectivity to specific methylation states. Nonetheless, if one wants to be able to exogenously interfere with the histone code using small-molecule therapeutics, one would need to understand the finest details of how lysine methylation states are recognized by their respective effector modules. Over the past few years, several research groups aimed to close this gap (Hughes et al., 2007; Lu et al., 2009; Eisert and Waters, 2011; Gao et al., 2011; Seeliger et al., 2012). Since the goal here was not only to build an empirical, mathematically formal model of the phenomenon, but also to grasp the underlying physical principles, the methods used favored rigor and accuracy. Most studies made use of all-atom molecular dynamics with explicit solvent in combination with experimental techniques for determination of binding affinity and, in some studies, synthetic chemistry to produce the smallest possible effector binders and/or validate computational hypotheses.

The earliest study investigating physical principles of histone-effector interactions has been published by Lu et al. (2009). The overarching goal of the study was to quantitatively characterize various energetic contributions to the recognition of the chromodomain of heterochromatin-associated protein 1 (HP1) by its cognate substrate, histone H3 methylated at lysine 9 (H3K9Me3). The readout of H3K9Me3 by an aromatic cage of the HP1 chromodomain (leading to transcriptional repression and heterochromatin formation (Nakayama et al., 2001)) was by then probably the best-characterized histone-effector interaction. However, the question of what force is the major driver of this binding event was a matter of controversy. On the one hand, the initial putative hypothesis was that the chromodomain-H3K9Me3 binding is mainly driven by the hydrophobic effect. The rationale behind this hypothesis was quite simple. Since the only difference between a nonmethylated lysine and a trimethylated one are the three methyl groups (which can do nothing else but hydrophobic interactions), the chromodomain-H3K9Me3 binding is necessarily hydrophobicity-driven. On the other hand, a then-recent work by Hughes et al. (2007) has demonstrated, using a model peptide system, that the HP1 chromodomain binding to H3K9Me3 is energetically determined by cation–π interactions. To explicitly elucidate the structural specifics of this interaction, the authors made use of free-energy perturbation (FEP) simulations to calculate relative binding free energies among chromodomain–H3 complexes with varying methylation states at H3K9. Moreover, in order to dissect the energetic contributions to the binding free energy, the authors have also simulated the binding of a peptide with a t-butyl-norleucine (tBuNle) and methyl-norleucine (MeNle) unnatural side chains as neutral KMe3 analogs. It has been observed that the binding free energy of tBuNle was by 3.1 kcal/mol weaker than that of KMe3, which was consistent with experimental data about the significant role that cation–π interaction plays in the binding of the methylated H3 peptide to the HP1 chromodomain. Moreover, the H3K9Me3 mutation to another neutral analogue, MeNle, was also found to weaken the chromodomain-H3 binding by 2.1 kcal/mol. The latter result suggests that the electrostatic component explaining the difference between how the nonmethylated and trimethylated lysines recognize the aromatic cage is not more than 1.0 kcal/mol and, hence, cannot fully account for the methylation-state specific recognition. In the same time, the calculated free-energy difference between, respectively, H3tBuNle9 and H3MeNle9 binding to HP1 was 1.9 kcal/mol, which points out the importance of nonpolar interactions for the selectivity with respect to a particular methylation state. Overall, the FEP simulations by Lu et al. (2009), in agreement with Hughes et al. (2007), confirmed the essential role of cation–π interactions for the binding of a methylated histone peptide to the HP1 chromodomain. Nevertheless, these results also clearly indicate that nonpolar (van der Waals and hydrophobic) interactions are critical to distinguish between different methylation states (to the extent that the nonmethylated lysine is an extremely weak binder despite its positive charge).

A few years later the Waters group (that has earlier determined that the recognition of trimethyllysine by a chromodomain is not driven by the hydrophobic effect (Hughes et al., 2007)) has furthered the understanding of specific recognition of various methylation states by chromodomains. The particular interest of this study is that the authors have not only suggested a model of how a chromodomain's aromatic cage selects specific methylation states, but also made use of the knowledge acquired in the course of the study to express new chromodomain mutants with modified methylation-state specificities (Eisert and Waters, 2011). Although the article by Eisert and Waters (Eisert and Waters 2011) does not describe original computational contribution, the authors have critically assessed previous computational studies and integrated them in their initial hypotheses. Moreover, this study has considerably informed the subsequent computational work and hence is worth mentioning in this overview. The focus of the investigation was on the *Drosophila* HP1α (dHP1α) chromodomain. The particularity of this domain is that it equally well binds the marks H3K9Me2 and H3K9Me3 with only small preference for trimethyllysine (Jacobs and Khorasanizadeh, 2002), while showing a significantly weaker binding to H3K9 and H3K9Me. To find out the structural and energetic drivers of such selectivity profiles, various hydrophobic and polar mutations have been induced at position Glu52 of dHP1α, including Phe, Trp, Ile, Val, Asp, and Gln.

It was of particular interest to learn what would be the effect of E52F and E52W, mutations introducing a fourth aromatic side chain into the aromatic cage. The authors hypothesized that the latter mutations would favor the selectivity with respect to KMe3, while significantly diminishing hydrogen bonding to KMe2. In addition, two hydrophobic mutations, E52I and E52V, were tested to explore the role of van der Waals contacts in methylation-state specificity. Then, a fluorescence anisotropy assay was used to measure dissociation constants between the E52 mutants expressed and the H3K9Me2 or H3K9Me3 carboxyfluorescein-tagged peptides. The most remarkable finding of this study was that the E52Q mutant exhibited a significant degree of selectivity for KMe3, showing approximately 3.5-fold stronger binding to the latter ($K_d = 15\,\mu M$) compared to $K_d = 52\,\mu M$ observed for the dimethylated peptide. Interestingly, the diminished electrostatic interaction in case of the E52Q mutant did not affect the affinity of dHP1α to H3K9Me3. Overall, this study provides an extremely useful insight into the role of ionic interactions and hydrogen bonds in differentiation of methylation states and has implications regarding the evolutionary pressure for selectivity in this protein–protein interaction. It also suggests that the importance of electrostatic interactions in binding different methylation states depends on the depth of the aromatic cage, which seems to be a general method by which nature adjusts the selectivity for the various lysine methylation states and ultimately resulting in a high-resolution control of gene expression through subtle chemical changes on the lysine side chains.

Another intriguing case of selective effector recruitment to a specific methylation state involved the malignant brain tumor (MBT) domains. The MBT domains bind mono- and dimethylated histones equally well, while they do not have detectable affinity to non- and trimethylated states. On the one hand it seems that a methyl group greatly impacts the KMe binding to this domain, since adding a single methyl to a non- or dimethyllysine can induce/abolish the histone-MBT recognition. On the other hand, adding a methyl group to a monomethyllysine does not affect binding at all. To address this conundrum, Gao et al. initiated a multidisciplinary study involving FEP simulations, chemical synthesis, and isothermal titration calorimetry (Gao et al., 2011). In particular, the authors introduced a set of fragment-like MBT binders (which they called biophysical probes), which closely mimic a single lysine residue while conserving a level of binding affinity comparable to that of histone polypeptides. The biophysical probes synthesized for this study represented all KMe states and an unnatural pyrrolidine state. In addition, each synthesized probe was matched with a virtual nonpolar isomorph, in which the ε-nitrogen was replaced by an sp3 carbon.

The general design of the study was as follows: (1) the consistency of free-energy calculations and ITC measurements was expected to make sure that the simulation had adequately sampled the phase space of bound ligands; (2) the simulation trajectories were analyzed in order to infer structural details of MBT binding for each biophysical probe; and (3) virtual nonpolar isomorphs were used to determine the relative polar and nonpolar contribution to the histone-MBT recognition. Eventually, based on the simulation analyses, the authors have formulated an intuitive model of how MBT domains can equally well bind mono- and dimethyllysines, while ignoring other methylation states. This model suggests that the MBT lysine binding pocket has evolved as a compromise that equally nonoptimally fits to both mono- and dimethylation states. A direct corollary of this is that the aromatic cages featured by all methyl effectors were not necessarily optimized by the nature for their cognate substrates and hence an optimal binder should be individually designed to each pocket of the family. Moreover, despite the apparent similarity of the KMe pocket architectures in all KMe effectors (i.e., an aromatic cage + an aspartate side chain), the driving forces of the substrate-effector recognition may differ depending on the methylation state involved (Hughes et al., 2007; Lu et al., 2009; Eisert and Waters, 2011).

The work by Seeliger et al. approaches the question of substrate-effector interactions from a very different angle (Seeliger et al., 2012). The principal motivation of the authors was to help molecular biologists elaborate an appropriate technology for tracking cellular effects of the histone methylation. By then, tracking KMe effectors and their function in the chromatin context was limited by the lack of accessible experimental techniques for inducing histone methylation states into recombinant histones. Alkylation of genetically induced cysteine residues (with an introduction of appropriate methyl-ammonium groups) has appeared as a promising alternative to a native KMe induction (Simon et al., 2007). However, a γ-sulfur atom in the side chain induces significant changes to its overall geometry compared to a native lysine side chain. The objective of the study by Seeliger et al. was to investigate whether the cysteine-derived KMe analogs are appropriate for using them in cell-based studies of chromatin biology. To address this question, the authors designed a multidisciplinary approach based on extraction of binding free-energy differences between the native KMe and their respective cysteine-based analogs from a thermodynamic cycle (using all-atom FEP simulations). The resulting values were then used to correct the experimental affinities obtained for histone peptides. The experimental affinity measurements were conducted using isothermal titration calorimetry (ITC) and fluorescence polarization (FP). The study involved four structurally and functionally unrelated KMe effector domains: (1) the chromodomain of Su(Var)205; (2) the 3xMBT module of L3MBTL1; (3) the PHD finger of ING1; and (4) the pseudo Tudor domain of UHRF1. Eventually, the authors have demonstrated that their multidisciplinary approach allows the remarkably accurate determination of affinity changes resulting from replacements of native lysine side chains by their cysteine-based analogs. Their results clearly suggest that these affinity changes can be quite significant and hence result in misleading chromatin-wide observations. The authors hence recommend direct experimental evaluation of cysteine-based KMe analogs for each new effector of interest before using them in cell-based assays.

On the whole, the studies discussed in this section were a necessary step in understanding the general principles of various methylated histones by their respective effectors and created a solid basis to approach the next logical step: computer-aided identification and design of small-molecule ligands capable of efficient histone-competitive binding to KMe effectors.

6. CHEMICAL PROBES: IDENTIFICATION AND DESIGN

The challenge linked to the identification and design of small-molecule ligands to inhibit the effect of KMe effector proteins is dual. First, the small-molecule ligands are expected to interfere with protein–protein interactions—a standing challenge in drug discovery. Second, the KMe effector is novel target, for which no privileged chemical motifs have been identified yet and, hence, each new project has to start from the scratch. Moreover, it is not even sure yet if all KMe effectors are druggable at all. In such a situation, computational approaches are expected to play, in combination with medicinal chemistry and high-throughput screening (HTS), a crucial role in generating initial chemical hits as starting points for hit-to-lead projects.

The very first publication reporting on identification of small-molecule ligands for KMe effectors published by Kireev et al. (2010) focused on the MBT domain of L3MBTL1, an effector protein recruited in vivo by H4K20Me (Kalakonda et al., 2008). In this study, two complementary strategies have been applied to search for potential MBT binders in a database of ~6 million commercially available compounds. One of the strategies involved a substructural search for compounds that feature a mono- or dimethylated lysine side chain. Another strategy was a two-step protocol including (1) a structure-based pharmacophore search followed by (2) docking hits from the previous step into the binding pocket of the L3MBTL1's second MBT domain. A total of 51 virtual hits have been purchased and experimentally screened in the alpha-screen assay, of which 19 have been confirmed as actual MBT binders using Alpha-Screen, a proximity fluorescence-based assay (Wigle et al., 2010) with potencies in the low micromolar range. Moreover, the pyrrolidine group present in some confirmed virtual hits (from the pharmacophore/docking strategy) turned out to be the optimal MBT-binding moiety (affording three- to fivefold affinity gain compared to analogs featuring mono- or dimethylated amine moiety (Herold et al., 2011)). Some confirmed virtual hits inspired a follow-up chemistry program. The best ligands discovered in the course of the follow-up program antagonize native histone peptide binding, exhibiting fivefold stronger binding affinity to L3MBTL1 than its cognate histone peptide (Herold et al., 2011). The first X-ray structure of a small molecule UNC669 bound to L3MBTL1 was determined and showed a good agreement with previous computational work (Herold et al., 2011).

Of utmost importance for any structure-based VS study is a scoring function that enables an enrichment rate of at least 3% to 5%, so that if ~50 compounds are purchased for experimental testing at least one to three true hits would be identified. As described in the previous section, FEP simulations (Chipot and Pohorille, 2007) allow one to accurately assess the binding affinity between a KMe histone effector and a small-molecule ligand (Lu et al., 2009; Gao et al., 2011). However, in the early probe discovery steps, such as virtual screening and hit-to-lead progression, one might need more affordable free-energy assessment methods. And the main question is whether existing empirical free-energy predictors that have been developed and validated for a classic protein-ligand complex (in which ligand is almost fully buried and hydrogen bonds dominate the binding) are applicable to shallow, largely hydrophobic pockets of KMe effectors. To address this question, we performed a broad benchmark study of computationally affordable affinity predictors applied to a set of seven small-molecule ligands interacting with L3MBTL1. The study has revealed three top performers, far ahead of the other techniques, including two scoring functions, PMF04 (Muegge, 2005) and PLP (Gehlhaar et al., 1995), along with a simulation-based method MM-PB/SA (Rastelli et al., 2010). It should be noted that while PMF04 and PLP show excellent results on

L3MBTL1, their performance varies greatly from one protein target to another. Therefore, their general utility for the KMe histone code readers is still to be confirmed when more experimental data become available. Alternatively, MM-PB/SA is less empirical, with a thorough handling of solvation effects, and may therefore be considered a tool of choice.

PHD finger domains (that bind unmodified, methylated, or acetylated lysine side chains (Sanchez and Zhou, 2011)) are probably the KMe effectors with the strongest links to disease, including cancer, autoimmune, and developmental conditions (Baker et al., 2008). In particular, PHD fingers have been demonstrated to be subject to oncogenic translocations (van Zutven et al., 2006; Wang et al., 2007b; Gough et al., 2014). However, for a long time, they have not been in the focus of medicinal chemists or chemical biologists. The first study reporting small-molecule PHD antagonists by Wagner et al. (2012) was performed in two steps, with the first step including assay development and screening of a small-molecule library, and the second step involving molecular modeling, point mutations, and synthetic chemistry. The authors developed a high-throughput assay that uses a HaloTag fusion (Los et al., 2008) to the third PHD finger of JARID1A. The NIH Clinical Collection 1 was screened against JARID1A PHD3 in the HaloTag assay. This library contains 446 compounds, most of which are either FDA approved drugs or have been in clinical trials for a broad range of therapeutic indications. The NIH library screening has identified disulfiram, phenothiazine, amiodarone, and tegaserod to significantly interfere with the HaloTag signal. All hits have been confirmed in an orthogonal FP binding assay (in which they competed with a fluorescein-labeled H3K4me3 peptide for binding to JARID1A PHD3). The hits have shown relatively weak affinity to the JARID1A-PHD3 domain with K_d's ranging from 70 to 400 μM. Molecular modeling was then used to identify possible sites where inhibitors might bind. To this end the authors used the AutoDock algorithm (Morris et al., 2009) with the entire PHD structure included in the docking grid. These modeling studies suggested that Asp1624, Asp1629, and Trp1625A contributed important interactions for binding amiodarone analogs and tegaserod. This hypothesis was tested and confirmed using D1624A and D1629A and W1625A mutants (the D1629A mutation was particularly damaging). Based on the modeling studies, the authors synthesized four amiodarone derivatives, one of which, WAG-003, was approximately 10-fold more potent against JARID1A PHD3 than amiodarone (IC_{50} = 30 μM). However, despite a tangible improvement in potency compared to the initial hit, substantial progress is still to be made toward useful chemical probes (that are expected to be at least submicromolar binders).

During the last decade, fragment-based discovery (FBD) has been steadily gaining momentum as an approach combining hit identification and hit-to-lead progression, and a reputation as a strategy capable of tackling unsolvable problems (Baker, 2013). It not surprising then, that attempts have been made to apply the FBD strategy to the challenging target class of KMe effectors. In a recent publication Miller et al., 2014 have reported the first use of the fragment screening by NMR to identify small-molecule ligands binding to the PHD domain of Pygo2, a signaling protein involved in the Wnt pathway (Adachi et al., 2004). There are two closely related Pygo proteins (Pygo1 and Pygo2) and both are required for the elevated levels of TCF-dependent transcription in colorectal cancer cells due to the hyperactivated β-catenin (De la Roche et al., 2008). Therefore, the Pygo-BCL9 complex represents a promising therapeutic target that seems to be druggable due to relatively small protein–protein interfaces that could be blocked by a small-molecule binder. Such a binder would become a valuable chemical probe to further validate the Pygo-BCL9 complex as a drug target. The authors initiated the project by a structure-based virtual screening of 225,000 commercially available compounds for potential drug-like binders to the histone-binding surface of the Pygo-BCL9 complex (Fiedler et al., 2008). This screening resulted in three confirmed hits; however, all of them had very low solubility and ligand efficiency (LE) (Hopkins et al., 2014) and hence were not appropriate starting points for a follow-up structure–activity relationship (SAR) studies. In the next step, the authors screened the Maybridge rule-of-three (Ro3) library of 1000 chemical fragments for binding to the PHD-HD1 complex using 2D nuclear magnetic resonance (NMR). This screening study identified two closely related benzothiazole compounds binding to its rear surface of the complex. The affinity for the complex of one of the hits, determined by the titration, turned out to be quite low (3.1 mM). However, the calculated LE = 0.31 kcal/mol suggested a descent fit. The high solubility of the fragment hits made them suitable for cocrystallization with PHD–HD1, which revealed fragment's binding pose in a crevice on the PHD–HD1 interface, as has been previously hypothesized from the NMR data and modeling. Eventually, the overall wealth of NMR data and X-ray structures enabled the first round of SAR follow-up, screening hit-fragment analogs with chemical variations at two attachment sites of the benzothiazole bicyclic core. This SAR study allowed the authors to further improve the binding affinity to ~500 μM. This potency is still far from being enough for a valid chemical probe, but due to a small size and high LE of the current benzothiazole hits they can serve as promising templates for further chemical optimization.

As can be seen from the preceding discussion in this section, the main challenge yet to be addressed by in silico approaches in the area of KMe probe design is to achieve a binding affinity level consistent with what is usually expected from a probe or a drug candidate (i.e., in low nanomolar range). Most likely, such an improvement may only be possible through finding spots in the vicinity of KMe binding sites that might be simultaneously bound by the probe. Potential solutions may involve somewhat larger small molecules, such as nonnatural peptides or peptoids, as demonstrated in a 2014 study where nanomolar antagonists for the KMe effector Chromobox Homolog 7 (CBX7) have been designed (Simhadri et al., 2014). The authors of this study were interested in developing tool compounds to probe the activities of protein chromodomains, KMe readers associated with chromatin remodeling (Blus et al., 2011). Their specific focus was on a chromodomain subfamily of chromobox homologs CBX2, CBX4, CBX6, CBX7, and CBX8 (Kaustov et al., 2011). In particular, there is a substantial body of evidence that CBX7 might be a promising therapeutic target in certain forms of cancer (Scott et al., 2007; Shinjo et al., 2014). The authors' strategy consisted of using the smallest possible KMe3-containing peptides as a template for systematic computer-aided SAR studies. The FALKMe3S pentapeptide (a CBX7-binding motif from SETDB1) has been chosen as a starting point since it was known as the most potent CBX7 binder ($IC_{50} = 12\,\mu M$) (Kaustov et al., 2011). The authors then built a model of FALKMe3S bound to CBX7 using the published NMR structure of CBX7 in complex with its cognate substrate H3K27me3 (PDB code: 2L1B) as a template. The subsequent analysis of the protein–peptide interface has then prompted initial hypotheses on what side chains might be preferred at each sequence position. These hypotheses guided iterative chemical optimization (involving natural and unnatural side-chain substitutions) throughout the project. Eventually, the accurate affinity assessment by ITC has been performed for four compounds that have shown the highest potency in the FP assay. Two most potent compounds have been found to bind CBX7 with $K_d \approx 200\,nM$. Moreover, the most selective compound turned out to be ~10-fold selective with respect to CBX8 and ~400-fold selective with respect to CBX1, two close CBX7 homologs. Therefore, small peptides may be considered as a promising alternative to nonpeptide ligands due to a more straightforward path toward high-affinity binders. However, this avenue has its own caveats, such as poor cell penetration and a higher risk of metabolic instability in vivo (although these issues can be alleviated by means of computer modeling in the course of the chemical optimization).

7. STRUCTURAL MECHANICS OF METHYLLYSINE SIGNALING

In the previous sections we discussed canonical roles for molecular modelers in medicinal chemistry and chemical biology. In general, these roles consist of performing—in silico—conventional experimental tasks. That is, the modelers can virtually screen large molecular databases for inhibitors against a given target or to create a 3D model of an inexistent chemical compound and test whether it would bind to a protein target of interest. Sometimes, in addition to the roles of virtual chemists or virtual biochemists, modelers can also be helpful as virtual structural biologists. In the field of chromatin regulation, where multidomain functional units of histone effectors are often out of reach for conventional structural approaches, such computational contribution is particularly welcome. Below, we discuss one such study by Rothbart et al. (2013), in which molecular simulations helped the authors elucidate the roles of paired TTD and PHD domains of UHRF1 to engage methylated histone in a bivalent fashion. The E3 ubiquitin-protein ligase UHRF1 is a well-studied epigenetic regulator, bridging DNA methylation and chromatin modification (Rothbart et al., 2012) and implicated in hepatocellular carcinoma (Mudbhary et al., 2014). However, until very recently the structural mechanism of its interaction with methylated histones remained elusive. Interestingly, in one available structure (PDB: 2L3R (Nady et al., 2011)) the histone binding cleft of the TTD was occluded by the TTD–PHD linker, while in the other structure (PDB: 3ASK (Arita et al., 2012)) the histone was bound to both PHD and TTD domains in a bivalent fashion. To gain further insight into how the PHD and linker modulate the interaction of the TTD with the H3 tail, a series of microsecond-scale nonequilibrium molecular dynamics simulations (Dickson, 2011) for the UHRF1 TTD–PHD and UHRF1 TTD–linker has been performed.

A collective average of the free-energy landscape is depicted in Figure 3 with the position of the PHD in the crystal structure (PDB: 3ASK) (Arita et al., 2012) located at the origin of the map. Simulations indicated a rugged free-energy landscape created by nonspecific interactions between the PHD and the TTD–linker and suggested that the PHD is scanning the TTD surface rather than residing in a single conformation relative to the TTD. Overall, it appears that both PHD and linker play the role of a regulator in multivalent histone engagement by the dual TTD–PHD module. This model was corroborated by an FP assay measuring competition of, respectively, TTD or

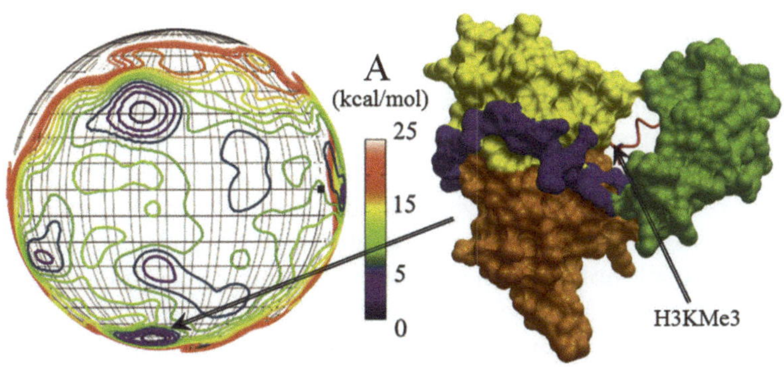

FIGURE 3

Estimate of the free-energy landscape in the TTD–PHD molecular dynamics simulation. The X-ray structure 3ASK including H3K9Me3 peptide (red), TTD (yellow-tan), PHD (green), and the TTD–PHD linker was used as a point of reference. Contours on the free energies are drawn every 4kβT, where kβ is the Boltzmann constant and T is the temperature.

TTD–linker with PHD for the $H3_{1-20}K9Me3$ peptide that demonstrated the capacity of a free PHD to perturb TTD–linker coordination of H3K9Me3, but not that of the isolated UHRF1 TTD. Overall, the simulations in this study helped to demonstrate that the linked TTD and PHD of UHRF1 operates as a functional unit in cells, providing a defined combinatorial readout of a heterochromatin signature within a single histone H3 tail that is essential for UHRF1-directed epigenetic inheritance of DNA methylation. These findings provided critical support for the histone code hypothesis, demonstrating that multivalent histone engagement plays a key role in driving a fundamental downstream biological event in chromatin.

8. CONCLUSIONS AND PERSPECTIVES

We hope that the studies and applications discussed in this chapter have definitely demonstrated how important various computational approaches can be in helping the chromatin research community better understand the function of KMe effectors and their role in disease. Ultimately, their continued contribution will enable validation of KMe effectors as new drug targets and guide the discovery of small-molecule therapeutics against these novel targets. However, in most studies published so far, the reported small-molecule hits show only weak to modest potency against therapeutically relevant KMe effectors, which would not be enough for a quality chemical probe or a drug candidate. Such a situation is consistent with the current structural data that do not provide evidence of large hydrophobic pockets capable of burying a typical drug-like molecule. Hence, the question whether KMe effectors are druggable targets remains open. To comprehensively address this issue, there are a few avenues to be explored. First, there is a record of success in fragment-based drug discovery, based on inducing ligand-binding pockets on the protein surface, which cannot be inferred from an apo-protein structure. A compelling example of this approach has been provided by Sun et al. (2012). Over NMR-based fragment screening for low-weight K-Ras binders, the authors have identified a highly efficient fragment, which binds by inducing a new pocket due to a major rearrangement of a tyrosine side chain. No applications of such a strategy to KMe effectors have been reported yet. Second, larger, though yet bioavailable, small molecules that would simultaneously bind more than one pocket probably need to be designed, using either in silico or NMR-based fragment strategy. Third, certain KMe effector domains can form either homo- or heterodimers, thus creating new potential binding pockets. Moreover, sometimes the domain dimerization can be induced by a ligand, as this is the case for L3MBTL3 3xMBT-dimers induced by a high-affinity probe UNCC1215 (James et al., 2013). Finally, probe/drug development against KMe effectors is yet to benefit from the target-family strategy. Since KMe effectors feature structurally and functionally related binding sites, it would be judicious to target them as a family rather than as independent targets. Such a strategy would focus on finding a reasonably small number of privileged chemical motifs that nonselectively bind to more than one effector domain. Such motifs can then be reused to initiate probe design for any family member. A desired selectivity can then be achieved by decorating privileged binding heads with diverse substituents. Then, all compounds synthesized are being systematically screened against a representative panel of KMe effector domains. The target-family strategy enables an efficient use of chemistry resources and favors serendipity, since a by-product of lead optimization for one target may turn out to be an optimal binder for another.

ACKNOWLEDGMENTS

The author thanks Bradley Dickson for providing a free-energy plot for Figure 3.

REFERENCES

Adachi, S., Jigami, T., Yasui, T., Nakano, T., Ohwada, S., Omori, Y., Sugano, S., Ohkawara, B., Shibuya, H., Nakamura, T., 2004. Role of a BCL9-related β-catenin-binding protein, B9L, in tumorigenesis induced by aberrant activation of Wnt signaling. Cancer Res. 64, 8496–8501.

Adams-Cioaba, M.A., Min, J., 2009. Structure and function of histone methylation binding proteins. Biochem. Cell Biol. 87, 93–105.

Arai, S., Miyazaki, T., 2005. Impaired maturation of myeloid progenitors in mice lacking novel polycomb group protein MBT-1. EMBO J. 24, 1863–1873.

Arita, K., Isogai, S., Oda, T., Unoki, M., Sugita, K., Sekiyama, N., Kuwata, K., Hamamoto, R., Tochio, H., Sato, M., Ariyoshi, M., Shirakawa, M., 2012. Recognition of modification status on a histone H3 tail by linked histone reader modules of the epigenetic regulator UHRF1. Proc. Natl. Acad. Sci. 109, 12950–12955.

Arrowsmith, C.H., Bountra, C., Fish, P.V., Lee, K., Schapira, M., 2012. Epigenetic protein families: a new frontier for drug discovery. Nat. Rev. Drug Discov. 11, 384–400.

Baker, L.A., Allis, C.D., Wang, G.G., 2008. PHD fingers in human diseases: disorders arising from misinterpreting epigenetic marks. Mutat. Res. Fundam. Mol. Mech. Mutagen. 647, 3–12.

Baker, M., 2013. Fragment-based lead discovery grows up. Nat. Rev. Drug Discov. 12, 5–7.

Ballaré, C., Lange, M., Lapinaite, A., Martin, G.M., Morey, L., Pascual, G., Liefke, R., Simon, B., Shi, Y., Gozani, O., Carlomagno, T., Benitah, S.A., Di Croce, L., 2012. Phf19 links methylated Lys36 of histone H3 to regulation of Polycomb activity. Nat. Struct. Mol. Biol. 19, 1257–1265.

Black, J.C., Van Rechem, C., Johnathan, R., 2012. Whetstine: histone lysine methylation dynamics: establishment, regulation, and biological impact. Mol. Cell 48, 491–507.

Blus, B.J., Wiggins, K., Khorasanizadeh, S., 2011. Epigenetic virtues of chromodomains. Crit. Rev. Biochem. Mol. Biol. 46, 507–526.

Bonasio, R., Lecona, E., Reinberg, D., 2009. MBT domain proteins in development and disease. Semin. Cell Dev. Biol. 21, 221–230.

Botuyan, M.V., Lee, J., Ward, I.M., Kim, J.-E., Thompson, J.R., Chen, J., Mer, G., 2006. Structural basis for the methylation state-specific recognition of histone H4-K20 by 53BP1 and Crb2 in DNA repair. Cell 127, 1361–1373.

Bouwman, P., Aly, A., Escandell, J.M., Pieterse, M., Bartkova, J., van der Gulden, H., Hiddingh, S., Thanasoula, M., Kulkarni, A., Yang, Q., Haffty, B.G., Tommiska, J., Blomqvist, C., Drapkin, R., Adams, D.J., Nevanlinna, H., Bartek, J., Tarsounas, M., Ganesan, S., Jonkers, J., 2010. 53BP1 loss rescues BRCA1 deficiency and is associated with triple-negative and BRCA-mutated breast cancers. Nat. Struct. Mol. Biol. 17, 688–695.

Bunnage, M.E., Chekler, E.L.P., Jones, L.H., 2013. Target validation using chemical probes. Nat. Chem. Biol. 9, 195–199.

Chi, P., Allis, C.D., Wang, G.G., 2010. Covalent histone modifications–miswritten, misinterpreted and mis-erased in human cancers. Nat. Rev. Cancer 10, 457–469.

Chipot, C., Pohorille, A., 2007. Free Energy Calculations: Theory and Applications in Chemistry and Biology. Springer, New York.

Collins, R.E., Northrop, J.P., Horton, J.R., Lee, D.Y., Zhang, X., Stallcup, M.R., Cheng, X., 2008. The ankyrin repeats of G9a and GLP histone methyltransferases are mono- and dimethyllysine binding modules. Nat. Struct. Mol. Biol. 15, 245–250.

Cooper, G.M., Hausman, R.E., 2013. The Cell: A Molecular Approach. Sinauer Associates, Sunderland, MA.

De la Roche, M., Worm, J., Bienz, M., 2008. The function of BCL9 in Wnt/β-catenin signaling and colorectal cancer cells. BMC Cancer 8, 199.

Dickson, B.M., 2011. Approaching a parameter-free metadynamics. Phys. Rev. E 84, 037701.

Eisert, R.J., Waters, M.L., 2011. Tuning HP1a chromodomain selectivity for di- and trimethyllysine. Chem. Bio. Chem 12, 2786–2790.

Ferguson, F.M., Fedorov, O., Chaikuad, A., Philpott, M., Muniz, J.R.C., Felletar, I., von Delft, F., Heightman, T., Knapp, S., Abell, C., Ciulli, A., 2013. Targeting low-druggability bromodomains: fragment based screening and inhibitor design against the BAZ2B bromodomain. J. Med. Chem. 56, 10183–10187.

Fiedler, M., Sánchez-Barrena, M.J., Nekrasov, M., Mieszczanek, J., Rybin, V., Müller, J., Evans, P., Bienz, M., 2008. Decoding of methylated histone H3 tail by the Pygo-BCL9 Wnt signaling complex. Mol. Cell 30, 507–518.

Filippakopoulos, P., Qi, J., Picaud, S., Shen, Y., Smith, W.B., Fedorov, O., Morse, E.M., Keates, T., Hickman, T.T., Felletar, I., Philpott, M., Munro, S., McKeown, M.R., Wang, Y., Christie, A.L., West, N., Cameron, M.J., Schwartz, B., Heightman, T.D., La Thangue, N., French, C.A., Wiest, O., Kung, A.L., Knapp, S., Bradner, J.E., 2010. Selective inhibition of BET bromodomains. Nature 468, 1067–1073.

Fischle, W., 2003. Molecular basis for the discrimination of repressive methyl-lysine marks in histone H3 by Polycomb and HP1 chromodomains. Genes Dev. 17, 1870–1881.

Flanagan, J.F., Mi, L.-Z., Chruszcz, M., Cymborowski, M., Clines, K.L., Kim, Y., Minor, W., Rastinejad, F., Khorasanizadeh, S., 2005. Double chromodomains cooperate to recognize the methylated histone H3 tail. Nature 438, 1181–1185.

Gao, C., Herold, J.M., Kireev, D., Wigle, T., Norris, J.L., Frye, S., 2011. Biophysical probes reveal a 'Compromise' nature of the methyl-lysine binding pocket in L3MBTL1. J. Am. Chem. Soc. 133, 5357–5362.

Gardner, K.E., Allis, C.D., Strahl, B.D., 2011. Operating on chromatin, a colorful language where context matters. J. Mol. Biol. 409, 36–46.

Gehlhaar, D.K., Verkhivker, G.M., Rejto, P.A., Sherman, C.J., Fogel, D.R., Fogel, L.J., Freer, S.T., 1995. Molecular recognition of the inhibitor AG-1343 by HIV-1 protease: conformationally flexible docking by evolutionary programming. Chem. Biol. 2, 317–324.

Goldberg, A.D., Allis, C.D., Bernstein, E., 2007. Epigenetics: a landscape takes shape. Cell 128, 635–638.

Gough, S.M., Lee, F., Yang, F., Walker, R.L., Zhu, Y.J., Pineda, M., Onozawa, M., Chung, Y.J., Bilke, S., Wagner, E.K., Denu, J.M., Ning, Y., Xu, B., Wang, G.G., Meltzer, P.S., Aplan, P.D., 2014. NUP98–PHF23 is a chromatin-modifying oncoprotein that causes a wide array of leukemias sensitive to inhibition of PHD histone reader function. Cancer Discov. 4, 564–577.

Guo, Y., Nady, N., Qi, C., Allali-Hassani, A., Zhu, H., Pan, P., Adams-Cioaba, M.A., Amaya, M.F., Dong, A., Vedadi, M., Schapira, M., Read, R.J., Arrowsmith, C.H., Min, J., 2009. Methylation-state-specific recognition of histones by the MBT repeat protein L3MBTL2. Nucleic Acids Res. 37, 2204–2210.

Hay, D.A., Fedorov, O., Martin, S., Singleton, D.C., Tallant, C., Wells, C., Picaud, S., Philpott, M., Monteiro, O.P., Rogers, C.M., Conway, S.J., Rooney, T.P.C., Tumber, A., Yapp, C., Filippakopoulos, P., Bunnage, M.E., Müller, S., Knapp, S., Schofield, C.J., Brennan, P.E., 2014. Discovery and optimization of small-molecule ligands for the CBP/p300 bromodomains. J. Am. Chem. Soc. 136, 9308–9319.

He, F., Umehara, T., Saito, K., Harada, T., Watanabe, S., Yabuki, T., Kigawa, T., Takahashi, M., Kuwasako, K., Tsuda, K., 2010. Structural insight into the zinc finger CW domain as a histone modification reader. Structure 18, 1127–1139.

Herold, J.M., Wigle, T.J., Norris, J.L., Lam, R., Korboukh, V.K., Gao, C., Ingerman, L.A., Kireev, D.B., Senisterra, G., Vedadi, M., Tripathy, A., Brown, P.J., Arrowsmith, C.H., Jin, J., Janzen, W.P., Frye, S.V., 2011. Small-molecule ligands of methyl-lysine binding proteins. J. Med. Chem. 54, 2504–2511.

Hopkins, A.L., Keseru, G.M., Leeson, P.D., Rees, D.C., Reynolds, C.H., 2014. The role of ligand efficiency metrics in drug discovery. Nat. Rev. Drug Discov. 13, 105–121.

Hoppmann, V., Thorstensen, T., Kristiansen, P.E., Veiseth, S.V., Rahman, M.A., Finne, K., Aalen, R.B., Aasland, R., 2011. The CW domain, a new histone recognition module in chromatin proteins. EMBO J. 30, 1939–1952.

Hughes, R.M., Wiggins, K.R., Khorasanizadeh, S., Waters, M.L., 2007. Recognition of trimethyllysine by a chromodomain is not driven by the hydrophobic effect. Proc. Natl. Acad. Sci. USA 104, 11184–11188.

Jacobs, S.A., Khorasanizadeh, S., 2002. Structure of HP1 chromodomain bound to a lysine 9-methylated histone H3 tail. Science 295, 2080–2083.

James, L.I., Barsyte-Lovejoy, D., Zhong, N., Krichevsky, L., Korboukh, V.K., Herold, J.M., MacNevin, C.J., Norris, J.L., Sagum, C.A., Tempel, W., Marcon, E., Guo, H., Gao, C., Huang, X.-P., Duan, S., Emili, A., Greenblatt, J.F., Kireev, D.B., Jin, J., Janzen, W.P., Brown, P.J., Bedford, M.T., Arrowsmith, C.H., Frye, S.V., 2013. Discovery of a chemical probe for the L3MBTL3 methyllysine reader domain. Nat. Chem. Biol. 9, 184–191.

Jungmichel, S., Stucki, M., 2010. MDC1: the art of keeping things in focus. Chromosoma 119, 337–349.

Kalakonda, N., Fischle, W., Boccuni, P., Gurvich, N., Hoya-Arias, R., Zhao, X., Miyata, Y., Macgrogan, D., Zhang, J., Sims, J.K., Rice, J.C., Nimer, S.D., 2008. Histone H4 lysine 20 monomethylation promotes transcriptional repression by L3MBTL1. Oncogene 27, 4293–4304.

Kaustov, L., Ouyang, H., Amaya, M., Lemak, A., Nady, N., Duan, S., Wasney, G.A., Li, Z., Vedadi, M., Schapira, M., Min, J., Arrowsmith, C.H., 2011. Recognition and specificity determinants of the human Cbx chromodomains. J. Biol. Chem. 286, 521–529.

Kireev, D., Wigle, T.J., Norris-Drouin, J., Herold, J.M., Janzen, W.P., Frye, S.V., 2010. Identification of non-peptide malignant brain tumor (MBT) repeat antagonists by virtual screening of commercially available compounds. J. Med. Chem. 53, 7625–7631.

Kondo, Y., Shen, L., Issa, J.-P.J., 2003. Critical role of histone methylation in tumor suppressor gene silencing in colorectal cancer. Mol. Cell. Biol. 23, 206–215.

Kouzarides, T., 2007. Chromatin modifications and their function. Cell 128, 693–705.

Kuo, A.J., Song, J., Cheung, P., Ishibe-Murakami, S., Yamazoe, S., Chen, J.K., Patel, D.J., Gozani, O., 2012. The BAH domain of ORC1 links H4K20me2 to DNA replication licensing and Meier-Gorlin syndrome. Nature 484, 115–119.

Lachner, M., O'Carroll, D., Rea, S., Mechtler, K., Jenuwein, T., 2001. Methylation of histone H3 lysine 9 creates a binding site for HP1 proteins. Nature 410, 116–120.

Lee, J.-S., Smith, E., Shilatifard, A., 2010. The language of histone crosstalk. Cell 142, 682–685.

Li, H., 2006. Molecular basis for site-specific read-out of histone H3K4me3 by the BPTF PHD finger of NURF. Nature 442, 91–95.

Li, H., Fischle, W., Wang, W., Duncan, E.M., Liang, L., Murakami-Ishibe, S., Allis, C.D., Patel, D.J., 2007. Structural basis for lower lysine methylation state-specific readout by MBT repeats of L3MBTL1 and an engineered PHD finger. Mol. Cell 28, 677–691.

Los, G.V., Encell, L.P., McDougall, M.G., Hartzell, D.D., Karassina, N., Zimprich, C., Wood, M.G., Learish, R., Ohana, R.F., Urh, M., 2008. HaloTag: a novel protein labeling technology for cell imaging and protein analysis. ACS Chem. Biol. 3, 373–382.

Lu, Z., Lai, J., Zhang, Y., 2009. Importance of charge independent effects in readout of the trimethyllysine mark by HP1 chromodomain. J. Am. Chem. Soc. 131, 14928–14931.

Margueron, R., Justin, N., Ohno, K., Sharpe, M.L., Son, J., Drury Iii, W.J., Voigt, P., Martin, S.R., Taylor, W.R., De Marco, V., 2009. Role of the polycomb protein EED in the propagation of repressive histone marks. Nature 461, 762–767.

Miller, T.C.R., Rutherford, T.J., Birchall, K., Chugh, J., Fiedler, M., Bienz, M., 2014. Competitive binding of a benzimidazole to the histone-binding pocket of the pygo PHD finger. ACS Chem. Biol. 9, 2864–2874.

Min, J., Allali-Hassani, A., Nady, N., Qi, C., Ouyang, H., Liu, Y., MacKenzie, F., Vedadi, M., Arrowsmith, C.H., 2007. L3MBTL1 recognition of mono- and dimethylated histones. Nat. Struct. Mol. Biol. 14, 1229–1230.

Moazed, D., 2011. Mechanisms for the inheritance of chromatin states. Cell 146, 510–518.

Morris, G.M., Huey, R., Lindstrom, W., Sanner, M.F., Belew, R.K., Goodsell, D.S., Olson, A.J., 2009. AutoDock4 and AutoDockTools4: automated docking with selective receptor flexibility. J. Comput. Chem. 30, 2785–2791.

Mudbhary, R., Hoshida, Y., Chernyavskaya, Y., Jacob, V., Villanueva, A., Fiel, M.I., Chen, X., Kojima, K., Thung, S., Bronson, Roderick T., Lachenmayer, A., Revill, K., Alsinet, C., Sachidanandam, R., Desai, A., SenBanerjee, S., Ukomadu, C., Llovet, J.M., Sadler, Kirsten C., 2014. UHRF1 overexpression drives DNA hypomethylation and hepatocellular carcinoma. Cancer Cell 25, 196–209.

Muegge, I., 2005. PMF scoring revisited. J. Med. Chem. 49, 5895–5902.

Musselman, C.A., Avvakumov, N., Watanabe, R., Abraham, C.G., Lalonde, M.-E., Hong, Z., Allen, C., Roy, S., Nuñez, J.K., Nickoloff, J., Kulesza, C.A., Yasui, A., Côté, J., Kutateladze, T.G., 2012a. Molecular basis for H3K36me3 recognition by the tudor domain of PHF1. Nat. Struct. Mol. Biol. 19, 1266–1272.

Musselman, C.A., Kutateladze, T.G., 2011. Handpicking epigenetic marks with PHD fingers. Nucleic Acids Res. gkr613.

Musselman, C.A., Lalonde, M.-E., Cote, J., Kutateladze, T.G., 2012b. Perceiving the epigenetic landscape through histone readers. Nat. Struct. Mol. Biol. 19, 1218–1227.

Nady, N., Lemak, A., Walker, J.R., Avvakumov, G.V., Kareta, M.S., Achour, M., Xue, S., Duan, S., Allali-Hassani, A., Zuo, X., Wang, Y.-X., Bronner, C., Chédin, F., Arrowsmith, C.H., Dhe-Paganon, S., 2011. Recognition of multivalent histone states associated with heterochromatin by UHRF1 protein. J. Biol. Chem. 286, 24300–24311.

Nakayama, J-i., Rice, J.C., Strahl, B.D., Allis, C.D., Grewal, S.I., 2001. Role of histone H3 lysine 9 methylation in epigenetic control of heterochromatin assembly. Science 292, 110–113.

Northcott, P.A., Nakahara, Y., Wu, X., Feuk, L., Ellison, D.W., Croul, S., Mack, S., Kongkham, P.N., Peacock, J., Dubuc, A., Ra, Y.-S., Zilberberg, K., McLeod, J., Scherer, S.W., Sunil Rao, J., Eberhart, C.G., Grajkowska, W., Gillespie, Y., Lach, B., Grundy, R., Pollack, I.F., Hamilton, R.L., Van Meter, T., Carlotti, C.G., Boop, F., Bigner, D., Gilbertson, R.J., Rutka, J.T., Taylor, M.D., 2009. Multiple recurrent genetic events converge on control of histone lysine methylation in medulloblastoma. Nat. Genet. 41, 465–472.

Owen, D.R., Trzupek, J.D., 2014. Epigenetic drugs that do not target enzyme activity. Drug Discov. Today Technol. 12, e29–e34.

Pena, P.V., 2006. Molecular mechanism of histone H3K4me3 recognition by plant homeodomain of ING2. Nature 442, 100–103.

Portela, A., Esteller, M., 2010. Epigenetic modifications and human disease. Nat. Biotech. 28, 1057–1068.

Prasanth, S.G., Shen, Z., Prasanth, K.V., Stillman, B., 2010. Human origin recognition complex is essential for HP1 binding to chromatin and heterochromatin organization. Proc. Natl. Acad. Sci. 107, 15093–15098.

Rastelli, G., Alberto, Degliesposti, G., Sgobba, M., 2010. Fast and accurate predictions of binding free energies using MM-PBSA and MM-GBSA. J. Comput. Chem. 31, 797–810.

Rivenbark, A.G., Coleman, W.B., Stahl, B.D., 2009. Histone methylation patterns in human breast cancer. FASEB J. 23 38.31.

Rothbart, S.B., Dickson, B.M., Ong, M.S., Krajewski, K., Houliston, S., Kireev, D.B., Arrowsmith, C.H., Strahl, B.D., 2013. Multivalent histone engagement by the linked tandem tudor and PHD domains of UHRF1 is required for the epigenetic inheritance of DNA methylation. Genes and Dev. 27, 1288–1298.

Rothbart, S.B., Krajewski, K., Nady, N., Tempel, W., Xue, S., Badeaux, A.I., Barsyte-Lovejoy, D., Martinez, J.Y., Bedford, M.T., Fuchs, S.M., Arrowsmith, C.H., Strahl, B.D., 2012. Association of UHRF1 with methylated H3K9 directs the maintenance of DNA methylation. Nat. Struct. Mol. Biol. 19, 1155–1160.

Ruthenburg, A.J., Allis, C.D., Wysocka, J., 2007. Methylation of lysine 4 on histone H3: intricacy of writing and reading a single epigenetic mark. Mol. Cell 25, 15–30.

Sanchez, R., Zhou, M.-M., 2011. The PHD finger: a versatile epigenome reader. Trends Biochem. Sci. 36, 364–372.

Scott, C.L., Gil, J., Hernando, E., Teruya-Feldstein, J., Narita, M., Martínez, D., Visakorpi, T., Mu, D., Cordon-Cardo, C., Peters, G., 2007. Role of the chromobox protein CBX7 in lymphomagenesis. Proc. Natl. Acad. Sci. 104, 5389–5394.

Seeliger, D., Soeroes, S., Klingberg, R., Schwarzer, D., Grubmüller, H., Fischle, W., 2012. Quantitative assessment of protein interaction with methyl-lysine analogues by hybrid computational and experimental approaches. ACS Chem. Biol. 7, 150–154.

Shinjo, K., Yamashita, Y., Yamamoto, E., Akatsuka, S., Uno, N., Kamiya, A., Niimi, K., Sakaguchi, Y., Nagasaka, T., Takahashi, T., 2014. Expression of chromobox homolog 7 (CBX7) is associated with poor prognosis in ovarian clear cell adenocarcinoma via TRAIL-induced apoptotic pathway regulation. Int. J. Cancer 135, 308–318.

Simhadri, C., Daze, K.D., Douglas, S.F., Quon, T.T.H., Dev, A., Gignac, M.C., Peng, F., Heller, M., Boulanger, M.J., Wulff, J.E., Hof, F., 2014. Chromodomain antagonists that target the polycomb-group methyllysine reader protein chromobox homolog 7 (CBX7). J. Med. Chem. 57, 2874–2883.

Simon, M.D., Chu, F., Racki, L.R., Cecile, C., Burlingame, A.L., Panning, B., Narlikar, G.J., Shokat, K.M., 2007. The site-specific installation of methyl-lysine analogs into recombinant histones. Cell 128, 1003–1012.

Strahl, B.D., Allis, C.D., 2000. The language of covalent histone modifications. Nature 403, 41–45.

Suganuma, T., Workman, J.L., 2011. Signals and combinatorial functions of histone modifications. Annu. Rev. Biochem. 80, 473–499.

Sun, Q., Burke, J.P., Phan, J., Burns, M.C., Olejniczak, E.T., Waterson, A.G., Lee, T., Rossanese, O.W., Fesik, S.W., 2012. Discovery of small molecules that bind to K-Ras and inhibit Sos-Mediated activation. Angew. Chem. 124, 6244–6247.

Taverna, S.D., Li, H., Ruthenburg, A.J., Allis, C.D., Patel, D.J., 2007. How chromatin-binding modules interpret histone modifications: lessons from professional pocket pickers. Nat. Struct. Mol. Biol. 14, 1025–1040.

Vallaster, M., Vallaster, C.D., Wu, S.M., 2012. Epigenetic mechanisms in cardiac development and disease. Acta Biochim. Biophys. Sin. 44, 92–102.

van Zutven, L.J., Önen, E., Velthuizen, S.C., van Drunen, E., von Bergh, A.R., van den Heuvel-Eibrink, M.M., Veronese, A., Mecucci, C., Negrini, M., de Greef, G.E., 2006. Identification of NUP98 abnormalities in acute leukemia: JARID1A (12p13) as a new partner gene. Genes Chromosomes Cancer 45, 437–446.

Wagner, E.K., Nath, N., Flemming, R., Feltenberger, J.B., Denu, J.M., 2012. Identification and characterization of small molecule inhibitors of a plant homeodomain finger. Biochemistry 51, 8293–8306.

Wang, G.G., Allis, C.D., Chi, P., 2007a. Chromatin remodeling and cancer, part I: covalent histone modifications. Trends Mol. Med. 13, 363–372.

Wang, G.G., Cai, L., Pasillas, M.P., Kamps, M.P., 2007b. NUP98-NSD1 links H3K36 methylation to Hox-A gene activation and leukaemogenesis. Nat. Cell Biol. 9, 804–812.

Wang, Y., Reddy, B., Thompson, J., Wang, H., Noma, K.-i., Yates, J.R., Jia, S., 2009. Regulation of Set9-mediated H4K20 methylation by a PWWP domain protein. Mol. Cell 33, 428–437.

Weiss, W.A., Taylor, S.S., Shokat, K.M., 2007. Recognizing and exploiting differences between RNAi and small-molecule inhibitors. Nat. Chem. Biol. 3, 739–744.

Wigle, T.J., Copeland, R.A., 2013. Drugging the human methylome: an emerging modality for reversible control of aberrant gene transcription. Curr. Opin. Chem. Biol. 17, 369–378.

Wigle, T.J., Herold, J.M., Senisterra, G.A., Vedadi, M., Kireev, D.B., Arrowsmith, C.H., Frye, S.V., Janzen, W.P., 2010. Screening for inhibitors of low-affinity epigenetic peptide-protein interactions: an AlphaScreen-based assay for antagonists of methyl-lysine binding proteins. J. Biomol. Screen 15, 62–71.

Wu, H., Zeng, H., Lam, R., Tempel, W., Amaya, M.F., Xu, C., Dombrovski, L., Qiu, W., Wang, Y., Min, J., 2011. Structural and histone binding ability characterizations of human PWWP domains. PLoS One 6, e18919.

Yap, K.L., Li, S., Muñoz-Cabello, A.M., Raguz, S., Zeng, L., Mujtaba, S., Gil, J., Walsh, M.J., Zhou, M.-M., 2010. Molecular interplay of the noncoding RNA ANRIL and methylated histone H3 lysine 27 by polycomb CBX7 in transcriptional silencing of INK4a. Mol. Cell 38, 662–674.

Yun, M., Wu, J., Workman, J.L., Li, B., 2011. Readers of histone modifications. Cell Res. 21, 564–578.

CHAPTER 11

Structure-Based Design and Computational Studies of Sirtuin Inhibitors

Berin Karaman[1], Manfred Jung[2], Wolfgang Sippl[1]

[1]*Institute of Pharmacy, Martin-Luther University of Halle-Wittenberg, Halle/Saale, Germany;*
[2]*Institute of Pharmaceutical Sciences, Albert-Ludwigs-University of Freiburg, Freiburg, Germany*

1. INTRODUCTION

Epigenetic regulation generally refers to "heritable but reversible changes in gene expression that occur without alterations in the underlying DNA sequence" (Dhanak, 2012). The DNA together with the histone proteins form the chromatin, which is involved in the regulation of gene expression. Histones are modified by post-translational modifications such as acetylation, methylation, phosphorylation, ubiquitinilation, crotonylation, among others that occur on amino acid residues located at histone tails. The histone modifications result in either gene activation or gene silencing (Jenuwein and Allis, 2001). The acetylation states of histone tails are modulated by opposing activities of two enzymes, namely histone acetyltransferases (HATs) and histone deacetylases (HDACs). Removal of the acetyl group from the ε-amino group of acetylated lysine residues by HDACs restores the positive charge on the lysine residue and allows the electrostatic interaction with the negatively charged DNA backbone causing a condensed/inactive chromatin structure (Jenuwein and Allis, 2001). The attempts to modulate the activities of HDACs with small molecules go back to the 1990s when it was discovered that cells treated with the HDAC inhibitor Trapoxin showed hyperacetylated histones (Kijima et al., 1993; Taunton et al., 1996). Small molecule inhibitors of some HDAC isoforms have already been shown to be of great value in cancer chemoprevention and therapy, for example, Vorinostat and Romidepsin were approved for the treatment of cutaneous T-cell lymphoma (Mottamal et al., 2015). Moreover, the design of selective HDACs inhibitors would be of great value to decipher the modes of action of HDACs and to understand the crosstalk between genetic and epigenetic partners upon changes in the acetylation status of histone tails. Thus, modulating the signaling pathways of HDACs remains a major focus and a challenging task in the field of drug discovery and development.

CONTENTS

Epi-Informatics. http://dx.doi.org/10.1016/B978-0-12-802808-7.00011-3

2. SIRTUINS, INSIGHTS INTO THEIR CHEMICAL MECHANISM, CHARACTERISTIC FEATURES, AND SUBSTRATES

Until today, 18 different members of HDACs have been identified and grouped into four classes, based on their homology to yeast histone deacetylases (de Ruijter et al., 2003; Thiagalingam et al., 2003). Class I, II, and IV HDACs, or the so-called classical histone deacetylases, remove the acetyl moiety on lysine residues in the presence of a Zn^{2+} ion via a charge-relay system (de Ruijter et al., 2003). In contrast, class III HDACs, sirtuins, are dependent on nicotinamide adenine dinucleotide (NAD^+) as a cofactor to catalyze the deacetylation reaction (Sauve, 2010). Sirtuins exist in a broad phylum of life ranging from bacteria to eukaryotes. Phylogenetic analyses of 60 sirtuins from a diverse array of organisms pointed out that eukaryotic sirtuins can be grouped into five branches (classes I–IV and U), where mammalian sirtuins (Sirt1-7) have representatives from all main branches, classes I–IV.

The deacetylation reaction catalyzed by sirtuins involves an ADPR-peptidyl-imidate intermediate, resulting in the production of acetyllysine, nicotinamide, and 2′-O-acetyl-ADP-ribose (OAADPr) (Sauve, 2010; Sauve and Youn, 2012). Emerging data acquired from biochemical and structural studies of sirtuins increased the understanding of the order of substrate binding and product release. Detailed kinetic analyses showed that (1) the acetylated substrate binds prior to NAD^+ and (2) the nicotinamide portion of NAD^+ is released as a first product followed by OAADPr and the deacetylated lysine (Borra et al., 2004). Ternary complexes of sirtuins provided evidence that NAD^+ can bind in a suitable conformation for the catalytic reaction, usually referred to as the productive conformation, in the presence of either the substrate (Du et al., 2011; Hoff et al., 2006) or molecules that mimic the peptide binding, for example, PEG (Avalos et al., 2005, 2004). In addition, NAD^+ was also shown in crystal structures to adopt various nonproductive conformations that do not allow the binding of the substrate in the acetyllysine channel (Avalos et al., 2005, 2004; Min et al., 2001). Collectively, these findings suggest that substrate binding induces the NAD^+ binding in a productive manner, which is followed by the attack of the carbonyl group of acetyllysine to the C1′ of the NAD^+-ribose resulting in the release of nicotinamide and the formation of C1′-O-alkylamidate species. In the next step, the 2′-hydroxyl group of the O-alkylamidate is activated with a conserved histidine residue in the active site and attacks the C1′-O-alkylamidate to afford the 1′,2′-cyclic intermediate. Finally, a base-activated water molecule conducts the formation of the final products: the deacetylated peptide and a mixture of 2′- and 3′-OAADPr (Figure 1).

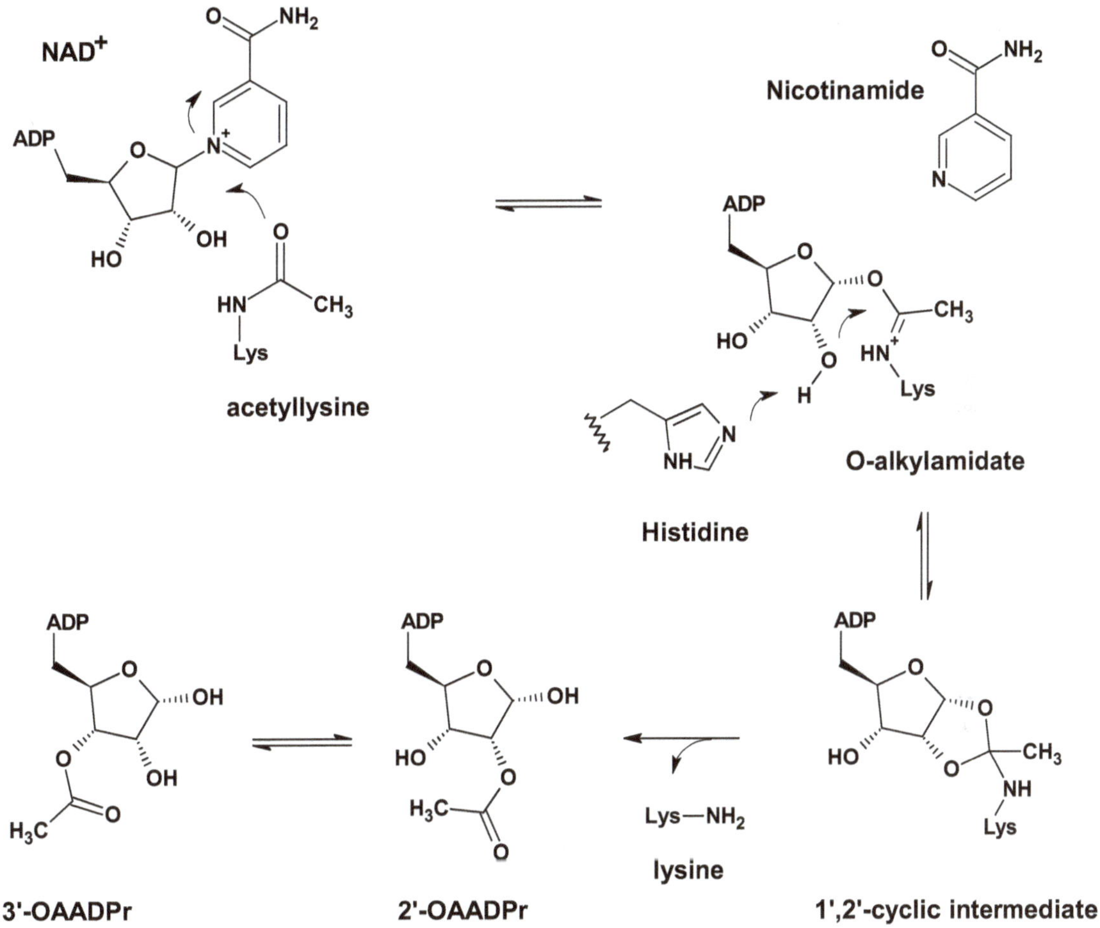

FIGURE 1
Catalytic mechanism of sirtuins.

In order to elucidate the details of the catalytic mechanism, substrate recognition and modulation with small molecules, several high-resolution structures of sirtuins have been determined. Crystal structures of sirtuins derived from prokaryotes (Avalos et al., 2005, 2004; Cosgrove et al., 2006; Hawse et al., 2008; Hoff et al., 2006; Min et al., 2001; Sun et al., 2002; Zhao et al., 2003, 2004) as well as human sirtuin subtypes—Sirt1 (Davenport et al., 2014; Zhao et al., 2013), Sirt2 (Finnin et al., 2001; Moniot et al., 2013; Rumpf et al., 2015; Yamagata et al., 2014), Sirt3 (Bao et al., 2014; Disch et al., 2013; Gertz et al., 2013, 2012; Jin et al., 2009b; Nguyen et al., 2013a,b; Szczepankiewicz et al., 2012; Wu et al., 2013), Sirt5 (Du et al., 2011; Gertz et al., 2012; Schuetz et al., 2007; Szczepankiewicz et al., 2012; Zhou et al., 2012), and Sirt6 (Jiang et al., 2013)—have provided insights into the overall structure of sirtuins. So far no 3D structures of human Sirt4 and Sirt7 isoforms are available. All sirtuins share a highly conserved catalytic core of ~275 amino acids and N- and C-terminal flanking extensions that vary in length, sequence, and disorder propensity. Genealogical studies have suggested that the terminal segments of sirtuins are involved in the regulation of functional activities and the recognition of other proteins (Costantini et al., 2013). The catalytic core is composed of two globular subdomains connected by four loops and an active site cleft formed between the two subdomains: a large Rossmann-fold domain and a smaller and structurally more diverse zinc-binding domain (Figure 2). The Rossmann-fold domain consists of six parallel β-strands that form a central β-sheet and several α-helices that pack against the β-sheets. Whereas, the zinc-binding domain comprises an α-helical bundle and a characteristic zinc-binding motif Cys—X_{2-4}—Cys—X_{15-40}—Cys—X_{2-4}—Cys (Moniot et al., 2012). The zinc ion is far away from the active site and does not directly participate in the catalysis. However, replacing the cysteine residues of the zinc-binding domain with

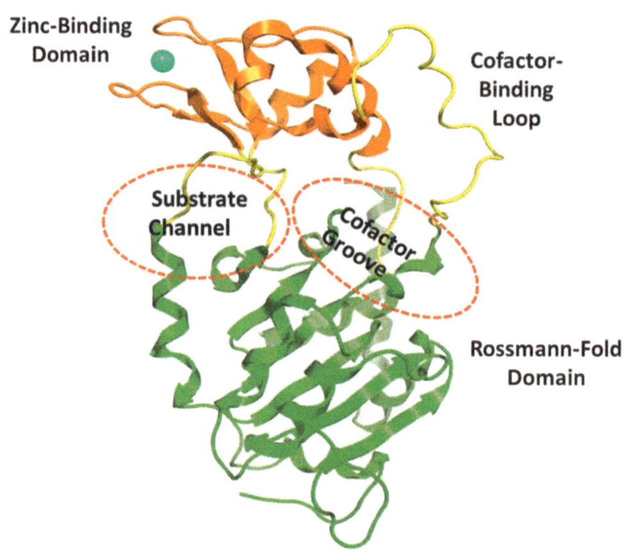

FIGURE 2

Overall structure of sirtuins, Sirt2 (PDB ID; 1J8F) as an example. The structure of Sirt2 is shown as a cartoon model. The zinc-binding domain and Rossman-fold domain are colored in orange and green, respectively. The four loops (including the cofactor-binding loop) that connect the subdomains are colored in yellow. Substrate and cofactor binding pockets are indicated by red-circle dashed lines. The zinc ion is shown as a cyan sphere.

alanine residues or using the zinc chelator o-phenanthroline abolished the deacetylase activity of sirtuins in vitro (Min et al., 2001).

The acetyllysine peptide substrate and NAD^+ cofactor bind to the active tunnel at the interface of two domains. Upon binding of the substrate or a pseudo-substrate, the zinc-binding domain rotates toward the Rossmann-fold domain and induces a closure of the active cleft. This closure of both domains helps to correctly orientate the acetyllysine in the hydrophobic tunnel, allowing the formation of an H-bond between the Nε atom of lysine and the backbone carbonyl of a conserved valine residue (Val233, Sirt2 numbering) as well as further van der Waals interactions with hydrophobic residues. Recently, Rumpf et al. (2015) showed that sirtuins can also adopt an locked-open conformation. It was shown that upon binding of the highly selective Sirt2 inhibitor, SirReal2, no closure of the substrate channel was observed even in the presence of a substrate. This newly discovered conformation of Sirt2 resembles an intermediate state between the apo, open conformation, and substrate-bound closed conformation.

Variations in the amino acid residues of the substrate channel in some sirtuins are shown to grant different modifications of lysine to be recognized as substrates. It was demonstrated that Sirt5 has a weak deacetylase activity but a robust lysine desuccinylase and demalonylase activity (Du et al., 2011). The structure of Sirt5 in complex with NAD^+ and a succinyl-lysine containing peptide has also been determined, showing that two nonconserved residues, Tyr102 and Arg105, in the back of the Sirt5 catalytic pocket form H-bond and salt–bridge interactions with the carboxylate group of the succinyl-lysine residue (Du et al., 2011). In addition, Jiang et al. (2013) showed that Sirt6 accepts long-chain fatty acyl groups as lysine modifications even more efficiently than a small acetyl group. The ternary structure of Sirt6 in complex with a myristoyl-lysine peptide and ADPR supported that Sirt6 possesses a large hydrophobic tunnel that can accommodate peptides with long-chain fatty acyl modifications on lysine residue (Jiang et al., 2013). Recently, Sirt2 was also shown to recognize fatty-acid-acetylated lysine residues using a photo-crosslinking chemistry approach (Liu et al., 2015). Interestingly, Sirt2 could remove the myristoyl group on Lys9 from the histone H3 peptide (H3K9myr) more efficiently than Sirt6. Moreover, Sirt2 showed a tighter binding to H3K9myr peptide than its known in vitro substrate, Lys9-acetylated histone (H3K9ac). More recently, Teng et al. (2015) have reported the crystal structure of Sirt2 in complex with such a thiomyristoyl peptide, confirming that the hydrophobic and flexible acyl pocket of Sirt2 allows the binding of long-chain fatty acyl groups. Collectively, these results suggest that sirtuins can be considered as lysine deacylases instead of lysine deacetylases and their role in different protein acyl lysine modifications still has to be fully elucidated.

3. CRYSTAL STRUCTURES OF SIRTUINS WITH INHIBITORS AND ACTIVATORS

3.1 Suramin and Analogs

Originally developed for the treatment of human African trypanosomiasis (sleeping sickness) (Dressel, 1961; Steverding, 2010), Suramin (Figure 3) is a well-known antipurinergic mediator. Today, Suramin is also used for the treatment of onchocerciasis (Babalola, 2011) and its potential therapeutic effects against numerous other diseases including cancer, AIDS, Hepatitis B infection (Voogd et al., 1993), and autism spectrum disorders (ASDs) (Naviaux et al., 2014) have been extensively investigated.

In 2003, attempts to find sirtuin modulators led Howitz et al. (2003) to discover that Suramin is an inhibitor of Sirt1. Soon after, the X-ray structure of Sirt5 in complex with Suramin was solved and shed light on the binding of Suramin to sirtuins (Schuetz et al., 2007). Interestingly, Suramin binds to the active cleft located between the substrate and cofactor binding pockets of two sirtuin monomers (Figure 4). Despite its known preference for purinergic binding sites, Suramin does not occupy the adenosine binding pocket of Sirt5. In a study by Trapp et al. (2007), a series of about 30 Suramin analogs were tested as sirtuin inhibitors and their structure–activity relationships (SARs) for Sirt1 and Sirt2 were investigated by means of docking studies. Suramin was found to be a potent Sirt1 (IC_{50} = 297 nM) and Sirt2 (IC_{50} = 1150 nM) inhibitor. The most potent Sirt1 inhibitor in the series was the analog NF675 (Figure 3; IC_{50} = 93 nM) whereas the most potent Sirt2 inhibitor was the chlorinated analog NF258 (Figure 3; IC_{50} = 407 nM). Docking studies suggested that Suramin analogs are likely to bind to Sirt2 in a similar fashion as observed for Sirt5.

3.2 Ex-527

Ex-527 (Figure 3), now named Selisistat, has reached Phase II clinical trials for Huntington's disease (Arrowsmith et al., 2012; Donmez and Outeiro, 2013; Solomon et al., 2006; Zhang et al., 2011). Ex-527 and several other indole derivatives were discovered as potent Sirt1 inhibitors in a high-throughput screen in 2005 (Napper et al., 2005). Ex-527 was shown to be >200 fold more selective for Sirt1 (IC_{50} = 0.098 μM) than its close homologs Sirt2 (IC_{50} = 19.6 μM) and Sirt3 (IC_{50} = 48.7 μM) and showed no inhibitory effect on class I/II HDACs and NAD glycohydrolase at 100 μM (Napper et al., 2005). Interestingly, only the S-isomer of Ex-527 (named Ex-243) was found to be responsible for the inhibition of the Sirt1-dependent deacetylation activity, whereas the R-isomer (Ex-242) was found to be inactive on Sirt1. Next, Zhao et al. (2013) determined the first ternary crystal structure of Sirt1 in complex with an Ex-527 analog (Figure 5) referred to as compound 35 (Figure 3). A closer look at the interactions of compound 35 reveals that the carboxamide group of the cyclohepta[*b*]

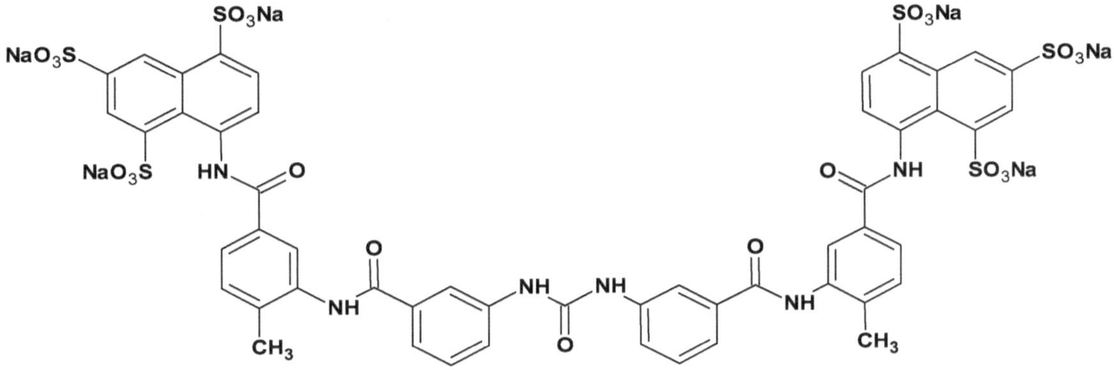

Suramin

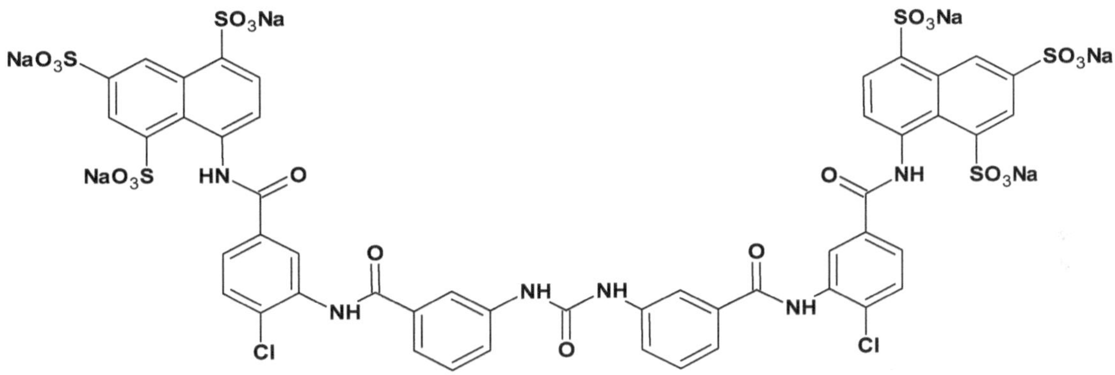

NF258

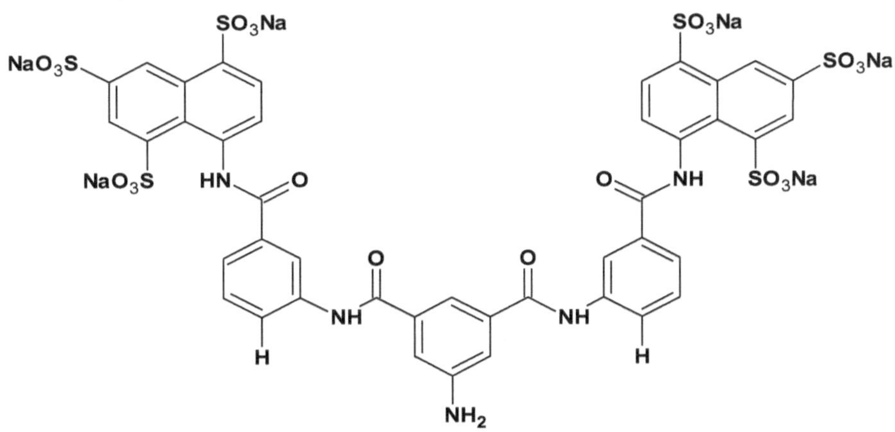

NF675

FIGURE 3
Molecular structures of sirtuin activators/inhibitors mentioned in the text.

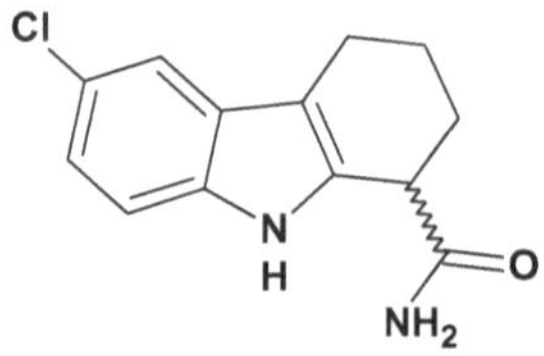

Ex-527

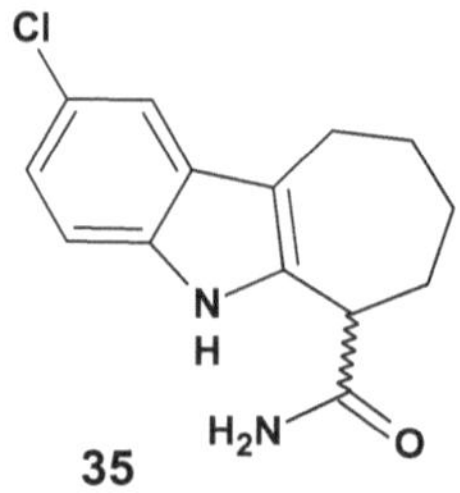

35

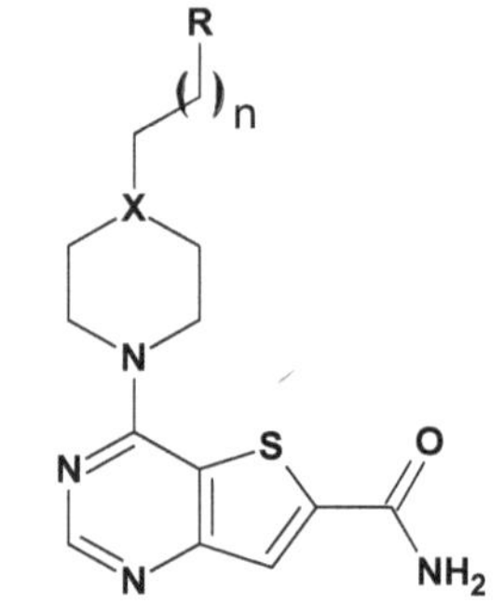

Thieno[3,2-*d*]pyrimidine-6-carboxamide derivatives

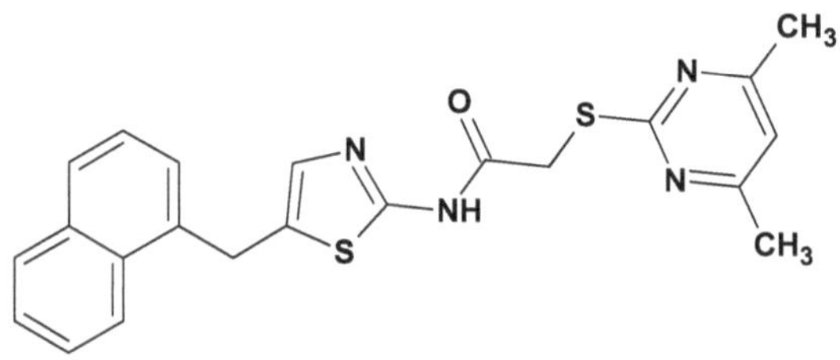

SirReal2

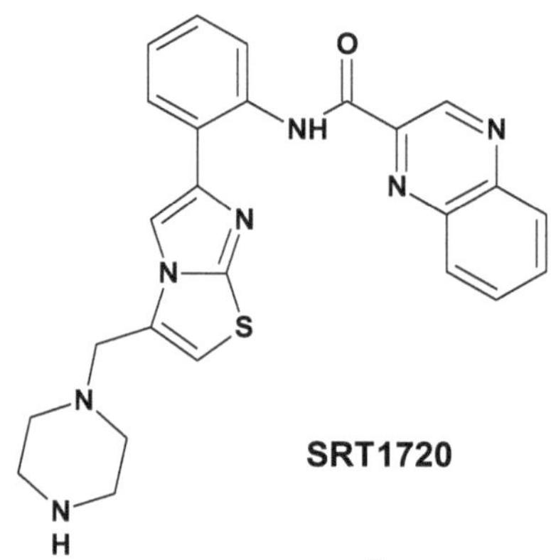

SRT1720

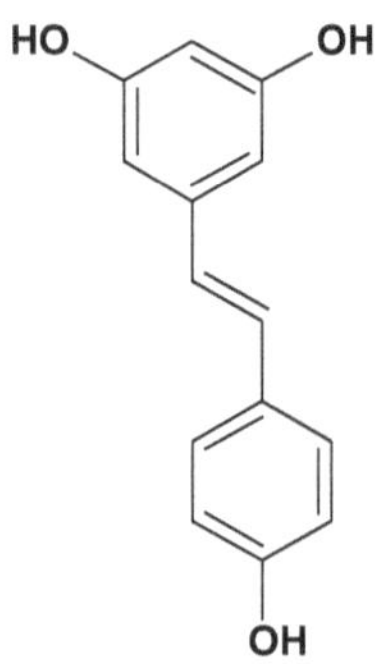

Resveratrol

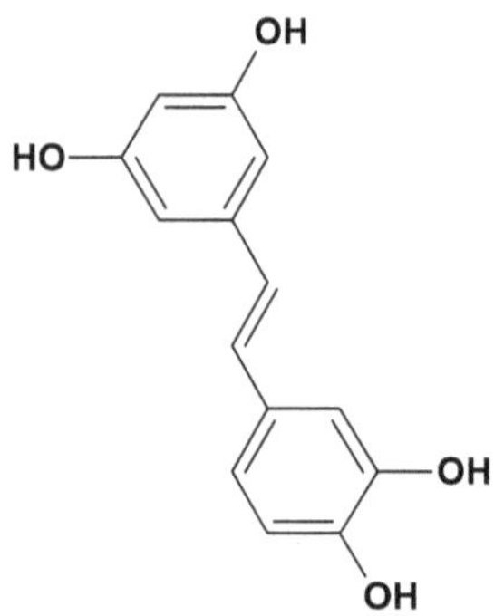

Piceatannol

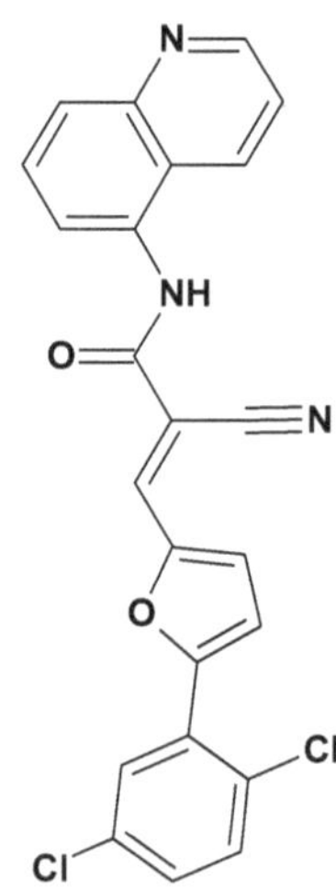

AGK-2

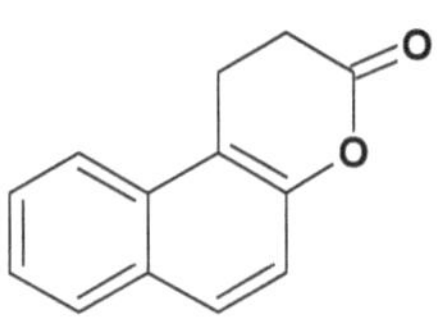

Splitomicin

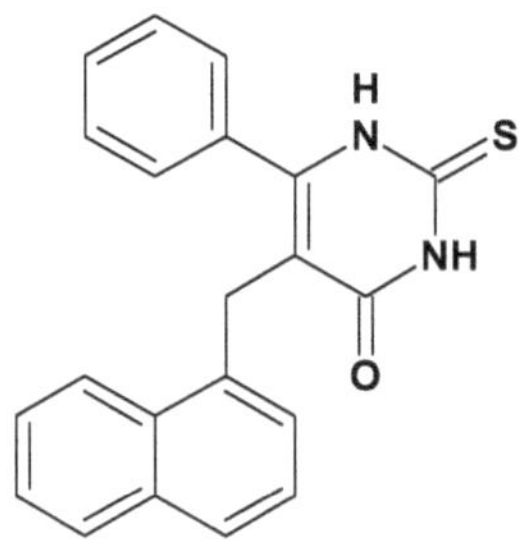

Cambinol

FIGURE 3 Cont'd

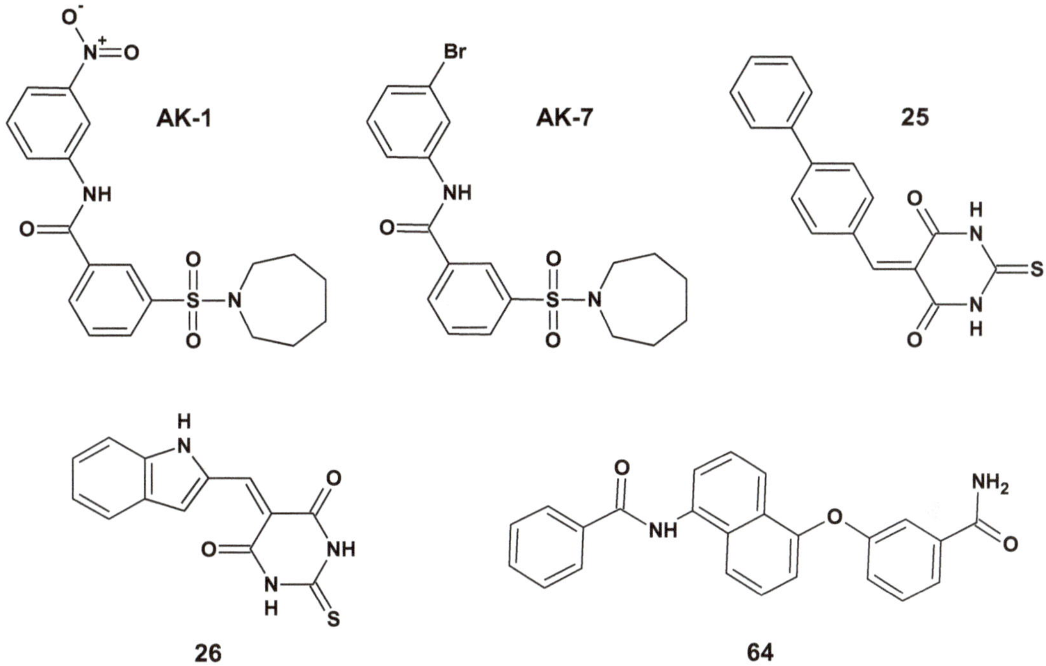

FIGURE 3 Cont'd

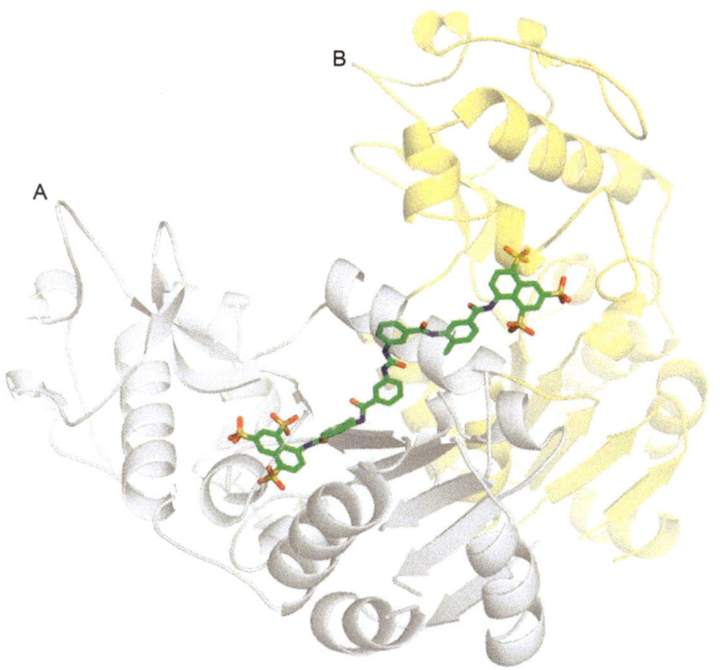

FIGURE 4

Crystal structure of Sirt5 in complex with Suramin (PDB ID; 2NYR). The structure of Sirt5 is presented as cartoon model (A chain, white; B chain, yellow). The bound inhibitor Suramin is shown as green sticks.

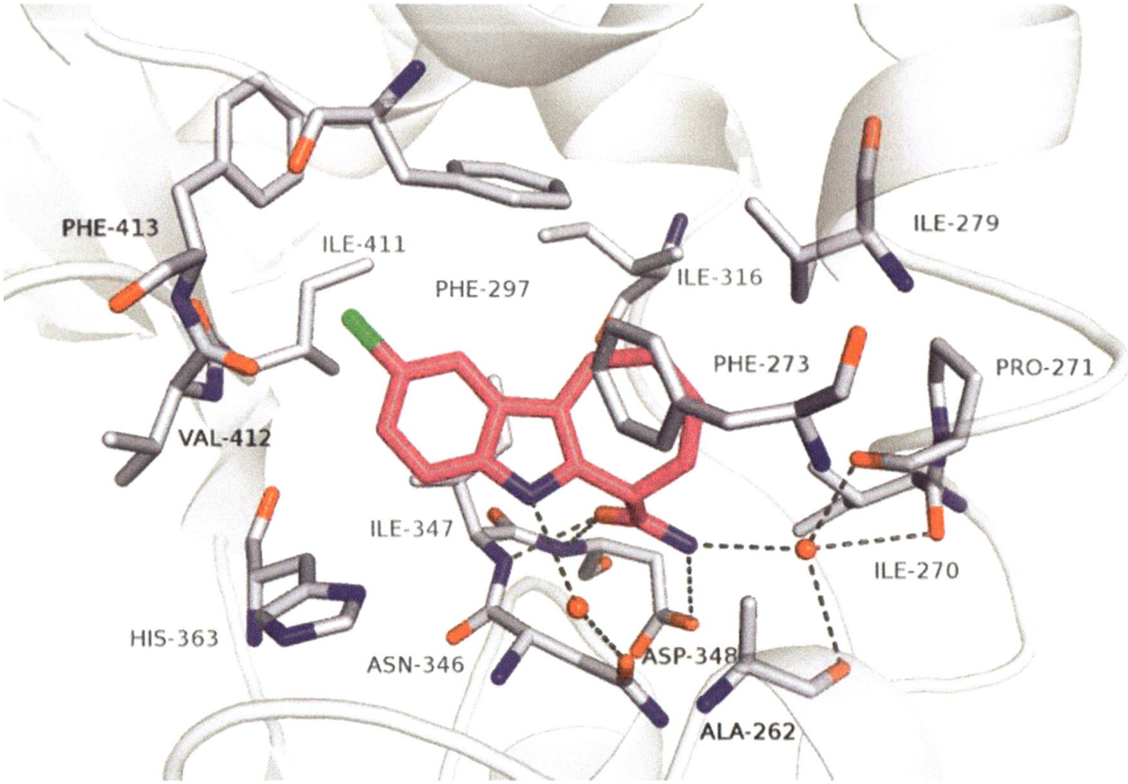

FIGURE 5

Crystal structure of Sirt1 in complex with NAD⁺ and the Ex-527 analog 35 (PDB ID; 4I5I). The structure of Sirt1 is presented as a white cartoon model and 35 is shown as pink sticks. NAD⁺ is not shown for clarity. Only the interacting protein residues are shown as white sticks, the bridging water molecules are displayed as red balls, and hydrogen bonds are depicted as gray dashed lines.

indole core mimics the carboxamide group of NAD⁺ (Avalos et al., 2005). Several crystal structures of Ex-527 in complex with Sirt3 and Sir2Tm were determined (Gertz et al., 2013). Consistent with the previous inhibition data reported by Napper et al. (2005), only the *S*-enantiomer of Ex-527 (Ex-243) was found to be bound in the crystal structures. Moreover, competition experiments and binding assays were performed using peptide substrates and NAD⁺ to better understand the inhibition mechanism of Ex-527. Collectively, these results indicated that Ex-527 binding to sirtuins occurs after the formation of the imidate-intermediate and the release of nicotinamide. The blocking of the formation of the 2′-*O*-acetyl-ADP ribose sirtuin complex is suggested to be the most relevant mechanism for the inhibitory activity of Ex-527. In light of these findings the authors speculated that Ex-527 inhibition mechanism is dependent on trapping sirtuin in a closed conformation where no product release is possible.

3.3 ELT Inhibitors

Recently, Disch et al. (2013) discovered a series of thieno[3,2-*d*]pyrimidine-6-carboxamide derivatives (Figure 3) as potent Sirt1-3 inhibitors using Encoded Library Technology (ELT) (Clark et al., 2009). The reported compounds were shown to inhibit Sirt3 with an IC_{50} in the range between 0.0032 and >50 µM. The compounds showed a similar inhibitory effect on Sirt1 (with an IC_{50} in the range between 0.0036 and >50 µM) and Sirt2 (with an IC_{50} in the range between 0.0011 and >50 µM). X-ray structures of Sirt3 in complex with three representative inhibitors from this series were resolved, enabling further exploration of SARs. Visual analysis of the Sirt3-inhibitor complexes revealed that the compounds bind in a similar fashion in the catalytic pocket, occupying both the acetyl lysine substrate channel and the nicotinamide C-pocket (which overlaps with the NAD^+ cofactor binding pocket). A closer look at the interactions reveals that the carboxamide group of the thieno[3,2-*d*]pyrimidine-6-carboxamide core mimics the corresponding group of NCA and the sirtuin inhibitor Ex-527 (Figure 6).

3.4 SirReals

In order to identify novel potent sirtuin inhibitors, Rumpf et al. (2015) recently screened a compound library using an in vitro assay for human Sirt1-3 with a fluorophore-labeled acetyl-lysine derivative as substrate (Heltweg et al., 2002). In this screening campaign, a series of compounds bearing an aminothiazole scaffold was discovered as potent Sirt2-selective inhibitors. Interestingly, these inhibitors—called SirReals (Sirtuin rearranging ligands)—were found to rearrange the Sirt2 binding pocket and behave as allosteric inhibitors. Among those, SirReal2 (Figure 3) was found to be highly selective, strongly inhibiting Sirt2 (IC_{50} of 140 nM), while only a slight inhibition of Sirt1 (22% inhibition at 100 µM) and Sirt6 (19% inhibition at 200 µM) could be detected. Moreover, no significant effect was seen against the sirtuin isoforms Sirt3-5 and the classical HDACs: HDAC1 and HDAC6. Sirt2-selective in vitro inhibition and binding was further confirmed by conducting HPLC-based conversion and thermal stability assays, while cellular assays confirmed the Sirt2 activity and selectivity in vivo. An additional stabilization of the Sirt2-SirReal2 complex could be achieved in the presence of either substrate or NAD^+ in the thermal stability assay.

By solving the X-ray structure of SirReals with Sirt2, it was shown that the isoform-selective inhibition by these inhibitors is derived from a ligand-induced structural rearrangement of the active site. SirReal2 binds to a highly hydrophobic pocket adjacent to the zinc-binding domain and does not prevent the binding of NAD^+ in its productive conformation. The binding conformation of SirReal2 shows an ideal complementary shape with the active cleft of Sirt2. The naphthyl moiety of SirReal2 occludes the acetyl-lysine channel and thus

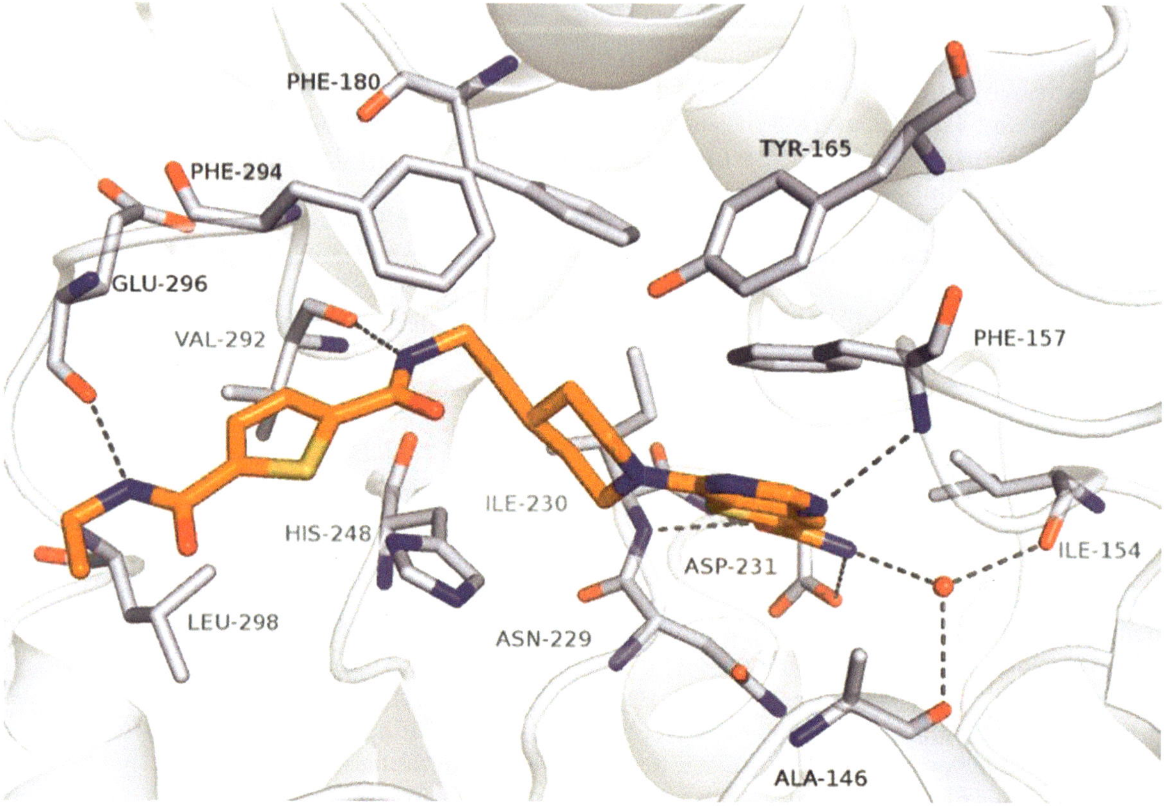

FIGURE 6

Crystal structure of Sirt3 in complex with the ELT inhibitor 11c (PDB ID; 4JSR). The structure of Sirt3 is presented as a white cartoon model and ELT 11c is shown as orange sticks. Only the interacting protein residues are shown as white sticks, the bridging water molecule is displayed as a red ball, and hydrogen bonds are depicted as gray dashed lines.

prevents the correct orientation of the substrate peptide in the channel (Figure 7). On the other hand, the dimethylpyrimidine substituent induces the formation of a pocket located deep inside the catalytic cavity above the C-pocket and has not been observed so far. This pocket was claimed to be the selectivity pocket for Sirt2.

3.5 Sirtuin Activators: Resveratrol and Others

In 2003, Howitz et al. (2003) identified a number of small molecules including Piceatannol and Resveratrol (Figure 3) as the first-known activators of Sirt1 using a fluorescent deacetylation assay with a fluorophore-labeled peptide as substrate. Although Resveratrol has sparked a great interest due to its potential chemo-preventive effects on the treatment of age-related diseases, cancer, metabolic disorders, and cardiovascular diseases (Camins et al., 2010), the mechanism

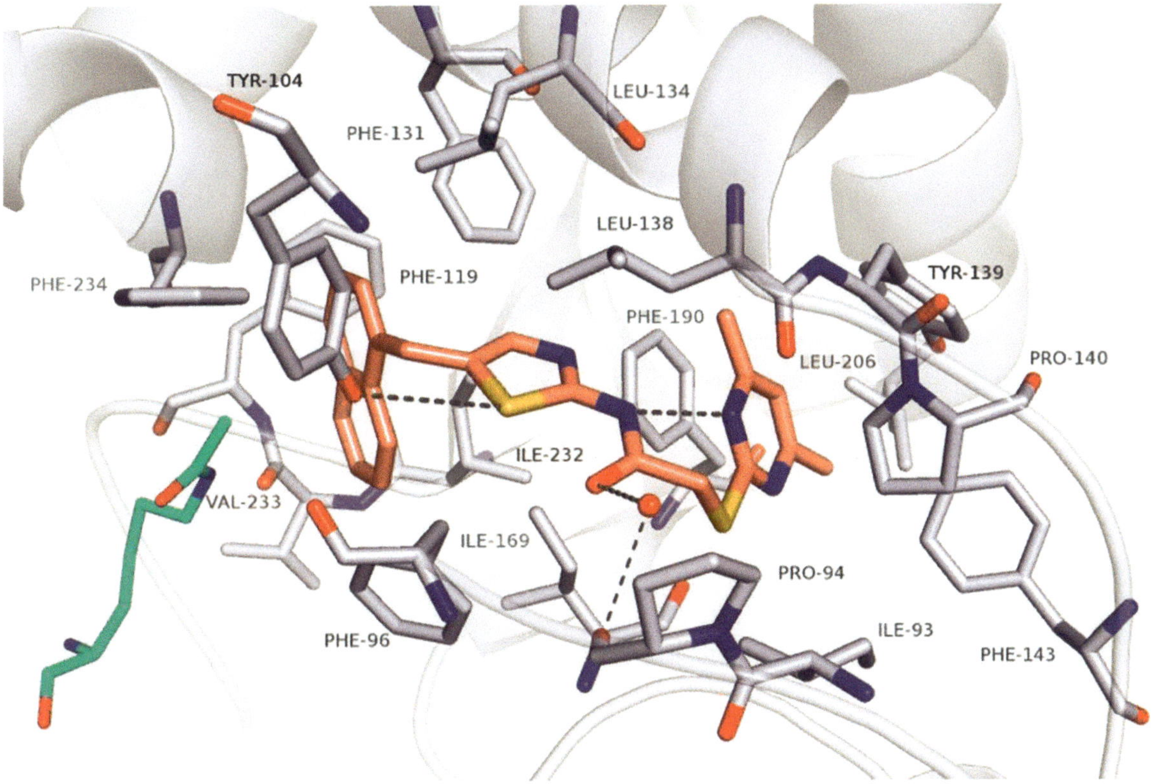

FIGURE 7

Crystal structure of Sirt2 in complex with SirReal2 and acetyllysine peptide substrate derived from histone H3 (PDB ID; 4RMH). The structure of Sirt2 is presented as a white cartoon model. SirReal2 and the acetyllysine peptide are shown as salmon and cyan sticks, respectively. Only the interacting protein residues are shown as white sticks, the bridging water molecule is displayed as a red ball, and hydrogen bonds are depicted as gray dashed lines.

behind the pharmacological benefits of Resveratrol has not been clarified. Several in vivo/in vitro studies have demonstrated that Resveratrol inhibits/activates a wide range of enzymes and is involved in various signaling pathways and it was also shown to indirectly regulate Sirt1 activation (Pirola and Frojdo, 2008). Speculations surrounding Resveratrol and Sirt1 have intensified upon the observation that Resveratrol could stimulate Sirt1 activation only with substrates carrying a fluorophore on their C-terminal end (Borra et al., 2005; Kaeberlein et al., 2005; Pacholec et al., 2010). In order to elucidate the Sirt1 activation mechanism with sirtuin activating compounds (STACs), Dai et al. (2010) performed isothermal titration calorimetry (ITC) experiments and fluorescence binding assays using both fluorophore-labeled and nonlabeled peptide substrates. Surprisingly, some compounds could accelerate Sirt1-dependent deacetylation of specific nonlabeled peptides without a fluorophore. Later, Gertz et al. (2012) discovered

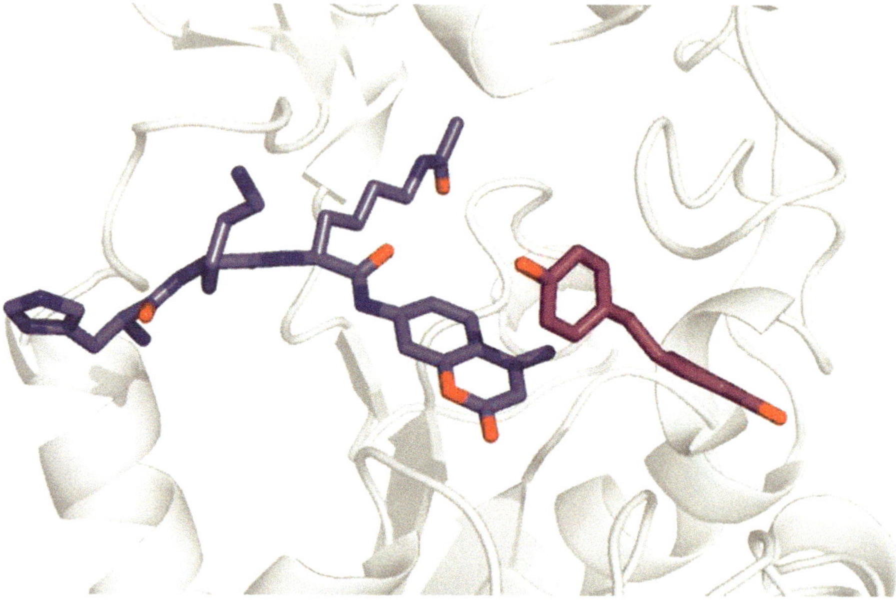

FIGURE 8

Crystal structure of Sirt5 in complex with Resveratrol and FdL-1 peptide substrate (PDB ID; 4HDA). The structure of Sirt5 is presented as a white cartoon model. Resveratrol and the FdL-1 peptide are shown as purple and slate sticks, respectively.

that Resveratrol and Piceatannol activate Sirt5 besides Sirt1; however, they both inhibit Sirt3. Ternary structures of Sirt5 in complex with a fluorogenic peptide substrate (FdL-1 peptide) and Resveratrol (Figure 8), and Sirt3 in complex with Piceatannol and FdL-1 peptide provided first evidences for a direct interaction between Resveratrol-like compounds and the coumarin fluorophore (Gertz et al., 2012). Interestingly, Resveratrol was shown to stimulate Sirt5-dependent deacetylation against fluorophore-free peptides and full-length proteins such as Peroxiredoxin 1 and Cytochrome c but not against the p53-Lys382 peptide in agreement with the previous data reported (Dai et al., 2010).

Meanwhile, Wu et al. (2013) published a new class of Sirt1 activators bearing a diaryl-acylhydrazone moiety. In addition, they cocrystallized the compounds with Sirt3 and the substrate FdL-1 peptide. Docking studies, ITC, and enzyme kinetic experiments were combined with mutagenesis results to elucidate the probable binding mode of these compounds. It was concluded that diaryl-acylhydrazones bind in the hydrophobic pocket formed by the side chains of the substrate peptide and the residues from the Rossmann-fold domain of Sirt1.

Remarkable changes were observed in sirtuin activity upon slight modifications of the Resveratrol structure. A recent study conducted by Nguyen et al. (2013a) revealed that 4′-bromo-resveratrol has potent inhibitory activity against Sirt1 (unlike its parent Sirt1-activator: Resveratrol) and Sirt3 isoforms. Crystal structures of Sirt3 in complex with 4′-bromo-resveratrol were determined in the existence of both a physiological target of Sirt3, acetyl-CoA-synthetase 2 (ACS2), and the FdL substrate. Two different binding modes of 4′-bromo-resveratrol were observed upon using different substrates. Moreover, only a weak inhibitory effect for 4′-bromo-resveratrol on Sirt5 desuccinylase activity could be obtained using succinylated fluorophore-labeled peptide substrate, indicating the potential of Resveratrol-like isoform-selective sirtuin inhibitors.

Attempts to find more potent Sirt1 activators led Sirtris Pharmaceuticals to identify unique small molecules that are structurally unrelated to Resveratrol (Milne et al., 2007). The activity of these compounds were evaluated by detecting the concentration of a compound needed to increase enzyme activity by 50% ($EC_{1.5}$) and the percentage of maximum activation (m. a.) achieved at the highest doses of the compound tested. Several compounds, including SRT1720 (Figure 3; $EC_{1.5}=0.16\,\mu M$ and m. a. = 781%) exhibited a remarkable increase in Sirt1 activation in a fluorescence polarization assay against a p53-derived peptide (Milne et al., 2007). SRT1720 treatment in mice models exhibited improved insulin sensitivity and glucose tolerance presenting its potential therapeutic effectiveness in the treatment of metabolic disorders such as type 2 diabetes (Milne et al., 2007). Initially recognized as a potent Sirt1 activator, SRT1720 was shown to possess an inhibitory effect against mouse Sirt3 isoform (Jin et al., 2009a). Competition experiments implied uncompetitive inhibition of Sirt3 with SRT1720 against NAD^+ and competitive inhibition in the presence of the substrate peptide. Recently, the determined ternary structure of Sirt3 in complex with SRT1720 has established the binding mode of this compound (Nguyen et al., 2013b). While the piperazine ring and the imidazothiazole group of SRT1720 occupy the acetyllysine channel, the rest of the compound is situated above the nicotinamide ribose and the nicotinamide group of NAD^+. The binding of SRT1720 in the C-pocket is mediated by π-stacking interactions between its quinoxaline ring and the nicotinamide moiety of NAD^+. Thus, the Sirt3/SRT1720 structure supported the reported competition between SRT1720 and the peptide substrate and explained the noncompetitive binding of the compound in the presence of NAD^+ (Jin et al., 2009a; Nguyen et al., 2013b).

3.6 Sirt5 CPS-1-Derived Peptidic Inhibitors

Sirt5 was also found to remove octanoyl or decanoyl residues from peptides (Feldman et al., 2013). In particular, Sirt5 was recently shown to be an effective demalonylase/desuccinylase in vitro (Du et al., 2011; Park et al., 2013) and in vivo (Park et al., 2013; Peng et al., 2011). Du et al. (2011) were able to demonstrate that the change from acetyl to succinyl residues in three different model peptide sequences increased the catalytic efficiencies (k_{cat}/K_M) between 75- and 1000-fold but individual K_M and k_{cat} values could not be determined for acetylated substrates. The analysis of Sirt5 crystal structure identified an interaction of the carboxylate group of the succinylated peptide lysine with the Sirt5 side chains of Arg105 and Tyr102, indicating a mechanism for substrate acyl discrimination (Du et al., 2011). Interestingly, these residues also cause isoforms and substrate acyl specific inhibitor effects (Fischer et al., 2012; Maurer et al., 2012). The effects of substrate lysine modifications were investigated in a more systematic way and peptide-based Sirt5 inhibitors were developed through acyl-group modifications (Roessler et al., 2014). A carbamoyl phosphate synthetase I (CPS1)-derived sequence was used as the parent peptide (Fischer et al., 2012).

In the Sirt5 crystal structure, Arg105 adopts different conformations, suggesting that larger acyl residues might be able to interact with Tyr102 and Arg105. Therefore, CPS1 substrate analogs were developed that are acylated at the lysine side chain with oxalyl, malonyl, succinyl, glutaryl, adipoyl, pimeloyl, and suberoyl residues. In addition, the head groups were further chemically modified by introducing double bonds, substituents, and nonacidic head groups. The X-ray structures of zebrafish Sirt5, which is highly homologous to the less reliably crystallizing human enzyme, in complex with some of the modified peptides, were solved. The crystallized peptides show a consistent salt bridge between the carboxyl group of the acyl moiety and Arg101 as well as a hydrogen bond to Tyr98, corresponding to Arg105 and Tyr98 in human Sirt5, respectively. In comparison to the succinyl moiety, the glutaryl- and adipoyl-modifications show an increasingly twisted conformation, allowing very similar positions for the distal carboxyl group despite different chain lengths. Several sirtuin inhibitors, such as Ex-527 (Gertz et al., 2013) and 4′-bromo-resveratrol (Zhao et al., 2003), were shown to bind to the nicotinamide accommodating C-site, indicating it as an attractive pocket for increasing the binding affinity. Succinyl-peptides with bulkier acyl substituents were therefore developed, in order to generate steric hindrance for the binding of NAD^+. A potent inhibitor was derived by introducing a benzylcarbamate group at C3 of the succinyl chain. The X-ray structure of this modified peptide showed that the terminal phenyl ring is indeed blocking the C-site of Sirt5 (Figures 9 and 10). The derived CPS-1 peptides represent useful templates for the further development of potent small molecule Sirt5 inhibitors, which are still lacking.

4. COMPUTATIONAL STUDIES OF SIRTUIN MODULATORS

Computer-based methods can be grouped into two classes, structure-based and ligand-based approaches. Structure-based approaches depend heavily on the availability of three-dimensional structures (3D) of the target biomolecule.

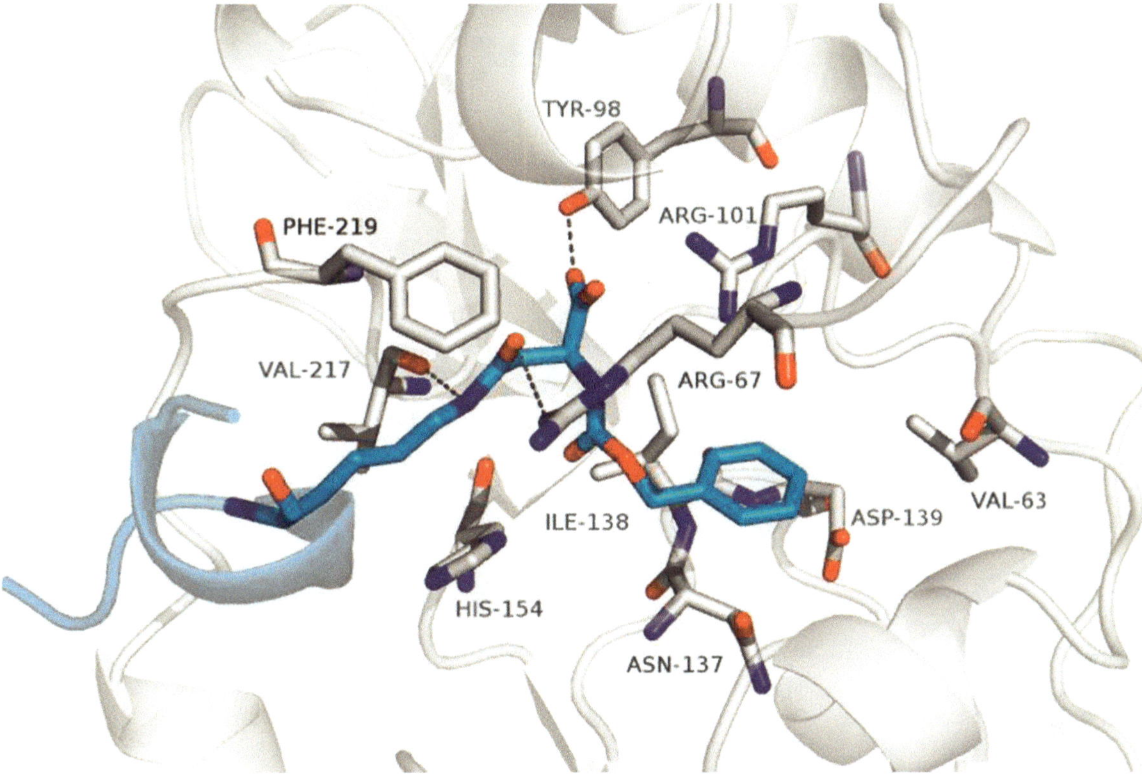

FIGURE 9
Crystal structure of zebrafish Sirt5 in complex with inhibitor 5i (PDB ID; 4UUA). The structure of Sirt5 is presented as a white cartoon model. Only the 3S-Z-amino succinylated lysine part of the inhibitor 5i is shown as blue sticks and the rest of the molecule is presented as blue cartoon. Protein residues involved in the interaction are shown as white sticks and hydrogen bonds are depicted as gray dashed lines.

Ideally, an X-ray structure of the macromolecular target has been already solved complexed with a lead compound. However, if no 3D structure is available, homology modeling (also known as comparative protein structure modeling) can be used to predict a 3D model for a protein of unknown structure (the target) from proteins that share sequence similarity with known structures (templates) (Cavasotto and Phatak, 2009; Marti-Renom et al., 2000). The inhibitor/activator compound of interest can be docked into a particular binding pocket to rationalize the probable interactions with the protein. Current molecular docking tools are able to model the interactions in the putative binding site and can easily screen large numbers of molecules and rank them according to their predicted affinities (Moitessier et al., 2008; Warren et al., 2006; Yuriev et al., 2011). On the other hand, ligand-based approaches can be applied when a lead structure has already been determined. Large compound databases can be filtered according to the chemical and/or electrostatic

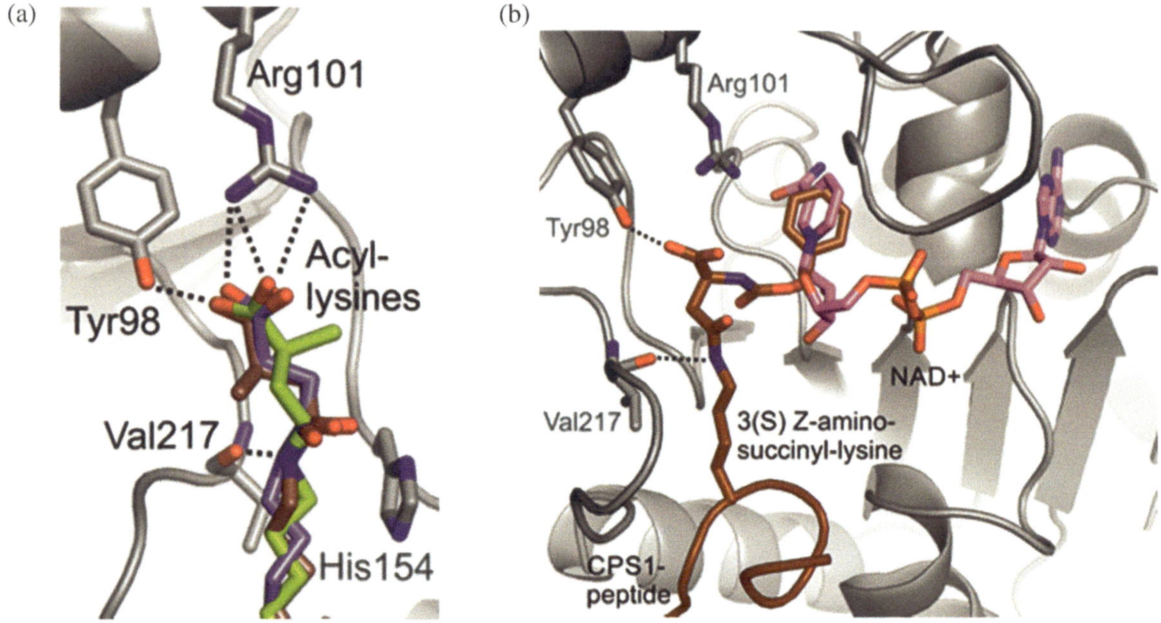

FIGURE 10

(a) Crystal structures of zSirt5 active site with superposed CPS-1 peptides containing modified acyl-groups. The protein is shown only once for clarity (zSirt5/5a complex; gray). Hydrogen bonds between active-site residues and substrate peptides' acylated lysine are indicated by dotted lines. (b) Crystal structure of zSirt5 active site with a CPS-1 peptide acting as inhibitor (brown). NAD^+ is shown in pink. The protein is shown only once for clarity (zSirt5/5i complex; gray). Hydrogen bonds between active-site residues and substrate peptides' acylated lysine are indicated by dotted lines.

similarities calculated for a set of known actives or via fitting a pharmacophore model (Bender and Glen, 2005; Hevener et al., 2012; Lower et al., 2011; Tan et al., 2010; Tresadern et al., 2009). Besides traditional scoring methods, which usually do not take solvation effects or protein flexibility into account, more sophisticated and demanding calculations of binding free energy (BFE), for example, linear interaction energy (LIE) (Aqvist et al., 1994), molecular mechanics/Poisson-Boltzmann (Generalized Born) surface area (MM-PB(GB) SA) (Srinivasan et al., 1998), free-energy perturbation (FEP), and thermodynamic integration (TI) (Kirkwood, 1935; Straatsma and McCammon, 1991) have been developed.

4.1 Docking Studies

Until today, several sirtuin inhibitors/activators have been identified by well-established in silico methods. Molecular docking has been the most extensively used computational tool to predict the potential binding interactions of structurally known small molecules and thus explain the inhibition/activation mechanisms of sirtuins. Moreover, when a lead structure is known, docking studies are often used to optimize potency and/or selectivity of compounds of interest. Splitomicin (Figure 3) was one of the first small molecules that has been identified as an inhibitor of yeast sirtuins using a phenotypic screening (Bedalov et al., 2001). In their studies, Neugebauer et al. (2008) performed molecular docking to characterize the binding interactions of splitomicin derivatives, which showed micromolar range of inhibition against human Sirt2 isoform. Since biological assays were carried out for racemates, both stereoisomers (R and S) were considered for the docking study. The authors demonstrated that only the R-enantiomer of β-phenylsplitomicin derivatives could show favorable interactions with Sirt2 structure. The R-isomers were predicted to bind in the C-pocket, making water-mediated H-bond interactions with Gln167. Moreover, the β-phenyl ring of this series of compounds is making sandwich interactions with the two aromatic residues Phe119 and His187 and by blocking the acetyllysine pocket that sirtuin substrate binds. A novel scaffold that shows structural similarity with sirtinol and splitomicin was discovered by Heltweg et al. (2006) and named Cambinol (Figure 3). Cambinol showed inhibitory activity against Sirt1 ($IC_{50}=56\,\mu M$) and Sirt2 ($IC_{50}=59\,\mu M$) and exhibited an antitumor effect in mice models (Heltweg et al., 2006). Based on docking studies and binding free energy calculations performed on Sirt2 crystal structure (Finnin et al., 2001) and a Sirt1 homology model (Sirt2 structure was used as a template), Medda et al. (2011, 2009) suggested that Cambinol analogs interact with the C-pocket residues and occupy the acetyllysine pocket. Docking results demonstrated that while the thiouracil ring of Cambinol forms H-bonds with Gln167 and His187 (Sirt2 numbering), the β-naphthol group makes π–π interactions with His187 and Phe119 (Sirt2 numbering) and the rest of the compound occupies the nicotinamide exchange pocket. Authors could rationalize the structure activity/selectivity relationship between Cambinol analogs and sirtuins well with this proposed binding model. The same binding region and a similar interaction was also described for other inhibitors of sirtuins such as splitomicin derivatives (Neugebauer et al., 2008) and indolinones (Huber et al., 2010), demonstrating the versatility of this binding model.

Aside from predicting the correct orientation of small molecules within the sirtuins, docking has also been exerted in virtual screening studies to find new active compounds. In some recent work (Pulla et al., 2014), a parallel target-based virtual screening was carried out that identified novel Sirt1 inhibitors and activators with diverse scaffolds from a commercially available database. In attempts to find new Sirt1 activators, the same research group performed docking studies on a previously reported homology model of Sirt1 (Autiero et al., 2009) based on the assumption that Resveratrol and other STACs activate Sirt1 via binding to its allosteric site. On the other hand, the authors used the available crystal structure of Sirt1 in complex with the inhibitor 35 (Figure 3) and NAD+ (PDB ID; 4I5I) (Zhao et al., 2013) to discover novel Sirt1 inhibitors. The authors utilized a consecutive docking approach where the precision of the scoring function was increased subsequently. Compounds that showed similar binding interactions as already reported sirtuin inhibitors (Splitomicin, Sirtinol, Ex-527 and 35) and the activator Resveratrol were selected based on the docking score. Hits were then evaluated for their biological effects on Sirt1 activity. Compounds from the inhibitor class showed Sirt1 inhibition with an IC_{50} range of 16.35 to 52.92 μM. On the other hand, compounds from the activator class showed moderate activity at 10 μM ($EC_{1.5}$ values were not calculated). The authors could demonstrate that fast and computationally less expensive methods such as target-based virtual screening with simple scoring functions are useful to find novel sirtuin inhibitors.

4.2 Ligand-Based Screening Studies

Ligand-based virtual screening represents a time-saving and efficient method to identify novel sirtuin inhibitors, based on the assumption that compounds sharing the same chemical features are likely to act in a similar manner. Outeiro et al. (2007) identified the two Sirt2 inhibitors AGK-2 and AK-1 (Figure 2) and analyzed the interaction with Sirt2 by docking studies. AGK-2, which inhibits Sirt2 with an IC_{50} of 3.5 μM, shows neuroprotective effects in central diseases such as Parkinson's disease. Taylor et al. (2011) could identify the first brain-permeable Sirt2 inhibitor, named AK-7 (Figure 3), by performing a substructure-based search on the Chembridge database using the AK-1 scaffold (Figure 3).

Uciechowska et al. (2008) reported a novel approach to target sirtuins and identify new inhibitors that could form similar binding interactions as the already reported Cambinol analogs (Medda et al., 2011, 2009). In this context, a large database of commercially available compounds (Chembridge database) was first filtered via fingerprint calculation to retrieve only molecules similar to Cambinol (Heltweg et al., 2006). Subsequently, hit compounds were docked into Sirt2 (PDB ID; 1J8F) (Finnin et al., 2001). A successful combination of ligand-based and target-based virtual screening (VS) methods led the authors to discover a number of thiobarbiturate and barbiturate derivatives as novel sirtuin inhibitors. Thiobarbiturates shared a similar binding mode as observed in case of Cambinol and showed a nonselective inhibitory effect against sirtuin isoforms. In this series of compounds, the most potent Sirt1 inhibitor, compound 26 (Figure 3), and the most potent Sirt2 inhibitor, compound 25 (Figure 3), showed an IC_{50} value of 5.9 μM and 8.7 μM, respectively. Binding free energies obtained by MM-PBSA calculations for thiobarbiturate and barbiturate derivatives showed good agreement with the experimental data.

Furthermore, fragment-based drug design has been used as an alternative computational tool as opposed to traditional virtual screening approaches for the identification of novel sirtuin modulators. Recently, Cui et al. (2014) carried out a fragment-based approach to identify selective and potent Sirt2 inhibitors. The authors could successfully combine the chemical fragments of Suramin and nicotinamide, which resulted in the design of (5-benzamidonaphthalen-1/2-yloxy) nicotinamide derivatives. Among these, compound 64 (Figure 3) showed nanomolar inhibitory activity against Sirt2 (IC_{50} = 0.0483 μM) and 200- to 900-fold selectivity over Sirt1 (IC_{50} = 12.0 μM) and Sirt3 (IC_{50} = 44.2 μM). The authors presented two potential binding modes based on docking studies of compound 64 into the Sirt2 crystal structure (PDB ID; 1J8F) (Finnin et al., 2001). While the nicotinamide moiety binds into the nicotinamide binding pocket and forms H-bonds with the conserved residues Ile169 and Asp170 in both docking poses, the naphthalene group makes π–π interactions either with Phe96 and Phe119 or Phe119 and His187 in model 1 and model 2, respectively. Considering the sufficient data on small molecule inhibitors of sirtuins, growing fragments of known binders of sirtuins into lead compounds represents an advantageous technique with growing interest.

4.3 Design of Pseudopeptidic Inhibitors

In addition to small molecules, designing pseudopeptidic inhibitors for sirtuins has spurred great interest in the past few years. In 2006, Fatkins et al. (2006) presented the potential of N^{ε}–thioacetyllysine containing pseudopeptides as sirtuin inhibitors. Based on this discovery, Huhtiniemi et al. (2011) designed several pseudopeptidic inhibitors by attaching various fragments on the C- and N-terminal ends of the thioacetyllysine residue. Pseudopeptides were virtually generated using a small size fragment library and carboxylic acid containing fragments that were obtained commercially. Then, docking calculations of virtual libraries were carried out using the Sirt3 crystal structure (Jin et al., 2009b). Based on the docking results, the authors proposed that H-bonds observed with residues Val292, Gly295, Glu296, and Glu325 (Sirt3 numbering) are essential for the correct orientation of pseudopeptidic compounds. This binding hypothesis facilitated the discovery of a number of new pseudopeptides, which showed potent Sirt1-3 inhibitory activity in in vitro assays and some could inhibit cell growth in breast and lung cancer cell lines (Huhtiniemi et al., 2011; Mellini et al., 2013). Therefore, the authors could demonstrate the effectiveness of fragment-based drug design in assessing pseudopeptidic inhibitors with druglike properties.

4.4 QSAR Studies

In order to develop more selective and potent pseudopeptidic sirtuin inhibitors it is important to determine the quantitative structure–activity relationship (QSAR) between the structural features of a series of compounds and the biological activity data. Kokkonen et al. (2014) reported the generation of the first 3D-QSAR model using the comparative molecular field analysis (CoMFA) technique for already published substrate-based Sirt1 inhibitors (Huhtiniemi et al., 2011, 2010; Kiviranta et al., 2009; Mellini et al., 2013). In their work, the authors either docked the pseudopeptidic inhibitors into the Sirt1 structure (Zhao et al., 2013) or aligned them with a reference inhibitor to obtain the most biologically relevant conformation. The authors demonstrated that the CoMFA contour map was consistent with the steric and electrostatic properties of the Sirt1 substrate binding site, and the predicted biological activities of the newly synthesized compounds were in good agreement with the experimental data.

Many substrate peptides that contain different protein acyl lysine modifications have been reported lately for sirtuins (Du et al., 2011; Jiang et al., 2013; Liu et al., 2015). The acetyllysine binding pocket of sirtuins can accommodate peptides with long-chain fatty acyl modifications on lysine residue. Therefore, it would be intriguing to combine these modifications with already known thioacetyllated pseudopeptides (Huhtiniemi et al., 2011; Kokkonen et al., 2014; Mellini et al., 2013). This combination should aid in identifying novel pseudopeptidic sirtuin inhibitors with high selectivity and potency. In addition, such attempts if utilized with fragment-based drug discovery will possibly lead to a better exploration of the acetyllysine pocket limitations. Moreover, 3D-QSAR models can be adapted to such studies to predict the biological activities of next-generation compounds and optimize them in a quantitative fashion.

Due to the high flexibility of the substrate and cofactor binding pockets observed in sirtuins, the structure-based optimization of inhibitors is still challenging. In a recent effort, different molecular modeling techniques such as docking, homology modeling, and binding free energy calculations were integrated to provide predictive QSAR models for the aforementioned thieno[3,2-*d*]pyrimidine-6-carboxamide derivatives that were shown to possess potent inhibitory activities against Sirt1-3 (Karaman and Sippl, 2015). A significant correlation between binding free energies derived from MM-GBSA calculations and in vitro data was found for all three sirtuin subtypes. The obtained results show that binding free energy calculations using a small number of snapshots retrieved from molecular dynamic simulations give a better agreement with the experimental data than docking scores. The established protocol appears to be useful for discovering novel sirtuin inhibitors and can be applied as a postdocking filter for virtual screening campaigns of larger databases.

5. CONCLUSION

Sirtuins are interesting drug targets due to the reported links of sirtuin activity to the pathogenesis of cancer, HIV, metabolic disorders, parasitic diseases (Lancelot et al., 2013), or neurodegeneration. Proof of cellular inhibition and especially selectivity in vivo has yet to be established for many sirtuin inhibitors. Additionally, for some inhibitors, sirtuin inhibition has been shown by hyperacetylation in a concentration range where the phenotypic response was observed but other mechanisms may be operative (e.g., kinase inhibition). While a substantial amount of progress has been made in the discovery of novel inhibitor scaffolds by the usage of structure- and computer-based approaches, two major steps must still be completed to validate sirtuins as drug targets. First, selective sirtuin inhibitors must be developed in order to properly identify the exact targets of the seven human sirtuin proteins. Second, active and cell-permeable druglike compounds will need to be developed from selective sirtuin inhibitors for rigorous testing in animal models.

ACKNOWLEDGMENTS

M.J. thanks the Deutsche Forschungsgemeinschaft (DFG) for funding (Project P10 within RTG1976).

REFERENCES

Aqvist, J., Medina, C., Samuelsson, J.E., 1994. A new method for predicting binding affinity in computer-aided drug design. Protein Eng. 7 (3), 385–391.

Arrowsmith, C.H., Bountra, C., Fish, P.V., Lee, K., Schapira, M., 2012. Epigenetic protein families: a new frontier for drug discovery. Nat. Rev. Drug Discov. 11 (5), 384–400.

Autiero, I., Costantini, S., Colonna, G., 2009. Human sirt-1: molecular modeling and structure-function relationships of an unordered protein. PLoS One 4 (10), e7350.

Avalos, J.L., Bever, K.M., Wolberger, C., 2005. Mechanism of sirtuin inhibition by nicotinamide: altering the NAD(+) cosubstrate specificity of a Sir2 enzyme. Mol. Cell 17 (6), 855–868.

Avalos, J.L., Boeke, J.D., Wolberger, C., 2004. Structural basis for the mechanism and regulation of Sir2 enzymes. Mol. Cell 13 (5), 639–648.

Babalola, O.E., 2011. Ocular onchocerciasis: current management and future prospects. Clin. Ophthalmol. 5, 1479–1491.

Bao, X., Wang, Y., Li, X., Li, X.M., Liu, Z., Yang, T., Wong, C.F., Zhang, J., Hao, Q., Li, X.D., 2014. Identification of 'erasers' for lysine crotonylated histone marks using a chemical proteomics approach. Elife 3.

Bedalov, A., Gatbonton, T., Irvine, W.P., Gottschling, D.E., Simon, J.A., 2001. Identification of a small molecule inhibitor of Sir2p. Proc. Natl. Acad. Sci. U.S.A. 98 (26), 15113–15118.

Bender, A., Glen, R.C., 2005. A discussion of measures of enrichment in virtual screening: comparing the information content of descriptors with increasing levels of sophistication. J. Chem. Inf. Model 45 (5), 1369–1375.

Borra, M.T., Langer, M.R., Slama, J.T., Denu, J.M., 2004. Substrate specificity and kinetic mechanism of the Sir2 family of NAD^+-dependent histone/protein deacetylases. Biochemistry 43 (30), 9877–9887.

Borra, M.T., Smith, B.C., Denu, J.M., 2005. Mechanism of human SIRT1 activation by resveratrol. J. Biol. Chem. 280 (17), 17187–17195.

Camins, A., Sureda, F.X., Junyent, F., Verdaguer, E., Folch, J., Pelegri, C., Vilaplana, J., Beas-Zarate, C., Pallas, M., 2010. Sirtuin activators: designing molecules to extend life span. Biochim. Biophys. Acta 1799 (10–12), 740–749.

Cavasotto, C.N., Phatak, S.S., 2009. Homology modeling in drug discovery: current trends and applications. Drug Discov. Today 14 (13–14), 676–683.

Clark, M.A., Acharya, R.A., Arico-Muendel, C.C., Belyanskaya, S.L., Benjamin, D.R., Carlson, N.R., Centrella, P.A., Chiu, C.H., Creaser, S.P., Cuozzo, J.W., Davie, C.P., Ding, Y., Franklin, G.J., Franzen, K.D., Gefter, M.L., Hale, S.P., Hansen, N.J.V., Israel, D.I., Jiang, J., Kavarana, M.J., Kelley, M.S., Kollmann, C.S., Li, F., Lind, K., Mataruse, S., Medeiros, P.F., Messer, J.A., Myers, P., O'Keefe, H., Oliff, M.C., Rise, C.E., Satz, A.L., Skinner, S.R., Svendsen, J.L., Tang, L., van Vloten, K., Wagner, R.W., Yao, G., Zhao, B., Morgan, B.A., 2009. Design, synthesis and selection of DNA-encoded small-molecule libraries. Nat. Chem. Biol. 5 (9), 647–654.

Cosgrove, M.S., Bever, K., Avalos, J.L., Muhammad, S., Zhang, X., Wolberger, C., 2006. The structural basis of sirtuin substrate affinity. Biochemistry 45 (24), 7511–7521.

Costantini, S., Sharma, A., Raucci, R., Costantini, M., Autiero, I., Colonna, G., 2013. Genealogy of an ancient protein family: the Sirtuins, a family of disordered members. BMC Evol. Biol. 13, 60.

Cui, H., Kamal, Z., Ai, T., Xu, Y., More, S.S., Wilson, D.J., Chen, L., 2014. Discovery of potent and selective sirtuin 2 (SIRT2) inhibitors using a fragment-based approach. J. Med. Chem. 57 (20), 8340–8357.

Dai, H., Kustigian, L., Carney, D., Case, A., Considine, T., Hubbard, B.P., Perni, R.B., Riera, T.V., Szczepankiewicz, B., Vlasuk, G.P., Stein, R.L., 2010. SIRT1 activation by small molecules: kinetic and biophysical evidence for direct interaction of enzyme and activator. J. Biol. Chem. 285 (43), 32695–32703.

Davenport, A.M., Huber, F.M., Hoelz, A., 2014. Structural and functional analysis of human SIRT1. J. Mol. Biol. 426 (3), 526–541.

de Ruijter, A.J., van Gennip, A.H., Caron, H.N., Kemp, S., van Kuilenburg, A.B., 2003. Histone deacetylases (HDACs): characterization of the classical HDAC family. Biochem. J. 370 (Pt 3), 737–749.

Dhanak, D., 2012. Cracking the code: the promise of epigenetics. ACS Med. Chem. Lett. 3 (7), 521–523.

Disch, J.S., Evindar, G., Chiu, C.H., Blum, C.A., Dai, H., Jin, L., Schuman, E., Lind, K.E., Belyanskaya, S.L., Deng, J., Coppo, F., Aquilani, L., Graybill, T.L., Cuozzo, J.W., Lavu, S., Mao, C., Vlasuk, G.P., Perni, R.B., 2013. Discovery of thieno[3,2-*d*]pyrimidine-6-carboxamides as potent inhibitors of SIRT1, SIRT2, and SIRT3. J. Med. Chem. 56 (9), 3666–3679.

Donmez, G., Outeiro, T.F., 2013. SIRT1 and SIRT2: emerging targets in neurodegeneration. EMBO Mol. Med. 5 (3), 344–352.

Dressel, J., 1961. The discovery of Germanin by Oskar Dressel and Richard Kothe. J. Chem. Educ. 38 (12), 620.

Du, J., Zhou, Y., Su, X., Yu, J.J., Khan, S., Jiang, H., Kim, J., Woo, J., Kim, J.H., Choi, B.H., He, B., Chen, W., Zhang, S., Cerione, R.A., Auwerx, J., Hao, Q., Lin, H., 2011. Sirt5 is a NAD-dependent protein lysine demalonylase and desuccinylase. Science 334 (6057), 806–809.

Fatkins, D.G., Monnot, A.D., Zheng, W., 2006. Nepsilon-thioacetyl-lysine: a multi-facet functional probe for enzymatic protein lysine Nepsilon-deacetylation. Bioorg. Med. Chem. Lett. 16 (14), 3651–3656.

Feldman, J.L., Baeza, J., Denu, J.M., 2013. Activation of the protein deacetylase SIRT6 by long-chain fatty acids and widespread deacylation by mammalian sirtuins. J. Biol. Chem. 288 (43), 31350–31356.

Finnin, M.S., Donigian, J.R., Pavletich, N.P., 2001. Structure of the histone deacetylase SIRT2. Nat. Struct. Biol. 8 (7), 621–625.

Fischer, F., Gertz, M., Suenkel, B., Lakshminarasimhan, M., Schutkowski, M., Steegborn, C., 2012. Sirt5 deacylation activities show differential sensitivities to nicotinamide inhibition. PLoS One 7 (9), e45098.

Gertz, M., Fischer, F., Nguyen, G.T., Lakshminarasimhan, M., Schutkowski, M., Weyand, M., Steegborn, C., 2013. Ex-527 inhibits Sirtuins by exploiting their unique NAD^+-dependent deacetylation mechanism. Proc. Natl. Acad. Sci. U.S.A. 110 (30), E2772–E2781.

Gertz, M., Nguyen, G.T., Fischer, F., Suenkel, B., Schlicker, C., Franzel, B., Tomaschewski, J., Aladini, F., Becker, C., Wolters, D., Steegborn, C., 2012. A molecular mechanism for direct sirtuin activation by resveratrol. PLoS One 7 (11), e49761.

Hawse, W.F., Hoff, K.G., Fatkins, D.G., Daines, A., Zubkova, O.V., Schramm, V.L., Zheng, W., Wolberger, C., 2008. Structural insights into intermediate steps in the Sir2 deacetylation reaction. Structure 16 (9), 1368–1377.

Heltweg, B., Gatbonton, T., Schuler, A.D., Posakony, J., Li, H., Goehle, S., Kollipara, R., Depinho, R.A., Gu, Y., Simon, J.A., Bedalov, A., 2006. Antitumor activity of a small-molecule inhibitor of human silent information regulator 2 enzymes. Cancer Res. 66 (8), 4368–4377.

Heltweg, B., Dequiedt, F., Verdin, E., Jung, M., 2002. Nonisotopic substrate for assaying both human zinc and NAD^+-dependent histone deacetylases. Anal. Biochem. 319, 42–48.

Hevener, K.E., Mehboob, S., Su, P.C., Truong, K., Boci, T., Deng, J., Ghassemi, M., Cook, J.L., Johnson, M.E., 2012. Discovery of a novel and potent class of F. tularensis enoyl-reductase (FabI) inhibitors by molecular shape and electrostatic matching. J. Med. Chem. 55 (1), 268–279.

Hoff, K.G., Avalos, J.L., Sens, K., Wolberger, C., 2006. Insights into the sirtuin mechanism from ternary complexes containing NAD^+ and acetylated peptide. Structure 14 (8), 1231–1240.

Howitz, K.T., Bitterman, K.J., Cohen, H.Y., Lamming, D.W., Lavu, S., Wood, J.G., Zipkin, R.E., Chung, P., Kisielewski, A., Zhang, L.L., Scherer, B., Sinclair, D.A., 2003. Small molecule activators of sirtuins extend Saccharomyces cerevisiae lifespan. Nature 425 (6954), 191–196.

Huber, K., Schemies, J., Uciechowska, U., Wagner, J.M., Rumpf, T., Lewrick, F., Suss, R., Sippl, W., Jung, M., Bracher, F., 2010. Novel 3-arylideneindolin-2-ones as inhibitors of NAD^+-dependent histone deacetylases (sirtuins). J. Med. Chem. 53 (3), 1383–1386.

Huhtiniemi, T., Salo, H.S., Suuronen, T., Poso, A., Salminen, A., Leppanen, J., Jarho, E., Lahtela-Kakkonen, M., 2011. Structure-based design of pseudopeptidic inhibitors for SIRT1 and SIRT2. J. Med. Chem. 54 (19), 6456–6468.

Huhtiniemi, T., Suuronen, T., Lahtela-Kakkonen, M., Bruijn, T., Jaaskelainen, S., Poso, A., Salminen, A., Leppanen, J., Jarho, E., 2010. N(epsilon)-Modified lysine containing inhibitors for SIRT1 and SIRT2. Bioorg. Med. Chem. 18 (15), 5616–5625.

Jenuwein, T., Allis, C.D., 2001. Translating the histone code. Science 293 (5532), 1074–1080.

Jiang, H., Khan, S., Wang, Y., Charron, G., He, B., Sebastian, C., Du, J., Kim, R., Ge, E., Mostoslavsky, R., Hang, H.C., Hao, Q., Lin, H., 2013. SIRT6 regulates TNF-alpha secretion through hydrolysis of long-chain fatty acyl lysine. Nature 496 (7443), 110–113.

Jin, L., Galonek, H., Israelian, K., Choy, W., Morrison, M., Xia, Y., Wang, X., Xu, Y., Yang, Y., Smith, J.J., Hoffmann, E., Carney, D.P., Perni, R.B., Jirousek, M.R., Bemis, J.E., Milne, J.C., Sinclair, D.A., Westphal, C.H., 2009a. Biochemical characterization, localization, and tissue distribution of the longer form of mouse SIRT3. Protein Sci. 18 (3), 514–525.

Jin, L., Wei, W., Jiang, Y., Peng, H., Cai, J., Mao, C., Dai, H., Choy, W., Bemis, J.E., Jirousek, M.R., Milne, J.C., Westphal, C.H., Perni, R.B., 2009b. Crystal structures of human SIRT3 displaying substrate-induced conformational changes. J. Biol. Chem. 284 (36), 24394–24405.

Kaeberlein, M., McDonagh, T., Heltweg, B., Hixon, J., Westman, E.A., Caldwell, S.D., Napper, A., Curtis, R., DiStefano, P.S., Fields, S., Bedalov, A., Kennedy, B.K., 2005. Substrate-specific activation of sirtuins by resveratrol. J. Biol. Chem. 280 (17), 17038–17045.

Karaman, B., Sippl, W., 2015. Docking and binding free energy calculations of sirtuin inhibitors. Eur. J. Med. Chem. 93, 584–598.

Kijima, M., Yoshida, M., Sugita, K., Horinouchi, S., Beppu, T., 1993. Trapoxin, an antitumor cyclic tetrapeptide, is an irreversible inhibitor of mammalian histone deacetylase. J. Biol. Chem. 268 (30), 22429–22435.

Kirkwood, J.G., 1935. Statistical mechanics of fluid mixtures. J. Chem. Phys. 3 (5), 300–313.

Kiviranta, P.H., Suuronen, T., Wallen, E.A., Leppanen, J., Tervonen, J., Kyrylenko, S., Salminen, A., Poso, A., Jarho, E.M., 2009. N(epsilon)-thioacetyl-lysine-containing tri-, tetra-, and pentapeptides as SIRT1 and SIRT2 inhibitors. J. Med. Chem. 52 (7), 2153–2156.

Kokkonen, P., Mellini, P., Nyrhila, O., Rahnasto-Rilla, M., Suuronen, T., Kiviranta, P., Huhtiniemi, T., Poso, A., Jarho, E., Lahtela-Kakkonen, M., 2014. Quantitative insights for the design of substrate-based SIRT1 inhibitors. Eur. J. Pharm. Sci. 59, 12–19.

Lancelot, J., Caby, S., Dubois-Abdesselem, F., Vanderstraete, M., Trolet, J., Oliveira, G., Bracher, F., Jung, M., Pierce, R.J., 2013. Schistosomamansonisirtuins: characterization and potential as chemotherapeutic targets. PLoS Negl. Trop. Dis. 7, e2428.

Liu, Z., Yang, T., Li, X., Peng, T., Hang, H.C., Li, X.D., 2015. Integrative chemical biology approaches for identification and characterization of "Erasers" for fatty-acid-acylated lysine residues within proteins. Angew. Chem. Int. Ed. Engl. 54 (4), 1149–1152.

Lower, M., Geppert, T., Schneider, P., Hoy, B., Wessler, S., Schneider, G., 2011. Inhibitors of Helicobacter pylori protease HtrA found by 'virtual ligand' screening combat bacterial invasion of epithelia. PLoS One 6 (3), e17986.

Marti-Renom, M.A., Stuart, A.C., Fiser, A., Sanchez, R., Melo, F., Sali, A., 2000. Comparative protein structure modeling of genes and genomes. Annu. Rev. Biophys. Biomol. Struct. 29, 291–325.

Maurer, B., Rumpf, T., Scharfe, M., Stolfa, D.A., Schmitt, M.L., He, W., Verdin, E., Sippl, W., Jung, M., 2012. Inhibitors of the NAD(+)-dependent protein desuccinylase and demalonylase Sirt5. ACS Med. Chem. Lett. 3 (12), 1050–1053.

Medda, F., Joseph, T.L., Pirrie, L., Higgins, M., Slawin, A.M.Z., Lain, S., Verma, C., Westwood, N.J., 2011. N1-Benzyl substituted cambinol analogues as isozyme selective inhibitors of the sirtuin family of protein deacetylases. MedChemComm 2 (7), 611–615.

Medda, F., Russell, R.J., Higgins, M., McCarthy, A.R., Campbell, J., Slawin, A.M., Lane, D.P., Lain, S., Westwood, N.J., 2009. Novel cambinol analogs as sirtuin inhibitors: synthesis, biological evaluation, and rationalization of activity. J. Med. Chem. 52 (9), 2673–2682.

Mellini, P., Kokkola, T., Suuronen, T., Salo, H.S., Tolvanen, L., Mai, A., Lahtela-Kakkonen, M., Jarho, E.M., 2013. Screen of pseudopeptidic inhibitors of human sirtuins 1-3: two lead compounds with antiproliferative effects in cancer cells. J. Med. Chem. 56 (17), 6681–6695.

Milne, J.C., Lambert, P.D., Schenk, S., Carney, D.P., Smith, J.J., Gagne, D.J., Jin, L., Boss, O., Perni, R.B., Vu, C.B., Bemis, J.E., Xie, R., Disch, J.S., Ng, P.Y., Nunes, J.J., Lynch, A.V., Yang, H., Galonek, H., Israelian, K., Choy, W., Iffland, A., Lavu, S., Medvedik, O., Sinclair, D.A., Olefsky, J.M., Jirousek, M.R., Elliott, P.J., Westphal, C.H., 2007. Small molecule activators of SIRT1 as therapeutics for the treatment of type 2 diabetes. Nature 450 (7170), 712–716.

Min, J., Landry, J., Sternglanz, R., Xu, R.M., 2001. Crystal structure of a SIR2 homolog-NAD complex. Cell 105 (2), 269–279.

Moitessier, N., Englebienne, P., Lee, D., Lawandi, J., Corbeil, C.R., 2008. Towards the development of universal, fast and highly accurate docking/scoring methods: a long way to go. Br. J. Pharmacol. 153 (Suppl. 1), S7–S26.

Moniot, S., Schutkowski, M., Steegborn, C., 2013. Crystal structure analysis of human Sirt2 and its ADP-ribose complex. J. Struct. Biol. 182 (2), 136–143.

Moniot, S., Weyand, M., Steegborn, C., 2012. Structures, substrates, and regulators of mammalian sirtuins—opportunities and challenges for drug development. Front. Pharmacol. 3, 16.

Mottamal, M., Zheng, S., Huang, T.L., Wang, G., 2015. Histone deacetylase inhibitors in clinical studies as templates for new anticancer agents. Molecules 20 (3), 3898–3941.

Napper, A.D., Hixon, J., McDonagh, T., Keavey, K., Pons, J.F., Barker, J., Yau, W.T., Amouzegh, P., Flegg, A., Hamelin, E., Thomas, R.J., Kates, M., Jones, S., Navia, M.A., Saunders, J.O., DiStefano, P.S., Curtis, R., 2005. Discovery of indoles as potent and selective inhibitors of the deacetylase SIRT1. J. Med. Chem. 48 (25), 8045–8054.

Naviaux, J.C., Schuchbauer, M.A., Li, K., Wang, L., Risbrough, V.B., Powell, S.B., Naviaux, R.K., 2014. Reversal of autism-like behaviors and metabolism in adult mice with single-dose antipurinergic therapy. Transl. Psychiatry 4, e400.

Neugebauer, R.C., Uchiechowska, U., Meier, R., Hruby, H., Valkov, V., Verdin, E., Sippl, W., Jung, M., 2008. Structure-activity studies on splitomicin derivatives as sirtuin inhibitors and computational prediction of binding mode. J. Med. Chem. 51 (5), 1203–1213.

Nguyen, G.T., Gertz, M., Steegborn, C., 2013a. Crystal structures of Sirt3 complexes with 4′-bromo-resveratrol reveal binding sites and inhibition mechanism. Chem. Biol. 20 (11), 1375–1385.

Nguyen, G.T., Schaefer, S., Gertz, M., Weyand, M., Steegborn, C., 2013b. Structures of human sirtuin 3 complexes with ADP-ribose and with carba-NAD^+ and SRT1720: binding details and inhibition mechanism. Acta Crystallogr. D. Biol. Crystallogr. 69 (Pt 8), 1423–1432.

Outeiro, T.F., Kontopoulos, E., Altmann, S.M., Kufareva, I., Strathearn, K.E., Amore, A.M., Volk, C.B., Maxwell, M.M., Rochet, J.C., McLean, P.J., Young, A.B., Abagyan, R., Feany, M.B., Hyman, B.T., Kazantsev, A.G., 2007. Sirtuin 2 inhibitors rescue alpha-synuclein-mediated toxicity in models of Parkinson's disease. Science 317 (5837), 516–519.

Pacholec, M., Bleasdale, J.E., Chrunyk, B., Cunningham, D., Flynn, D., Garofalo, R.S., Griffith, D., Griffor, M., Loulakis, P., Pabst, B., Qiu, X., Stockman, B., Thanabal, V., Varghese, A., Ward, J., Withka, J., Ahn, K., 2010. SRT1720, SRT2183, SRT1460, and resveratrol are not direct activators of SIRT1. J. Biol. Chem. 285 (11), 8340–8351.

Park, J., Chen, Y., Tishkoff, D.X., Peng, C., Tan, M., Dai, L., Xie, Z., Zhang, Y., Zwaans, B.M., Skinner, M.E., Lombard, D.B., Zhao, Y., 2013. SIRT5-mediated lysine desuccinylation impacts diverse metabolic pathways. Mol. Cell 50 (6), 919–930.

Peng, C., Lu, Z., Xie, Z., Cheng, Z., Chen, Y., Tan, M., Luo, H., Zhang, Y., He, W., Yang, K., Zwaans, B.M., Tishkoff, D., Ho, L., Lombard, D., He, T.C., Dai, J., Verdin, E., Ye, Y., Zhao, Y., 2011. The first identification of lysine malonylation substrates and its regulatory enzyme. Mol. Cell Proteomics 10 (12) M111.012658.

Pirola, L., Frojdo, S., 2008. Resveratrol: one molecule, many targets. IUBMB Life 60 (5), 323–332.

Pulla, V.K., Alvala, M., Sriram, D.S., Viswanadha, S., Sriram, D., Yogeeswari, P., 2014. Structure-based drug design of small molecule SIRT1 modulators to treat cancer and metabolic disorders. J. Mol. Graph Model 52, 46–56.

Roessler, C., Nowak, T., Pannek, M., Gertz, M., Nguyen, G.T., Scharfe, M., Born, I., Sippl, W., Steegborn, C., Schutkowski, M., 2014. Chemical probing of the human sirtuin 5 active site reveals its substrate acyl specificity and peptide-based inhibitors. Angew. Chem. Int. Ed. Engl. 53 (40), 10728–10732.

Rumpf, T., Schiedel, M., Karaman, B., Roessler, C., North, B.J., Lehotzky, A., Olah, J., Ladwein, K.I., Schmidtkunz, K., Gajer, M., Pannek, M., Steegborn, C., Sinclair, D.A., Gerhardt, S., Ovadi, J., Schutkowski, M., Sippl, W., Einsle, O., Jung, M., 2015. Selective Sirt2 inhibition by ligand-induced rearrangement of the active site. Nat. Commun. 6, 6263.

Sauve, A.A., 2010. Sirtuin chemical mechanisms. Biochim. Biophys. Acta 1804 (8), 1591–1603.

Sauve, A.A., Youn, D.Y., 2012. Sirtuins: NAD(+)-dependent deacetylase mechanism and regulation. Curr. Opin. Chem. Biol. 16 (5–6), 535–543.

Schuetz, A., Min, J., Antoshenko, T., Wang, C.L., Allali-Hassani, A., Dong, A., Loppnau, P., Vedadi, M., Bochkarev, A., Sternglanz, R., Plotnikov, A.N., 2007. Structural basis of inhibition of the human NAD+-dependent deacetylase SIRT5 by suramin. Structure 15 (3), 377–389.

Solomon, J.M., Pasupuleti, R., Xu, L., McDonagh, T., Curtis, R., DiStefano, P.S., Huber, L.J., 2006. Inhibition of SIRT1 catalytic activity increases p53 acetylation but does not alter cell survival following DNA damage. Mol. Cell Biol. 26 (1), 28–38.

Srinivasan, J., Cheatham, T.E., Cieplak, P., Kollman, P.A., Case, D.A., 1998. Continuum solvent studies of the stability of DNA, RNA, and phosphoramidate–DNA helices. J. Am. Chem. Soc. 120 (37), 9401–9409.

Steverding, D., 2010. The development of drugs for treatment of sleeping sickness: a historical review. Parasit. Vectors 3 (1), 15.

Straatsma, T.P., McCammon, J.A., 1991. Multiconfiguration thermodynamic integration. J. Chem. Phys. 95 (2), 1175–1188.

Sun, D., Chen, Z.W., Mathews, F.S., Davidson, V.L., 2002. Mutation of alphaPhe55 of methylamine dehydrogenase alters the reorganization energy and electronic coupling for its electron transfer reaction with amicyanin. Biochemistry 41 (47), 13926–13933.

Szczepankiewicz, B.G., Dai, H., Koppetsch, K.J., Qian, D., Jiang, F., Mao, C., Perni, R.B., 2012. Synthesis of carba-NAD and the structures of its ternary complexes with SIRT3 and SIRT5. J. Org. Chem. 77 (17), 7319–7329.

Tan, L., Batista, J., Bajorath, J., 2010. Computational methodologies for compound database searching that utilize experimental protein-ligand interaction information. Chem. Biol. Drug Des. 76 (3), 191–200.

Taunton, J., Hassig, C.A., Schreiber, S.L., 1996. A mammalian histone deacetylase related to the yeast transcriptional regulator Rpd3p. Science 272 (5260), 408–411.

Taylor, D.M., Balabadra, U., Xiang, Z., Woodman, B., Meade, S., Amore, A., Maxwell, M.M., Reeves, S., Bates, G.P., Luthi-Carter, R., Lowden, P.A., Kazantsev, A.G., 2011. A brain-permeable small molecule reduces neuronal cholesterol by inhibiting activity of sirtuin 2 deacetylase. ACS Chem. Biol. 6 (6), 540–546.

Teng, Y.-B., Jing, H., Aramsangtienchai, P., He, B., Khan, S., Hu, J., Lin, H., Hao, Q., 2015. Efficient demyristoylase activity of SIRT2 revealed by kinetic and structural studies. Sci. Rep. 5.

Thiagalingam, S., Cheng, K.H., Lee, H.J., Mineva, N., Thiagalingam, A., Ponte, J.F., 2003. Histone deacetylases: unique players in shaping the epigenetic histone code. Ann. N.Y. Acad. Sci. 983, 84–100.

Trapp, J., Meier, R., Hongwiset, D., Kassack, M.U., Sippl, W., Jung, M., 2007. Structure-activity studies on suramin analogues as inhibitors of NAD+-dependent histone deacetylases (sirtuins). ChemMedChem 2 (10), 1419–1431.

Tresadern, G., Bemporad, D., Howe, T., 2009. A comparison of ligand based virtual screening methods and application to corticotropin releasing factor 1 receptor. J. Mol. Graph Model 27 (8), 860–870.

Uciechowska, U., Schemies, J., Neugebauer, R.C., Huda, E.M., Schmitt, M.L., Meier, R., Verdin, E., Jung, M., Sippl, W., 2008. Thiobarbiturates as sirtuin inhibitors: virtual screening, free-energy calculations, and biological testing. ChemMedChem 3 (12), 1965–1976.

Voogd, T.E., Vansterkenburg, E.L., Wilting, J., Janssen, L.H., 1993. Recent research on the biological activity of suramin. Pharmacol. Rev. 45 (2), 177–203.

Warren, G.L., Andrews, C.W., Capelli, A.M., Clarke, B., LaLonde, J., Lambert, M.H., Lindvall, M., Nevins, N., Semus, S.F., Senger, S., Tedesco, G., Wall, I.D., Woolven, J.M., Peishoff, C.E., Head, M.S., 2006. A critical assessment of docking programs and scoring functions. J. Med. Chem. 49 (20), 5912–5931.

Wu, J., Zhang, D., Chen, L., Li, J., Wang, J., Ning, C., Yu, N., Zhao, F., Chen, D., Chen, X., Chen, K., Jiang, H., Liu, H., Liu, D., 2013. Discovery and mechanism study of SIRT1 activators that promote the deacetylation of fluorophore-labeled substrate. J. Med. Chem. 56 (3), 761–780.

Yamagata, K., Goto, Y., Nishimasu, H., Morimoto, J., Ishitani, R., Dohmae, N., Takeda, N., Nagai, R., Komuro, I., Suga, H., Nureki, O., 2014. Structural basis for potent inhibition of SIRT2 deacetylase by a macrocyclic peptide inducing dynamic structural change. Structure 22 (2), 345–352.

Yuriev, E., Agostino, M., Ramsland, P.A., 2011. Challenges and advances in computational docking: 2009 in review. J. Mol. Recognit. 24 (2), 149–164.

Zhang, F., Wang, S., Gan, L., Vosler, P.S., Gao, Y., Zigmond, M.J., Chen, J., 2011. Protective effects and mechanisms of sirtuins in the nervous system. Prog. Neurobiol. 95 (3), 373–395.

Zhao, K., Chai, X., Marmorstein, R., 2003. Structure of the yeast Hst2 protein deacetylase in ternary complex with 2′-O-acetyl ADP ribose and histone peptide. Structure 11 (11), 1403–1411.

Zhao, K., Harshaw, R., Chai, X., Marmorstein, R., 2004. Structural basis for nicotinamide cleavage and ADP-ribose transfer by NAD(+)-dependent Sir2 histone/protein deacetylases. Proc. Natl. Acad. Sci. U.S.A. 101 (23), 8563–8568.

Zhao, X., Allison, D., Condon, B., Zhang, F., Gheyi, T., Zhang, A., Ashok, S., Russell, M., MacEwan, I., Qian, Y., Jamison, J.A., Luz, J.G., 2013. The 2.5 A crystal structure of the SIRT1 catalytic domain bound to nicotinamide adenine dinucleotide (NAD^+) and an indole (EX527 analogue) reveals a novel mechanism of histone deacetylase inhibition. J. Med. Chem. 56 (3), 963–969.

Zhou, Y., Zhang, H., He, B., Du, J., Lin, H., Cerione, R.A., Hao, Q., 2012. The bicyclic intermediate structure provides insights into the desuccinylation mechanism of human sirtuin 5 (SIRT5). J. Biol. Chem. 287 (34), 28307–28314.

CHAPTER 12

Drug Repurposing for Epigenetic Targets Guided by Computational Methods

J. Jesús Naveja[1,3], Alfonso Dueñas-González[2], José L. Medina-Franco[3]
[1]Facultad de Medicina, PECEM, Universidad Nacional Autónoma de México, Mexico City, México; [2]Unidad de Investigación Biomédica en Cáncer, Instituto de Investigaciones Biomédicas, Universidad Nacional Autónoma de México and Instituto Nacional de Cancerología, Mexico City, México; [3]Facultad de Química, Departamento de Farmacia, Universidad Nacional Autónoma de México, Mexico City, México

1. INTRODUCTION

1.1 Basis of Drug Repurposing

Drug repurposing (or drug repositioning) is the process of finding a new medical indication for a drug. The drug might be currently approved for another use, withdrawn because of adverse effects, or not accepted for failing to prove efficacy (Ashburn and Thor, 2004; Turner et al., 2011). Although many cases of drug repurposing emerged from serendipity by discovering beneficial side effects of drugs for patients in a regime, numerous efforts are arising to perform drug repurposing deliberately with more systematic approaches.

Known facts that make drug repurposing possible are the polypharmacology of small molecules (that allows a drug to have an activity on more than one target) and the connections among metabolic pathways (so that changes in the activity in a given target can affect indirectly the overall activity of other targets) (Ma et al., 2013).

Polypharmacology of small molecules can not only cause adverse effects but also lead to drug repurposing. Size, polarity, and concentration are major features upon which a drug's promiscuity depends (Medina-Franco et al., 2013b; Merino et al., 2010). Polypharmacology can be largely exploited, since a drug hits on average six to 13 targets (Vogt and Mestres, 2010). Nonetheless, very promiscuous drugs may have a considerable number of undesired effects. Therefore, many strategies to predict polypharmacology have been developed (Medina-Franco et al., 2013b; Hopkins, 2009; Merino et al., 2010; Vogt and Mestres, 2010). Also, metabolic pathways can be exploited to repurpose drugs given that many diseases involve multiple genes, whose products may be involved in multiple pathways (Issa et al., 2013).

CONTENTS

Epi-Informatics. http://dx.doi.org/10.1016/B978-0-12-802808-7.00012-5

1.2 Benefits of Drug Repurposing versus Traditional Drug Discovery

Despite the large investments and efforts performed so far for developing new drugs (an average of $1.78 billion dollars and 13.5 years from discovery to commercialization), only 20 to 30 of them are approved for their clinical use each year in the United States. This represents only 11% of drugs that enter to clinical trials (Li and Jones, 2012). This is true for the traditional approach of one drug–one indication.

Drug repurposing is a promising strategy in pharmaceutics, since it reduces considerably the resources needed for developing any given therapy for a disease and magnifies the probability that the drug enters from Phase I to market for the new indication (Temesi et al., 2014). Indeed, taking this into consideration, along with the fact that drug repurposing skips the need for initial target discovery, lead optimization, preclinical development, and early clinical safety testing, approximately 50% of the cost may be reduced to get a drug approved for a new indication (Paul and Lewis-Hall, 2013). As a consequence, early repositioning is emerging for developing novel therapies as a result of more efficiently connecting compounds that are safe and have an acceptable bioavailability with plausible targets. This has favorable economical effects by enhancing (1) a wider use of a drug that has been already approved, and (2) the finding of possible targets for molecules that have evidence of safety and bioavailability but failed at demonstrating efficacy treating the disease for which it was developed (Temesi et al., 2014). The US National Institutes of Health (NIH) has started to release funding for projects focused on drug repurposing with different aims, ranging from finding drugs to achieve smoking cessation, to drugs capable of slowing Alzheimer's disease (AD) progression. Similar cases had also appeared in the United Kingdom (Wadman, 2012, 2013; Nair, 2013). However, the renewal of this support to repurposed drugs depends on the results of these pioneer works (Wadman, 2012).

The effect of drug repurposing goes beyond the reduction on the resources needs, provided that it is also a tool capable of finding treatments for rare and neglected diseases, for which it is not economically plausible to develop a treatment (Ekins et al., 2011; Muthyala, 2011; Sardana et al., 2011).

1.3 Representative Examples of Successful Drug Repurposing Campaigns

Drug repurposing has already found several successful cases (Dobrosotskaya et al., 2011; Mukhopadhyay et al., 2002; Nygren and Larsson, 2014; Pantziarka et al., 2014). Representative examples are summarized in Table 1. Sildenafil is a primary example of a repurposed drug. Sildenafil was developed as a treatment for angina pectoris, but clinical trials could not prove its efficacy. However, the side

Table 1 Examples of Drugs that Underwent or Are Being Repurposed and Their Characteristic Features

Drug	Original Indication	New Indication	Repurposing Approach	Current Development Stage	Efficacy	Toxicity	References
Mebendazole	Helminthes infection	Cancer, especially MACC[a] and RMCC[b]	Literature-based	Phases I and II	Proved in vitro and in two patients	Low; adverse effects are rare	Pantziarka et al. (2014); Mukhopadhyay et al. (2002); Nygren and Larsson (2014); Dobrosotskaya et al. (2011)
Sildenafil	Intended for angina (lack of efficacy)	ED[c], PF[d], PAH[e]	ED: serendipity; PF and PAH literature-based	Approved (Phase IV)	ED: 82% efficacy; PF and PAH: Improves clinical outcomes (reduces vascular resistance index in 32.5%)	Uncommon, mild, and transient adverse effects	Ghofrani et al. (2002); Ashburn and Thor (2004); Boolell et al. (1996); Corbin et al. (2005)
Thalidomide	Discontinued pregnancy anti-emetic (teratogenesis)	MM[f] and ENL[g]	ENL: serendipity; MM: literature-based	Phase IV (approved)	ENL: 92–99% efficacy; MM: 32% total rate response	Low (when not used during pregnancy)	Ashburn and Thor (2004); Gupta et al. (2013); Singhal et al. (1999); Zhou et al. (2013)
Ketoconazole	Antifungal	Cushing's syndrome	Finding of adrenal inhibition: serendipity; Application: literature-based	Approved (Phase IV)	50% efficacy in ectopic ACTH secretion and nearly 100% in adrenal carcinoma	Common, mild adverse effects; rarely, liver dysfunction	Pont et al. (1982); Tritos and Biller (2012)
Lonafarnib	Intended for cancer (lack of efficacy)	Hutchinson–Gilford progeria syndrome	Literature-based	Phase II clinical trials completed	Improves clinical outcomes (36% for weight gain; 76% increment in bone density; 44% low frequency hearing improvement)	Common mild adverse effects; can cause liver toxicity	Wadman (2012); Capell et al. (2005); Gordon et al. (2012)
Itraconazole	Antifungal	Angiogenesis inhibition in BCC[h] and SCLC[i]	HTS[j] with 2604 FDA-approved drugs	Phase II clinical trials completed	Reduced 45% cell proliferation and 24% tumor area	Moderate; can cause fatigue and congestive heart failure	Ekins et al. (2011); Chong et al. (2007); Aftab et al. (2011); Rudin et al. (2013); Kim et al. (2014)

Continued...

Table 1 Examples of Drugs that Underwent or Are Being Repurposed and Their Characteristic Features *Continued*

Drug	Original Indication	New Indication	Repurposing Approach	Current Development Stage	Efficacy	Toxicity	References
Saracatinib	Intended for cancer treatment (lack of efficacy)	Alzheimer	Literature-based	Currently in Phase Ib clinical trials	Not reported	Adverse effects are common and go from fatigue to adrenal insufficiency or hypoxia	Gucalp et al. (2011); Nygaard et al. (2014)
Acrisorcine	Antifungal	Antimalarial	HTS of 4000 compounds against the *Plasmodium falciparum* protein Hsp90, followed by cellular study	Preclinical studies	IC_{50} in nanomolar range for both a susceptible strain and MDR[k] strain (0.0513±0.0237 and 0.0544±0.0071, respectively)	Adverse effects are rare	Ekins et al. (2011); Ramdhave et al. (2013); Njoroge et al. (2014)

[a]*MACC: metastatic adrenocortical carcinoma.*
[b]*RMCC: refractory metastatic colon cancer.*
[c]*ED: erectile dysfunction.*
[d]*PF: pulmonary fibrosis.*
[e]*PAH: pulmonary arterial hypertension.*
[f]*MM: multiple myeloma.*
[g]*ENL: erythema nodosum leprosum.*
[h]*BCC: basal cell carcinoma.*
[i]*SCLC: small-cell lung cancer.*
[j]*HTS: high-throughput screening.*
[k]*MDR: multidrug-resistant.*

effects of the drug—unusually strong and maintained erections—did not pass unnoticed. The drug was approved by the FDA in 1998 for the treatment of erectile dysfunction, since in clinical trials it demonstrated to exert a positive effect in more than 80% of the patients (Boolell et al., 1996; Ashburn and Thor, 2004). It was then found that the target of sildenafil, phosphodiesterase type 5 (PDE5), was not only highly expressed in the penis, but also in lungs, making it a good prospect for treating pulmonary arterial hypertension (Corbin et al., 2005). However, clinical trials with sildenafil for pulmonary fibrosis and hypertension, and even its FDA approval, were performed before this fact was known (Ghofrani et al., 2002; Corbin et al., 2005). Thus, sildenafil was repositioned twice.

There are many other examples of drug repositioning by serendipity, like thalidomide, a teratogenic drug that is effective to treat erythema nodosum leprosum (Singhal et al., 1999; Gupta et al., 2013; Zhou et al., 2013), a very painful skin manifestation of leprosy for which this drug is the only cure; or antifungals, which inhibit steroidogenesis and can be used for treating diseases with excessive steroid production (Pont et al., 1982; Tritos and Biller, 2012). There is also a great interest in finding drugs against aggressive types of cancer through repurposing (Chong et al., 2007; Aftab et al., 2011; Rudin et al., 2013; Kim et al., 2014).

As stated before, neglected and orphan diseases can also benefit from drug repurposing. Indeed, a remarkable example of drug repositioning was for progeria, a rare disease with a very bad prognosis and no cure to date (Wadman, 2012). It was found that the progerin (a mutant of laminin A protein) accumulates in the nuclear membrane and leads to morphological abnormalities of the nucleus in interphase due to keeping its farnesyl group (Wu et al., 2014). Hence, the inhibition of farnesylation has become a pharmacological target, and Phase I studies with lonafarnib have demonstrated improvement in clinical outcomes for patients (Capell et al., 2005; Gordon et al., 2012). Many more examples are emerging for developing therapies for rare or neglected diseases through drug repurposing, and this includes potential treatments for Chagas' disease, trypanosomiasis, leishmaniasis, malaria, and many more (Ekins et al., 2011; Ramdhave et al., 2013; Njoroge et al., 2014). Computational (also called in silico) approaches are becoming an important part of this process (Ekins et al., 2011; Lotharius et al., 2014).

Although the repurposing of sildenafil to pulmonary arterial hypertension was not initially based on the specific knowledge that PDE5 is concentrated in lung tissue, it was performed as an application of the known facts of the pathogenesis and the drug's effect in PDE5. This general method is called literature-based approach, which is illustrated in Table 1. The literature-based approach is, undoubtedly, a more comprehensive way of performing drug repurposing than mere serendipity. In the next section, we will describe the pros and cons of the different strategies that aim for drug repurposing, from serendipity to more systematic approaches based on the literature, to the most systematic computational strategies that can be conveniently integrated with traditional data mining strategies.

2. DRUG REPURPOSING STRATEGIES

2.1 Noncomputational Approaches

As discussed earlier, most of the current repurposed drugs emerged from serendipity. This is why it is imperative to keep a record of the side effects of a drug, whether they are adverse or not, since that information can be helpful either for identifying a drug's toxicity or for determining the potential use of the drug for a new indication. That is how serendipity works.

Given the benefit obtained from the first repurposed drugs, more systematized approaches have been developed to obtain the maximum number of indications for a drug. Knowing the mechanism of action for a drug, as well as the physiopathology of many diseases, could help to simply speculate about whether the drug will have more than one possible indication. Inversely, unraveling the physiopathology of a new disease while knowing the mechanism of action of many drugs will lead to just associate a drug that could counteract the physiopathology. These approaches are clustered in the group of literature-based discovery strategies and some examples of drugs that fall into this category are summarized in Table 1.

In vitro screening of compounds was among the first approaches intended to systematize drug discovery. The evolution of this screening goes from low-throughput (analyzing a set consisting of a few compounds), to ultra-high-throughput screening (that can yield results for up to 100,000 compounds a day) (Hertzberg and Pope, 2000; Sundberg, 2000). Probably, the main limitation of this methodology is the high costs of the required robotics equipment and the rest of infrastructure, as well as the number of false positives. Nonetheless, systematic experimental screening is a valid and attractive manner of conducting drug repurposing, since even the whole cellular system may be tested by performing in vitro assays with live cells (Chong et al., 2007; Sundberg, 2000).

While literature-based discovery may become a "data archeology" endeavor (in particular if the data is hidden or not very well organized), computational approaches are playing a key role to transform data archeology into a systematic data mining approach (Deftereos et al., 2011). Figure 1 shows drug repurposing strategies discussed throughout this chapter. In the following paragraphs, data mining and other computational methods for systematic drug repurposing will be discussed.

2.2 Computational Methods

Even though there are many recent and integrating computational approaches, it is essential to highlight the most successful or cost-efficient, since a detailed

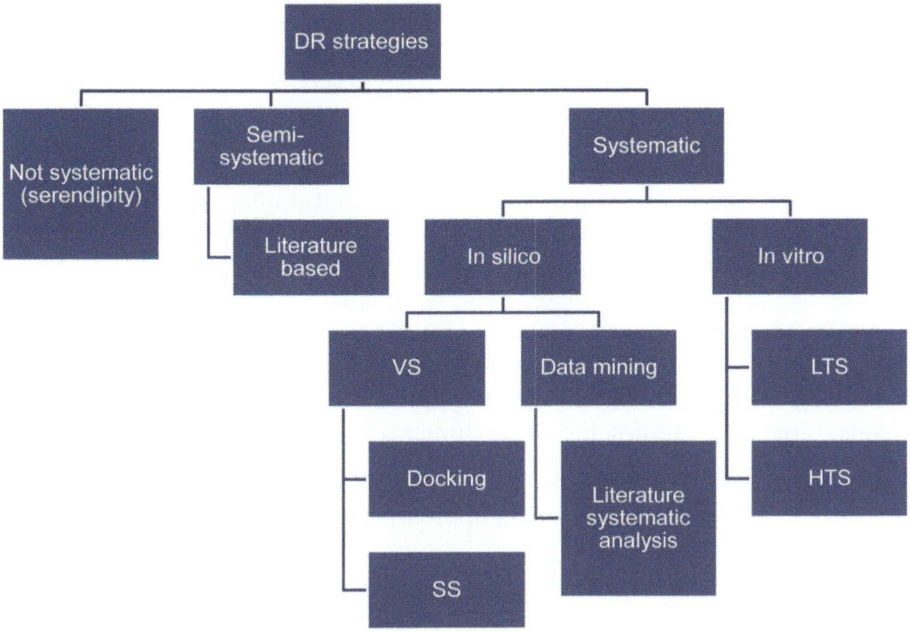

FIGURE 1

Schematic view of the most commonly used drug repurposing strategies. Abbreviations: DR: drug repurposing; VS: virtual screening; SS: similarity searching; LTS: low-throughput screening; HTS: high-throughput screening.

description of the methods regarding drug repurposing out of the field of epigenetics goes beyond the scope of this chapter.

Among the first, but still widely used, approaches to conduct drug repurposing is in silico virtual screening (VS). Virtual screening has two major approaches: structure- and ligand-based. Automated molecular docking and similarity searching (SS) are representative examples of each approach, respectively. Virtual screening by docking involves the scanning of chemical compounds (usually obtained from chemical databases), which are evaluated for the probability of binding to a molecular target using a scoring function that determines the ligand affinity to the target (Cheng et al., 2012; Danishuddin and Khan, 2015; Wang et al., 2012). For drug repurposing applications, the chemical databases can be collections of approved drugs such as DrugBank (Wishart et al., 2008) or compounds in the Therapeutic Target Database (Qin et al., 2014). Rather than being a firmly settled strategy, molecular docking is still an active area of research. The most recent approaches to conduct docking with machine learning are relegating the past ones in the sense that now the nonlinearity of noncovalent bonds is taken into account, and the fact that they are not strictly additive (Kinnings et al., 2011). Nonetheless, controversy in this field is growing, since it has also been found that a more precise chemical description of the protein–ligand complexes does not necessarily lead to a more accurate prediction of binding affinities, and there is not a concluding answer for why this happens (Ballester et al., 2014).

Similarity searching is an alternative manner of conducting VS. This approach emerges from the similarity principle, assuming that similar ligands would exert comparable effects on targets (Ma et al., 2013). In drug repurposing the basic assumption is that a drug can be repurposed if it is similar to one with a proved activity against a target. Similarity searching is closely related to the concept of structure–activity relationships (Medina-Franco et al., 2009, 2013a; Medina-Franco, 2012; Medina-Franco and Aguayo-Ortiz, 2013; Swamidass, 2011) and can be expanded and can find direct applications to identify compounds with similar bioactivity profiles, including similar side effects (Lounkine et al., 2012).

Computational methods emerged as a strategy to systematize drug discovery, and are strongly based on the one-target approach (Reaume, 2011). Nevertheless, as discussed in the section "Introduction," the application of these approaches need to be taken with caution as many of them seem to be too reductionist (Medina-Franco et al., 2013b; Pujol et al., 2010; Reaume, 2011). For this to be taken into account, numerous strategies are either implementing a data mining approach to establish new targets, to drugs from coupling drug networks, to disease networks extracted from the literature (Azmi and Mohammad, 2014; Carrella et al., 2014; Emig et al., 2013; Frijters et al., 2010; Hu and Bajorath, 2013; Kinnings et al., 2010; Klipp et al., 2010; Larocque et al., 2014; Loging et al., 2011; Medina-Franco and Aguayo-Ortiz, 2013; Pujol et al., 2010; Vogt and Mestres, 2010; Wu et al., 2013a,b; Zanzoni et al., 2009), or turning to phenotypic (although highly analytical) studies through connectivity maps, which can also (but not necessarily) incorporate gene expression data (Chung et al., 2014; Do et al., 2010; Li et al., 2009; Liu et al., 2013). There are also networks' analysis trying to characterize metabolic pathways in order to identify drugs acting not only on a single target, but on a whole network of proteins (Chang et al., 2010; Duarte et al., 2007; Knight et al., 2010; Raman and Chandra, 2008; Schneider and Klabunde, 2013; Xie et al., 2011). Interestingly, some data mining approaches that perform high-throughput literature analysis can unravel previously unknown niches for drug repurposing by associating a drug and a disease only based on the analysis of the frequency of concurrence of specific terms in the scientific texts (Choi et al., 2013; Deftereos et al., 2011; Eichborn et al., 2011; Lekka et al., 2011; Persidis et al., 2004). As can be seen so far, informatics allows multiple manners for conducting drug repurposing. However, there are key steps that should be taken into account while developing such strategies, in order to create more rationalized approaches: (1) consider whether the method is centered on the drug (finding new indications) or on a disease (e.g., repurposing drugs for neglected diseases); (2) choosing a strategy that can be either computational or not; and (3) calibrating the tool with known examples of repurposed drugs. This entire process is depicted in Figure 2.

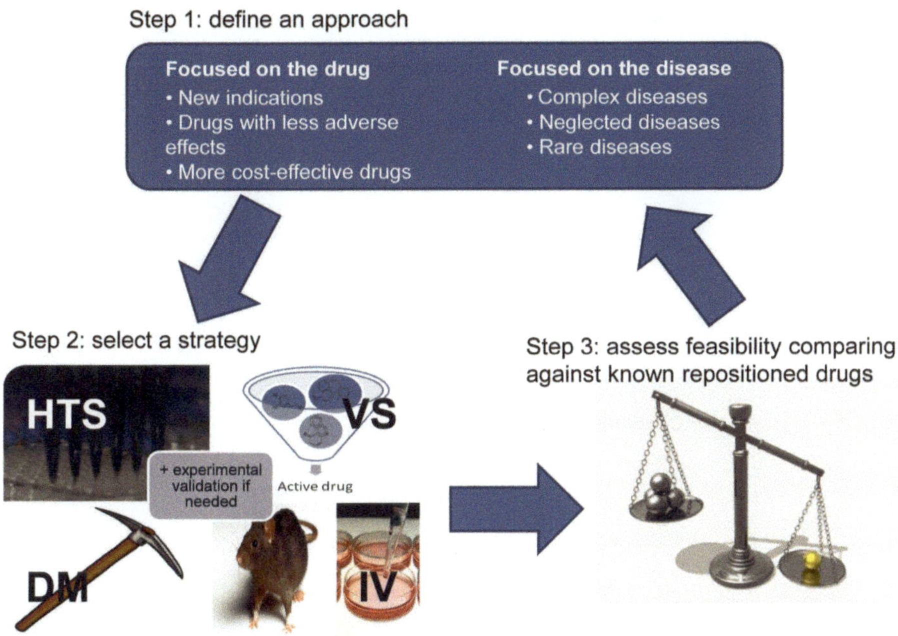

FIGURE 2

Steps for a rationalized drug repurposing. Abbreviations: HTS: high-throughput screening; DM: data mining; VS: virtual screening; IV: in vivo/in vitro methods. *Adapted from Liu et al. (2013).*

3. DRUG REPURPOSING IN EPIGENETICS

3.1 It Works! Examples of Approved Drugs as Ligands of Epigenetic Targets

Epigenetic alterations are being studied as a hallmark of many diseases, for example, cancer, inflammatory disease, obesity, cardiovascular disorders, mental disorders, and dementia (Cacabelos, 2014). Thus, it is expected that drugs with epigenetic effects could represent a novel therapy for many clinical disorders. Interestingly, the current FDA-approved epigenetic therapies are examples of repurposed drugs, and it is estimated that 5% of these drugs have an epigenetic effect (Lötsch et al., 2013). Figure 3 depicts the chemical structure of representative examples of compounds with an effect on epigenetic targets, either currently FDA-approved or under clinical development.

The pioneer drug in the field is 5-azacytidine (Figure 3). It was first proposed as an anticancer drug because of the antimetabolic effects when tested in cancer cells in vitro. Although the drug was also found to be an effective chemotherapeutic for acute myeloid leukemia, the toxicity displayed by 5-azacytidine eventually made other therapies preferable (Christman, 2002). However, the finding that 5-azacytidine and 5-aza-2′-deoxycytidine (decitabine) inhibited

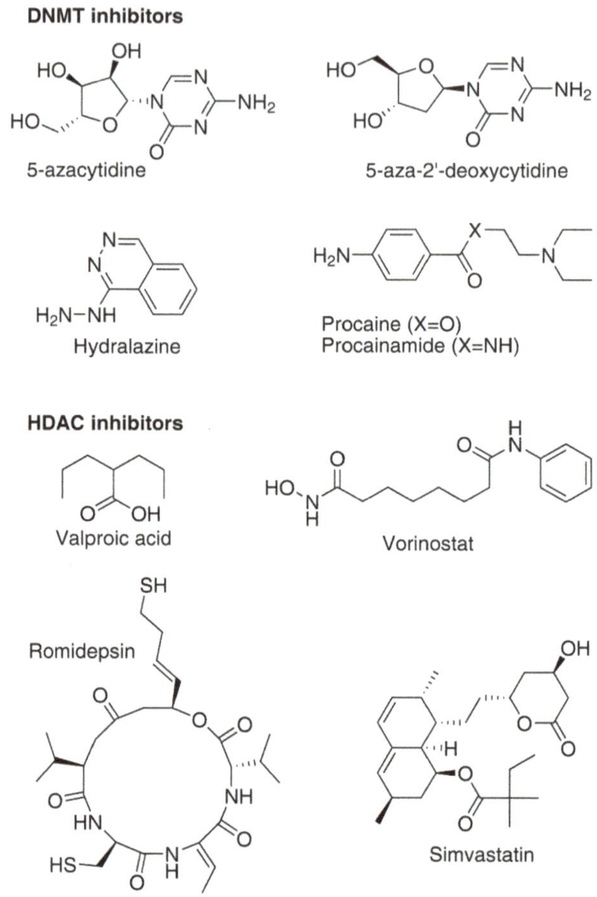

FIGURE 3
Representative molecules with epigenetic activity.

DNA methylation after its incorporation into DNA strands in vivo led to new uses for these drugs in either cancers with a key epigenetic silencing or myelodysplastic syndromes (Christman, 2002; Fenaux et al., 2009; Kantarjian et al., 2006; Mummaneni and Shord, 2014). Hydralazine, formerly an antihypertensive, is a repurposed drug that affects DNA methylation, and it has been found to be a cost-effective and nontoxic therapy against advanced cervical cancer and myelodisplastic syndromes when combined with magnesium valproate (a second repurposed drug, vide infra) (Candelaria et al., 2010; Soto et al., 2014; Dueñas-Gonzalez et al., 2014). The combination of hydralazine and magnesium valproate has also been shown to reverse imatinib resistance in patients with chronic myeloid leukemia (Cervera et al., 2012). Procaine and procainamide (Figure 3) are also notable examples of repurposed drugs in epigenetics. It has been demonstrated that procaine, a local anesthetic, and procainamide, an antiarrhytmic drug, may inhibit DNA methylation, causing reactivation of silenced genes such as RARβ, GSTP1 and WIF-1, whose promoter is commonly hypermethylated in lung cancer cells (Jablons, 2009). These data correlate well with the hypomethylated DNA found in hepatoma cells treated with procaine and then control cells (Tada et al., 2007). Therefore, procaine and procainamide are candidates that are worth investigation for treating hepatocellular carcinoma, a common malignancy associated with a high mortality (Anestopoulos et al., 2015), as well as for other cancers.

Another very important group of pharmacological agents with epigenetic effects are HDACi (histone deacetylase inhibitors). HDACi are capable of generating a pleiotropic effect: they induce apoptosis, arrest the cell cycle, immunomodulate and inhibit angiogenesis and metastasis (Ma et al., 2012; Sigalotti et al., 2007). Therefore, these are drugs with an increasing use among the chemotherapeutic agents. For example, drugs like vorinostat (Figure 3) are currently in several clinical trials against different types of cancer (Wagner et al., 2010; Nebbioso et al., 2012). Valproic acid (Figure 3), an anticonvulsant, was found to be a HDACi, and is a representative example of a successful repurposed drug in epigenetics, since it is now known as an active treatment of resistant-to-treatment myelodysplastic syndrome, multiple myeloma, and solid malignancies (Wagner et al., 2010). Indeed, other drugs, some of them FDA-approved and without an apparent relationship with epigenetics, have also been found to exert effects in either methylation or acetylation. Statins, compounds used in the chemotherapy of hypercholesterolemia, have been shown to inhibit histone deacetylation (Lin et al., 2008). A study by Nowakowski et al. (2010) found that statins may produce a benefit in patients with follicular Hodgkins lymphoma. Moreover, a clinical study testing the combination of gefitinib and simvastatin against gefitinib alone suggested that patients with advanced non–small-cell lung cancer and resistance to gefitinib may benefit from the combination (Han et al., 2011). These effects of statins on cancer may be linked to the previously mentioned fact that they affect histone deacetylation (Barker and Yahalom, 2010).

On the other hand, epigenetic changes are not exclusive of cancer, but they can be found in many diseases, such as asthma, type 2 diabetes, metabolic disorders, autoimmune disease, and brain disorders, including Huntington's disease, autism, and AD (Szyf, 2009; Cacabelos, 2014; Caraci et al., 2013). As mentioned by Cacabelos and Torrellas (2014), drugs with epigenetic targets can be a therapy for AD. This assumption is based on studies that correlate neuropathological hallmarks of the disease (i.e., senile plaques, neurofibrillary tangles) with alteration in DNA methylation, histone modifications, chromatin remodeling, and noncoding RNA dysregulation. The polygenic and multifactorial nature of AD had made it difficult to find an effective therapy. Nevertheless, there are no clinical trials yet for testing epigenetic drugs in the context of AD (Cacabelos and Torrellas, 2014).

Interestingly, it has been speculated that some of the persistent adverse effects of drugs may be due to epigenetic modifications (Csoka and Szyf, 2009). Identifying drugs causing this type of adverse effect may lead to characterizing more drugs with epigenetic effects that would be further studied, looking forward to their repositioning.

3.2 Current Efforts to Reposition Drugs Using Computational Approaches

As summarized by Knapp and Weinmann (Knapp and Weinmann, 2013), increasing attention is being held on finding drugs that exert an effect on epigenetic targets. HTS is the most common methodology among the current studies in the field, and the majority of them are aiming toward inhibition of bromodomains (particularly BRD4), histone methyltransferases (HMTs), histone deacetylases (HDACs), and histone demethylases (HDMT). Although HTS has exhibited, in general, a very low hit rate, there are only a few research groups using different approaches, like computational methods, aiming to achieve a more efficient search of active compounds against epigenetic factors. Though, as depicted in Figure 4(a) with data from Scopus® (as per February 2015), a low but rapidly increasing number of papers about drug repurposing has been published during the last seven years. Figure 4(b) illustrates the proportion of these publications that are original articles, review articles, book chapters, and notes. The next subsection of this chapter is intended to introduce a few (since there are not many) concrete examples of drugs that are being studied with computational methods for their repurposing as modifiers of epigenetic targets.

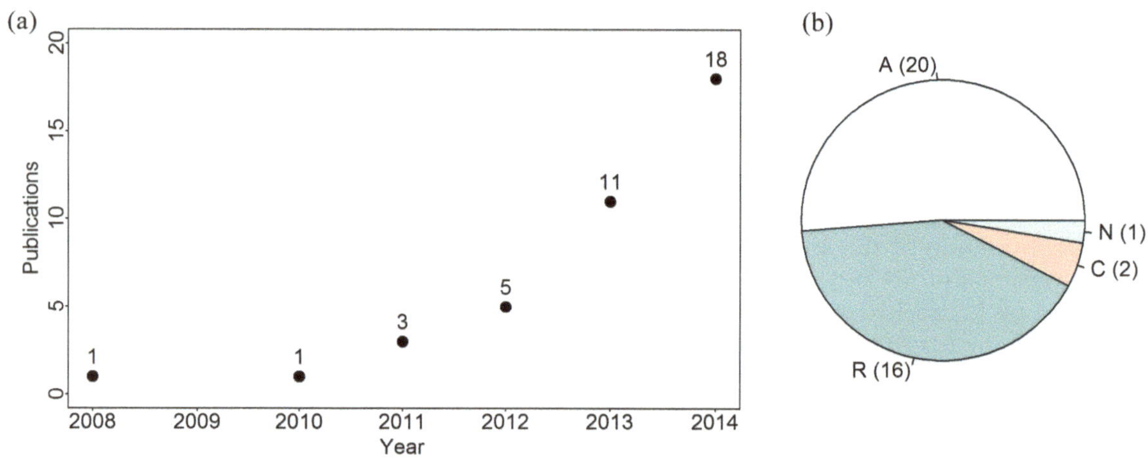

FIGURE 4

Number of publications about drug repurposing and epigenetics by: year (a) and type of publication (b). Abbreviations: A: article; R: review; C: conference paper; N: note. *Data from Scopus (as per March 2015).*

3.2.1 Case Study 1: Olsalazine as a Novel Hypomethylating Agent

In a recent VS approach for identifying DNA hypomethylating agents, Méndez-Lucio et al. (2014) uncovered olsalazine, a drug that has been previously approved by the FDA as an anti-inflammatory agent. Using a previously recognized hypomethylating agent as reference molecule (query) (NSC14778), a SS approach was conducted comparing the structure of the reference with DrugBank database, which contained 1582 FDA-approved compounds at the time of the study. The Tanimoto Score, which is included in the Overlay Chemical Structures (ROCS) package from OpenEye Scientific Software, was used as metric to calculate the similarity values. Interestingly, only 19 of the drugs enlisted in DrugBank had a significantly high similarity value. The significance was determined if a compound had a similarity value at least two standard deviations (SD) above the mean of the dataset. In this case, the similarity mean of the database was 0.62 and SD = 0.15, so that values of 0.922 and over are significant. Despite the general low similarity between the reference molecule and approved drugs in DrugBank, olsalazine showed a similarity value of 1.032 with NSC14778. Olsalazine was chosen as a good candidate for DNMT1 inhibition due to its high structural similarity to the query compound (Figure 5) after a visual comparison of both structures. Olsalazine is an anti-inflammatory agent currently approved for its use against inflammatory bowel disease (i.e., Crohn's disease and ulcerative colitis). Further, an in vivo study with HeLa cells adapted to report gene expression visually through Green fluorescent protein was calibrated with 5-azacytidine for identifying hypomethylation. This method has been described to accurately report DNMT1 (responsible of maintaining methylation patterns) activity. Olsalazine was tested by this method, and the results pointed toward a hypomethylation pattern. Interestingly, olsalazine was well tolerated by cells, a feature not observable with 5-azacytidine treatments.

A docking study of olsalazine focused on DNMT1 and DNMT3b (the enzymes against which NSC14778 exhibit detectable activity) was performed, in order to explore the probable mechanism of action of olsalazine at a molecular level. In order to conduct the docking study, olsalazine was optimized in silico using the MMFF94x force field, and Gasteiger–Marsilli atomic charges were selected in MGLTools 1.5.4. The torsional root and branches were chosen with the criteria implemented in this software, which allows flexibility of

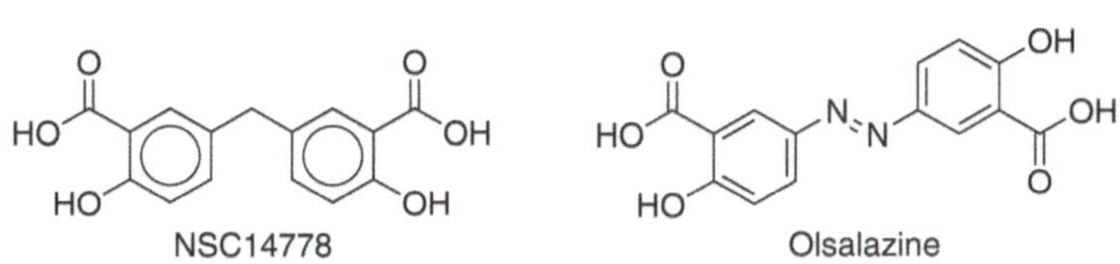

FIGURE 5
Structure of NSC14778 (query) and olsalazine (repurposing candidate).

all the rotatable bonds of the ligand. Although the crystallographic structure of DNMT1 was readily available, it was necessary to use a validated homology model of DNMT3b generated based on the crystal structure of DNMT3a. The conformational search was conducted following a Lamarckian genetic algorithm. More details on the docking settings may be consulted elsewhere (Méndez-Lucio et al., 2014). The results of the docking assay were consistent with the experimental observations, since olsalazine and NSC14778 showed very similar predicted binding energies (around −9 and −10 kcal/mol for both enzymes), and also similar binding pockets. The model proposed that the ligand–enzyme complex is mainly stabilized by hydrogen bonds between the carboxylic acids within olsalazine structure and Ser111, Gly112, Arg157, and Arg193 (in the case of DNMT1), and Arg731, Arg733, Asp582, Gly587, and Cys651 (in DNMT3b, thus preventing the nucleophilic attack of this residue on the target cytosine). Collectively, this integrative approach assesses the theoretical and experimental epigenetic character of a drug initially developed for a different indication. The full characterization of olsalazine as an anticancer molecule remains to be done.

3.2.2 *Case Study 2: Pentamidine and Amitriptyline for Renal Cancer*

The approach followed by Zerbini et al. (2014) is a notable example of drug repurposing in epigenetics. Although they were not directly searching for an inhibitor of an epigenetic target, they performed an Individualized Bioinformatics Analysis (IBA, developed by themselves) of the transcriptome of patients with clear cells renal carcinoma (ccRC) at advanced stage. Therefore, their approach is directly linked with epigenetics. In an earlier work, the authors of that work set gene signatures of progression and metastasis for ccRC (Jones et al., 2005). But the aim of their work of 2014 was to identify the 100 most overexpressed and the 100 most underexpressed genes in the individual transcriptome of the tumor of 21 patients and then, through a Gen Set Enrichment Analysis (GSEA) of the Connectivity Maps (C-MAP) in the literature, find the drugs that have been reported to counteract this alterated expression, thus theoretically reversing the altered phenotype to normality. They found the overlap of the top misregulated genes to be around 30% to 40% from one patient to other, hence demonstrating the existence of both a significant overlap and divergence in the expression profiles.

Although they analyzed the transcriptome of each patient rather than the one unified, the drugs suggested by the analysis were a small subset of drugs shared by most of the patients, with some significant exceptions that probably reflected the different pathways altered. The consensus top-scoring drugs (those with a negative enrichment correlation between −0.7 and −1.0 in more than 50% of the patients) were selected to be tested in vitro and in vivo. The list included pentamidine (antiparasitic drug), amitriptyline (antidepressant), oligomycin

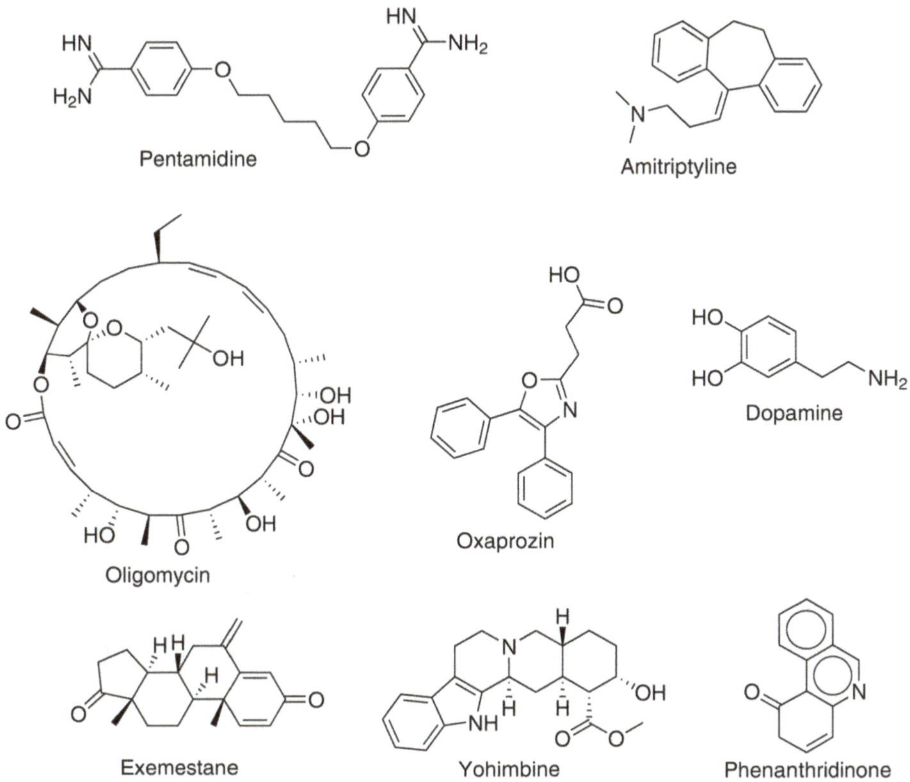

FIGURE 6

Candidate drugs for repurposing in renal cancer.

(antibiotic), yohimbine (alpha-2-adrenergic blocker), phenanthridinone (immunosuppressive), oxaprozin (NSAID), dopamine (vasoconstrictor), and exemestane (aromatase inhibitor) (Figure 6). Indeed, these eight drugs were tested against renal cancer and noncancerous cell lines. Five of these drugs exhibited a strongly incremented rate of apoptosis in cancer cells, whereas they did not affect the survival of normal cells. They also demonstrated that the status of the VHL gene (whose mutation is known for causing ccRC in patients with von Hippel-Lindau syndrome) was strongly associated with the response. Actually, the best responses were in cells deficient in VHL. Moreover, it was demonstrated that combination of these drugs (particularly those including pentamidine and/or amitriptyline) sometimes had a synergistic effect in apoptosis induction. Furthermore, the authors of that work reported results of an in vivo model by ccRC xenograft tumors injected in immunodeficient mice treated with amitriptyline, oxaprozin, or pentamidine. Only pentamidine showed a significant reduction in tumor growth and extension in the mice survival. Finally, a ccRC cell line was treated with pentamidine, to assess whether the C-MAP-based prediction of the reversion of the cancer cells' transcriptome to normality by pentamidine was accurate. Although the gene signature of normal renal cells and ccRC cells treated with pentamidine can be accurately distinguished by a hierarchical clustering analysis, the genes downregulated by pentamidine were significantly associated with pathways involved in metastasis, cell signaling, and differentiation, as a posterior pathway enrichment analysis indicated.

Since the drugs tested by Zerbini et al. were modifiers of the gene expression signature of cells, it must not be surprising that these drugs were found to have epigenetic effects, probably even as inhibitors of DNMTs, HDACs, BRDs, and such, and this probability rises if the following facts are taken into account: (1) it has been found that genes encoding enzymes histone demethylases and methyltransferases have a higher prevalence of mutation in ccRC cells (Dalgliesh et al., 2010); (2) amitriptyline, one of the most active drugs in the in vitro assays by Zerbini et al., has been intended to be repurposed as an anticancer drug before since it has shown apoptosis induction in cells of many cancer types; the anticancer effect of this drug has been linked with its identification as a HDACi (Mao et al., 2011); and (3) human HDACi has been studied, with promising results, against *Leishmania donovani*, since this microorganism is highly dependent on HDACs for its metabolism. Interestingly, pentamidine is considered a standard antileishmania agent, although its precise mechanism of action is not already known (Sodji et al., 2014). However, it must be noted that Zerbini et al. did identify plausible mechanisms of action for pentamidine in the literature, and they did not include any epigenetic modifier as a possible target. In addition, to our knowledge there are no reports of the enzymatic inhibitory activity of pentamidine against HDACs.

3.2.3 Case Study 3: Ribavirin as a Cancer Drug

A recent study used SS to identify histone methyltransferase inhibitors among approved drugs (De la Cruz-Hernandez et al., 2015). Ribavirin, which is a nucleosidic drug approved for the treatment of infection with hepatitis C virus, turned out to resemble to a high extent with 3-deazaneplanocin A, a known inhibitor of SAM-dependent methyltransferase that targets the degradation of EZH2 and leads to apoptosis in various malignancies. Similarity was assessed by MACCS/T in this study.

The biological tests of ribavirin were performed against various cancer cell lines (i.e., breast, brain, cervical, colon, and prostate cancer), and it was found that ribavirin was not only capable of inhibiting the expression of EZH2 but of other two epigenetic targets linked with cancer. It was shown that ribavirin cytotoxicity was higher for the breast, prostate, and brain cancer cell lines, although it also inhibited colon cancer cell lines to a considerable extent. These results indicate that SS based on fingerprints for drug repurposing in epigenetics is feasible and reliable.

3.2.4 Novel Computational Methodologies for Drug Repurposing against Epigenetic Targets

While some examples of computational approaches for conducting drug repurposing for epigenetic targets have been already discussed, it is important to emphasize important features, particularly those concerning VS.

In order to identify novel DNMT inhibitors, Kuck et al. (2010b) performed a VS based on a multistep docking approach that began with more than a quarter million compounds from the NCI database. Subsequent steps (that included a lead-like selection, high-throughput docking, and parallel docking) led to 24 consensus hits and experimental validation of the computationally obtained data against DNMT1 and DNMT3B in enzymatic assays. Optimization studies have been performed on NSC 319,745, one of the computational hits validated experimentally, leading to specific and more active molecules (particularly against DNMT3B). Nanaomycin A was also identified in the study by Kuck et al. (2010) as a possible DNMT3B inhibitor. Experimental studies showed that nanaomycin A is able to inhibit DNMT3B selectively in vitro. Furthermore, this natural product compound showed an ability to induce demethylation and reactivation of silenced tumor suppressor genes of human tumors (tested in colon, lung, and bone marrow cancer cell lines) (Kuck et al., 2010a). A molecular dynamics study of nanaomycin A revealed structural insights about its mode of action in DNMT3B, particularly about the long lasting interactions with Arg731, Arg733, Arg832, and the catalytic Cys 651 (involved in the methylation of DNA). In that study it was shown that S-adenosyl-methionine (SAM) may enhance the protein–ligand interaction (Caulfield and Medina-Franco, 2011). Also, DNMTi has been proposed to act through either covalent or noncovalent mechanisms (Medina-Franco and Caulfield, 2011).

Similarity searching, a different computational approach, was conducted by Medina-Franco and Yoo (2013). In that study, the assessment of chemical structures was performed with Tanimoto (T) and Tanimoto substructure (T_{SS}) measures of similarity between queries—or reference compounds, usually exhibiting a demonstrated activity against DNMTs—and published chemical databases. Visible heterogeneity was found among databases, regarding the similarity, but there were identified approved drugs with both a high T score and promising docking results. This study was based on a prior chemoinformatic characterization of compounds libraries, which demonstrated that collections of DNMTi share features with approved drugs and molecules from other libraries (Yoo and Medina-Franco, 2011a).

Modeling pharmacophores is another computational strategy that has been applied to DNMTi. Although the best-scoring pharmacophoric feature from a selection of 14 DNMTi was found to be a negative charge interacting closely with Ser1229, Gly1230, and Arg1311, four other features were significant in a computational model incorporating a homology-based reconstruction of DNMT (Yoo and Medina-Franco, 2011b). Despite the fact that different DNMTi may have different binding modes, they match three or more of the pharmacophore features of that model (Medina-Franco et al., n.d.; Kabro et al., 2013; Valente et al., 2014; Medina-Franco and Yoo, 2013; Castellano et al., 2011).

After the publication of the crystallographic structure of DNMT1, computational studies previously performed with homology models are repeated, finding a good agreement between the models previously published with the homology models and the actual crystallographic structure (Medina-Franco and Yoo, 2013).

In a separate study and using a crystallographic structure, Medina-Franco et al., (2014) elucidated the binding mode of SW155246 with DNMT1. SW155246 is an active molecule identified through an HTS of over 180,000 molecules (Kilgore et al., 2013). It was found that SW155246 binds to both the catalytic and cofactor-binding sites, and it represents an activity cliff, since a subtle modification of its hydroxyl group turns into a loosing of a hydrogen bond shared with Asn1267 in the catalytic site, thus helping to explain, at the molecular level, the loss of activity (Medina-Franco et al., 2014).

Another possible application of computational methods is in the visualization of potential optimizations of apparently weak DNMTi. Docking and molecular dynamics studies conducted with hydralazine has suggested that the molecule is a prospect for optimization, since although it may enter to the catalytic site in DNMT1, a relatively small number of interactions are formed between the enzyme and the drug (Singh et al., 2009).

Finally, computational methods provide the possibility of assessing large libraries containing natural products. Medina-Franco et al. (2010) have performed in silico VS of natural product collections found in the ZINC database with the aim of finding nonnucleosidic inhibitors with distinct chemical scaffold and decreased toxicity. The experimental validation of the VS hits needs to be conducted.

3.2.5 Drug Repurposing in Epigenetics by Similarity Searching Virtual Screening

In order to further illustrate the concepts and pragmatic features of SS, a representative example of a VS aiming to repurpose drugs in epigenetics is described in this section. One of the most important features of similarity-based VS is the definition of the queries—the compounds that are taken as a reference, mainly because of their proven activity against a target—and the chemical databases—the lists of molecules with an unknown activity against the target of interest. Since drug repurposing,

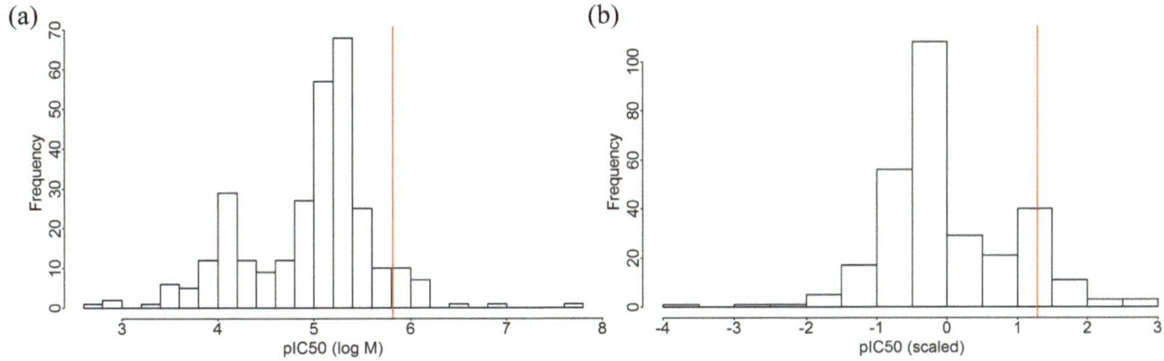

FIGURE 7

Selection of queries for virtual screening based on the distribution of the activity data. (a) pIC_{50}, the distribution seems bimodal; (b) Scaled pIC_{50} (mean = 0, sd = 1), this approximately normal distribution makes the selection easier. Vertical red lines represent the threshold.

as it was discussed before, aims to find new indications of drugs that are currently approved or in advanced clinical stages, the screening database has to be a collection of drugs approved for clinical use such as DrugBank. In order to select the queries, 296 compounds were selected from ChEMBL with reported activity against DNMT1. The most active compounds were further used as references for the VS. In order to select the most active compounds reported in ChEMBL, the distribution of activities was considered: selection of compounds with a t-score of 0.90 or more leads to a selection of 19 queries. The distribution of activities is depicted in Figure 7. Figure 7(a) shows the distribution of the pIC50 values and Figure 7(b) shows the distribution of the scaled values. Both graphs indicate the thresholds above which the activities were considered as significantly high. The selection and the corresponding figures were produced in RStudio 0.98.953 (R Development Core Team, 2008).

3.2.5.1 Selection of Queries

A diverse set of active compounds was selected. To this end, a similarity matrix of the queries was computed using MACCS keys (166-bits, as implemented in Molecular Operating Environment) and the Tanimoto coefficient. Six compounds with high similarity values to the rest of the queries (e.g., ≥0.85) were identified and excluded from the queries set. The remaining 13 compounds were selected for the similarity-based VS of DrugBank.

3.2.5.2 Similarity Searching

The similarity search was performed using the Tanimoto coefficient and two fingerprint representations, MACCS keys, and extended connectivity fingerprints (ECFPs) radius 4 as implemented in MayaChemTools

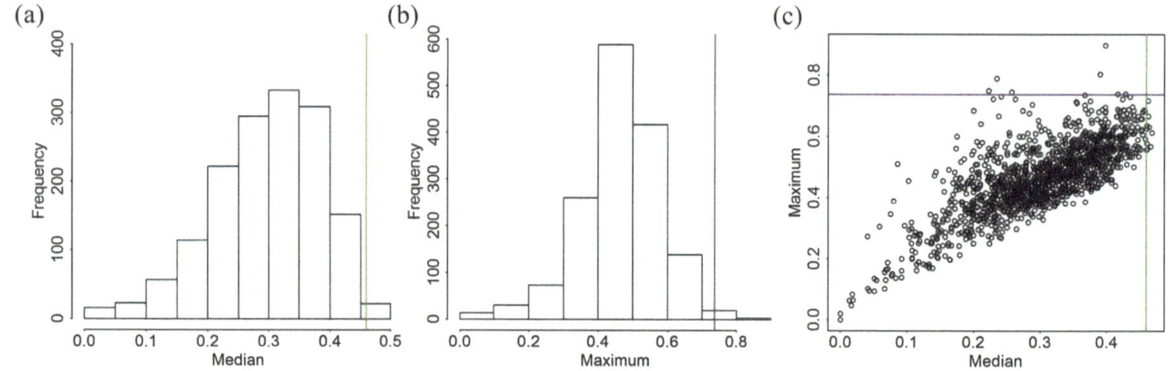

FIGURE 8

Selection of repurposing candidates by either high median of similarity with queries (green lines: cut point for median) (a); or maximum values of similarity (blue lines: cut point for maximum) (b). Scatter plot in (c) depicts the thresholds in the multifusion similarity map. Tanimoto scores with MACCS fingerprints were used to assess similarity.

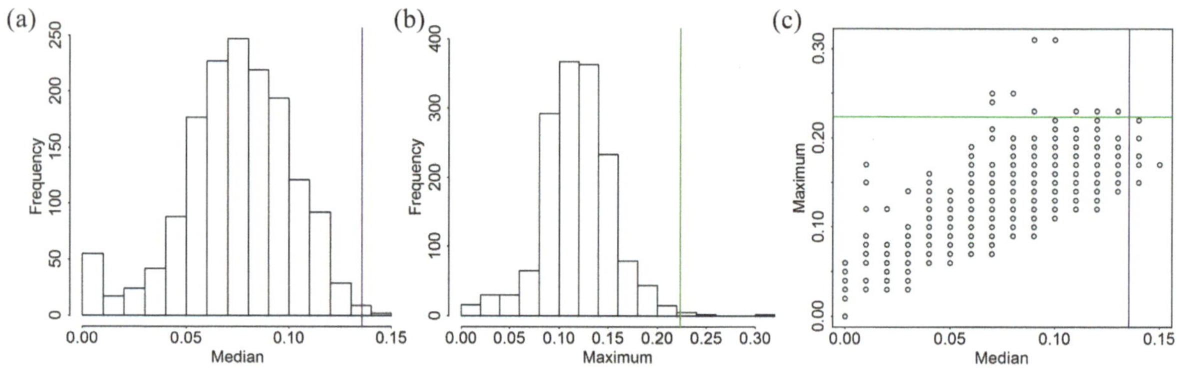

FIGURE 9

Selection of repurposing candidates by either high median of similarity with queries (blue lines: cut point for median) (a); or maximum values of similarity (green lines: cut point for maximum) (b). The plot in (c) depicts the process in two dimensions. In (c) there is high overlap of data points of maximum and median values in the central area of the graph (see also Figure 10). Tanimoto scores with ECFPs were used to assess similarity.

(http://www.mayachemtools.org/). Based on the principles of data fusion (Arany et al., 2013; Chen et al., 2010), the maximum value and the median of similarity were employed using RStudio as parallel methods for finding suitable candidates for repurposing. The corresponding multifusion similarity plots (Medina-Franco et al., 2007) obtained with MACCS keys and ECFPs are shown in Figures 8 and 9, respectively. Figure 10 shows a density plot of the multifusion similarity map presented in Figure 9(c), since a high overlap of points was found. This situation may suggest that ECFPs do not discriminate

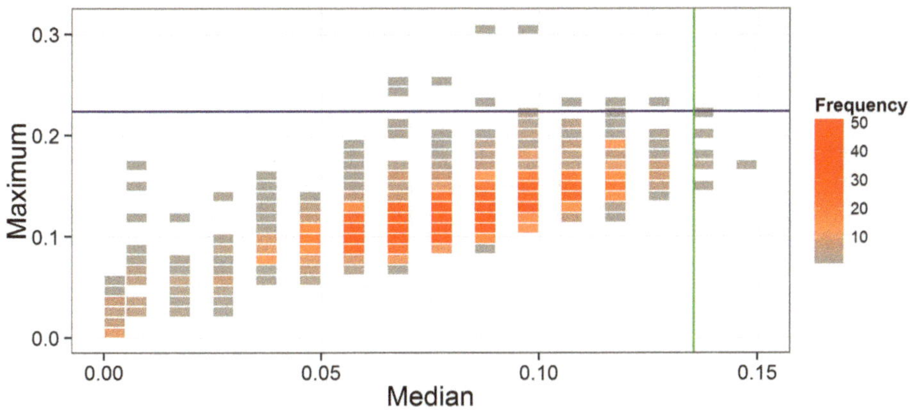

FIGURE 10

A density plot with the multifusion similarity plot depicted in Figure 9(c). Note the high superposition of points. The cut point for maximum is represented by the green line, while the cut point for median is represented by the blue line.

properly among compounds with intermediate similarity values. Here a neighborhood diameter of 4 was used, while it is possible to assess molecules with neighborhood diameters of 6, though with a small difference in the results (Duan et al., 2010; Cereto-Massagué et al., 2015). Median was preferred for this study as a central tendency measure over mean because it is less sensitive to extreme values, and so molecules with high similarity with a few queries, but very low with the majority are less probable to appear in both maximum and median groups. It is not yet known whether, in practice, molecules that are similar with many different queries have a higher probability to be active compounds.

3.2.4.3 Hit Selection

The screening compounds with higher values of maximum or median of Tanimoto scores to the queries were selected, leading to distinct and unexpected VS hits (Figure 11). Many of them already were found to be anticancer drugs, and this was attributed mainly to the presence of nucleosidic structures among the queries. Hence, a second round of the analysis was done excluding nucleosidic structures from the queries (Figure 12). It was possible to identify consensus hits, the drugs that consistently showed high similarity value for both searches with and without nucleoside analogs included in the queries (e.g., mebendazole). Of course, the experimental validation is the next logical step after this representative example that is focused on showing an example of a computational methodology that can be easily implemented.

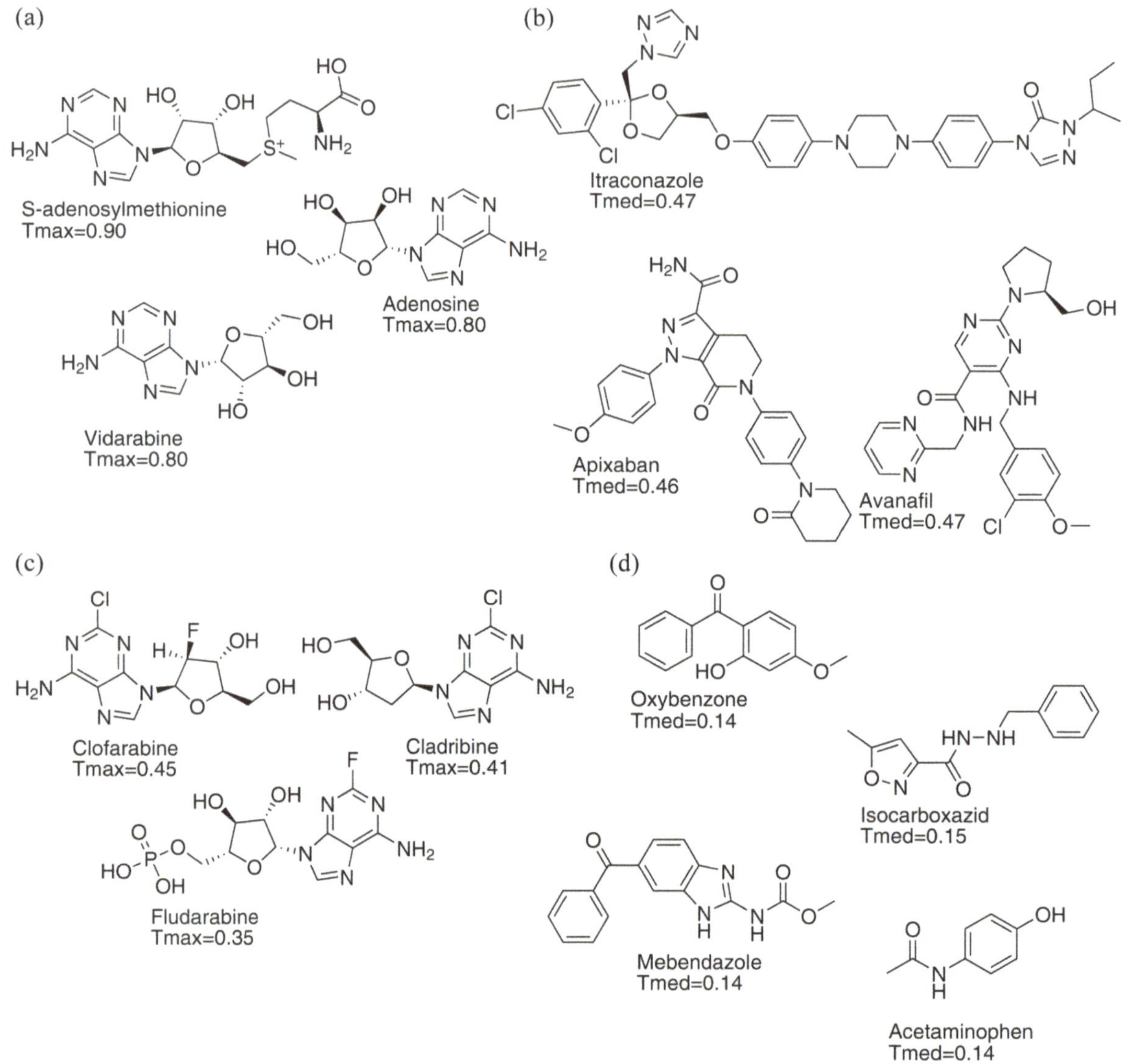

FIGURE 11

SS hits obtained by similarity search including nucleosidic queries. Many of the hits are drugs already approved against cancer. (a) Drugs with the highest maximum MACCS/T; (b) drugs with the highest median of MACCS/T; (c) drugs with the highest maximum ECFPs/T; (d) drugs with the highest median ECFPs/T. Tmax, maximum Tanimoto score value among queries; Tmed, median of Tanimoto score values of queries.

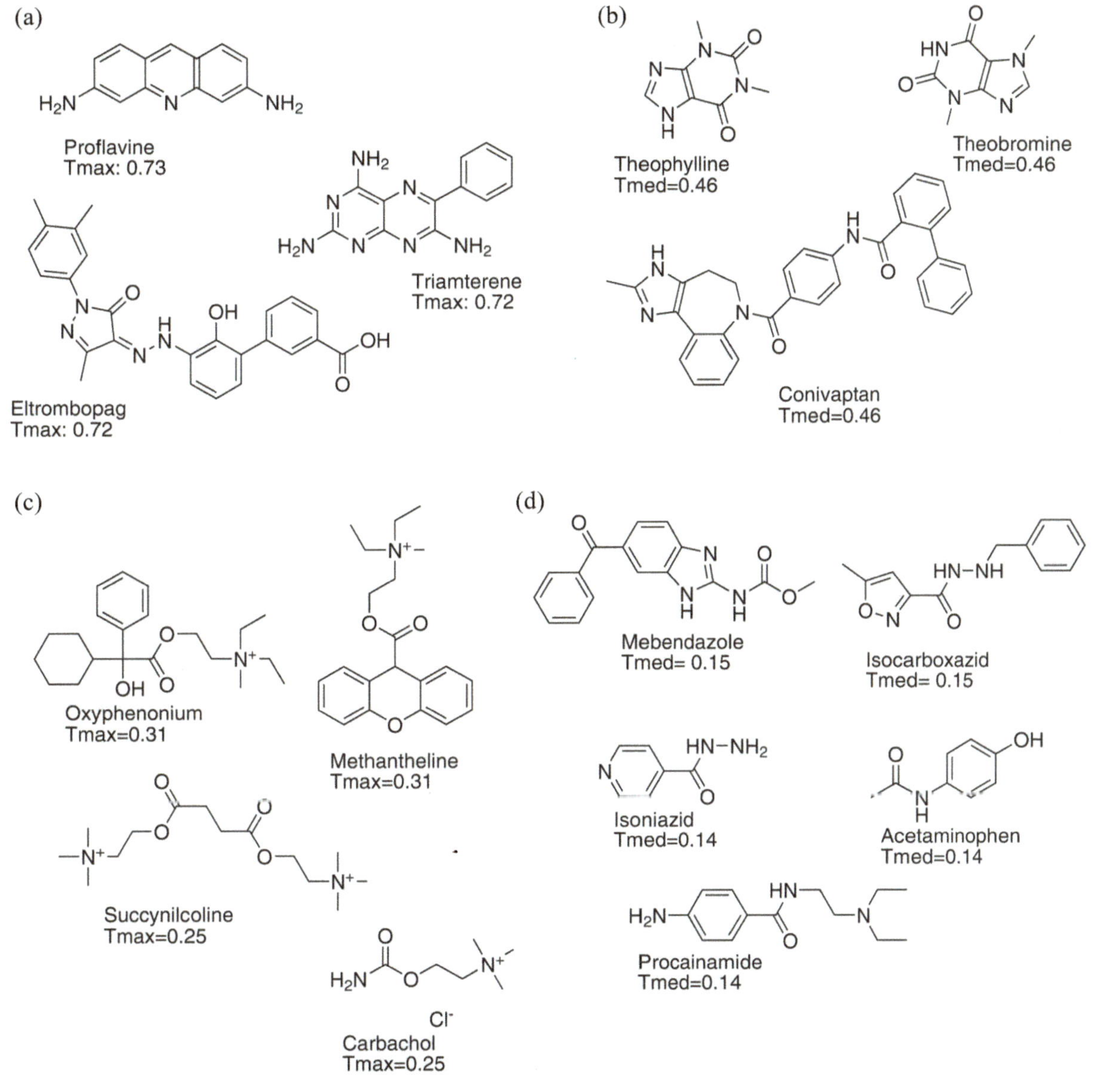

FIGURE 12

Molecules obtained by similarity search excluding nucleosidic queries. (a) Drugs with the highest maximum MACCS/T; (b) drugs with the highest median of MACCS/T; (c) drugs with the highest maximum ECFPs/T; (d) drugs with the highest median ECFPs/T. Tmax, maximum Tanimoto score value among queries; Tmed, median of Tanimoto score values of queries.

4. CONCLUSIONS

Drug repositioning (or repurposing) is the process of using a drug for an indication different from the initial indication. This is a complex approach that, in principle, may be predicted and carried out, at least in part, with a systematic method. Virtual screening is one of the systematic methods for conducting drug repurposing with the aid of computational methods. Virtual screening can be conducted typically using molecular docking (including sequential docking strategies combining high-throughput docking with consensus scoring) or SS approaches. Given the increasing amount of data available, data mining approaches based in the literature are gaining significant strength, and are also a valid method to outperform drug positioning through a systematic approach. It must be noted that in every case, experimental validation is compulsory, since the actual systematic methods are not 100% accurate. Thus far, DNMTs are the most explored epigenetic targets for drug repositioning, although there is still a considerable lack of understanding.

Diverse VS strategies have been proposed and it remains to be seen which one (or the best combinations) is (are) the best. Highlighting the research perspectives within the area, there are many doubts remaining about which are the best in silico models for predicting the activity of a compound against DNMTs. In the concise example of similarity search, it is important to assess different indicators, such as maximum values versus central tendency measures, to find a more reliable way of performing drug repurposing. Also, the exploration of DNMTi from the biological to the clinical points of view is necessary, in order to assess the feasibility of their incorporation as therapies in diseases with an epigenetic basis. The increasing attention on epigenetic targets as an opportunity for repurposing drugs provides a high expectation of epi-drugs.

ACKNOWLEDGMENTS

We thank the assistance of Mariana Giralda Herrera Núñez in assembling the data set of DNMT inhibitors used in this work. This work was supported by the National Autonomous University of Mexico (UNAM), grant PAIP 5000-9163 to J.L.M-F. Authors are also grateful to the Department of Pharmacy, School of Chemistry, UNAM for the support.

REFERENCES

Aftab, B.T., Dobromilskaya, I., Liu, J.O., Rudin, C.M., 2011. Itraconazole inhibits angiogenesis and tumor growth in non-small cell lung cancer. Cancer Res. 71, 6764–6772.

Anestopoulos, I., Voulgaridou, G.P., Georgakilas, A.G., Franco, R., Pappa, A., Panayiotidis, M.I., 2015. Epigenetic therapy as a novel approach in hepatocellular carcinoma. Pharmacol. Ther. 145, 103–119.

Arany, Á., Bolgár, B., Balogh, B., Antal, P., Mátyus, P., 2013. Multi-aspect candidates for repositioning: data fusion methods using heterogeneous information sources. Curr. Med. Chem. 20, 95–107.

Ashburn, T.T., Thor, K.B., 2004. Drug repositioning: identifying and developing new uses for existing drugs. Nat. Rev. Drug Discov. 3, 673–683.

Azmi, A.S., Mohammad, R.M., 2014. Rectifying cancer drug discovery through network pharmacology. Future Med. Chem. 6, 529–539.

Li, Y.Y., Jones, S.J., 2012. Drug repositioning for personalized medicine. Genome Med. 4, 27.

Ballester, P.J., Schreyer, A., Blundell, T.L., 2014. Does a more precise chemical description of protein–ligand complexes lead to more accurate prediction of binding affinity? J. Chem. Inf. Model. 54, 944–955.

Barker, C.A., Yahalom, J., 2010. Histone deacetylase inhibition as a mechanism for the therapeutic effect of statins (3-hydroxy-3-methylglutaryl-coenzyme a reductase inhibitors) in follicular lymphoma. J. Clin. Oncol. 28, e426.

Boolell, M., Gepi-Attee, S., Gingell, J.C., Allen, M.J., 1996. Sildenafil, a novel effective oral therapy for male erectile dysfunction. Br. J. Urol. 78, 257–261.

Cacabelos, R., 2014. Epigenomic networking in drug development: from pathogenic mechanisms to pharmacogenomics. Drug Dev. Res. 75, 348–365.

Cacabelos, R., Torrellas, C., 2014. Epigenetic drug discovery for Alzheimer's disease. Expert Opin. Drug Discov. 9, 1059–1086.

Candelaria, M., Herrera, A., Labardini, J., González-Fierro, A., Trejo-Becerril, C., Taja-Chayeb, L., et al., 2010. Hydralazine and magnesium valproate as epigenetic treatment for myelodysplastic syndrome. preliminary results of a phase-ii trial. Ann. Hematol. 90, 379–387.

Capell, B.C., Erdos, M.R., Madigan, J.P., Fiordalisi, J.J., Varga, R., Conneely, K.N., et al., 2005. Inhibiting farnesylation of progerin prevents the characteristic nuclear blebbing of hutchinson-gilford progeria syndrome. Proc. Natl. Acad. Sci. U.S.A. 102, 12879–12884.

Caraci, F., Leggio, G.M., Drago, F., Salomone, S., 2013. Epigenetic drugs for Alzheimer's disease: hopes and challenges. Br. J. Clin. Pharmacol. 75, 1154–1155.

Carrella, D., Napolitano, F., Rispoli, R., Miglietta, M., Carissimo, A., Cutillo, L., et al., 2014. Mantra 2.0: an online collaborative resource for drug mode of action and repurposing by network analysis. Bioinformatics 30, 1787–1788.

Castellano, S., Kuck, D., Viviano, M., Yoo, J., López-Vallejo, F., Conti, P., et al., 2011. Synthesis and biochemical evaluation of δ2-isoxazoline derivatives as DNA methyltransferase 1 inhibitors. J. Med. Chem. 54, 7663–7677.

Caulfield, T., Medina-Franco, J.L., 2011. Molecular dynamics simulations of human DNA methyltransferase 3B with selective inhibitor nanaomycin A. J. Struct. Biol. 176, 185–191.

Cereto-Massagué, A., Ojeda, M.J., Valls, C., Mulero, M., Garcia-Vallvé, S., Pujadas, G., 2015. Molecular fingerprint similarity search in virtual screening. Methods 71, 58–63.

Cervera, E., Candelaria, M., López-Navarro, O., Labardini, J., Gonzalez-Fierro, A., Taja-Chayeb, L., et al., 2012. Epigenetic therapy with hydralazine and magnesium valproate reverses imatinib resistance in patients with chronic myeloid leukemia. Clin. Lymphoma Myeloma Leuk. 12, 207–212.

Chang, R.L., Xie, L., Xie, L., Bourne, P.E., Palsson, B.Ø., 2010. Drug off-target effects predicted using structural analysis in the context of a metabolic network model. PLoS Comput. Biol. 6, e1000938.

Chen, B., Mueller, C., Willett, P., 2010. Combination rules for group fusion in similarity-based virtual screening. Mol. Inform. 29, 533–541.

Cheng, T., Li, Q., Zhou, Z., Wang, Y., Bryant, S.H., 2012. Structure-based virtual screening for drug discovery: a problem-centric review. AAPS J. 14, 133–141.

Choi, J., Kim, K., Song, M., Lee, D., 2013. Generation and application of drug indication inference models using typed network motif comparison analysis. BMC Med. Inform. Decis. Mak. 13 (Suppl. 1), S2.

Chong, C.R., Xu, J., Lu, J., Bhat, S., Sullivan, D.J., Liu, J.O., 2007. Inhibition of angiogenesis by the antifungal drug itraconazole. ACS Chem. Biol. 2, 263–270.

Christman, J.K., 2002. 5-azacytidine and 5-aza-2′-deoxycytidine as inhibitors of DNA methylation: mechanistic studies and their implications for cancer therapy. Oncogene 21, 5483–5495.

Chung, F.-H., Chiang, Y.-R., Tseng, A.-L., Sung, Y.-C., Lu, J., Huang, M.-C., et al., 2014. Functional module connectivity map (FMCM): a framework for searching repurposed drug compounds for systems treatment of cancer and an application to colorectal adenocarcinoma. PloS One 9, e86299.

Corbin, J.D., Beasley, A., Blount, M.A., Francis, S.H., 2005. High lung pde5: a strong basis for treating pulmonary hypertension with pde5 inhibitors. Biochem. Biophys. Res. Commun. 334, 930–938.

De la Cruz-Hernandez, E., Medina-Franco, J., Trujillo, J., Chavez-Blanco, A., Dominguez-Gomez, G., Perez-Cardenas, E., et al., 2015. Ribavirin as a tri-targeted antitumor repositioned drug. Oncol. Rep. 33 (5), 2384–2392.

Csoka, A.B., Szyf, M., 2009. Epigenetic side-effects of common pharmaceuticals: a potential new field in medicine and pharmacology. Med. Hypotheses 73, 770–780.

Dalgliesh, G.L., Furge, K., Greenman, C., Chen, L., Bignell, G., Butler, A., et al., 2010. Systematic sequencing of renal carcinoma reveals inactivation of histone modifying genes. Nature 463, 360–363.

Danishuddin, M., Khan, A.U., 2015. Structure based virtual screening to discover putative drug candidates: necessary considerations and successful case studies. Methods 71, 135–145.

Deftereos, S.N., Andronis, C., Friedla, E.J., Persidis, A., Persidis, A., 2011. Drug repurposing and adverse event prediction using high-throughput literature analysis. Wiley Interdiscip. Rev. Syst. Biol. Med. 3, 323–334.

Dobrosotskaya, I.Y., Hammer, G.D., Schteingart, D.E., Maturen, K.E., Worden, F.P., 2011. Mebendazole monotherapy and long-term disease control in metastatic adrenocortical carcinoma. Endocr. Pract. 17, e59–e62.

Duan, J., Dixon, S.L., Lowrie, J.F., Sherman, W., 2010. Analysis and comparison of 2D fingerprints: insights into database screening performance using eight fingerprint methods. J. Mol. Graph. Model. 29, 157–170.

Duarte, N.C., Becker, S.A., Jamshidi, N., Thiele, I., Mo, M.L., Vo, T.D., et al., 2007. Global reconstruction of the human metabolic network based on genomic and bibliomic data. Proc. Natl. Acad. Sci. U.S.A. 104, 1777–1782.

Dueñas-Gonzalez, A., Coronel, J., Cetina, L., González-Fierro, A., Chavez-Blanco, A., Taja-Chayeb, L., 2014. Hydralazine–valproate: a repositioned drug combination for the epigenetic therapy of cancer. Expert Opin. Drug Metab. Toxicol. 10, 1433–1444.

Do, L.H., Esteves, F.F., Karten, H.J., Bier, E., 2010. Booly: a new data integration platform. BMC Bioinform. 11, 513.

von Eichborn, J., Murgueitio, M.S., Dunkel, M., Koerner, S., Bourne, P.E., Preissner, R., 2011. Promiscuous: a database for network-based drug-repositioning. Nucleic Acids Res. 39, D1060–D1066.

Ekins, S., Williams, A.J., Krasowski, M.D., Freundlich, J.S., 2011. In silico repositioning of approved drugs for rare and neglected diseases. Drug Discov. Today 16, 298–310.

Emig, D., Ivliev, A., Pustovalova, O., Lancashire, L., Bureeva, S., Nikolsky, Y., et al., 2013. Drug target prediction and repositioning using an integrated network-based approach. PloS One 8, e60618.

Li, J., Zhu, X., Chen, J.Y., 2009. Building disease-specific drug-protein connectivity maps from molecular interaction networks and pubmed abstracts. PLoS Comput. Biol. 5, e1000450.

Fenaux, P., Mufti, G.J., Hellstrom-Lindberg, E., Santini, V., Finelli, C., Giagounidis, A., et al., 2009. Efficacy of azacitidine compared with that of conventional care regimens in the treatment of higher-risk myelodysplastic syndromes: a randomised, open-label, phase III study. Lancet Oncol. 10, 223–232.

Frijters, R., van Vugt, M., Smeets, R., van Schaik, R., de Vlieg, J., Alkema, W., 2010. Literature mining for the discovery of hidden connections between drugs, genes and diseases. PLoS Comput. Biol. 6, e1000943.

Ghofrani, H.A., Wiedemann, R., Rose, F., Schermuly, R.T., Olschewski, H., Weissmann, N., et al., 2002. Sildenafil for treatment of lung fibrosis and pulmonary hypertension: a randomised controlled trial. Lancet 360, 895–900.

Gordon, L.B., Kleinman, M.E., Miller, D.T., Neuberg, D.S., Giobbie-Hurder, A., Gerhard-Herman, M., et al., 2012. Clinical trial of a farnesyltransferase inhibitor in children with hutchinson–gilford progeria syndrome. Proc. Natl. Acad. Sci. U.S.A. 109, 16666–16671.

Gucalp, A., Sparano, J.A., Caravelli, J., Santamauro, J., Patil, S., Abbruzzi, A., et al., 2011. Phase II trial of saracatinib (azd0530), an oral src-inhibitor for the treatment of patients with hormone receptor-negative metastatic breast cancer. Clin. Breast Cancer 11, 306–311.

Gupta, S.C., Sung, B., Prasad, S., Webb, L.J., Aggarwal, B.B., 2013. Cancer drug discovery by repurposing: teaching new tricks to old dogs. Trends Pharmacol. Sci. 34, 508–517.

Han, J.-Y., Lee, S.-H., Yoo, N.J., Hyung, L.S., Moon, Y.J., Yun, T., et al., 2011. A randomized phase II study of gefitinib plus simvastatin versus gefitinib alone in previously treated patients with advanced non–small cell lung cancer. Clin. Cancer Res. 17, 1553–1560.

Hertzberg, R.P., Pope, A.J., 2000. High-throughput screening: new technology for the 21st century. Curr. Opin. Chem. Biol. 4, 445–451.

Hopkins, A.L., 2009. Drug discovery: predicting promiscuity. Nature 462, 167–168.

Hu, Y., Bajorath, J., 2013. Compound promiscuity: what can we learn from current data? Drug Discov. Today 18, 644–650.

Issa, N.T., Kruger, J., Byers, S.W., Dakshanamurthy, S., 2013. Drug repurposing a reality: from computers to the clinic. Expert Rev. Clin. Pharmacol. 6, 95–97.

Jablons, D.M., 2009. Procaine and procainamide inhibit the Wnt canonical pathway by promoter demethylation of wif-1 in lung cancer cells. Oncol. Rep. 22.

Jones, J., Otu, H., Spentzos, D., Kolia, S., Inan, M., Beecken, W.D., et al., 2005. Gene signatures of progression and metastasis in renal cell cancer. Clin. Cancer Res. 11, 5730–5739.

Kabro, A., Lachance, H., Marcoux-Archambault, I., Perrier, V., Doré, V., Gros, C., et al., 2013. Preparation of phenylethylbenzamide derivatives as modulators of dnmt3 activity. MedChemComm 4, 1562–1570.

Kantarjian, H., Issa, J.-P.J., Rosenfeld, C.S., Bennett, J.M., Albitar, M., DiPersio, J., et al., 2006. Decitabine improves patient outcomes in myelodysplastic syndromes. Cancer 106, 1794–1803.

Kilgore, J.A., Du, X., Melito, L., Wei, S., Wang, C., Chin, H.G., et al., 2013. Identification of DNMT1 selective antagonists using a novel scintillation proximity assay. J. Biol. Chem. 288, 19673–19684.

Kim, D.J., Kim, J., Spaunhurst, K., Montoya, J., Khodosh, R., Chandra, K., et al., 2014. Open-label, exploratory phase II trial of oral itraconazole for the treatment of basal cell carcinoma. J. Clin. Oncol. 49 (9525), 2013.

Kinnings, S.L., Liu, N., Tonge, P.J., Jackson, R.M., Xie, L., Bourne, P.E., 2011. A machine learning-based method to improve docking scoring functions and its application to drug repurposing. J. Chem. Inf. Model. 51, 408–419.

Kinnings, S.L., Xie, L., Fung, K.H., Jackson, R.M., Xie, L., Bourne, P.E., 2010. The mycobacterium tuberculosis drugome and its polypharmacological implications. PLoS Comput. Biol. 6, e1000976.

Klipp, E., Wade, R.C., Kummer, U., 2010. Biochemical network-based drug-target prediction. Curr. Opin. Biotechnol. 21, 511–516.

Knapp, S., Weinmann, H., 2013. Small-molecule modulators for epigenetics targets. ChemMedChem 8, 1885–1891.

Knight, Z.A., Lin, H., Shokat, K.M., 2010. Targeting the cancer kinome through polypharmacology. Nat. Rev. Cancer 10, 130–137.

Kuck, D., Caulfield, T., Lyko, F., Medina-Franco, J.L., 2010a. Nanaomycin a selectively inhibits DNMT3b and reactivates silenced tumor suppressor genes in human cancer cells. Mol. Cancer Ther. 9, 3015–3023.

Kuck, D., Singh, N., Lyko, F., Medina-Franco, J.L., 2010b. Novel and selective DNA methyltransferase inhibitors: docking-based virtual screening and experimental evaluation. Bioorg. Med. Chem. 18, 822–829.

Larocque, M., Chénard, T., Najmanovich, R., 2014. A curated c. difficile strain 630 metabolic network: prediction of essential targets and inhibitors. BMC Syst. Biol. 8, 117.

Lekka, E., Deftereos, S.N., Persidis, A., Persidis, A., Andronis, C., 2011. Literature analysis for systematic drug repurposing: a case study from biovista. Drug Discov. Today Ther. Strateg. 8, 103–108.

Lin, Y.-C., Lin, J.-H., Chou, C.-W., Chang, Y.-F., Yeh, S.-H., Chen, C.-C., 2008. Statins increase p21 through inhibition of histone deacetylase activity and release of promoter-associated HDAC1/2. Cancer Res. 68, 2375–2383.

Liu, Z., Fang, H., Reagan, K., Xu, X., Mendrick, D.L., Slikker Jr., W., et al., 2013. In silico drug repositioning – what we need to know. Drug Discov. Today 18, 110–115.

Loging, W., Rodriguez-Esteban, R., Hill, J., Freeman, T., Miglietta, J., 2011. Cheminformatic/bioinformatic analysis of large corporate databases: application to drug repurposing. Drug Discov. Today Ther. Strateg. 8, 109–116.

Lotharius, J., Gamo-Benito, F.J., Angulo-Barturen, I., Clark, J., Connelly, M., Ferrer-Bazaga, S., et al., 2014. Repositioning: the fast track to new anti-malarial medicines? Malar. J. 13, 143.

Lötsch, J., Schneider, G., Reker, D., Parnham, M.J., Schneider, P., Geisslinger, G., et al., 2013. Common non-epigenetic drugs as epigenetic modulators. Trends Mol. Med. 19, 742–753.

Lounkine, E., Keiser, M.J., Whitebread, S., Mikhailov, D., Hamon, J., Jenkins, J.L., et al., 2012. Large-scale prediction and testing of drug activity on side-effect targets. Nature 486, 361–367.

Medina-Franco, J.L., Yoo, J., 2013. Molecular modeling and virtual screening of DNA methyltransferase inhibitors. Curr. Pharm. Des. 19, 2138–2147.

Ma, D.-L., Chan, D.S.-H., Leung, C.-H., 2013. Drug repositioning by structure-based virtual screening. Chem. Soc. Rev. 42, 2130–2141.

Mao, X., Hou, T., Cao, B., Wang, W., Li, Z., Chen, S., et al., 2011. The tricyclic antidepressant amitriptyline inhibits d-cyclin transactivation and induces myeloma cell apoptosis by inhibiting histone deacetylases: in vitro and in silico evidence. Mol. Pharmacol. 79, 672–680.

Ma, X., Ezzeldin, H.H., Diasio, D.R.B., 2012. Histone deacetylase inhibitors. Drugs 69, 1911–1934.

Medina-Franco, J.L., 2012. Interrogating novel areas of chemical space for drug discovery using chemoinformatics. Drug Dev. Res. 73, 430–438.

Medina-Franco, J.L., Aguayo-Ortiz, R., 2013. Progress in the visualization and mining of chemical and target spaces. Mol. Inform. 32, 942–953.

Medina-Franco, J.L., Caulfield, T., 2011. Advances in the computational development of DNA methyltransferase inhibitors. Drug Discov. Today 16, 418–425.

Medina-Franco, J.L., Edwards, B.S., Pinilla, C., Appel, J.R., Giulianotti, M.A., Santos, R.G., et al., 2013a. Rapid scanning structure–activity relationships in combinatorial data sets: identification of activity switches. J. Chem. Inf. Model. 53, 1475–1485.

Medina-Franco, J.L., Giulianotti, M.A., Welmaker, G.S., Houghten, R.A., 2013b. Shifting from the single to the multitarget paradigm in drug discovery. Drug Discov. Today 18, 495–501.

Medina-Franco, J.L., Giulianotti, M.A., Yu, Y., Shen, L., Yao, L., Singh, N., 2009. Discovery of a novel protein kinase B inhibitor by structure-based virtual screening. Bioorg. Med. Chem. Lett. 19, 4634–4638.

Medina-Franco, J.L., López-Vallejo, F., Kuck, D., Lyko, F., 2010. Natural products as DNA methyltransferase inhibitors: a computer-aided discovery approach. Mol. Divers. 15, 293–304.

Medina-Franco, J.L., Maggiora, G.M., Giulianotti, M.A., Pinilla, C., Houghten, R.A., 2007. A similarity-based data-fusion approach to the visual characterization and comparison of compound databases. Chem. Biol. Drug Des. 70, 393–412.

Medina-Franco, J.L., Méndez-Lucio, O., Dueñas-González, A., Yoo, J., n. d. Discovery and development of DNA methyltransferase inhibitors using in silico approaches. Drug Discov. Today.

Medina-Franco, J.L., Méndez-Lucio, O., Yoo, J., 2014. Rationalization of activity cliffs of a sulfonamide inhibitor of DNA methyltransferases with induced-fit docking. Int. J. Mol. Sci. 15, 3253–3261.

Medina-Franco, J.L., Yoo, J., 2013. Docking of a novel DNA methyltransferase inhibitor identified from high-throughput screening: insights to unveil inhibitors in chemical databases. Mol. Divers. 17, 337–344.

Méndez-Lucio, O., Tran, J., Medina-Franco, J.L., Meurice, N., Muller, M., 2014. Toward drug repurposing in epigenetics: olsalazine as a hypomethylating compound active in a cellular context. ChemMedChem 9, 560–565.

Merino, A., Bronowska, A.K., Jackson, D.B., Cahill, D.J., 2010. Drug profiling: knowing where it hits. Drug Discov. Today 15, 749–756.

Mukhopadhyay, T., Sasaki, J., Ramesh, R., Roth, J.A., 2002. Mebendazole elicits a potent antitumor effect on human cancer cell lines both in vitro and in vivo. Clin. Cancer Res. 8, 2963–2969.

Mummaneni, P., Shord, S.S., 2014. Epigenetics and oncology. Pharmacotherapy 34, 495–505.

Muthyala, R., 2011. Orphan/rare drug discovery through drug repositioning. Drug Discov. Today Ther. Strateg. 8, 71–76.

Nair, P., 2013. Second act: drug repurposing gets a boost as academic researchers join the search for novel uses of existing drugs. Proc. Natl. Acad. Sci. U.S.A. 110, 2430–2432.

Nebbioso, A., Carafa, V., Benedetti, R., Altucci, L., 2012. Trials with "epigenetic" drugs: an update. Mol. Oncol. 6, 657–682.

Njoroge, M., Njuguna, N.M., Mutai, P., Ongarora, D.S.B., Smith, P.W., Chibale, K., 2014. Recent approaches to chemical discovery and development against malaria and the neglected tropical diseases human african trypanosomiasis and schistosomiasis. Chem. Rev. 114, 11138–11163.

Nowakowski, G.S., Maurer, M.J., Habermann, T.M., Ansell, S.M., Macon, W.R., Ristow, K.M., et al., 2010. Statin use and prognosis in patients with diffuse large b-cell lymphoma and follicular lymphoma in the rituximab era. J. Clin. Oncol. 28, 412–417.

Nygaard, H.B., van Dyck, C.H., Strittmatter, S.M., 2014. Fyn kinase inhibition as a novel therapy for Alzheimer's disease. Alzheimers Res. Ther. 6, 8.

Nygren, P., Larsson, R., 2014. Drug repositioning from bench to bedside: tumour remission by the antihelmintic drug mebendazole in refractory metastatic colon cancer. Acta Oncol. 53, 427–428.

Pantziarka, P., Bouche, G., Meheus, L., Sukhatme, V., Sukhatme, V.P., 2014. Repurposing drugs in oncology (ReDo)—mebendazole as an anti-cancer agent. Ecancermedicalscience. 8, 442.

Paul, S.M., Lewis-Hall, F., 2013. Drugs in search of diseases. Sci. Transl. Med. 5, 186fs18.

Persidis, A., Deftereos, S., Persidis, A., 2004. Systems literature analysis. Pharmacogenomics 5, 943–947.

Pont, A., Williams, P.L., Loose, D.S., Feldman, D., Reitz, R.E., Bochra, C., et al., 1982. Ketoconazole blocks adrenal steroid synthesis. Ann. Intern. Med. 97, 370–372.

Pujol, A., Mosca, R., Farrés, J., Aloy, P., 2010. Unveiling the role of network and systems biology in drug discovery. Trends Pharmacol. Sci. 31, 115–123.

Qin, C., Zhang, C., Zhu, F., Xu, F., Chen, S.Y., Zhang, P., et al., 2014. Therapeutic target database update 2014: a resource for targeted therapeutics. Nucleic Acids Res. 42, D1118–D1123.

Raman, K., Chandra, N., 2008. *Mycobacterium tuberculosis* interactome analysis unravels potential pathways to drug resistance. BMC Microbiol. 8, 234.

R Development Core Team, 2008. R: A Language and Environment for Statistical Computing. R Foundation for Statistical Computing, Vienna, Austria.

Reaume, A.G., 2011. Drug repurposing through nonhypothesis driven phenotypic screening. Drug Discov. Today Ther. Strateg. 8, 85–88.

Rudin, C.M., Brahmer, J.R., Juergens, R.A., Hann, C.L., Ettinger, D.S., Sebree, R., et al., 2013. Phase 2 study of pemetrexed and itraconazole as second-line therapy for metastatic nonsquamous non-small-cell lung cancer. J. Thorac. Oncol. 8, 619–623.

Ramdhave, A.S., Patel, D., Ramya, I., Nandave, M., Kharkar, P.S., 2013. Targeting heat shock protein 90 for malaria. Mini Rev. Med. Chem. 13, 1903–1920.

Sardana, D., Zhu, C., Zhang, M., Gudivada, R.C., Yang, L., Jegga, A.G., 2011. Drug repositioning for orphan diseases. Brief. Bioinform. 12, 346–356.

Schneider, H.-C., Klabunde, T., 2013. Understanding drugs and diseases by systems biology? Bioorg. Med. Chem. Lett. 23, 1168–1176.

Sigalotti, L., Fratta, E., Coral, S., Cortini, E., Covre, A., Nicolay, H.J.M., et al., 2007. Epigenetic drugs as pleiotropic agents in cancer treatment: biomolecular aspects and clinical applications. J. Cell Physiol. 212, 330–344.

Singhal, S., Mehta, J., Desikan, R., Ayers, D., Roberson, P., Eddlemon, P., et al., 1999. Antitumor activity of thalidomide in refractory multiple myeloma. N. Engl. J. Med. 341, 1565–1571.

Singh, N., Dueñas-González, A., Lyko, F., Medina-Franco, J.L., 2009. Molecular modeling and molecular dynamics studies of hydralazine with human dna methyltransferase 1. ChemMedChem 4, 792–799.

Sodji, Q., Patil, V., Jain, S., Kornacki, J.R., Mrksich, M., Tekwani, B.L., et al., 2014. The antileishmanial activity of isoforms 6- and 8-selective histone deacetylase inhibitors. Bioorg. Med. Chem. Lett. 24, 4826–4830.

Soto, H., Sanchez, K., Escobar, J.Y., Constanzo, A., Fernandez, Z., Melendez, C., 2014. Cost-effectiveness analysis of hydralazine and magnesium valproate lp associated with treatment for adult patients with metastatic recurrent or persistent cervical cancer in Mexico. Value Health 17, A639.

Sundberg, S.A., 2000. High-throughput and ultra-high-throughput screening: solution- and cell-based approaches. Curr. Opin. Biotechnol. 11, 47–53.

Swamidass, S.J., 2011. Mining small-molecule screens to repurpose drugs. Brief. Bioinform. 12, 327–335.

Szyf, M., 2009. Epigenetics, DNA methylation, and chromatin modifying drugs. Annu. Rev. Pharmacol. Toxicol. 49, 243–263.

Tada, M., Imazeki, F., Fukai, K., Sakamoto, A., Arai, M., Mikata, R., et al., 2007. Procaine inhibits the proliferation and DNA methylation in human hepatoma cells. Hepatol. Int. 1, 355–364.

Temesi, G., Bolgár, B., Arany, A., Szalai, C., Antal, P., Mátyus, P., 2014. Early repositioning through compound set enrichment analysis: a knowledge-recycling strategy. Future Med. Chem. 6, 563–575.

Tritos, N.A., Biller, B.M.K., 2012. Advances in medical therapies for Cushing's syndrome. Discov. Med. 13, 171–179.

Turner, T.J., Bascomb, N., Maki, J., Rao, N., Young, F.S., 2011. Adopting orphaned drugs: developing multidrug regimens using generic drugs. Drug Discov. Today Ther. Strateg. 8, 117–121.

Valente, S., Liu, Y., Schnekenburger, M., Zwergel, C., Cosconati, S., Gros, C., et al., 2014. Selective non-nucleoside inhibitors of human dna methyltransferases active in cancer including in cancer stem cells. J. Med. Chem. 57, 701–713.

Vogt, I., Mestres, J., 2010. Drug-target networks. Mol. Inform. 29, 10–14.

Wadman, M., 2012. New cures sought from old drugs. Nature 490, 15.

Wadman, M., 2013. NIH gambles on recycled drugs. Nature 499, 263–264.

Wagner, J.M., Hackanson, B., Lübbert, M., Jung, M., 2010. Histone deacetylase (HDAC) inhibitors in recent clinical trials for cancer therapy. Clin. Epigenetics 1, 117–136.

Wang, J.-C., Chu, P.-Y., Chen, C.-M., Lin, J.-H., 2012. Idtarget: a web server for identifying protein targets of small chemical molecules with robust scoring functions and a divide-and-conquer docking approach. Nucleic Acids Res. 40, W393–W399.

Wishart, D.S., Knox, C., Guo, A.C., Cheng, D., Shrivastava, S., Tzur, D., et al., 2008. Drugbank: a knowledgebase for drugs, drug actions and drug targets. Nucleic Acids Res. 36, D901–D906.

Wu, C., Gudivada, R.C., Aronow, B.J., Jegga, A.G., 2013a. Computational drug repositioning through heterogeneous network clustering. BMC Syst. Biol. 7 (Suppl. 5), S6.

Wu, Z., Wang, Y., Chen, L., 2013b. Network-based drug repositioning. Mol. Biosyst. 9, 1268–1281.

Wu, D., Flannery, A.R., Cai, H., Ko, E., Cao, K., 2014. Nuclear localization signal deletion mutants of lamin a and progerin reveal insights into lamin a processing and emerin targeting. Nucleus 5, 66–74.

Xie, L., Evangelidis, T., Xie, L., Bourne, P.E., 2011. Drug discovery using chemical systems biology: weak inhibition of multiple kinases may contribute to the anti-cancer effect of nelfinavir. PLoS Comput. Biol. 7, e1002037.

Yoo, J., Medina-Franco, J.L., 2011a. Chemoinformatic approaches for inhibitors of DNA methyltransferases: comprehensive characterization of screening libraries. Comput. Mol. Biosci. 1, 7–16.

Yoo, J., Medina-Franco, J.L., 2011b. Homology modeling, docking and structure-based pharmacophore of inhibitors of dna methyltransferase. J. Comput. Aided Mol. Des. 25, 555–567.

Zanzoni, A., Soler-López, M., Aloy, P., 2009. A network medicine approach to human disease. FEBS Lett. 583, 1759–1765.

Zerbini, L.F., Bhasin, M.K., de Vasconcellos, J.F., Paccez, J.D., Gu, X., Kung, A.L., et al., 2014. Computational repositioning and preclinical validation of pentamidine for renal cell cancer. Mol. Cancer Ther. 13, 1929–1941.

Zhou, S., Wang, F., Hsieh, T.-C., Wu, J.M., Wu, E., 2013. Thalidomide–a notorious sedative to a wonder anticancer drug. Curr. Med. Chem. 20, 4102–4108.

CHAPTER 13

Computational Structure–Activity Relationship Studies of Epigenetic Target Inhibitors

Oscar Méndez-Lucio
Centre for Molecular Informatics, Department of Chemistry, University of Cambridge, Cambridge, UK

1. INTRODUCTION

Epigenetics has been established as a major driver of the phenotype as it regulates when and which genes should be expressed without altering the DNA sequence (Berger et al., 2009). This gene expression regulation is mediated by covalent modifications either directly in the DNA or in proteins such as histones. The effect of this regulation can be very dynamic or very stable depending on the epigenetic modifications (Reik, 2007). The proteins responsible to create, maintain, or delete these epigenetic modifications can be roughly divided depending on its function in writers (add a chemical group covalently), erasers (remove a chemical group), and readers (bind to a modified histone) (Muller and Brown, 2012). There are several reviews that explain in detail the effect and mechanism of each of the epigenetic modifications (Arrowsmith et al., 2012; Portela and Esteller, 2010).

Due to the importance of epigenetic modifications, many of the proteins involved in them have been adopted as potential drug targets for treating cancer, CNS indications, diabetes, inflammation, among other diseases (Best and Carey, 2010). Currently, the FDA has approved several drugs with epigenetic targets (also known as epi-drugs), including Azacitadine and Decitabine, DNA methyltransferase (DNMT) inhibitors approved for the treatment of myelodysplastic syndrome; and Vorinostat and Romidepsin, histone deacetylases (HDAC) inhibitors for the treatment of T-cell lymphoma (Ho et al., 2013; Mack, 2010). Although epigenetic modulators are very attractive targets to pharmaceutical companies as they are involved in many different diseases, only few projects reach successful stages. This is exemplified by the number of clinical trials (Falahi et al., 2014; Nebbioso et al., 2012), where only few new compounds have been able to reach clinical phase III (DeWoskin and Million, 2013).

CONTENTS

Epi-Informatics. http://dx.doi.org/10.1016/B978-0-12-802808-7.00013-7

The wide range of epigenetic targets together with the large chemical diversity of inhibitors makes the use of computational methods necessary to guide the identification of new drug candidates. Computational methods used in drug discovery are divided into two large groups: target-based and ligand-based. Target-based drug discovery uses the target three-dimensional (3D) structure in order to study the conformations that ligands adopt in the binding sites, for example, molecular docking and molecular dynamics. Although the crystal structures of many epigenetic targets (such as DNMT, HDACs, and bromodomains) have already been reported, many others are still unknown, limiting the application of these methods. On the other hand, ligand-based drug discovery is based on the chemical structures of known inhibitors and rely mainly on molecular similarity. These methods are especially useful when the 3D structure of the target is unknown and are more suitable to study structure–activity relationships (SARs) of ligands. Some examples of ligand-based methods include similarity searching, quantitative structure–activity relationships (QSAR), proteochemometric modeling, and pharmacophore modeling, among others.

The main goal of this chapter is to discuss how computational methods have been useful to study and analyze SARs of epigenetic targets inhibitors. In order to do that, a brief introduction of the computational methods commonly used to study SARs will be given in the second section, just after this introduction. In the third section I will be discussing novel applications of computational methods to navigate SARs of some epigenetic targets including HDACs, DNMTs, and bromodomains. The fourth section will include a discussion of the future developments and challenges faced by computational methods specifically when modeling epigenetic target selectivity. The fifth and last section presents a summary and general conclusions of this chapter.

2. COMPUTATIONAL METHODS TO STUDY STRUCTURE–ACTIVITY RELATIONSHIPS

Computational methods have been extensively used to study and predict ligand–target interactions. There is an enormous range of different computational methods to analyze different kinds of SARs. In order to choose the more suitable method, we need to know what kind of information we want to know and if we have the appropriate data requirements for that. In this section, I will present a brief description of the most common computational methods used to analyze SAR data. Nonetheless, more detailed information can be found in the references given at the end of this chapter.

2.1 Quantitative Structure–Activity Relationships

It has been more that 50 years since the publication of the first QSAR by Hansch et al. (1962). They reported for the first time the relationship between biological activity and structural parameters, such as log P and Hammett's σ, using a computational approach. Since then, the field of QSAR has suffered an enormous evolution, which includes the use of more complex descriptors, larger data sets, and sophisticated machine learning techniques. However, the main idea behind this technique is still the same, finding the relationship between a readout (i.e., biological activity, reactivity, etc.) and the structure of a set of molecules that can help us to rationalize the data and to predict the behavior of new compounds.

In practical terms, there are three fundamental elements needed for any QSAR study: a data set, molecular descriptors, and a mathematical method. An initial set of molecules to train the model is the most critical requirement to obtain good performance and predictive power. In the last years, the increased amount of chemogenomic data has facilitated the generation of more robust and better quality QSAR models. Most of this bioactivity data has been compiled in different databases such as ChEMBL (Gaulton et al., 2012), Binding Database (Liu et al., 2007), MOAD (Benson et al., 2008; Hu et al., 2005), and EpiDBase (Loharch et al., 2015), among others. These databases, together with DrugBank (Wishart et al., 2006) and PubChemBioAssay (Wang et al., 2012), represent a large repository with experimental data suitable for QSAR studies. It is important to keep in mind that in an ideal situation, all molecules should be tested in the same assay under the same conditions; otherwise possible measurement variations should be taken into account.

Another crucial step in the development of QSAR models is to encode chemical structures into a format that captures suitable information for modeling. At early stages of QSAR modeling physicochemical properties of the compounds were preferred for this objective. The perfect example of this is the use of log P and Hammett's σ by Hansch et al. (1962). Nowadays, there is a wide variety of different molecular descriptors, some of them designed specifically for structure–activity modeling, such as extended-connectivity fingerprints (ECFP) (Rogers and Hahn, 2010). A more comprehensive compilation of chemical descriptors can be found in excellent reviews (Xue and Bajorath, 2000) or in the Handbook of Molecular Descriptors (Todeschini and Consonni, 2008). Once having the data set encoded into appropriate chemical descriptors, the last critical step is to choose the best mathematical method for modeling. Historically, linear methods such as multiple linear regression and partial least squares (PLS) have been used to generate QSAR models. Recently, more sophisticated machine learning techniques have been applied to capture in a detailed manner the SARs of large data sets. Some of the machine learning techniques used in QSAR modeling are support vector machines (SVM), random forest (RF), decision trees, and *k*-nearest neighbors (*k*NN) to name a few. All of these methods and their application to drug discovery are reviewed in detail in an exhaustive review by Lavecchia (2015).

Apart from the three basic points mentioned before, it is important to define the domain of applicability and apply a rigorous validation procedure (Eriksson et al., 2003). However it would be impossible to cover in detail all aspects needed to generate a trustworthy and useful model in this section. Nevertheless, in-depth reviews on QSAR modeling, the application, and future approaches have been published before (Cherkasov et al., 2014; Dehmer et al., 2012).

2.2 Activity Landscape Modeling and Activity Cliffs

Although medicinal and computational chemists have been using QSAR modeling for more than 50 years, it is not the only computational method suited to analyze SARs. Recently, Gerald Maggiora (2006) proposed to conceptualize SAR data as an "activity landscape". Following this idea, a continuous SAR would be represented by gently rolling hills, whereas a discontinuous SAR would be comparable to a rugged landscape. In this topological metaphor it is possible to find atypical cases where a pair of molecules with high similarity presents a large difference in potency, namely "activity cliffs." In general, activity cliffs are common in hit-to-lead or lead optimization processes, and their identification gives important information of small chemical modifications associated with an increased or decreased activity (Stumpfe et al., 2014). It has been discussed that the presence of activity cliffs in a database can be regarded as good or bad depending on the final end of the project (Cruz-Monteagudo et al., 2014). For instance, they give structural information of chemical features associated with large changes in potency, which can speed up a medicinal chemistry project or guide the optimization of a ligand. On the other hand, they represent a challenge for computational methods that are based on a similarity principle (which assumes a continuous SAR), as these usually fail when predicting activity cliffs.

Different useful concepts have risen in order to facilitate the activity landscape analysis and modeling. One of those is activity cliff hot spots, which highlight the "interacting protein atoms directly in the active site of the target protein," which are associated with the formation of activity cliffs (Seebeck et al., 2011). The identification of activity cliff hot spots facilitates the structure-based interpretation of activity cliffs. In other words, they help to identify which regions in the binding site are associated with the large changes in potency presented by the ligands. An alternative way to identify activity cliff hot spots is by the use of protein–ligand fingerprints (PLIFs) as was shown recently using a database of kinase–ligand complexes (Méndez-Lucio et al., 2015). A different way to analyze activity cliffs is by finding the activity cliff generators, which are defined as "molecules with high probability to form activity cliffs with structurally similar molecules tested in the same biological assay" (Medina-Franco et al., 2014; Méndez-Lucio et al., 2012b). The analysis of all activity cliff generators in a database facilitates the interpretation of the chemical features associated with activity cliffs. Bajorath and his research team have written in-depth reviews on the advances in activity cliff analysis (Stumpfe and Bajorath, 2012; Stumpfe et al., 2014).

In the last years, the range of methods to model activity landscapes has increased, allowing the users to choose the most appropriate for a specific data. For example, structure–activity similarity (SAS) maps (Shanmugasundaram and Maggiora, 2001) and dual–activity difference (DAD) maps (Méndez-Lucio et al., 2012a; Perez-Villanueva et al., 2011) have been used for a qualitative description of large databases. In a similar way, the Structure–Activity Landscape Index (SALI) (Guha and Van Drie, 2008) and Structure–Activity Relationship Index (SARI) (Peltason and Bajorath, 2007) offer a more quantitative description of the activity landscape. Additionally, there have been efforts to predict the presence of activity cliffs in a database by using machine learning techniques (Heikamp et al., 2012), predicting the SALI (Guha, 2012b) or through conditional probabilistic approaches (Santos et al., 2013). Many of these methods have been extensively reviewed in two papers by Medina-Franco (2012) and Guha (2012a).

2.3 Proteochemometric Modeling

Proteochemometric (PCM) modeling can be conceptualized as an extension of QSAR modeling. This is a novel technique that simultaneously analyzes both chemical space (i.e., ligands) and biological space (i.e., protein target) (van Westen et al., 2011). Including all this information in the same model allows navigation of a multitarget SAR by the quantitative evaluation of ligand and target structural similarities. Similarly to QSAR, many different descriptors can be used to encode the compound information, namely physicochemical descriptors, topological descriptors, molecular interaction fields, and so on. Additionally, this technique requires the use of protein information, which can be encoded by descriptors based on the amino acidic sequence or on the 3D shape of the binding site (if the 3D structure of the protein is available). PCM modeling has been applied to analyze the multitarget SAR of different protein families such as kinases (Fernandez et al., 2010; Lapins and Wikberg, 2010; Subramanian et al., 2013), proteases (Ain et al., 2014), cytochrome P450 (Lapins et al., 2013), adenosine receptors (van Westen et al., 2012), to name a few. All the specific requirements, techniques, and new applications of PCM modeling have been nicely reviewed elsewhere (Cortes-Ciriano et al., 2015; van Westen et al., 2011).

2.4 Similarity Search

Similarity searching can be defined as the action of selecting compounds from a database that present high structural similarity to a query (Downs and Willett, 2007). Contrary to substructure searching, here the query generally is the entire structure of the compound that is encoded into chemical descriptors, which are compared to the descriptors of the molecules in the database. As a result, the user obtains a list in which compounds are ranked based on the similarity they present to the query molecule (Willett et al., 1998). Based on the similarity principle one could expect that the most similar compounds (i.e., top-ranked molecules in the result list) would present similar proprieties to the query compound. The computational efficiency of this method makes it a good first option to screen or filter large molecular databases. An advantage of the similarity searching approach is that it captures the global similarity of the molecules. Nevertheless, in some cases this technique can overlook those small chemical features that can dramatically increase or decrease the biological activity of a molecule (e.g., a hydrogen bond donor) (Kubinyi, 1998). A recent review by Stumpfe and Bajorath (2011) gives a very complete overview of the principles, methods, and applications of similarity search.

2.5 Pharmacophore Modeling

The term pharmacophore is defined as "a set of structural features in a molecule that is recognized at a receptor site and is responsible for the molecule's biological activity" (Gund, 1977). This set of structural features is often shared by a group of compounds that bind to the same protein target. The main goal of the pharmacophore modeling is to find this recognition pattern shared by a set of known ligands. The result of this particular type of modeling is a set of individual features (e.g., hydrogen-bond donors or acceptors, positively or negatively charged groups, hydrophobic regions, aromatic rings, etc.) that are usually arranged in a specific 3D conformation. The pharmacophore obtained from the computational analysis is a helpful tool that can be used to screen chemical libraries with high efficiency. Additionally, this technique could be applied for the prediction of ligand binding modes or ligand 3D alignments. It is important to note that contrary to similarity searching, pharmacophore modeling centers on the local similarity between compounds (Sanders et al., 2012). In other words, it captures the small chemical features in the molecule directly associated with biological activity. More information regarding theory, software, and applications of pharmacophore modeling can be found in excellent reviews previously published (Leach et al., 2010; Van Drie, 2013).

3. ANALYZING SARs OF EPIGENETIC TARGETS USING COMPUTATIONAL TOOLS

Proteins involved in the epigenetic machinery have been attractive targets for drug discovery. However, the absence of crystallographic structures for these proteins at early stages of drug development limited the use of molecular dynamics or docking techniques. On the other hand, this helped ligand-based approaches to gain relevance for the SAR analysis and the design of new epigenetic target inhibitors. From the last section it can be seen that there are numerous different methods that can be used to understand the SAR of a set of compounds. However, the election of the best method to navigate the SAR will always depend on the database in study. Most of these methods have been applied to understand the SARs of different epigenetic target inhibitors. This section shows real-life examples of how computational methods have been applied successfully to study the SARs on epigenetic targets and will focus on three of the most studied epigenetic protein families: histone deacetylases, DNA methyltransferases (DNMT), and bromodomains.

3.1 Histone Deacetylases Inhibitors

The main function of histone deacetylases (HDACs) is to remove acetyl groups from the acetylated lysines located in the histone tails. Consequently, the positive charge of the amino group in the lysines interacts with the negative charge in the DNA phosphates, promoting chromatin condensation. As a result, gene silencing is observed since the transcription machinery cannot access gene-promoter regions (Sterner and Berger, 2000). Based on this, several histone deacetylases inhibitors (HDACIs) have been designed in order to increase the transcription during gene-silencing diseases (e.g., cancer) (Cress and Seto, 2000). HDACIs can be divided into five categories based on their chemical structures (Figure 1): short-chain aliphatic carboxylates, hydroxamic acids, benzamides, ketones, and cyclic peptides with a pendant functional group (Drummond et al., 2005). Currently, the US Food and Drug Administration (FDA) has approved the use of four HDACIs: Vorinostat (SAHA) and Istodax (romidepsin) for the treatment of cutaneous T cell lymphoma (CTCL), Belinostat for peripheral T-cell lymphoma (PTCL), and more recently Panobinostat for multiple myeloma (Guha, 2015). Nevertheless, several research groups in industry and academia are in constant search for novel, more potent, and selective HDACIs, many of them making use of computational methods.

The paper published by Tang et al. (2009) represents a good example of how computational methods are valuable tools to identify new HDAC1 inhibitors and analyze their structures. They generated consensus QSAR models from 59 chemically diverse HDAC1 inhibitors. Molconn Z.05 software (version 4.05; eduSoft, LC) was employed to calculate 262 chemically relevant topological descriptors for each molecule, which were used together with *k*NN and SVM algorithms to generate QSAR models. Having good predictive power ($q^2_{LOO\_cv} = 0.81$, $R^2_{test} = 0.80$ and $\text{RMSE}_{\text{test}} = 0.38$ for the best *k*NN model; $q^2_{LOO\_cv} = 0.94$, $R^2_{test} = 0.81$ and $\text{RMSE}_{\text{test}} = 0.51$ for the best SVM model), consensus models of both techniques were used for the virtual screening of 9.5×10^6 molecules. In "the consensus models the predicted activities from individual models (70 *k*NN and 17 SVM models) were averaged to yield a consensus pIC_{50} value"

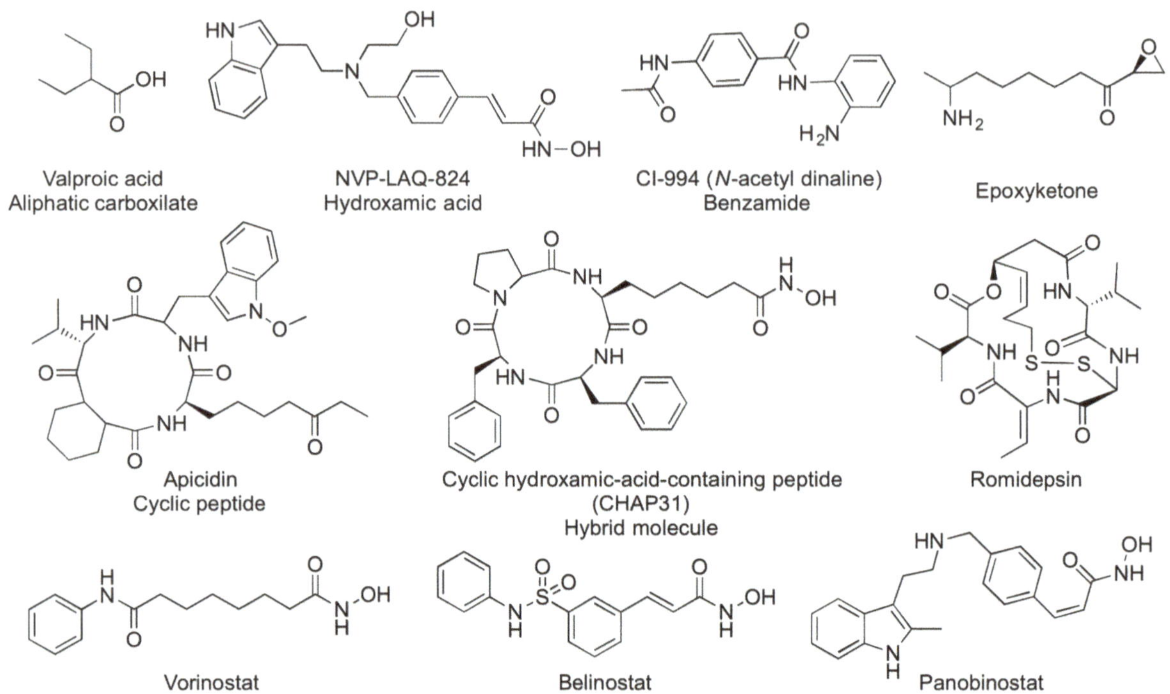

FIGURE 1

Chemical structures of selected HDAC inhibitors including the four FDA-approved drugs. Some of these compounds are classified in different groups based on their structure and present diverse Zinc-binding groups.

(Tang et al., 2009). In the end, 45 hits were obtained from the consensus virtual screening; two of them were reported previously as potential HDAC1 inhibitors (Remiszewski et al., 2002; Wang et al., 2004). The other four compounds were validated as HDACIs, from which three resulted in HDAC1 inhibitors (with pIC_{50} of 4, 5.9 and 6), whereas the fourth compound was confirmed as an HDAC6 inhibitor. Interestingly, two of these compounds are FDA approved drugs, namely, Bufexamac, which is used as an anti-inflammatory agent; and Roxatidine, a competitive histamine H_2 antagonist. Additionally, after analyzing the structure of the 45 hits, the authors of this work realized that the model was able to identify chemical features associated with HDACIs that were not present in the original training set, such as triple bonds and the 3-bromo-4-hydroxy-phenyl group. Also, they found that unsaturated bonds in the linker region of HDACIs were frequently observed in the 45 hit compounds. With this finding, the authors proposed that unsaturated bonds could be an interesting alternative for the linker region since they are likely to reduce conformational freedom and hence decrease the net entropy of the system during the ligand binding process.

Another example is the excellent work published by Hou et al. (2015) where they use a different method to find new HDAC8 inhibitors. In this paper they reported a high-sensitivity pharmacophore model for HDAC8 inhibitors. They started from the assumption that a proper description of the zinc-binding group (ZBG) in HDACIs could increase the accuracy of the pharmacophore model when searching for inhibitors of HDAC1-11 (which are zinc-dependent enzymes). In previous pharmacophore models (Chen et al., 2008, 2009; Thangapandian et al., 2010), ZBGs were usually represented by hydrogen donor/acceptor features, which results in a high false positive rate and/or a low specificity. In order to avoid this common problem, Hou et al. undertook the task to design a specific feature for the ZBG present in HDACIs based on 17 known ZBGs and 22 computational predicted ZBGs. The final model was derived from the binding conformation of six compounds and consists of an aromatic ring feature, a hydrophobic feature, and the ZBG feature. Interestingly, the ZBG-based model overperformed several molecular docking protocols and other pharmacophore models without the ZBG feature in discriminating 134 HDAC8 inhibitors from 10,000 decoy compounds. After using the ZBG-based pharmacophore to screen a library of 150,000 drug-like compounds the authors of this work were able to identify three HDAC8 inhibitors. It is worth noting that the biological activities exhibited by these inhibitors (IC_{50} from 1.8 – 1.9 μM) are comparable to the IC_{50} of SAHA (4.3 μM). Moreover, one of these compounds (H8-A5) presented 16- and 20-fold selectivity for HDAC8 over HDAC1 and HDAC4, respectively, and antiproliferative activity in the MDA-MB-231 cell line (IC_{50} = 3.6 μM). These results represent a very successful application of pharmacophore modeling to drug discovery.

As noted by Hou et al., the ZBG of HDACIs is a critical feature that can be associated with gain or loss of biological activity. In fact, many research groups are using experimental and computational techniques to find new ZBGs that can increase the activity of HDACIs (Botta et al., 2011; Chen et al., 2013; Li and Woster, 2015; Patil et al., 2013). In this regard, the work of Liu et al. (2010) gains great relevance since they used an SVM classification model to identify novel ZBG useful for HDACIs. They trained their model on 702 hydroxamate HDACIs and 64,334 putative non-HDACIs encoded using 98 molecular descriptors. The final model (with a sensitivity and specificity above 99% in the 5-fold cross-validation) was used to screen PubChem and MDDR databases, where they identified over 40% of the reported HDACIs. A substructure analysis uncovered seven different types of ZBG in the hit compounds: sulfonamides, cyano containing groups, isothiocyanate, hydroxypyrones and its analogs, heterocyclic ketones, nitro group, and five-member heterocyclic compounds (e.g., thiadiazoles). Some of these chemicals groups are already used as ZBGs, for example, the nitro group, exhibited by some carboxypeptidase A inhibitors (Wang et al., 2008).

A structure-based alternative to analyze ZBGs was proposed by Chen et al. (2013). In a recent publication they used a two-step procedure to evaluate the binding affinity of potential ZBGs. In the first stage, they calculate the energy difference (ΔE) between the bound and free ligand using density functional theory (DFT) and the M05-2X/6-31+G* method only taking into account the binding site of HDAC2. Compounds with a favorable binding energy where submitted to a QM/MM calculation using the same functional together with the AMBER99 force field. After evaluating 48 fragments, the model was able to detect known ZBGs (e.g., hydroxamates), but also predicted new substructures as potential ZBGs: α-aminoalcohols, guaiacolate and 2-(aminomethyl) pyridines. This paper represents an original example of how structure-based methods can also be used to navigate the SAR of HDACIs.

As mentioned before, different computational techniques have been applied to the SAR analysis of HDACIs. Unfortunately, not all the relevant examples can be mentioned in this section. Nevertheless, Table 1 lists some of the recent (from 2012 to 2015) computational analyses that take HDACIs as subjects of

Table 1 Examples of Computational SAR Analyses on HDACIs from 2012 to 2015

References	HDAC Isoform	Data Set Size	Chemical Class	Method	Validation Results
Nair et al. (2012)	HDAC2	Training set: 38 Test set: 15 Pharmacophore: 19	Benzimidazole acrylamides, imidazole acrylamides	3D-QSAR, Pharmacophore modeling	$R^2_{training} = 0.99$ $R^2_{test} = 0.75$ $RMSE_{test} = 0.25$
Xiang et al. (2012)	HDAC2	Training set: 43 Test set: 7	Benzimidazole and imidazole inhibitors	3D-QSAR (COMFA, COMSIA)	$R^2_{training} = 0.98$ $R^2_{test} = 0.61$ $q^2_{LOO\_cv} = 0.80$
Li et al. (2014)	HDAC8	Training set: 51 Test set: 10	Hydroxamic acid analogues	3D-QSAR (COMFA, COMSIA)	$R^2_{training} = 0.99$ $q^2_{LOO\_cv} = 0.70$ $R^2_{test} = 0.82$
Kokkonen et al. (2014)	SIRT1	Training set: 65 Test set: 14	Modified lysine	3D-QSAR (COMFA, COMSIA)	$R^2_{training} = 0.86$ $R^2_{test} = 0.84$
Gupta et al. (2014)	HDAC class I	Training set: 11 Test set: 33 Decoy: 299	Hydroxamic acid derivatives, γ-lactam ring, bis-oxo benzo[b]thiophene analogues, etc.	Pharmacophore modeling	AUC = 0.78 Specif = 0.94 TPR = 0.969
Chuang et al. (2015)	SIRT2	Training set: 29 Test set: 17	2-Anilinobenzamide analogues	3D-QSAR (COMFA)	$R^2_{training} = 0.99$ $q^2_{LOO\_cv} = 0.88$ $R^2_{test} = 0.80$
Hou et al. (2015)	HDAC8	Training set: 6 Test set: 134 Decoy: 10,000	Inhibitors extracted from HDAC8 crystal structures	Pharmacophore modeling	$EF_{1\%} = 67.2$ $EF_{2\%} = 50$ $EF_{5\%} = 20$

study. Information of QSAR models published before 2012 can be found in a comprehensive review by Pontiki and Hadjipavlou-Litina (2012). Some interesting patterns can be observed in Table 1. For example, most of the recent research uses 3D-QSAR or pharmacophore models, which take into account the 3D structure of the ligand. Now is common to find computational studies on both zinc-dependent (HDAC1-11) and NAD-dependent (SIRTs family) HDACs. They usually represent high predictive models (q^2 or R^2_{test} > 0.7), nonetheless only a few of them reported experimental validation. Interestingly, the chemical diversity of the compounds increased compared to the studies published before 2012 when most of them were hydroxamic acid derivatives. Following this trend, it would not be surprising in the future to find new computational analyses describing the SARs of novel classes of HDACIs.

3.2 DNA Methyltransferase Inhibitors

One of the long-term mechanisms for gene expression regulation is DNA methylation (Reik, 2007). This process is mediated by DNA methyltransferases (DNMTs), a family of enzymes that add a methyl group to the carbon-5 position of cytosine residues (Vilkaitis et al., 2001). The inhibition of DNMTs by small molecules has been suggested as a strategy for the treatment of several pathologies including cancer (Baylin and Chen, 2005; Gros et al., 2012), age-related macular degeneration (AMD) (Kwa and Thrimawithana, 2014), depression (Sales et al., 2011; Xing et al., 2014), and some addictions (Anier et al., 2010; Han et al., 2010; Warnault et al., 2013), among others. To date, two DNMT inhibitors (DNMTIs) have been approved by the FDA (5-azacitidne and Decitabine) for the treatment of myelodysplastic syndrome despite their substantial cellular and clinical toxicity (Stresemann and Lyko, 2008). Apart from these molecules, many other DNMTIs have been reported in scientific literature and are generally divided into five groups: nucleoside analogs, natural products, repurposed drugs, hits from virtual screening, and compounds from lead optimization and high-throughput screening (HTS) hits. Some examples of DNMTIs are shown in Figure 2(a).

During the past years, numerous hypomethylating compounds have been identified, many of them by the use of computational methods and virtual screening (Medina-Franco and Caulfield, 2011; Medina-Franco et al., 2015). Surprisingly, robust and predictive QSAR models are not very common to study the SAR of DNMTIs. The principal reason for this is the high interassay variability and poor reproducibility in the existing biological assays (Yoo and Medina-Franco, 2011b). As a result, pharmacophore modeling and similarity searching, together with structure-based techniques, have positioned as the preferred methods for the computational analysis and identification of DNMTIs. In fact, a recent publication (Yoo and Medina-Franco, 2011c) reports a pharmacophore model derived from the predicted binding conformations

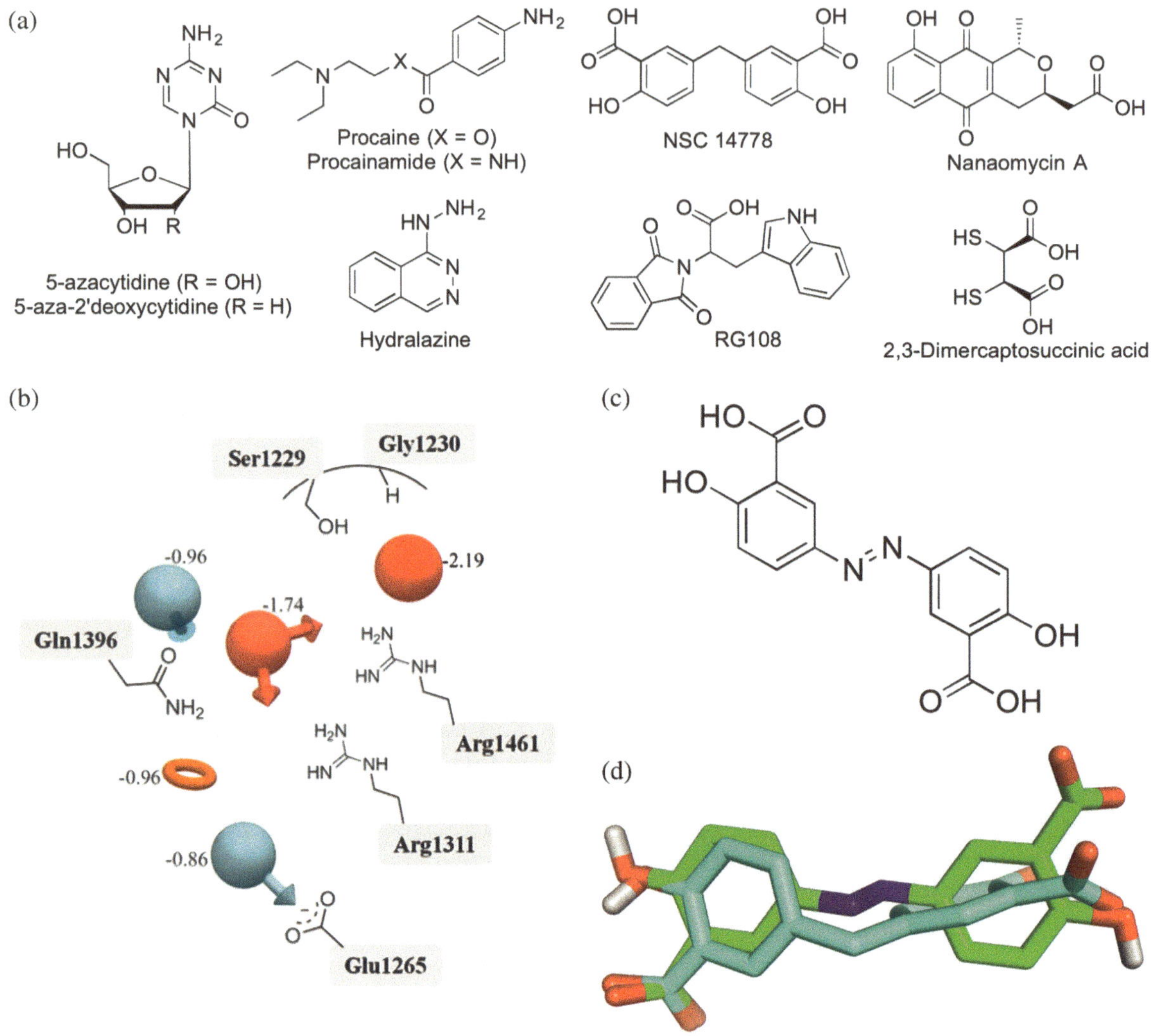

FIGURE 2

(a) Chemical structures of selected DNMT inhibitors. (b) Pharmacophore hypothesis published by Yoo and Medina-Franco (2011c) consisting of a negative charge (red sphere), a hydrogen bond acceptor (pink sphere), an aromatic ring (orange ring), and a hydrogen donor (blue sphere). *(Adapted from Yoo and Medina-Franco (2011c). Copyright (2011) John Wiley & Sons, Inc.)* (c) Chemical structure of olsalazine; (d) overlap of the structures of olsalazine (green) and NSC14778 (cyan). *Adapted from Méndez-Lucio et al. (2014).*

of 14 DNMTIs. The authors of this work used ligand–protein energetic terms to rank the importance of each of the pharmacophoric features. The final model contained four features: (1) a negative charge close to the side chains of Ser1229, Gly1230, and Arg1311; (2) a hydrogen bond acceptor adjacent to side chains of Arg1311 and Arg1416; (3) an aromatic ring that stabilizes the conformation of the ligands; and (4) a hydrogen donor site nearby the side chain of Glu1265 (Figure 2(b)). Interestingly, some of these features lie in close proximity to amino acids that take part in the catalytic mechanism, for example, Glus1265, Arg1311, and Arg1416. As one could expect, both FDA-approved drugs and other nucleoside analogs match the five pharmacophoric features, whereas many of the nonnucleoside inhibitors match at least one of the features. It is worth mentioning that other DNMTIs discovered independently also agree with this pharmacophoric model. This is the case of trimethylaurintricarboxylic acid, a DNMT1 inhibitor, which matches the ring, hydrogen bond acceptor and donor features (Yoo and Medina-Franco, 2012). Apart from identifying novel compounds during the virtual screening of compound libraries, this model is useful in guiding the optimization of hits. In general, this pharmacophoric model provides valuable information to understand the SAR of DNMTIs.

The large number of molecular databases combined with different computational methods also has accelerated the discovery of DNMTIs. As a proof of this, now it is common to perform virtual screening campaigns using large databases that usually result in the discovery of novel DNMTIs (e.g., RG108, NSC14778, and NSC319745) (Kuck et al., 2010; Siedlecki et al., 2006). In an effort to better understand the chemical properties of possible DNMTIs, Yoo and Medina-Franco (2011a) performed a chemoinformatic analysis of the databases commonly used for virtual screening. This analysis suggests that the DNMT-focused library, the traditional Chinese medicine (TCM) and the Specs databases present similar physicochemical properties to drugs. An important result that contributes to the SAR analysis of these compounds is the identification of chemical scaffolds associated with DNMTIs. In this case, the most common scaffolds in the DNMT-focused library contain a benzopyrrol or a benzodiazole ring, whereas a flavone scaffold was frequent in the TCM library. This was not surprising, as many flavonoids have been identified as DNMTIs before (Lee et al., 2005).

More recently, a similarity search for DNMTIs was conducted in one of these databases (i.e., DrugBank) using compound NSC14778 as a template (Méndez-Lucio et al., 2014). The DrugBank database contains data of 1600 FDA-approved small molecules, 160 FDA-approved biotech drugs, 89 nutraceuticals, and over 6000 experimental compounds (Wishart et al., 2006). As a result of this study, Olsalazine (Figure 2(c)), an anti-inflammatory agent, was identified and validated as a DNA hypomethylating agent possibly acting in one or more DNMTs. Interesting SAR information can be retrieved from this study; for example, the importance of the salicylate group for ligand binding. The salicylate is a frequent chemical feature among DNMTIs, not only present in NSC14778 and olsalazine, but also in other inhibitors such as trimethylaurintricarboxylic acid and aurintricarboxylic acid. The predicted binding conformations of these compounds show that the carboxylic group of the salicylate forms hydrogen bonds with Arg1311 and/or Arg1461, whereas the hydroxyl forms a hydrogen bond with Glu1265 (Kuck et al., 2010; Méndez-Lucio et al., 2014; Yoo and Medina-Franco, 2012). Not surprisingly, these interactions also match the hydrogen bond donor and acceptor features of the pharmacophore reported by Yoo and Medina-Franco (2011c).

3.3 Bromodomains Inhibitors

Bromodomains are considered some of the epigenetic readers that recognize marks in proteins of the epigenetic machinery (Filippakopoulos and Knapp, 2014). Specifically, bromodomains bind to acetylated lysine (AcK) located in the histone tails, which are susceptible to different post-translational modifications (Sạnchez and Zhou, 2009). In general, these conserved domains contain around 110 amino acids that always are found as part of a more complex protein structure (Hewings et al., 2012). The human proteome comprises 61 bromodomains that form part of 46 proteins, hence some proteins contain more than one bromodomain isoform (Filippakopoulos et al., 2012). The function of many bromodomains is still not completely understood. However, some of them, like the bromo and extra-terminal (BET) family of bromodomains, have been used as drug targets for male contraception (Matzuk et al., 2012), and to treat cancer (Dawson et al., 2011, 2012), obesity, and other diseases (Denis, 2010). In this regard, many bromodomain inhibitors have been designed as cancer drugs, but many others were developed as compound tools to understand better the biological function of bromodomains. In general, bromodomain inhibitors can be clustered based on the chemical group that mimics an AcK, which is a key feature for ligand binding. Some examples of AcK mimetic groups are triazolodiazepine, isoxazole, l-acyltetrahydroquinoline, and others shown in Figure 3(a) (Gallenkamp et al., 2014; Hewings et al., 2012; Muller and Knapp, 2014).

Since the resolution of the first crystal structure of a human bromodomain in 1999 (Dhalluin et al., 1999), many research groups have opted for a structure-based strategy to design bromodomain inhibitors. However, the numerous ligand–bromodomain structures reported so far have also facilitated the design of different pharmacophore models. For instance, Bamborough et al. (2012) reported a pharmacophore model based on crystal structures of BRD2 bound to compound I-BET762 (also known as GSK525762A) and the 4-phenyl 3,5-dimethyl isoxazole fragment. This pharmacophore model consists of four features: (1) a hydrogen bond acceptor at the nitrogen atom of the isoxasole ring, (2) a steric feature on the methyl group at position 5 of the isoxazole, (3) an aromatic feature on the phenyl ring of the fragment, and (4) a lipophilic feature at the pendant phenyl ring of I-BET762 (Figure 3(b)). The first two pharmacophoric

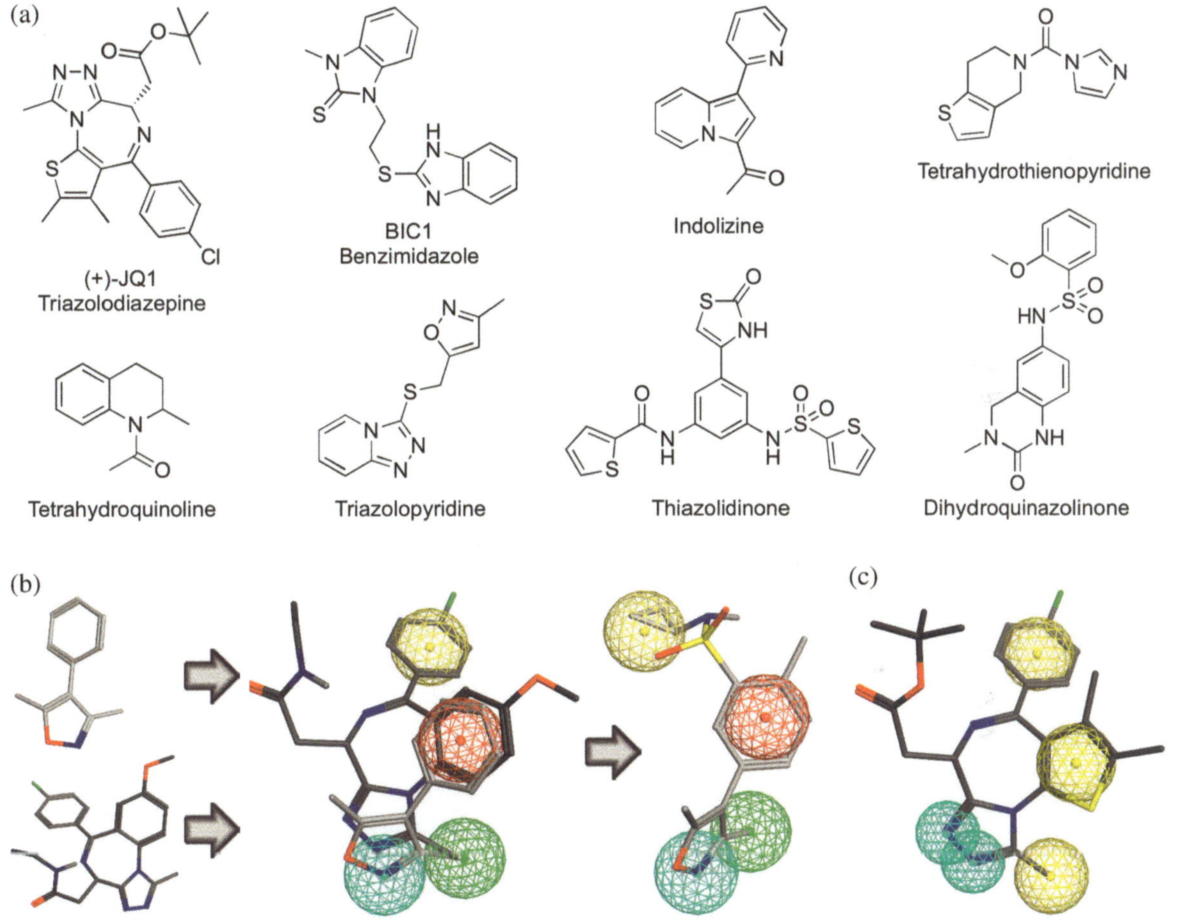

FIGURE 3

(a) Examples of bromodomain inhibitors exhibiting different chemical groups that mimic an acetylated lysine. (b) Schematic representation of the work published by Bamborough et al. (2012) showing the pharmacophore derived from I-BET762 and the 4-phenyl 3,5-dimethyl isoxazole fragment resulting in a series isoxazole analogs with a sulfonamide attached at meta position of the phenyl ring. (c) Pharmacophoric hypothesis used by Vidler et al. (2013) based on compound (+)-JQ1. In panels b and c blue = hydrogen bond acceptor, green = methyl feature, yellow = hydrophobic/lipophilic feature, and red = aromatic feature.

features are located on chemical groups that mimic the properties of an AcK in order to facilitate molecular recognition. Moreover, the lipophilic feature was added to identify substituents that interact with the WPF shelf, a region formed by residues 97–99 in BRD2 that is known to bind aromatic groups (Chung et al., 2011). These four pharmacophoric features were used to search a database of commercially available compounds, resulting in several hits. A set of isoxazole analogs with a sulfonamide attached at meta position of the phenyl ring were found to be up to 100-fold more potent than the initial fragment. Contrary to the initial isoxazole fragment, the hit compounds match the four pharmacophoric features. The results obtained in this study highlight the importance of having these four pharmacophoric features in order to design potent inhibitors.

In a more recent publication, Vidler et al. (2013) followed a similar approach in order to find bromodomain inhibitors with novel AcK mimetic groups. As part of their workflow, the authors performed docking, pharmacophore, shape, and 2D fingerprint (ECFP4) searches on different chemical databases. The pharmacophore model was based on the crystal structure of (+)-JQ1, a potent BRD4 inhibitor. Interestingly, the model used in this study is very similar to one reported by Bamborough et al.; that is, the WPF shelf, the ZA channel, and the methyl group covered by one hydrophobic feature each and two hydrogen bond acceptor features at the triazole ring (Figure 3(c)). The three ligand-based searches were followed by a docking protocol to ensure that the hit compounds fit in the restricted shape of the BRD4 biding site. As a result of this elegant workflow, six new inhibitors were found and used as the starting point for optimization projects. It is important to emphasize that the new inhibitors contain four new scaffolds that act as AcK mimetic, namely, chloropyridines, triazolopyrimidines, quinolines, and pyrrolopyridines. Other groups have reported pharmacophore models developed using more compounds but reaching to similar pharmacophore hypothesis (Devi et al., 2015).

4. OVERCOMING SELECTIVITY CHALLENGES IN THE COMPUTATIONAL DESIGN OF EPI-DRUGS

Although different inhibitors have been reported for most of the epigenetic targets, there are still many challenges to tackle in this field. One of the most important challenges in epigenetic drug discovery is the need for isoform and family selective inhibitors (Mack, 2010). Like many other cancer targets, epigenetic proteins are also involved in normal biological processes. Inhibiting all members of an epigenetic target class by an unselective inhibitor results in an inefficient control of the biological outcome. For example, a pan-epi-drug can reactivate hundreds of genes, some of them not as beneficial as expected (e.g., oncogenes) (Guha, 2015). Recently, specific cancer types have been associated with abnormal function of individual epigenetic targets (Deng et al., 2009; Pasqualucci et al., 2011). This suggests that selective inhibitors will keep their therapeutic effect, possibly with a reduced number of side effects for the patient. In addition to the therapeutic benefit, selective inhibitors will help to reveal the specific biological functions of each epigenetic target (Erdmann et al., 2015).

From the computational viewpoint, modeling selectivity profiles of a set of compounds is not straightforward. The complexity arises from the fact that selectivity depends on the chemical features present in the ligand but also on the amino acidic sequence and 3D conformation of the target. Usually, multitarget models are preferred to model selectivity, as the activity of a single compound needs to be predicted against a set of proteins. One option is to generate a set of independent QSAR models, each of them to evaluate the activity of the ligand against an individual target (Zhao et al., 2013). An alternative method is to include both target and ligand information into the same model, as in the case of PCM, which regularly performs better than individual QSAR models (see Section 2.3). In this regard, the work of Wu et al. (2012) gains relevance as they generated a PCM model capable to predict selectivity patterns among different HDAC isoforms. In order to do this they trained an SVM model with activity data (IC_{50}) of 1275 compound-HDAC pairs. This data contained information for HDAC2, HDAC4, HDAC6, HDAC7, and HDAC8. They were able to obtain high correlation during the internal cross-validation procedure ($q^2_{cv}=0.75$) and also to predict a test set ($R^2_{test}=0.72$). As explained before, PCM modeling is able to take into account differences and similarities among the target proteins included in the model. In this case, Wu et al. showed that using 3D structure similarity of the targets as a descriptor showed improved results compared to the use of sequence similarity descriptors. On the other hand, no differences in the predictive power were observed when using two different compound descriptors (i.e., General Descriptors and Drug-Like Index). Interestingly, this model was tested on an external set of 10 known ligands and was able to distinguish between pan-HDAC inhibitors, class I-specific inhibitors, and class II specific inhibitors. This model is very useful in finding antitumor drugs that target a single HDAC and hence present fewer side effects.

Following a similar idea, Silvestri et al. (2012) generated several structure-based 3D QSAR models. With these models they were able to identify the ligand-protein interactions needed for ligand binding. These models include activity data of 94 compound-HDAC pairs corresponding to 15 different inhibitors and 12 HDAC isoforms. All compounds were docked in the respective HDAC to generate a putative 3D conformation used to calculate 3D descriptors. They apply a modified version of the comparative binding energy analysis (Ortiz et al., 1995) named COMBINEr, which uses AutoDock's AutoGrid to calculate interaction energies (Ballante et al., 2012). With this, they were able to correlate the pIC_{50} data of the inhibitors with the electrostatic and van der Waals interaction energies. The best model presented a good internal predictability with a $q^2_{LOO\_cv}=0.76$ and $SDEP_{cv}=0.81$ in the Leave-One-Out (LOO) and in the 5-fold cross-validation. The use of 3D descriptors together with PLS facilitated the interpretation of the model by

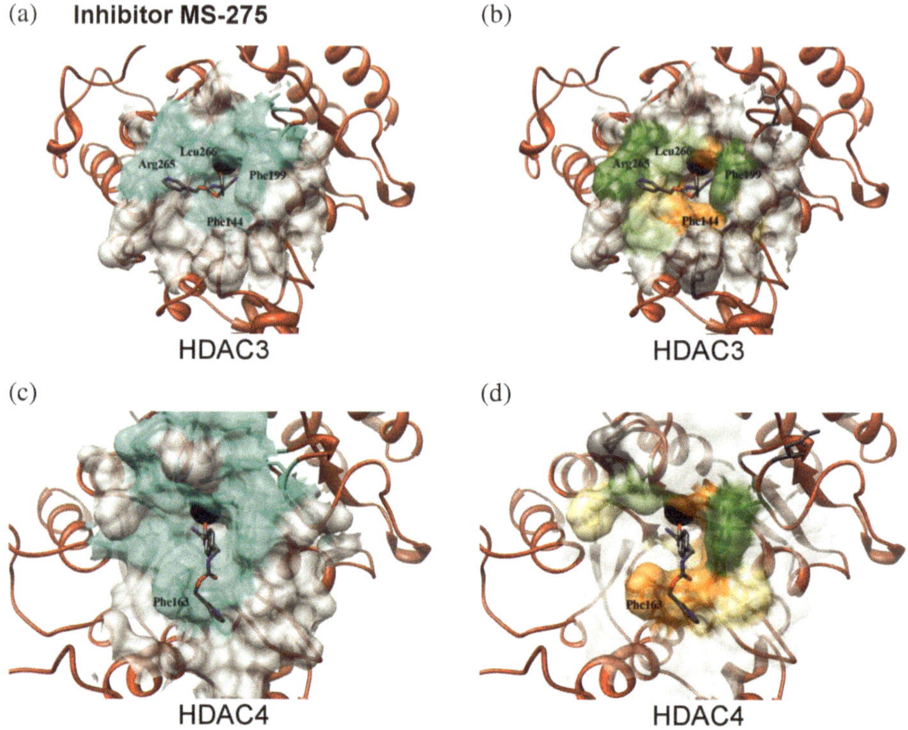

FIGURE 4

Compound MS-275 bound to HDAC3 and HDAC4. Panels a and c depict the ELE filed, whereas b and d depict the DRY field used in the calculations. For the ELE field, light blue refers to areas with poor positive contribution to the activity, whereas red areas represent a negative contribution. For the DRY field, positive contributions are indicated in green and negative contributions in yellow. Dark gray surfaces indicate zero contribution. *Reprinted with permission from Silvestri et al. (2012). Copyright 2012 American Chemical Society.*

showing the influence of each amino acid in the binding site to ligand binding (Figure 4). As a result, Silvestri et al. were able to identify five residues (His254, His294, Asp392, Phe/Tyr263, Met/Lys/Leu401 following the amino acid numbering used in this paper) that are of major importance in modulating the inhibitory activity against all HDACs. Moreover, with the COMBINEr method they were able to explain isoform selectivity with the information of only 34 residues in the binding site. Figure 4 shows a graphical representation of compound MS-275 bound to HDAC3 and HDAC4, highlighting the amino acids that influence the binding of this compound. Overall, similar strategies to model multitarget activity seem to be rapidly advancing to predict intrafamily selectivity of epi-drugs.

5. CONCLUSIONS

As seen throughout this chapter, epigenetic targets have captured the attention of medicinal chemists for the last decades. One attractive characteristic of these targets is that they can be useful to treat several, and in some cases unrelated, diseases. The need for novel and diverse therapies, especially to treat cancer, has accelerated the FDA approval for many epi-drugs, despite the numerous side effects presented by some of them. On this subject, computational methods have served as a useful tool to accelerate the discovery and improvement of epi-drugs. The rapid development and disclosure of new data sets, together with the increased number of reported crystallographic structures, has facilitated the use of computational methods. As mentioned at the beginning of this chapter, many of these methods are suited for the systematic study and mining of large datasets, giving as a result important SAR information. In addition to this valuable information, in many cases they also help to identify compounds that can be employed as starting points of medicinal chemistry projects.

There are still many challenges in the epigenetic therapy field. Among these, selectivity, drug combinations, and target validation are of special importance. The new generation of epi-drugs is expected to be more selective and with defined drug targets. In order to achieve this, it will be necessary to make direct or indirect use of computational methods at some point of the drug development. For this, computational methods will need to adapt to new types of data (e.g., drug combinations) and to evolve to model more complex interactions (selectivity).

Although it would be impossible to cover all the computational studies performed to discover new epi-drugs, the aim of this chapter was to give a general overview of recent efforts taken in this field. It is intriguing that many of the examples showed in this chapter discovered new chemical entities that will be used to generate analogs, which are very likely to be fed back into a computational method at some point. Based on this successful iterative process, it can be anticipated that computational methods will be applied in the discovery of inhibitors for new epigenetic targets. Moreover, they will continue as key components of epigenetic drug discovery projects in the years to come.

ACKNOWLEDGMENTS

The author thanks Stephanie Ashenden for proofreading the manuscript and for insightful comments. O. M-L. acknowledges CONACyT (No. 217442/312933) and the Cambridge Overseas Trust for funding. This work was supported by a scholarship from the Secretariat of Public Education and the Mexican government.

REFERENCES

Ain, Q.U., Mendez-Lucio, O., Ciriano, I.C., Malliavin, T., van Westen, G.J.P., Bender, A., 2014. Modelling ligand selectivity of serine proteases using integrative proteochemometric approaches improves model performance and allows the multi-target dependent interpretation of features. Integr. Biol. 6 (11), 1023–1033.

Anier, K., Malinovskaja, K., Aonurm-Helm, A., Zharkovsky, A., Kalda, A., 2010. DNA methylation regulates cocaine-induced behavioral sensitization in mice. Neuropsychopharmacology 35 (12), 2450–2461.

Arrowsmith, C.H., Bountra, C., Fish, P.V., Lee, K., Schapira, M., 2012. Epigenetic protein families: a new frontier for drug discovery. Nat. Rev. Drug Discov. 11 (5), 384–400.

Ballante, F., Musmuca, I., Marshall, G., Ragno, R., 2012. Comprehensive model of wild-type and mutant HIV-1 reverse transciptases. J. Comput. Aided Mol. Des. 26 (8), 907–919.

Bamborough, P., Diallo, H., Goodacre, J.D., Gordon, L., Lewis, A., Seal, J.T., et al., 2012. Fragment-based discovery of bromodomain inhibitors part 2: optimization of phenylisoxazole sulfonamides. J. Med. Chem. 55 (2), 587–596.

Baylin, S.B., Chen, W.Y., 2005. Aberrant gene silencing in tumor progression: implications for control of cancer. Cold Spring Harb. Symp. Quant. Biol. 70, 427–433.

Benson, M.L., Smith, R.D., Khazanov, N.A., Dimcheff, B., Beaver, J., Dresslar, P., et al., 2008. Binding MOAD, a high-quality protein–ligand database. Nucleic Acids Res. 36 (Suppl. 1), D674–D678.

Berger, S.L., Kouzarides, T., Shiekhattar, R., Shilatifard, A., 2009. An operational definition of epigenetics. Genes Dev. 23 (7), 781–783.

Best, J.D., Carey, N., 2010. Epigenetic therapies for non-oncology indications. Drug Discov. Today 15 (23–24), 1008–1014.

Botta, C.B., Cabri, W., Cini, E., De Cesare, L., Fattorusso, C., Giannini, G., et al., 2011. Oxime amides as a novel zinc binding group in histone deacetylase inhibitors: synthesis, biological activity, and computational evaluation. J. Med. Chem. 54 (7), 2165–2182.

Chen, K., Xu, L., Wiest, O., 2013. Computational exploration of zinc binding groups for HDAC inhibition. J. Org. Chem. 78 (10), 5051–5055.

Chen, Y., Li, H., Tang, W., Zhu, C., Jiang, Y., Zou, J., et al., 2009. 3D-QSAR studies of HDACs inhibitors using pharmacophore-based alignment. Eur. J. Med. Chem. 44 (7), 2868–2876.

Chen, Y.-D., Jiang, Y.-J., Zhou, J.-W., Yu, Q.-S., You, Q.-D., 2008. Identification of ligand features essential for HDACs inhibitors by pharmacophore modeling. J. Mol. Graph. Model. 26 (7), 1160–1168.

Cherkasov, A., Muratov, E.N., Fourches, D., Varnek, A., Baskin, I.I., Cronin, M., et al., 2014. QSAR modeling: where have you been? where are you going to? J. Med. Chem. 57 (12), 4977–5010.

Chuang, Y.-C., Chang, C.-H., Lin, J.-T., Yang, C.-N., 2015. Molecular modelling studies of sirtuin 2 inhibitors using three-dimensional structure-activity relationship analysis and molecular dynamics simulations. Mol. Biosyst. 11 (3), 723–733.

Chung, C.-W., Coste, H., White, J.H., Mirguet, O., Wilde, J., Gosmini, R.L., et al., 2011. Discovery and characterization of small molecule inhibitors of the BET family bromodomains. J. Med. Chem. 54 (11), 3827–3838.

Cortes-Ciriano, I., Ain, Q.U., Subramanian, V., Lenselink, E.B., Mendez-Lucio, O., Ijzerman, A.P., et al., 2015. Polypharmacology modelling using proteochemometrics (PCM): recent methodological developments, applications to target families, and future prospects. MedChemComm 6 (1), 24–50.

Cress, W.D., Seto, E., 2000. Histone deacetylases, transcriptional control, and cancer. J. Cell. Physiol. 184 (1), 1–16.

Cruz-Monteagudo, M., Medina-Franco, J.L., Pérez-Castillo, Y., Nicolotti, O., Cordeiro, M.N.D.S., Borges, F., 2014. Activity cliffs in drug discovery: Dr Jekyll or Mr Hyde? Drug Discov. Today 19 (8), 1069–1080.

Dawson, M.A., Kouzarides, T., Huntly, B.J.P., 2012. Targeting epigenetic readers in cancer. N. Engl. J. Med. 367 (7), 647–657.

Dawson, M.A., Prinjha, R.K., Dittmann, A., Giotopoulos, G., Bantscheff, M., Chan, W.-I., et al., 2011. Inhibition of BET recruitment to chromatin as an effective treatment for MLL-fusion leukaemia. Nature 478 (7370), 529–533.

Dehmer, M., Varmuza, K., Bonchev, D., 2012. Statistical Modelling of Molecular Descriptors in QSAR/QSPR, vol. 2. Wiley-VCH Verlag GmbH & Co. KGaA, Weinheim, Germany.

Deng, T., Kuang, Y., Wang, L., Li, J., Wang, Z., Fei, J., 2009. An essential role for DNA methyltransferase 3a in melanoma tumorigenesis. Biochem. Biophys. Res. Commun. 387 (3), 611–616.

Denis, G.V., 2010. Bromodomain coactivators in cancer, obesity, type 2 diabetes, and inflammation. Discov. Med. 10 (55), 489–499.

Devi, S.T., Tandon, H., Ampasala, D.R., 2015. Identification of potent bromodomain4 (BRD4) inhibitors by energy-pharmacophore based virtual screening to target BRD4-NUT midline carcinoma. Int. J. Pharm. Sci. 7 (4), 77–84.

DeWoskin, V.A., Million, R.P., 2013. The epigenetics pipeline. Nat. Rev. Drug Discov. 12 (9), 661–662.

Dhalluin, C., Carlson, J.E., Zeng, L., He, C., Aggarwal, A.K., Zhou, M.-M., et al., 1999. Structure and ligand of a histone acetyltransferase bromodomain. Nature 399 (6735), 491–496.

Downs, G.M., Willett, P., 2007. Similarity searching in databases of chemical structures. Reviews in Computational Chemistry, vol. 7. John Wiley & Sons, Inc. 1–66.

Drummond, D.C., Noble, C.O., Kirpotin, D.B., Guo, Z., Scott, G.K., Benz, C.C., 2005. Clinical development of histone deacetylase inhibitors as anticancer agents. Annu. Rev. Pharmacol. Toxicol. 45, 495–528.

Erdmann, A., Halby, L., Fahy, J., Arimondo, P.B., 2015. Targeting DNA methylation with small molecules: what's next? J. Med. Chem. 58 (6), 2569–2583.

Eriksson, L., Jaworska, J., Worth, A.P., Cronin, M.T., McDowell, R.M., Gramatica, P., 2003. Methods for reliability and uncertainty assessment and for applicability evaluations of classification- and regression-based QSARs. Environ. Health Perspect. 111 (10), 1361–1375.

Falahi, F., van Kruchten, M., Martinet, N., Hospers, G., Rots, M., 2014. Current and upcoming approaches to exploit the reversibility of epigenetic mutations in breast cancer. Breast Cancer Res. 16 (4), 412.

Fernandez, M., Ahmad, S., Sarai, A., 2010. Proteochemometric recognition of stable kinase inhibition complexes using topological autocorrelation and support vector machines. J. Chem. Inf. Model. 50 (6), 1179–1188.

Filippakopoulos, P., Knapp, S., 2014. Targeting bromodomains: epigenetic readers of lysine acetylation. Nat. Rev. Drug Discov. 13 (5), 337–356.

Filippakopoulos, P., Picaud, S., Mangos, M., Keates, T., Lambert, J.-P., Barsyte-Lovejoy, D., et al., 2012. Histone recognition and large-scale structural analysis of the human bromodomain family. Cell 149 (1), 214–231.

Gallenkamp, D., Gelato, K.A., Haendler, B., Weinmann, H., 2014. Bromodomains and their pharmacological inhibitors. ChemMedChem 9 (3), 438–464.

Gaulton, A., Bellis, L.J., Bento, A.P., Chambers, J., Davies, M., Hersey, A., et al., 2012. ChEMBL: a large-scale bioactivity database for drug discovery. Nucleic Acids Res. 40 (D1), D1100–D1107.

Gros, C., Fahy, J., Halby, L., Dufau, I., Erdmann, A., Gregoire, J.-M., et al., 2012. DNA methylation inhibitors in cancer: recent and future approaches. Biochimie 94 (11), 2280–2296.

Guha, M., 2015. HDAC inhibitors still need a home run, despite recent approval. Nat. Rev. Drug Discov. 14 (4), 225–226.

Guha, R., 2012a. Exploring structure–activity data using the landscape paradigm. WIREs Comput. Mol. Sci. 2 (6), 829–841.

Guha, R., 2012b. Exploring uncharted territories: predicting activity cliffs in Structure–Activity landscapes. J. Chem. Inf. Model. 52 (8), 2181–2191.

Guha, R., Van Drie, J.H., 2008. Structure–Activity landscape Index: identifying and quantifying activity cliffs. J. Chem. Inf. Model. 48 (3), 646–658.

Gund, P., 1977. Three-dimensional pharmacophoric pattern searching. In: Hahn, F., Kersten, H., Kersten, W., Szybalski, W. (Eds.), Progress in Molecular and Subcellular Biology, vol. 5. Springer Berlin Heidelberg, pp. 117–143.

Gupta, N., Sitwala, N., Patel, K., 2014. Pharmacophore modelling, validation, 3D virtual screening, docking, design and in silico ADMET simulation study of histone deacetylase class-1 inhibitors. Med. Chem. Res. 23 (11), 4853–4864.

Han, J., Li, Y., Wang, D., Wei, C., Yang, X., Sui, N., 2010. Effect of 5-aza-2-deoxycytidine microinjecting into hippocampus and prelimbic cortex on acquisition and retrieval of cocaine-induced place preference in C57BL/6 mice. Eur. J. Pharmacol. 642 (1–3), 93–98.

Hansch, C., Maloney, P.P., Fujita, T., Muir, R.M., 1962. Correlation of biological activity of phenoxyacetic acids with Hammett substituent constants and partition coefficients. Nature 194 (4824), 178–180.

Heikamp, K., Hu, X., Yan, A., Bajorath, J., 2012. Prediction of activity cliffs using support vector machines. J. Chem. Inf. Model. 52 (9), 2354–2365.

Hewings, D.S., Rooney, T.P.C., Jennings, L.E., Hay, D.A., Schofield, C.J., Brennan, P.E., et al., 2012. Progress in the development and application of small molecule inhibitors of Bromodomain–Acetyl-lysine interactions. J. Med. Chem. 55 (22), 9393–9413.

Ho, A.S., Turcan, S., Chan, T.A., 2013. Epigenetic therapy: use of agents targeting deacetylation and methylation in cancer management. Onco Targets Ther. 6, 223–232.

Hou, X., Du, J., Liu, R., Zhou, Y., Li, M., Xu, W., et al., 2015. Enhancing the sensitivity of pharmacophore-based virtual screening by incorporating customized ZBG features: a case study using histone deacetylase 8. J. Chem. Inf. Model. 55 (4), 861–871. http://dx.doi.org/10.1021/ci500762z.

Hu, L., Benson, M.L., Smith, R.D., Lerner, M.G., Carlson, H.A., 2005. Binding MOAD (Mother of all databases). Proteins 60 (3), 333–340.

Kokkonen, P., Mellini, P., Nyrhilä, O., Rahnasto-Rilla, M., Suuronen, T., Kiviranta, P., et al., 2014. Quantitative insights for the design of substrate-based SIRT1 inhibitors. Eur. J. Pharm. Sci. 59, 12–19.

Kubinyi, H., 1998. Similarity and dissimilarity: a medicinal chemist's view. Perspect. Drug Discov. Des. 9-11 (0), 225–252.

Kuck, D., Singh, N., Lyko, F., Medina-Franco, J.L., 2010. Novel and selective DNA methyltransferase inhibitors: docking-based virtual screening and experimental evaluation. Bioorg. Med. Chem. 18 (2), 822–829.

Kwa, F.A.A., Thrimawithana, T.R., 2014. Epigenetic modifications as potential therapeutic targets in age-related macular degeneration and diabetic retinopathy. Drug Discov. Today 19 (9), 1387–1393.

Lapins, M., Wikberg, J., 2010. Kinome-wide interaction modelling using alignment-based and alignment-independent approaches for kinase description and linear and non-linear data analysis techniques. BMC Bioinf. 11 (1), 339.

Lapins, M., Worachartcheewan, A., Spjuth, O., Georgiev, V., Prachayasittikul, V., Nantasenamat, C., et al., 2013. A unified proteochemometric model for prediction of inhibition of cytochrome P450 isoforms. PLoS One 8 (6), e66566.

Lavecchia, A., 2015. Machine-learning approaches in drug discovery: methods and applications. Drug Discov. Today 20 (3), 318–331.

Leach, A.R., Gillet, V.J., Lewis, R.A., Taylor, R., 2010. Three-dimensional pharmacophore methods in drug discovery. J. Med. Chem. 53 (2), 539–558.

Lee, W.J., Shim, J.-Y., Zhu, B.T., 2005. Mechanisms for the inhibition of DNA methyltransferases by tea catechins and bioflavonoids. Mol. Pharmacol. 68 (4), 1018–1030.

Li, X., Zhang, Y., Zhang, L., Xu, Y., Xu, W., 2014. 3D QSAR and docking studies of a series of histone deacetylase inhibitors. Med. Chem. Res. 23 (5), 2229–2241.

Li, Y., Woster, P.M., 2015. Discovery of a new class of histone deacetylase inhibitors with a novel zinc binding group. MedChemComm 6 (4), 613–618.

Liu, T., Lin, Y., Wen, X., Jorissen, R.N., Gilson, M.K., 2007. BindingDB: a web-accessible database of experimentally determined protein–ligand binding affinities. Nucleic Acids Res. 35 (Suppl. 1), D198–D201.

Liu, X.H., Song, H.Y., Zhang, J.X., Han, B.C., Wei, X.N., Ma, X.H., et al., 2010. Identifying novel type ZBGs and nonhydroxamate HDAC inhibitors through a SVM based virtual screening approach. Mol. Inf. 29 (5), 407–420.

Loharch, S., Bhutani, I., Jain, K., Gupta, P., Sahoo, D.K., Parkesh, R., 2015. EpiDBase: a manually curated database for small molecule modulators of epigenetic landscape. Database 2015. http://dx.doi.org/10.1093/database/bav1013.

Mack, G.S., 2010. To selectivity and beyond. Nat. Biotech. 28 (12), 1259–1266.

Maggiora, G.M., 2006. On outliers and activity cliffs – why QSAR often disappoints. J. Chem. Inf. Model. 46 (4), 1535.

Matzuk, M.M., McKeown, M.R., Filippakopoulos, P., Li, Q., Ma, L., Agno, J.E., et al., 2012. Small-molecule inhibition of BRDT for male contraception. Cell 150 (4), 673–684.

Medina-Franco, J.L., 2012. Scanning structure–activity relationships with structure–activity similarity and related maps: from consensus activity cliffs to selectivity switches. J. Chem. Inf. Model. 52 (10), 2485–2493.

Medina-Franco, J.L., Caulfield, T., 2011. Advances in the computational development of DNA methyltransferase inhibitors. Drug Discov. Today 16 (9–10), 418–425.

Medina-Franco, J.L., Méndez-Lucio, O., Dueñas-González, A., Yoo, J., 2015. Discovery and development of DNA methyltransferase inhibitors using in silico approaches. Drug Discov. Today 20 (5), 569–577. http://dx.doi.org/10.1016/j.drudis.2014.12.007.

Medina-Franco, J.L., Méndez-Lucio, O., Martinez-Mayorga, K., 2014. Chapter one – the interplay between molecular modeling and chemoinformatics to characterize protein–ligand and protein–protein interactions landscapes for drug discovery. In: Tatyana, K.-C. (Ed.), Advances in Protein Chemistry and Structural Biology, vol. 96. Academic Press, pp. 1–37.

Méndez-Lucio, O., Kooistra, A.J., Graaf, C.d., Bender, A., Medina-Franco, J.L., 2015. Analyzing multitarget activity landscapes using protein–ligand interaction fingerprints: interaction cliffs. J. Chem. Inf. Model. 55 (2), 251–262.

Méndez-Lucio, O., Pérez-Villanueva, J., Castillo, R., Medina-Franco, J.L., 2012a. Activity landscape modeling of PPAR ligands with dual-activity difference maps. Bioorg. Med. Chem. 20 (11), 3523–3532.

Méndez-Lucio, O., Pérez-Villanueva, J., Castillo, R., Medina-Franco, J.L., 2012b. Identifying activity cliff generators of PPAR ligands using SAS maps. Mol. Inf. 31 (11–12), 837–846.

Méndez-Lucio, O., Tran, J., Medina-Franco, J.L., Meurice, N., Muller, M., 2014. Toward drug repurposing in epigenetics: olsalazine as a hypomethylating compound active in a cellular context. Chem. Med. Chem. 9 (3), 560–565.

Muller, S., Brown, P.J., 2012. Epigenetic chemical probes. Clin. Pharmacol. Ther. 92 (6), 689–693.

Muller, S., Knapp, S., 2014. Discovery of BET bromodomain inhibitors and their role in target validation. MedChemComm 5 (3), 288–296.

Nair, S.B., Teli, M.K., Pradeep, H., Rajanikant, G.K., 2012. Computational identification of novel histone deacetylase inhibitors by docking based QSAR. Comput. Biol. Med. 42 (6), 697–705.

Nebbioso, A., Carafa, V., Benedetti, R., Altucci, L., 2012. Trials with 'epigenetic' drugs: an update. Mol. Oncol. 6 (6), 657–682.

Ortiz, A.R., Pisabarro, M.T., Gago, F., Wade, R.C., 1995. Prediction of drug binding affinities by comparative binding energy analysis. J. Med. Chem. 38 (14), 2681–2691.

Pasqualucci, L., Dominguez-Sola, D., Chiarenza, A., Fabbri, G., Grunn, A., Trifonov, V., et al., 2011. Inactivating mutations of acetyltransferase genes in B-cell lymphoma. Nature 471 (7337), 189–195.

Patil, V., Sodji, Q.H., Kornacki, J.R., Mrksich, M., Oyelere, A.K., 2013. 3-Hydroxypyridin-2-thione as novel zinc binding group for selective histone deacetylase inhibition. J. Med. Chem. 56 (9), 3492–3506.

Peltason, L., Bajorath, J., 2007. SAR index: quantifying the nature of structure–activity relationships. J. Med. Chem. 50 (23), 5571–5578.

Perez-Villanueva, J., Santos, R., Hernandez-Campos, A., Giulianotti, M.A., Castillo, R., Medina-Franco, J.L., 2011. Structure-activity relationships of benzimidazole derivatives as antiparasitic agents: dual activity-difference (DAD) maps. Med. Chem. Comm. 2 (1), 44–49.

Pontiki, E., Hadjipavlou-Litina, D., 2012. Histone deacetylase inhibitors (HDACIs). Structure—activity relationships: history and new QSAR perspectives. Med. Res. Rev. 32 (1), 1–165.

Portela, A., Esteller, M., 2010. Epigenetic modifications and human disease. Nat. Biotech. 28 (10), 1057–1068.

Reik, W., 2007. Stability and flexibility of epigenetic gene regulation in mammalian development. Nature 447 (7143), 425–432.

Remiszewski, S.W., Sambucetti, L.C., Atadja, P., Bair, K.W., Cornell, W.D., Green, M.A., et al., 2002. Inhibitors of human histone deacetylase: synthesis and enzyme and cellular activity of straight chain hydroxamates. J. Med. Chem. 45 (4), 753–757.

Rogers, D., Hahn, M., 2010. Extended-connectivity fingerprints. J. Chem. Inf. Model. 50 (5), 742–754.

Sales, A.J., Biojone, C., Terceti, M.S., Guimarães, F.S., Gomes, M.V.M., Joca, S.R.L., 2011. Antidepressant-like effect induced by systemic and intra-hippocampal administration of DNA methylation inhibitors. Br. J. Pharmacol. 164 (6), 1711–1721.

Sanchez, R., Zhou, M.M., 2009. The role of human bromodomains in chromatin biology and gene transcription. Curr. Opin. Drug Discov. Devel. 12 (5), 659–665.

Sanders, M.P.A., McGuire, R., Roumen, L., de Esch, I.J.P., de Vlieg, J., Klomp, J.P.G., et al., 2012. From the protein's perspective: the benefits and challenges of protein structure-based pharmacophore modeling. MedChemComm 3 (1), 28–38.

Santos, R.G., Giulianotti, M.A., Houghten, R.A., Medina-Franco, J.L., 2013. Conditional probabilistic analysis for prediction of the activity landscape and relative compound activities. J. Chem. Inf. Model. 53 (10), 2613–2625.

Seebeck, B., Wagener, M., Rarey, M., 2011. From activity cliffs to target-specific scoring models and pharmacophore hypotheses. ChemMedChem 6 (9), 1630–1639.

Shanmugasundaram, V., Maggiora, G.M., 2001. Characterizing property and activity landscapes using an information-theoretic approach. In: Paper Presented at the 222nd ACS National Meeting, Chicago, IL, USA.

Siedlecki, P., Boy, R.G., Musch, T., Brueckner, B., Suhai, S., Lyko, F., et al., 2006. Discovery of two novel, small-molecule inhibitors of DNA methylation. J. Med. Chem. 49 (2), 678–683.

Silvestri, L., Ballante, F., Mai, A., Marshall, G.R., Ragno, R., 2012. Histone deacetylase inhibitors: structure-based modeling and isoform-selectivity prediction. J. Chem. Inf. Model. 52 (8), 2215–2235.

Sterner, D.E., Berger, S.L., 2000. Acetylation of histones and transcription-related factors. Microbiol. Mol. Biol. Rev. 64 (2), 435–459.

Stresemann, C., Lyko, F., 2008. Modes of action of the DNA methyltransferase inhibitors azacytidine and decitabine. Int. J. Cancer 123 (1), 8–13.

Stumpfe, D., Bajorath, J., 2011. Similarity searching. Wiley Interdiscip. Rev. Comput. Mol. Sci. 1 (2), 260–282.

Stumpfe, D., Bajorath, J., 2012. Exploring activity cliffs in medicinal chemistry. J. Med. Chem. 55 (7), 2932–2942.

Stumpfe, D., Hu, Y., Dimova, D., Bajorath, J., 2014. Recent progress in understanding activity cliffs and their utility in medicinal chemistry. J. Med. Chem. 57 (1), 18–28.

Subramanian, V., Prusis, P., Pietilä, L.-O., Xhaard, H., Wohlfahrt, G., 2013. Visually interpretable models of kinase selectivity related features derived from field-based proteochemometrics. J. Chem. Inf. Model. 53 (11), 3021–3030.

Tang, H., Wang, X.S., Huang, X.-P., Roth, B.L., Butler, K.V., Kozikowski, A.P., et al., 2009. Novel inhibitors of human histone deacetylase (HDAC) identified by QSAR modeling of known inhibitors, virtual screening, and experimental validation. J. Chem. Inf. Model. 49 (2), 461–476.

Thangapandian, S., John, S., Sakkiah, S., Lee, K.W., 2010. Ligand and structure based pharmacophore modeling to facilitate novel histone deacetylase 8 inhibitor design. Eur. J. Med. Chem. 45 (10), 4409–4417.

Todeschini, R., Consonni, V., 2008. Handbook of Molecular Descriptors. Wiley-VCH Verlag GmbH, Germany.

Van Drie, J.H., 2013. Generation of three-dimensional pharmacophore models. Wiley Interdiscip. Rev. Comput. Mol. Sci. 3 (5), 449–464.

Vidler, L.R., Filippakopoulos, P., Fedorov, O., Picaud, S., Martin, S., Tomsett, M., et al., 2013. Discovery of novel small-molecule inhibitors of BRD4 using structure-based virtual screening. J. Med. Chem. 56 (20), 8073–8088.

Vilkaitis, G., Merkienė, E., Serva, S., Weinhold, E., Klimašauskas, S., 2001. The mechanism of DNA Cytosine-5 Methylation: kinetic and mutational dissection of Hhal methyltransferase. J. Biol. Chem. 276 (24), 20924–20934.

Wang, D.-F., Wiest, O., Helquist, P., Lan-Hargest, H.-Y., Wiech, N.L., 2004. QSAR Studies of PC-3 cell line inhibition activity of TSA and SAHA-like hydroxamic acids. Bioorg. Med. Chem. Lett. 14 (3), 707–711.

Wang, S.-H., Wang, S.-F., Xuan, W., Zeng, Z.-H., Jin, J.-Y., Ma, J., et al., 2008. Nitro as a novel zinc-binding group in the inhibition of carboxypeptidase A. Bioorg. Med. Chem. 16 (7), 3596–3601.

Wang, Y., Xiao, J., Suzek, T.O., Zhang, J., Wang, J., Zhou, Z., et al., 2012. PubChem's BioAssay database. Nucleic Acids Res. 40 (D1), D400–D412.

Warnault, V., Darcq, E., Levine, A., Barak, S., Ron, D., 2013. Chromatin remodeling — a novel strategy to control excessive alcohol drinking. Transl. Psychiatry 3, e231.

van Westen, G.J.P., Wegner, J.K., Ijzerman, A.P., van Vlijmen, H.W.T., Bender, A., 2011. Proteochemometric modeling as a tool to design selective compounds and for extrapolating to novel targets. MedChemComm 2 (1), 16–30.

van Westen, G.J.P., van den Hoven, O.O., van der Pijl, R., Mulder-Krieger, T., de Vries, H., Wegner, J.K., et al., 2012. Identifying novel adenosine receptor ligands by simultaneous proteochemometric modeling of rat and human bioactivity data. J. Med. Chem. 55 (16), 7010–7020.

Willett, P., Barnard, J.M., Downs, G.M., 1998. Chemical similarity searching. J. Chem. Inf. Comput. Sci. 38 (6), 983–996.

Wishart, D.S., Knox, C., Guo, A.C., Shrivastava, S., Hassanali, M., Stothard, P., et al., 2006. DrugBank: a comprehensive resource for in silico drug discovery and exploration. Nucleic Acids Res. 34 (Suppl. 1), D668–D672.

Wu, D., Huang, Q., Zhang, Y., Zhang, Q., Liu, Q., Gao, J., et al., 2012. Screening of selective histone deacetylase inhibitors by proteochemometric modeling. BMC Bioinf. 13 (1), 212.

Xiang, Y., Hou, Z., Zhang, Z., 2012. Pharmacophore and QSAR studies to design novel histone deacetylase 2 inhibitors. Chem. Biol. Drug Des. 79 (5), 760–770.

Xing, B., Liu, P., Xu, W.-J., Xu, F.-Y., Dang, Y.-H., 2014. Effect of microinjecting of 5-aza-2-deoxycytidine into ventrolateral orbital cortex on depressive-like behavior in rats. Neurosci. Lett. 574, 11–14.

Xue, L., Bajorath, J., 2000. Molecular descriptors in chemoinformatics, computational combinatorial chemistry, and virtual screening. Comb. Chem. High Throughput Screen. 3 (5), 363–372.

Yoo, J., Medina-Franco, J.L., 2011a. Chemoinformatic approaches for inhibitors of DNA methyltransferases: comprehensive characterization of screening libraries. Comput. Mol. Biosci. 01 (01), 7–16.

Yoo, J., Medina-Franco, J.L., 2011b. Discovery and optimization of inhibitors of DNA methyltransferase as novel drugs for cancer therapy. In: Rundfeldt, C. (Ed.), Drug Development - A Case Study Based Insight into Modern Strategies. InTech, pp. 3–22.

Yoo, J., Medina-Franco, J.L., 2011c. Homology modeling, docking and structure-based pharmacophore of inhibitors of DNA methyltransferase. J. Comput. Aided Mol. Des. 25 (6), 555–567.

Yoo, J., Medina-Franco, J.L., 2012. Trimethylaurintricarboxylic acid inhibits human DNA methyltransferase 1: insights from enzymatic and molecular modeling studies. J. Mol. Model. 18 (4), 1583–1589.

Zhao, L., Xiang, Y., Song, J., Zhang, Z., 2013. A novel two-step QSAR modeling work flow to predict selectivity and activity of HDAC inhibitors. Bioorg. Med. Chem. Lett. 23 (4), 929–933.

CHAPTER 14

Role of Nutrition in Epigenetics and Recent Advances of In Silico Studies

Karina Martinez-Mayorga[1], **Carolina Peña Montes**[2]

[1]*Instituto de Química, Universidad Nacional Autónoma de México, Mexico City, México;* [2]*Facultad de Química, Universidad Nacional Autónoma de México, Mexico City, Mexico*

CONTENTS

1. NUTRITIONAL GENOMICS

The repercussion of the human genome project (HGP) on health has been evident and important in the last decade, especially in nutrition science (Williams, 2008). Consequently, a new discipline called nutritional genomics was created, which is the joint study of nutrition and genome including all the "-omics" genomics' derivate disciplines (transcriptomics, proteomics, and metabolomics). In 2014, the US Academy of Nutrition and Dietetics stated that nutritional genomics is the broad term encompassing nutrigenetics, nutrigenomics, and nutritional epigenomics, which provides insights into how interactions of diet and genotype affect the phenotype, including disease risk (Camp and Trujillo, 2014).

Two subdisciplines of the nutritional genomics are nutrigenomics and nutrigenetics. The word nutrigenetics was used for the first time in 1975 by Brennan (1977) and nutrigenomics was firstly used in 1999 by DellaPenna (1999). Nutrigenetics is the study of the phenotypic different responses to diet considering the genetic variability between individuals, and nutrigenomics is the discipline that studies the molecular mechanisms explaining that phenotypic different response to diet. Nutrigenomics focuses on studying how the nutrients regulate the expression of genes, how the polymorphisms affect the gene expression and regulation, and how these changes are interrelated with proteomics and metabolomics issues.

Epigenetics is an emergent concept, which can be defined as the study of changes that do not involve alterations to the genetic code but that can be passed down to future generations. The epigenome exerts its effect of gene control through a variety of mechanisms that physically block or reveal areas of the genome concerned with gene regulation (Thunders, 2015). In this context, nutritional epigenomics refers to the influence of the diet on changes in gene expression without changing the DNA sequence. At the molecular level,

Epi-Informatics. http://dx.doi.org/10.1016/B978-0-12-802808-7.00014-9

changes can occur under several mechanisms such as DNA methylation and chromatin remodeling, RNA transcripts and their encoded proteins, and the posttranslational modifications of histones. The primary role of epigenetic control is gene transcription silencing so they are not expressed when they are not needed. Additionally, epigenetic control has the function of managing the body's response in the most rapid and dynamic adaptive response to environmental stimuli (Figure 1).

Beyond the genetic traits selected naturally over many generations, it is also possible for changes in diet composition to affect gene expression and manifest as epigenetic modifications (epigenotype) just as we have seen before. However, it is possible also to manifest as changes in microbiome composition (phylotype). These epigenetic and microbiome modifications occur continuously as a real-time response to diet and other environmental factors. It is these layers of the human genome that we do have the power to change with dietary and lifestyle modifications.

It is known that our genes influence our diet, and nutrients—or the lack of them—affect gene initiation, expression, and progression. Thus, diet can be used as a strategy for disease prevention and control. It is also known that regional differences in food and culture have left their mark on our genome and that populations have adapted their diets to make the most of local resources. This has resulted in our current knowledge of the influence of diet on health. For instance, cardiovascular health might be the strongest application of nutrigenomics to date.

While the constituent parts of food might have been teased out, their impact on the body is only starting to be appreciated. As an example, genetic diversity of plant foods is a significant dietary feature, and it could play a role in chronic disease risk by influencing the epigenotype and phylotype in humans; however there is limited investigation in this field (Daniell and Ryan, 2012). Additionally, nutritional systems biology that assembles information from nutritional genomics also allows the studying of the interaction between nutrients, host metabolism, and microbiome (interactome).

2. DIET AND HEALTH

Nowadays, there are examples of diseases related to nutritional genomics; for example, the best-known case of nutrigenetics is phenylketonuria (PKU), caused by mutations in the phenylalanine hydroxylase gene. The primary treatment is a low phenylalanine diet (Camp and Trujillo, 2014). Table 1 shows examples of diseases related to nutrigenetics and nutrigenomics and their nutritional therapeutic intervention. In what follows, we will describe epigenetic factors in depth.

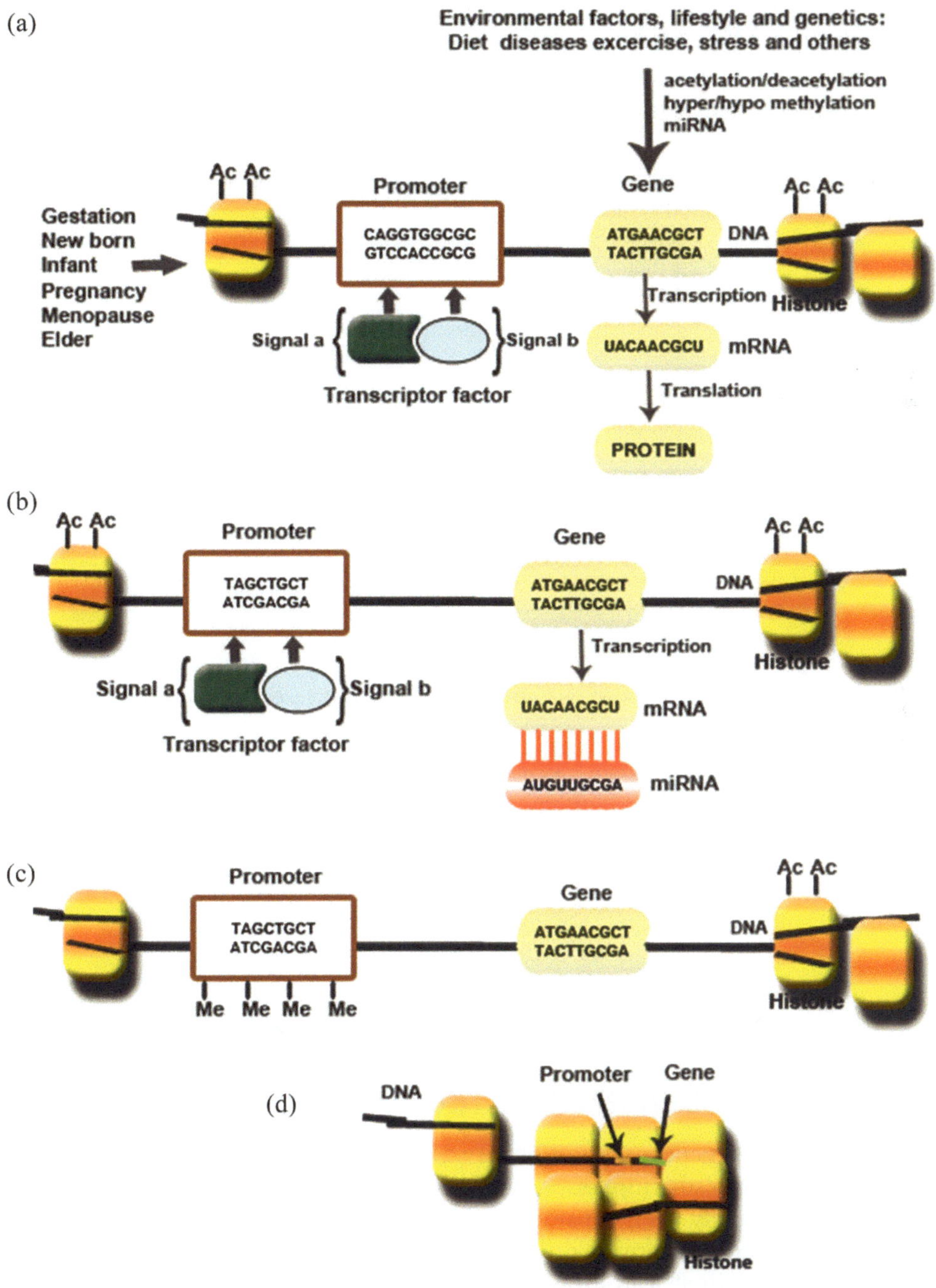

FIGURE 1
Protein production from an actively transcribed gene (a) and gene regulation by microRNAs (b), methyl groups (c), and histone deacetylation (d).

Table 1 Examples of Diseases Related to Nutrigenetics and Nutrigenomics and Their Nutritional Therapeutic Intervention

Disease	Etiology	Nutritional Therapeutics	References
Nutrigenetics			
Phenylketonuria	Mutations in the phenylalanine hydrolase (PAH) gene	Low phenylalanine diet	MacDonald (2011)
Cardiovascular disease	Single nucleotide polymorphism (SNP) of the methylenetetrahydrofolate reductase gene (MTHFR)	Folate-rich and methionine depletion diet	Yang (2012)
Coronary heart disease	Apolipoprotein E (ApoE) isoforms (E2, E3, and E4); E2 and E4 are dysfunctional	Dietary fat recommendations base on apoE genotype testing	Bennet (2007)
Obesity	Variations in fat mass- and obesity-associated (FTO) gene	Hypocaloric diet	Hetherington (2010)
Cancer colorectal	Vitamin D receptor (VDR) gene (Fok1) polymorphism is linked to increased colorectal cancer risk when calcium intakes are low	Higher consumption of milk/dairy products and high calcium intake diet	Hucharek (2009)
Nutrigenomics			
Obesity	Genes related to metabolism and insulin-like growth factor are increased	Hypocaloric diet	Kallio (2009)
Prostate cancer	RAS family oncogens are up-regulated which function as androgen receptor coactivator, and for which expression is increased in tumor tissues	Diet with 12% energy from fat	Ornish (2009)
Celiac disease (CD)	Gliadin sequence contains regions that play a special role in CD pathogenesis: exert a cytotoxic activity or immunomodulatory activity. Other regions trigger oxidative stress and induce the release of pro-inflammatory cytokines	The only sustainable treatment is a strict, lifelong gluten-free diet that avoids wheat, rye, barley, and related products	Ferretti (2012)

3. DISEASES RELATED TO EPIGENETIC CHANGES

Epigenetic modifications are defined as heritable but reversible changes in chromatin structure associated with alterations in gene expression. Epigenetics has recently been considered as a potential mechanism involved in the development of many disorders, including allergic, neurological (autism, Alzheimer), intestinal, cardiovascular, obesity, diabetes, and cancer. How food interacts and alters epigenetic mechanisms is still a work in progress. Food is a multivariable system, and composition of raw food may vary throughout the year and by geographical region. Chronic consumption of specific foods may affect epigenetic mechanisms. It has also been postulated that nutrients and environmental factors on early childhood affect individuals in adulthood. Moreover, cooking habits and portion preferences also have a direct influence on health. This scenario explains why the influence of diet on epigenetics has produced contradicting results, making it difficult to draw solid conclusions. Currently, an exciting research area is the identification of dietary components as useful modulators of the epigenome. It is well recognized that numerous bioactive dietary components mediate epigenetic modifications associated with the pathophysiology of several diseases. For example, it is known that polyphenols limit the development of cancers, cardiovascular diseases, neurodegenerative diseases, diabetes, obesity, and osteoporosis (Ferguson and Philpott, 2008).

However, much more research is still needed, in particular to define the clinical consequences of epigenetic alterations. Natural foods and bioactive food components have characteristics and functions that are similar to epigenetic inhibitors and therefore have potential in disease control and prevention. Food components and in particular micronutrients modulate gene expression at different levels, for example, fatty acids. Some vitamins and oligoelements function as signals to activate or repress key factors of gene expression and also they can bind to transcription factors or their cofactors and act as gene switches. An excess or deficiency in some micronutrients can directly damage DNA through chromosome or DNA breaks, nucleotide deletions, or changes and telomere shortening. Nutrients can also affect mRNA translation to protein. Other food components such as vitamins B12, B2, and B6, choline, and folic acid act on gene expression through epigenetics, which modify the genome by adding or removing molecular tags to DNA regions or histones (Constantin and Wahli, 2013). Multiple studies have shown that nutrients influence epigenetic regulation. Thus, bioactive compounds naturally occurring in plant foods may be useful for the manipulation of epigenetic profiles in health and disease.

3.1 Allergy

Food intake and environmental factors can alter gene transcription. The hypersensitivity to pollen, certain foods, or dust caused by allergic reactions are influenced by epigenetics. Progress has been made linking environmental pollution, tobacco smoke, and diet with epigenetic mechanisms (Tezza et al., 2013). A complete revision of epigenetics and the development of food allergy is given by Hong (Hong and Wang, 2012), where briefly it was suggested that early life is a critical window for the development of food allergies and that immune development is under epigenetic regulation. The kinetics and balance of CD4+ T-cell differentiation into TH1 or TH2 effector cells is critical to the development of food allergies, since a relatively TH2-skewed status predisposes an increased risk of allergy. A growing body of studies supports that TH2 cell differentiation is under epigenetic regulation. Notably, a recent study, conducted in 2759 participants, showed an epigenetic mediation of peanut allergy. The authors identified a peanut allergy specific loci in the HLADR and DQ gene region, they found that single nucleotide polymorphisms (SNPs) were associated with differential DNA methylation levels (Hong and Wang, 2012).

3.2 Aging Neurological Disorders

Abnormal epigenetic regulation is a hallmark in many pathological conditions such as Alzheimer's disease. Growing evidence supports the premise that environmental conditions and individual genetic susceptibility may increase the incidence and accelerate the onset of Alzheimer's disease. Alzheimer's disease is the most common cause of senile dementia. The significant role of epigenetic mechanisms for brain activities and neuropathology has become increasingly clear with the aim of being used for therapy (Bacalini et al., 2014). Epigenetic nutritional research has potential for Alzheimer's disease and may represent an opportunity to complement other interventions. The mechanisms involved in Alzheimer's disease, some of which may be epigenetically modulated by bioactive food, has been recently reviewed (Bacalini et al., 2014; Ford et al., 2011).

Dietary factors exert their effects on the brain by affecting molecular events related to the management of energy metabolism and synaptic plasticity. Energy metabolism influences neuronal function, neuronal signaling, and synaptic plasticity, ultimately affecting mental health. Epigenetic regulation of neuronal plasticity appears to be an important mechanism by which foods can prolong their effects on long-term neuronal plasticity.

Oxidative stress promotes damage to phospholipids present in the plasma membrane such as the omega-3 fatty acid docosahexenoic acid, disrupting neuronal signaling. Thus, dietary docosahexenoic acid (DHA) seems crucial for supporting plasma membrane function, interneuronal signaling, and cognition.

Brain foods such as omega-3 fatty acids provide structural material for plasma membranes, and other dietary components such as polyphenols provide support to selected energy metabolic events that assist specific aspects of synaptic functions. Dietary management may become a natural, noninvasive, and cost-effective therapeutic solution to maintaining a healthy brain and a strong defense system against some of the most common disorders in the world. In addition, based on current developments in the epigenetic field, it is likely that the effects of lifestyle on the brain have the capacity to influence the neurological health of future generations.

3.3 Cancer

Cancer is one of the leading causes of death worldwide; understanding the factors that contribute to cancer development may facilitate the development of strategies for cancer prevention and control. Cancer initiation and progression involves genetic and epigenetic alterations. The potential reversibility of epigenetic modifications makes them an attractive target for cancer prevention. Thus, epigenetics provides a link between the environment, diet, and cancer development. Proper food selection is imperative for better health and to avoid cancer and other diseases (Su et al., 2012). Multiple food components have been implicated in cancer risk and tumor behavior, with mixed results. Such findings suggest that not all individuals respond identically to a diet (Riscuta and Dumitrescu, 2012).

Recent studies provide evidence that dietary components may affect the process of carcinogenesis. Isothiocyanates from cruciferous vegetables (cauliflower, cabbage, and broccoli), diallyl sulfide (an organosulphur compound from garlic), isoflavone, phytosterole, folate, selenium, vitamin E, flavonoids, and dietary fibers may reduce the risk of cancer (Riscuta and Dumitrescu, 2012; Su et al., 2012; Supic et al., 2013; Verma, 2013).

A number of bioactive compounds from food have shown methylation-inhibitory and deacetylation-inhibitory properties. For example, nutrients such as folic acid supply methyl groups showing a protective effect in colon cancer.

The potential reversibility of epigenetic changes suggests that they could be modulated by nutrition and bioactive food compounds. Elucidating the impact of nutrition on epigenetic mechanisms may serve as a tool to predict an individual's susceptibility to cancer, provide dietary recommendations, or provide therapeutic applications of natural compounds against diseases such as cancer. There is limited information about tissue specificity and temporal aspects of dietary treatments. The optimal duration and the dose necessary for a chemopreventive effect require further studies (Supic et al., 2013; Verma, 2013).

It is important to mention that more research is needed on the nutrition of cancer patients and as consequence on the epigenetic factor. It is known that cancer patients present nutritional problems that are attributed to the disease itself, to the antineoplastic therapy, and to the psychological response of the patient to the cancer diagnosis. Cachexia (loss of body mass that cannot be reversed by eating more calories) can be caused by functional and/or mechanical problems. In the early stages of cancer, weight loss is associated with limited food intake, increased nutritional needs, malabsorption of substances, and their rapid passage through the gastrointestinal (GI) tract. In advanced stages, malnutrition is caused by alterations in metabolism. Eicosapentaenoic acid (EPA) is a compound included in supplement feeds. It is an essential long-chain polyunsaturated fatty acid of the n-3 fatty acid series. Studies suggest that EPA attenuates cancer-related wasting, improves immune function, halts or reverses weight loss, modulates the acute-phase response, and prolongs survival. The use of EPA reduced serum levels of C-reactive protein and interleukin

(IL-6) production by peripheral blood mononuclear cells (PBMCs) is seen in studies of patients with pancreatic cancer cachexia. Antioxidants may enhance the anticancer effects of chemotherapy by detoxifying reactive oxygen. For example, vitamin E was shown to prevent and treat chemotherapy-induced mucositis, while vitamin C reduced bleomycin-induced chromosomal damage in lymphocytes. Epigenetic regulation through these food compounds has to be elucidated (Bishop and Ferguson, 2015).

3.4 Intestinal

Although it is reasonable to believe that epigenetic mechanisms play a crucial role in normal gastrointestinal development (Waterland, 2006), this topic remains remarkably underinvestigated. Evidence for nutritional modulation of DNA methylation in the large intestine is being reported. However, there are gaps in our knowledge relating to other aspects of gastrointestinal epigenetics. The aging human colorectal mucosa develops aberrant patterns of DNA methylation that may contribute to its increasing vulnerability to cancer. Evidence suggests that age-dependent loss of global methylation, together with hypermethylation of CpG islands associated with cancer-related genes, may be influenced by nutritional and metabolic factors. Folates are essential for the maintenance of normal DNA methylation, and folate metabolism is known to modify epigenetic mechanisms under experimental conditions. Folates and other nutritional and metabolic factors appear as determinants of colorectal mucosal DNA methylation. Future studies should focus on the possibility that folic acid fortification may exert unforeseen effects on the human gastrointestinal epigenome.

3.5 Food Bioactive Compounds

A current aim of nutrigenetics is to personalize nutritional practices according to genetic variations that influence the way of digestion and metabolism of nutrients introduced with the diet. Nutrition in early life or in critical periods of development may have a role in modulating gene expression, and, therefore, have later effects on health. Epidemiological studies have clearly shown that whole-grain cereals can protect against obesity, diabetes, cardiovascular disease, and cancers. For example, the protective effects of fiber, Mg, and numerous bioactive compounds, especially those in the bran and germ, are well recognized. In addition, human breast milk is well known for its ability in preventing several acute and chronic diseases. The presumed epigenetic effects of human breast milk and components has been recently reviewed (Verduci et al., 2014). While evidence suggests that a direct relationship may exist of some components of human breast milk with epigenetic changes, the mechanisms involved are still unclear. Studies have to be conducted to clarify the actual role of human breast milk on genetic expression, in particular when linked to the risk of noncommunicable diseases, to potentially benefit the infant's health and later life (Verduci et al., 2014). The involvement of polyphenols in cell signaling and gene regulation, and of sulfur compounds, lignin and phytic acid, should be considered in antioxidant protection. Whole-grain wheat is also a rich source of methyl donors and lipotropes (methionine, betaine, choline, inositol, and folates) that may be involved in cardiovascular and/or hepatic protection, lipid metabolism, and DNA methylation. Potential protective effects of bound phenolic acids within the colon, of the B-complex vitamins on the nervous system and mental health, of oligosaccharides as prebiotics, of compounds associated with skeleton health, and of other compounds such as α-linolenic acid, policosanol, melatonin, phytosterols, and *para*-aminobenzoic acid also deserve to be studied in more depth. Benefits of nutrigenomics to study complex physiological effects of the whole-grain package, and promising ways for improving the nutritional quality of cereal products, have been discussed (Fardet, 2010).

4. COMPUTATIONAL STUDIES ON NUTRIGENOMICS AND FOOD CHEMICALS

4.1 Computational Methods for Database Analysis of Nutrigenomics Data

The multivariable nature of the large amount of data generated on genetic experiments requires statistical methods for analysis and processing. There are different types of data and analysis reported for databases of nutrigenomics studies. For instance, in 2006, García-Castro et al. reported on the use of ontology for nutrigenomics data. Two years later, Castle et al. reported on the analysis of electronic health records and remarked on the relevance of maintaining the privacy of the information of the individuals in the databases (Castle and DeBusk, 2008). The following year, in 2009, Harttig et al. developed and implemented a concept for data management and computing infrastructure that supports collaboration between nutrigenomics research. The program is called NUGO Information Network and is a decentralized distributed system for data exchange based on web technology. The central idea of this initiative is to promote the production of common databases instead of privet storing.

In this context, two important publications came out. Stumbo et al. (2010) recognized the need for web-software tools and data, to measure nutrient diet intake and physical activity for personalized health research. That same year Valavanis reported the use of neural networks in nutrigenomics to study the etiology behind obesity. In that work, 63 variables compressing genetic variations, nutrition, and gender were used to classify the normal versus overweight population. The mean accuracy of the developed model was 61.46% (Valavanis et al., 2010).

More recently, other methods have been implemented to explore and analyze nutrigenomics data. In 2013, Valour et al. derived a heuristic algorithm for gene selection on a nutrigenomics study. The method was developed for the search of transcriptome and metabolome connections. In that model, each gene subset is linked to the whole metabolic or clinical dataset that represents the biological phenotype of interest. The main outcome of this study is the selection of a smaller set of genes for biologists to further explore phenotyping. The algorithm was shown to work efficiently on three published datasets, resulting in meaningfully broadened gene networks (Valour et al., 2013).

The multivariable nature of gene expression patterns or food matrices calls for the use of methods to reduce dimensions. Principal component analysis (PCA) is a useful tool in the linear case. Microarray gene expression is nonlinear and Kernel representation offers an alternative to nonlinear functions. Kernel PCA reduces dimensionality and integrates multiple data sets. Kernel-based exploratory tools presented by the authors allows for data mining to extract nonlinear features while representing variables (Reverter et al., 2014).

With the aim of describing the potential applications of chemoinformatics tools in food chemistry, we have recently reported on the importance and variety of molecular descriptors and physicochemical properties that can be applied to the so-called Foodinformatics field (Martinez-Mayorga and Medina-Franco, 2013). On that study, we also surveyed representative databases with emphasis on food-related ones (Medina-Franco et al., 2012). Methods commonly used in molecular design such as similarity searching, pharmacophore modeling, quantitative structure–activity relationships (QSAR), and molecular docking can be applied in the food chemistry field; a revision of these concepts and applications has been reported (Martinez-Mayorga and Medina-Franco, 2009). These methodologies can be readily applied to epigenetic targets or lead molecules in the search of new compounds to modulate epigenetic processes.

Chemoinformatic analysis of flavor chemicals has been performed by our group (Martinez-Mayorga et al., 2011; Medina-Franco et al., 2012; Martinez-Mayorga et al., 2013) and by others (Sprous and Salemme, 2007). Flavor perception involves, among a number of physiological and psychological processes, the recognition of chemicals by olfactory and taste receptors. We have shown that flavor information contained in databases can be analyzed following standard chemoinformatics methods. Moreover, it can be envisioned that such analyses can be applied to epigenetic targets and epigenetic lead compounds, to explore compounds contained on nutrigenomics databases.

4.2 Molecular Modeling Studies of Bioactive Food Chemicals Interacting with Epigenetic Targets

The premise that bioactive food compounds are both therapeutically and nutritionally relevant has been tested and screening efforts are now widely employed. Using the principles of molecular similarity we identified compounds with potential mood-modulating properties. We showed the ability of certain GRAS molecules to inhibit histone deacetylase-1 (HDAC1), proposed as an important player in mood modulation. Two GRAS chemicals were identified as HDAC1 inhibitors in the micromolar range; the mood prescription drug used as reference in our similarity searches exhibits HDAC1 inhibition in the same nanomolar range. Additional studies on bioavailability, toxicity at higher concentrations, and off-target effects will show the practicality of this finding (Martinez-Mayorga et al., 2013).

Chemical constituents isolated from food and other natural sources can interfere with many biological targets in humans. Compounds like curcumin, genistein, plant and food polyphenols, resveratrol, and sulforaphane are able to modulate the biological activity of cell pathways and its functions in relation to metabolism and epigenetics. The exploration of natural and food compounds in cancer epigenetics and metabolism has been reviewed by Del Rio and Da Costa (2014). Understanding how food chemicals interact and modulate metabolism and epigenetics is an exciting and evolving research area. To this end, computational methods are valuable tools and it is expected that they will continue to be used in the search for epigenetic modulators.

ACKNOWLEDGMENTS

KMM acknowledges the Institute of Chemistry for financial support at UNAM for support.

REFERENCES

Bacalini, M.G., Friso, S., Olivieri, F., Pirazzini, C., Giuliani, C., Capri, M., Santoro, A., Franceschi, C., Garagnani, P., 2014. Present and future of anti-ageing epigenetic diets. Mech. Ageing Dev. 136, 101–115.

Bennet, A.M., Di Angelantonio, E., Ye, Z., Wensley, F., Dahlin, A., Ahlbom, A., Danesh, J., 2007. Association of apolipoprotein E genotypes with lipid levels and coronary risk. Jama 298 (11), 1300–1311.

Bishop, K.S., Ferguson, L.R., 2015. The Interaction between Epigenetics, Nutrition and the Development of Cancer. Nutrients 7 (2), 922–947.

Brennan, R.O., 1977. Nutrigenetics: New Concepts for Relieving Hypoglycemia. M. Evans & Co.

Camp, K.M., Trujillo, E., 2014. Position of the academy of nutrition and dietetics: nutritional genomics. J. Acad. Nutr. Diet. 114, 299–312.

Castle, D., DeBusk, R., 2008. The electronic health record, genetic information and patient privacy. J. Am. Diet. Assoc. 108, 1372–1374.

Constantin, N., Wahli, W., 2013. Nutrigenomic foods: what will be eating tomorrow? Nutrafoods 12, 3–12.

Daniell, E., Ryan, E.P., 2012. The nutrigenome and gut microbiome: chronic disease prevention with crop phytochemical diversity. In: Caliskan, M. (Ed.), The Molecular Basis of Plant Genetic Diversity. Intech. Available from: http://www.intechopen.com/books/the-molecular-basis-of-plant-genetic-diversity/the-nutrigenome-and-gut-microbiome-chronic-disease-prevention-with-crop-phytochemical-diversity.

Del Rio, A., Da Costa, F.B., 2014. Molecular approaches to explore natural and food-compound modulators in cancer epigenetics and metabolism. In: Martinez-Mayorga, K., Medina-Franco, J.L. (Eds.), Foodinformatics: Applications of Chemical Information to Food Chemistry. Springer, Switzerland.

DellaPenna, D., 1999. Nutritional genomics: manipulating plant micronutrients to improve human health. Science 285, 375–379.

Fardet, A., 2010. New hypothesis for the health-protective mechanism of whole-grain cereals: what is beyond fibre? Nutr. Res. Rev. 23, 65–134.

Ferguson, L.R., Philpott, M., 2008. Nutrition and mutagenesis. Annu. Rev. Nutr. 28, 13–29.

Ferretti, G., Bacchetti, T., Masciangelo, S., Saturni, L., 2012. Celiac disease, inflammation and oxidative damage: a nutrigenetic approach. Nutrients 4 (4), 243–257.

Ford, D., Ions, L.J., Alatawi, F., Wakeling, L.A., 2011. The potential role of epigenetic responses to diet in ageing. Proc. Nutr. Soc. 70, 374–384.

Garcia Castro, A., Rocca-Serra, P., Stevens, R., Taylor, C., Nashar, K., Ragan, M.A., Sansone, S.A., 2006. The use of concept maps during knowledge elicitation in ontology development processes-the nutrigenomics use case. BMC Bioinform. 7, 267.

Harttig, U., Travis, A.J., Rocca-Serra, P., Renkema, M., van Ommen, B., Boeing, H., 2009. Owner controled data exchange in nutrigenomic collaborations: the NuGO information network. Genes Nutr. 4, 113–122.

Hetherington, M.M., Joanne, E.C., 2010. Gene-environment interactions in obesity. Karger 63, 195–203.

Hong, X., Wang, X., 2012. Early life precursors, epigenetics, and the development of food allergy. Semin. Immunopathol. 34, 655–669.

Huncharek, M., Muscat, J., Kupelnick, B., 2009. Colorectal cancer risk and dietary intake of calcium, vitamin D, and dairy products: a meta-analysis of 26,335 cases from 60 observational studies. Nutr. Cancer 61 (1), 47–69.

Kallio, P., Tolppanen, A.M., Kolehmainen, M., Poutanen, K., Lindström, J., Tuomilehto, J., Uusitupa, M., 2009. Association of sequence variations in the gene encoding insulin-like growth factor binding protein 5 with adiponectin. International Journal of Obesity 33 (1), 80–88.

MacDonald, A., Ahring, K., Dokoupil, K., Gokmen-Ozel, H., Lammardo, A.M., Motzfeldt, K., Bélanger-Quintana, A., 2011. Adjusting diet with sapropterin in phenylketonuria: what factors should be considered? British Journal of Nutrition 106 (2), 175–182.

Martinez-Mayorga, K., Medina-Franco, J.L., 2009. Chemoinformatics-applications in food chemistry. In: Taylor, S. (Ed.), Advances in Food and Nutrition Research, vol. 58, pp. 33–55.

Martinez-Mayorga, K., Medina-Franco, J.L., 2013. FoodInformatics: applications of chemical information to food chemistry. In: Paper Presented at the 245th ACS National Meeting, New Orleans, LI, United States.

Martinez-Mayorga, K., Peppard, T.L., Yongye, A.B., Santos, R., Giulianotti, M., Medina-Franco, J.L., 2011. Characterization of a comprehensive flavor database. J. Chemometr. 25 (10), 550–560.

Martinez-Mayorga, k, Peppard, T.L., Lopez-Vallejo, F., Yongye, A.B., Medina-Franco, J.L., 2013. Systematic mining of generally recognized as safe flavor chemicals for bioactive compounds. J. Agric. Food Chem. 61, 7507–7514.

Medina-Franco, J.L., Martínez-Mayorga, K., Peppard, T.L., Del Rio, A., 2012. Chemoinformatic analysis of GRAS (Generally Recognized as Safe) flavor chemicals and natural products. PLoS One 7, e50798.

Ornish, D., 2009. Intensive lifestyle changes and health reform. The Lancet Oncology 10 (7), 638–639.

Reverter, F., Vegas, F., Oller, J.M., 2014. Kernel-PCA data integration with enhanced interpretability. BMC Syst. Biol. 8 (Suppl. 2), S6.

Riscuta, G., Dumitrescu, R.G., 2012. Nutrigenomics: Implications for breast and colon cancer prevention. Methods Mol. Biol. 863, 343–358.

Sprous, D.G., Salemme, F.R., 2007. A comparison of the chemical properties of drugs and FEMA/FDA notified GRAS chemical compounds used in the food industry. Food Chem. Toxicol. 45, 1419–1427.

Stumbo, P.J., Weiss, R., Newman, J.W., Pennington, J.A., Tucker, K.L., Wiesenfeld, P.L., Illner, A.-K., Klurfeld, D.M., Kaput, J., 2010. Web-enabled and improved software tools and data are needed to measure nutrient intakes and physical activity for personalized health research. J. Nutr. 140, 2104–2115.

Su, L.J., Mahabir, S., Ellison, G.L., McGuinn, L.A., Reid, B.C., 2012. Epigenetic contributions to the relationship between cancer and dietary intake of nutrient's bioactive food components, and environmental toxicants. Front. Genet. 2, 1–12.

Supic, G., Jagodic, M., Magic, Z., 2013. Epigenetics: a new link between nutrition and cancer. Nutr. Cancer 65, 781–792.

Tezza, G., Mazzei, F., Boner, A., 2013. Epigenetics of allergy. Early Hum. Dev. 89S1, S20–S210.

Thunders, M., 2015. Epigenetics: its understanding is crucial to a sustainable healthcare system. Healthcare 3, 194–204.

Valavanis, I.K., Mougiakakou, S.G., Grimaldi, K.A., Nikita, K.S., 2010. A multifactorial analysis of obesity as CVD risk factor: use of neural network based methods in a nutrigenetics context. BMC Bioinform. 11, 453.

Valour, D., Hue, I., Grimard, B., Valour, B., 2013. Gene selection heuristic algorithm for nutrigenomics studies. Physiol. Genom. 45, 615–628.

Verduci, E., Banderali, G., Barberi, S., Radaelli, G., Lops, A., Betti, F., Riva, E., Giovannini, M., 2014. Epigenetic effects of human breast milk. Nutrients 6, 1711–1724.

Verma, M., 2013. Cancer control and prevention: nutrition and epigenetics. Curr. Opin. Clin. Nutr. Metab. Care 16, 376–384.

Waterland, R.A., 2006. Epigenetic mechanisms and gastrointestinal development. J. Pediatr. 149, S137–S142.

Williams, S.F., 2008. Nutrigenomics: a new approach to personalized nutrition. Nutr. Perspect. 31, 9–16.

Yang, Q., Bailey, L., Clarke, R., Flanders, W.D., Liu, T., Yesupriya, A., Friedman, J.M., 2012. Prospective study of methylenetetrahydrofolate reductase (MTHFR) variant C677T and risk of all-cause and cardiovascular disease mortality among 6000 US adults. Am. J. Clin. Nutr. 95 (5), 1245–1253.

CHAPTER 15

The Road Ahead of the Epi-Informatics Field

José L. Medina-Franco[1], Jakyung Yoo[2]

[1]Facultad de Química, Departamento de Farmacia, Universidad Nacional Autónoma de México, Mexico City, México; [2]Life Science Research Institute, Daewoong Pharmaceutical Co., Ltd, Gyeonggi-do, Republic of Korea

1. INTRODUCTION

Epigenetic drug discovery is becoming a major research area not only in oncology but also for the treatment of other diseases associated with epigenetic alterations, particularly those influenced by the environment. For instance, DNA methylation has shown importance in Alzheimer's disease and psychiatric conditions such as depression, bipolar disorder, and schizophrenia. DNA methylation is also involved in autoimmune diseases and genetic disorders (Gros et al., 2012). A number of technological advances are being developed to screen compound databases and develop assays in high- or low-throughput mode (Zheng, 2015). Computational procedures commonly used in drug discovery are increasingly being used to develop molecules targeting epigenetic targets. Table 1 summarizes representative applications of computational approaches that have been applied to epigenetic drug and probe discovery. These applications can be roughly classified into two main research fields: molecular modeling and informatics. Such classification is very approximate since there are several instances where, for example, molecular modeling the three-dimensional structures of the targets and management of chemical information of the ligands play a significant role in the study. Typical examples of such cases are virtual screening and computer-aided drug repurposing.

In several cases, in particular in hit identification and development, computational approaches form part of a multidisciplinary effort working in synergy with methods such as in vitro and in vivo testing, chemical synthesis, and pharmacological evaluations, to name a few examples. In multidisciplinary projects, computational methods are usually implemented in iterative cycles that involve modeling and experimental testing. A typical example is the computational screening of compound databases followed by in vitro characterization of the activity of the hits compounds; based on the results of the experimental testing, new cycles of virtual screening (of new chemical

CONTENTS

Epi-Informatics. http://dx.doi.org/10.1016/B978-0-12-802808-7.00015-0

Table 1 Selected Computational Approaches to Epigenetic Drug and Probe Discovery

Application	Main Field
Simulations of protein–ligand interactions	Molecular modeling
Structural studies, simulations, and molecular modeling of epigenetic molecular targets	Molecular modeling
Computer-guided optimization of lead molecules	Molecular modeling
Design and identification of modulators of protein–protein interactions	Molecular modeling
Studies of the mechanism of action of active molecules at the structural level	Molecular modeling
Screening compound data sets (with emphasis on large screening collections), to identify distinct molecular scaffolds for later optimization	Molecular modeling and informatics
Computer-aided drug repurposing	Molecular modeling and informatics
Quantitative systematic studies of structure–activity relationships (SAR)	Informatics
Design of screening chemical libraries focused on epigenetic targets	Informatics
Organization and mining screening data, in particular large compound databases	Informatics

databases or more exhaustive searches of the same collection) can be made. As part of the refinement cycles new chemical libraries focused on epigenetic targets can be designed with the aid of chemoinformatic tools. Thus, after several cycles of computational studies and experimental testing, ligands with improved activity or better pharmacological profile are obtained.

Research networks and funding opportunities to create or maintain such networks play active roles to boost the progress of a research area. In epigenetics, examples of such initiatives are the NIH Roadmap Epigenomics Program (available at http://www.roadmapepigenomics.org/) and the Epigenome Roadmap (available at http://www.nature.com/collections/vbqgtr/), which open new drug development opportunities (Mullard, 2015). It is clear that computational methods play a fundamental role to conduct large-scale integrative analysis of reference epigenomes (8. Computational tools and methods, 2015).

A second example of a research network in epigenetic drug discovery is the Anti-Parasitic Drug Discovery in Epigenetics, A-ParaDDisE group that aims to "develop target-based strategies for the discovery of novel drug leads against epigenetic targets in schistosomiasis, leishmaniasis, Chagas disease, and malaria by focusing on key histone-modifying enzymes." In silico screening form part of A-ParaDDisE (accessible at http://a-paraddise.cebio.org/).

Emerging challenges and novel trends in drug discovery are also relevant in epigenetics. One such trend is the change in the paradigm of drug design from single target to multitarget drug design (Medina-Franco et al., 2013). The high complexity of epigenetics, with several enzymes and interactions participating in normal and disease processes, makes it very difficult or nearly impossible to predict how pharmacological interactions will produce favorable responses without producing epigenetic changes that could favor disease development (Medina-Franco, 2015b). As an approach to address this issue, Franci et al. have proposed the idea of epigenetic polypharmacolgy (Franci et al., 2010). In a rich paper, the authors remark that "An epigenetic poly-pharmacological approach targeting multiple chromatin-modifying enzymes may represent a 'smart' option to treat cancer versus the current view on the selective and single pharmacological targeting of epigenetic enzymes" (Franci et al., 2010).

Drug repurposing is a second area of intense research where the application of multidisciplinary efforts is required. As discussed elsewhere, this approach has proven to be useful in different therapeutic areas to find alternative applications of drugs initially applied for different indications. In silico methods are contributing to conduct drug repurposing in a systematic manner, giving rise to the area of so-called computer-aided drug repurposing (Bellera et al., 2015; Oprea et al., 2011). Examples of computer-aided drug repurposing for epigenetic targets are emerging.

The combination of the increased interest in epigenetic drug discovery with the constant advances in the development of computational methods is taking momentum. As discussed in this book, epi-informatics is an emerging area with several successful stories but also with challenges ahead, which are outlined and elaborated below. This chapter is organized in four major sections. After this introduction, Section 2 describes promising computational approaches that are being used that might be used alone or in combination with other strategies to further advance the development of epi-drugs. Section 3 presents representative challenges to be addressed in epigenetic drug discovery with emphasis on polypharmacology. Section 4 presents summary and conclusions.

2. PROMISING COMPUTATIONAL APPROACHES IN EPIGENETIC DRUG DISCOVERY

As discussed in the previous section, there are a number of computational methods that have been used successfully in epigenetic drug discovery. Representative molecular modeling and informatics strategies are summarized schematically in Figure 1.

Most if not all of these techniques can be further applied to refine current results and hypotheses derived for a given epigenetic target. Alternatively, the

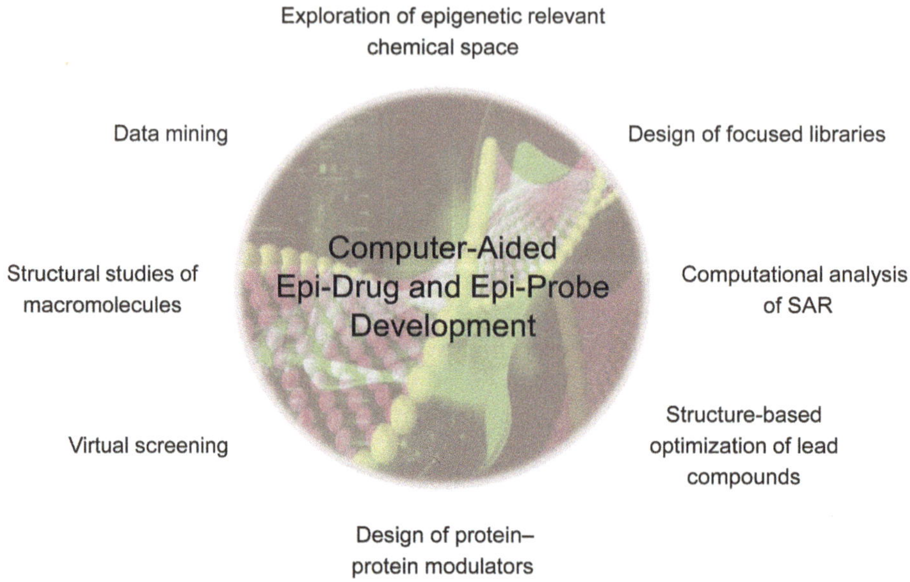

FIGURE 1
Integration of selected informatics applications for computer-aided epigenetic drug and probe discovery.

technique can be implemented for epigenetic targets different from the target for which the application was initially developed. Following, representative computational approaches are discussed in the context of epigenetic drug discovery. The theoretical background of each method has been discussed elsewhere and the interested reader is referred to the corresponding sources.

2.1 Epigenetic Target Fishing

Target fishing can be considered part of computational chemogenomics. Its major aim is to identify potential molecular targets for a given ligand (Rognan, 2010). In the context of chemogenomics, target fishing is the inverse but complementary approach to virtual screening, where the goal is to identify potential ligands for a given molecular target. Alternative uses of inverse docking are pharmacological profiling of chemical compounds, elucidation of the mode of action of active compounds, drug repurposing, and assessment of the toxicity profile (Cheng et al., 2011; Do et al., 2015).

A common strategy for target fishing is inverse docking, where a ligand molecule is docked across a panel of protein structures. Inverse docking, like any other docking approach, requires the availability of the three-dimensional structures of the molecular targets (typically obtained from crystallography, nuclear magnetic resonance, or homology modeling (Haupt and Schroeder, 2011)). Other common approaches, alternative to inverse docking, are data mining and similarity searching. In similarity searching, targets are represented by their ligands and the query molecules are compared with the known ligands. Based on the hypothesis that similar compounds have similar activities, with the exception of true activity cliffs (as opposed to artificial cliffs (Medina-Franco, 2013a)), it is expected that similar molecular structures will have similar biological activities. Thus, it is possible to relate the query compounds to their ligand targets by finding similar ligands to query structures. Tools for in silico target fishing have been recently reviewed (Cereto-Massagué et al., 2015).

Target fishing has been broadly used in natural product drug discovery to identify potential targets for a given compound (Do et al., 2015; Medina-Franco, 2013b). The same principles and in silico tools can be implemented for the systematic screening of the epigenetic-relevant biological space.

2.2 Structure-Guided Optimization of Lead Molecules

Structure-based drug design can be used to suggest structural modifications to ligand molecules to enhance the interactions with the molecular target and, hopefully, increase the biological activity. Crystallographic structures of ligand–receptor complexes or predicted binding models (e.g., obtained with molecular docking) can be the starting point of structure-based refinement. Fragment-based design (Murray and Rees, 2009) is also dependent on the structures of receptor–fragment complexes characterized by biophysical methods. Such information will guide the next steps to identify the most suitable fragments and guide the linking of those fragments. Evidently, such an approach is frequently performed in close collaborations with synthetic chemists who plan routes to link the fragments.

Structure-guided approaches have already contributed to optimize the activity of drugs targeting epigenetic targets such as DNMT inhibitors and bromodomains. This has been largely favored by the three-dimensional structural information available for epigenetic targets. Table 2 summarizes few examples of crystallographic structures of epigenetic targets available in the Protein Data Bank (Berman et al., 2000).

2.3 Computer-Aided Drug Repurposing

Drug repurposing is an attractive approach to speed up drug discovery and reduce the large costs associated with the drug discovery process. Of note, the large cost of drug development will impact directly the budget of the patients and may represent a significant burden in the access of the drug to the patient (Kaitin, 2010). As discussed elsewhere, drug repurposing has led to several successful stories of drugs that are being employed for an indication different from the original, including novel applications in oncology (Duenas-Gonzalez et al., 2008). Repositioning a drug can be conveniently achieved using a

Table 2 Representative Three-Dimensional Structures of Epigenetic Targets Available in the Protein Data Bank[a]

Epigenetic Target/Structure	Gene	PDB	Resolution (Å)
Reader			
Crystal structure analysis of the N-terminal bromodomain of human BRD2 complexed with acetylated histone H4 peptide	BRD2	2DVS	2.04
Crystal structure of the bromodomain 2 of human bromodomain containing protein 3 (BRD3)	BRD3	2OO1	1.70
Crystal structure of the first bromodomain of human Brd4 in complex with IBET inhibitor	BRD4	3P5O	1.60
Eraser			
The structure of HDAC1 in complex with the dimeric ELM2-SANT domain of MTA1 from the NuRD complex	HDAC 1	4BKX	3.00
Structure of human HDAC2 in complex with SAHA (vorinostat)	HDAC 2	4LXZ	1.85
Structure of HDAC3 bound to corepressor and inositol tetraphosphate	HDAC 3	4A69	2.06
Structure of HDAC4 catalytic domain bound to a hydroxamic acid inhibitor	HDAC 4	2VQM	1.80
Crystal structure of human HDAC6 in complex with ubiquitin	HDAC 6	3PHD	3.00
Crystal structure of HDAC8-inhibitor complex	HDAC 8	2V5X	2.25
Structure of NAD-dependent protein deacetylase sirtuin-1 (closed state, 1.85 A)	SIRT 1	4KXQ	1.85
Human SIRT2 histone deacetylase	SIRT 2	1J8F	1.70
Crystal structure of human SIRT3	SIRT 3	3GLS	2.70
Crystal structure and mechanism of human lysine-specific demethylase-1	KDM1A	2H94	2.90
Crystal structure of JMJD2A complexed with histone H3 peptide trimethylated at Lys9	KDM4A	2OQ6	2.00
Crystal structure of human JMJD2C catalytic domain	KDM4C	2XML	2.55
Crystal structure of JARID1A-PHD3 complexed with H3(1–9)K4me3 peptide	KDM5A	3GL6	1.90
Writer			
Structural basis for autoinhibition of the acetyltransferase activity of p300	EP300	4BHW	2.80
Crystal structure of the bromodomain of human PCAF	KAT2B	3GG3	2.25
Acetyltransferase domain of human HIV-1 Tat interacting protein, 60 kDa, isoform 3	KAT5	2OU2	2.30
Structure of the catalytic domain of human DOT1L, a non-SET domain nucleosomal histone methyltransferase	DOT1L	1NW3	2.50
Crystal structure of the EZH2 SET domain	EZH2	4MI5	2.00
SETD7 in complex with inhibitor (R)-PFI-2 and S-adenosyl-methionine	KMT7	4JLG	1.90
Crystal structure of protein arginine methyltransferase 3	PRMT3	3SMQ	2.00
Crystal structure of coactivator-associated arginine methyltransferase 1	PRMT4	4IKP	2.00
Crystal structure of human DNMT1(646–1600) in complex with DNA	DNMT1	3PTA	3.60
Structure of Dnmt3a-Dnmt3L C-terminal domain complex	DNMT3A	2QRV	2.89

[a]*All structures are for humans.*

blend of experimental and computational strategies. Indeed, as discussed in detail in this book, computational strategies are playing a key role to uncover epigenetic modulators among drugs currently targeted for other indications. For example, for DNMTs, similarity-based virtual screening led to the identification of the anti-inflammatory drug olsalazine as a distinct hypomethylating compound (Méndez-Lucio et al., 2014). In a follow-up study, olsalazine is being used as query to identify novel hypomethylating agents in databases of approved drugs.

A second recent example of a drug repurposing project guided by computational approaches is the identification of the antiviral drug ribavirin as inhibitor of histone methyltransferase zeste homolog 2 (EZH2) (De la Cruz-Hernandez et al., 2015). Ribavirin is a guanosine analog that is used to stop viral RNA synthesis and viral mRNA capping. Ribavirin is a known inhibitor of eIF4E and inosine 5′-phosphate dehydrogenase and also inhibits histone methyltransferase zeste homolog 2, EZH2. Similarity searching revealed that ribavirin has a high structural similarity to 3-deazaneplanocin A, a known inhibitor of EZH2. In that work, the growth inhibitory effects of ribavirin as well as its effects upon epigenetic enzymes were evaluated in various cancer cell lines. It was found that ribavirin decreased EZH2 expression, inhibited histone methyltransferase activity, and decreased H3K27 trimethylation (De la Cruz-Hernandez et al., 2015). It is expected that more systematic, computer-guided drug repurposing campaigns will be conducted not only for DNMT or histone methyltransferase but also for other epigenetic targets.

2.4 Structure–Activity Relationships (SAR) with Computational Methods

Data sets of molecules screened across epigenetic targets are rich sources to systematically analyze SARs in a descriptive and/or predictive manner. It has been discussed in the literature that before trying to build predictive models, it is convenient to understand first the structure-activity relationships (SAR) through systematic descriptive methods (Medina-Franco et al., 2015b). The detection of pairs of compounds with very similar chemical structures but very different biological activity—the so-called activity cliffs (Cruz-Monteagudo et al., 2014; Maggiora, 2006)—is a crucial step before attempting to develop predictive models where it is assumed linear SAR; that is, that similar structures will have similar activity. A typical example of an approach that often considers linear relationships is QSAR modeling. It has been shown that the presence of activity cliffs reduce the predictive ability of QSAR models (Guha and Van Drie, 2008).

Activity landscape modeling (ALM) is a distinct concept to analyze systematically the SAR of screening data sets using computational methods, mostly chemoinformatics (Bajorath, 2012). ALM can be regarded as a special case of a broader concept; that is, property landscape modeling (PLM) (Medina-Franco et al., 2012).

One of the major outcomes of ALM/PLM is the systematic identification of activity or, more general, property cliffs.

A second major application of ALM/PLM is the identification of scaffold hops or R hops. The first concept is applied to diverse sets, while the R hops is applied to analog series. These hops are compounds with different chemical structure; for example, different core scaffold or R-group substitutions/decorations but with comparable biological activity. The scaffold or R-hops, recently termed similarity cliffs (Iyer et al., 2013), are valuable in drug development since they provide a suitable alternative of lead compounds if issues emerge in the earlier preclinical development of a core scaffold (e.g., ADMET, intellectual property, or derivatization issues). As revised in this book, several QSAR studies have been developed for epigenetic target data sets (Chuang et al., 2015; Ebrahimi et al., 2013; Zhao et al., 2013). However, studies toward the identification of epigenetic activity cliffs and epigenetic scaffold/R-hops is still limited and represents promising research for further development.

2.4.1 Protein–Ligand Interaction Fingerprints

A convenient manner to conduct systematic studies of the SAR at the structural level (i.e., structure–protein–ligand interaction relationships) is through the understanding of protein–ligand interactions. Indeed, understanding protein–ligand interactions is at the core of molecular recognition and they are involved in several practical applications in drug discovery. Examples are docking, structure-based design, virtual screening, clustering of complexes, and structural interpretation of activity cliffs (Medina-Franco et al., 2014b). Protein–ligand interactions can be conveniently studied using protein–ligand interaction fingerprints (PLIFs), also called structural interaction fingerprints. Fingerprints are convenient representations to simplify the atomic coordinates of PLIFs (Desaphy et al., 2013). PLIFs are designed to "capture a 1D representation of the interactions between ligand and protein either in complexes of known structure or in docked poses" (Brewerton, 2008; Da and Kireev, 2014; Salentin et al., 2014). PLIFs are easy to generate and manipulate, and enable fast comparisons of a vast number of protein–ligand complexes (Desaphy et al., 2013). In epigenetics, PLIFs have great potential to facilitate the analysis of large amounts of data and evaluate if similar binding sites identically recognize similar ligands, if protein–ligand interaction patterns are conserved across target families, and if different ligand structures or substructures have the same interaction patterns with a single target (Desaphy et al., 2013). A recent review of PLIFs and their applications to identify hot spots is published elsewhere (Medina-Franco et al., 2014b).

2.5 Flexible Docking

Induced-fit docking and other computational techniques that account for protein flexibility (B-Rao et al., 2009), at least partially, will be of major significance in epigenetic drug discovery. Thus far, proof-of-concept studies have been published showing the relevance of flexible docking in modeling DNMT inhibitors and other epigenetic targets (Salo et al., 2013; Yoo et al., 2013). It is anticipated that advanced docking approaches or combination of rigid docking with more comprehensive optimization of protein–ligand complexes will result in more accurate binding models of epigenetic compounds. In turn, these models will have a positive impact on the better selection of candidates from a virtual screening or the improved analysis of structure-based design studies.

2.6 Virtual Screening

Virtual or in silico screening is a filtering approach to guide the identification of bioactive molecules. In order to potentiate the benefits, virtual screening should be conducted along with experimental characterization in an iterative manner. Although it is desirable to obtain highly potent compounds in the first iteration, one of the major goals is to identify compounds with novel chemical scaffolds that can be used as starting points in lead optimization programs or than can enter in new cycles of virtual/experimental screening. One or more computational procedures are used to filter (usually) large chemical databases and select compounds for experimental characterization.

Computational methods to conduct virtual screening can be divided into two major groups, depending on the structural information available, ligand-based and structure-based. Ligand-based approaches utilize structure–activity data from a set of known actives in order to identify candidate compounds for experimental evaluation. Structure-based approaches leverage the three-dimensional structure of a target. Comprehensive reviews of virtual screening including methods, successful applications, pitfalls, and workarounds are published elsewhere (Scior et al., 2012; Westermaier et al., 2015). Several successful applications have been reported for epigenetic targets (Chen et al., 2014; Kuck et al., 2010; Li et al., 2015; Siedlecki et al., 2006). In the next few years, an increased number of virtual screening studies, either structure- or ligand-based, for epigenetic drug discovery are anticipated.

2.6.1 Exploring Novel Regions of Chemical Space

One of the first criteria for a successful virtual screening campaign is the selection of the chemical databases to be searched. The selection of the screening collection will depend on the goals of the virtual screening. For example, diverse chemical libraries are more suitable to identify novel chemical scaffolds. General and diverse screening libraries are normally used in the first iteration of virtual/experimental screening. In the second and subsequent rounds of screenings, or, in general, during the optimization of lead compounds or further exploration of the SAR, focused libraries can be used. That is, compound collections with novel structures but within a confined region of the chemical space (Medina-Franco et al., 2014b).

During the last few years, chemical companies have been developing focused libraries for epigenetic drug discovery. Such libraries are available for purchase and screening. Examples of epigenetic-related libraries are:

- Epigenetics Compound Library, with 24 small molecule modulators with biological activity used for epigenetic research (http://www.selleckchem.com).
- Screen-Well® Epigenetics library, with 43 compounds with defined activity against enzymes that carry out epigenetic modification of lysine. It also includes DNA methylation inhibitors (http://www.enzolifesciences.com).
- Epigenetics Screening Library, with more than 90 molecules that are known to modulate the activity of methyltransferases, demethylases, histone acetyltransferases, histone deacetylases, and acetylated histone binding proteins. According to the library's vendor, the composition of this screening library varies somewhat depending upon inventory (http://www.caymanchem.com).

The chemical structures of the libraries are deposited on the company's websites and can be requested directly from the chemical vendors. Of note, focused libraries can be designed using computational approaches using the same principles and protocols implemented in successful virtual screening campaigns.

Virtual screening using molecular databases with high structural novelty may increase the chances to expand the medicinal chemistry relevant space. For example, virtual screening of natural product databases, natural product analogs, or combinatorial libraries with novel chemical structures (López-Vallejo et al., 2011) may identify hit compounds not easily identified in other general screening collections (López-Vallejo et al., 2012). Similarly, it has been discussed that screening libraries with increased molecular complexity may favor the identification of selective compounds (López-Vallejo et al., 2012; Lovering et al., 2009). This may be particularly attractive to develop epi-probes by means of identifying selective compounds for epigenetic targets and isoforms of the same epigenetic target.

2.6.2 Expanding the Set of Reference Compounds

In addition to the selection of the screening library, the definition of the reference molecules that will be used as queries is a crucial step in similarity-based virtual screening. Similarity searching has led to the identification of lead compounds, for instance, to identify distinct DNMT inhibitors. In a pilot study focused on drug repurposing, the small molecule NSC14778, previously identified from virtual screening (Kuck et al., 2010), was used as a query to search potential distinct DNMT inhibitors or hypomethylating agents in a database of approved drugs. The searching approach was based on a three-dimensional shape comparison. This strategy led to the identification of olsalazine as a novel hypomethylating compound (Méndez-Lucio et al., 2014). In that study, only one reference compound was used. One of the next logical steps is to expand the similarity searching in at least two directions, namely:

- Use olsalazine as a reference hypomethylating molecule to search structural related compounds with potential activity.
- Use other known DNMT inhibitors and/or hypomethylating molecules as queries.

The virtual screening can be conducted using a purely ligand-based approach (e.g., with two- and/or three-dimensional representations); a structure-based approach, taking advantage of the availability of crystallographic structures of DNMTs (Medina-Franco et al., 2015b); or a combination of ligand- and structure-based methods. As discussed in the previous section, virtually any compound collection can be used; for example, general screening collections, natural products, focused libraries, combinatorial libraries, or other databases of approved drugs. The same rationale can be applied to several other hits identified from virtual screening for epigenetic targets (Chen et al., 2014).

2.7 Combination of Computational Approaches

Given that a single theoretical model or approach does not capture all the details that successfully explain and/or predict a biological event (which is usually extremely complex), it is known that combining methods derived from different theoretical backgrounds deliver improved results. One example is the combination of structure-based and ligand-based methods. Data fusion in similarity searching (Willett, 2013) and consensus scoring (Feher, 2006) are good examples that show that the aggregation of multiple methods have better performance than the individual approaches. In the same context, consensus activity cliffs (Medina-Franco et al., 2009) and, more generally, consensus models of the activity landscape (Pérez-Villanueva et al., 2010) have been suggested as robust representations of SARs. In SAR analysis using the concept of activity landscapes, data aggregation is meant to prioritize the exploration of the consensus activity cliffs and other relevant data.

The combination of methods or protocols frequently occurs in virtual screening workflows where fast computational protocols are used first for the rapid filtering of chemical compounds, for example, filtering by drug- and lead-likeness empirical rules, structural flags associated with toxic compounds, pharmacophore models, or high-throughput docking. The fast protocols are followed by more time-consuming but presumably more accurate methods, for instance, multistep or cascade docking where compounds are selected and docked in a sequential manner using, in increasing order, more accurate and more time-consuming protocols.

Combining computational methodologies has already yielded significant results to further understand the activity of epigenetic targets. For instance, a pharmacophore model has been derived for DNMT1 inhibitors based on the binding modes predicted with docking (Yoo and Medina-Franco, 2011). Also, three-dimensional QSAR studies have been conducted along with docking for better understanding of SARs at the structural level (Li et al., 2014; Ragno et al., 2008).

3. EPI-INFORMATICS: THE ROAD AHEAD

Progress on epigenetic research will largely be favored by the successful implementation of computational approaches. Evidently, there are a wide number of research areas within epigenetics that involve the application of in silico procedures, ranging from target identification and validation as druggable targets, bioinformatic-based prediction of epigenetic target function, computational analysis of epigenetic data, and discovery and development of drug candidates, to name just few applications (Vellore and Baron, 2014; Wei et al., 2014).

As largely discussed in this book, there are several molecular modeling and chemoinformatic methods that have been used to date to identify novel epi-drugs and epi-probes. Applications cover a large range of structure and ligand-based methods that are used depending on the goals of the specific project. Thus, molecular docking and dynamics, quantum mechanics calculations, pharmacophore modeling, QSAR, ALM, molecular similarity, and virtual screening are examples of techniques and methods that are used (in more or less degree) to further advance epigenetic drug and probe discovery. It is expected that the scientific community will continue (and increase) to use these approaches in synergy with experimental methods. For example:

1. Combining computer-aided drug design with novel synthetic techniques to prepare the designed molecules in an efficient manner.
2. Virtual screening of novel chemical libraries including focused libraries.
3. Development of innovative workflows that combine computational methods, for instance, high-throughput virtual screening approaches followed by more intensive but more accurate protocols.

Evidently, advances and the progress in the technologies will have a fundamental role for the rapid evaluation of compounds suggested, for example, by virtual screening.

In this section of the manuscript, we will focus on a major drug design avenue that largely influences current drug discovery and it is expected to have a major impact on epigenetic drug and probe development.

3.1 Epigenetic Polypharmacology or Multiepigenetic Target Drug Design

As discussed above, epigenetics is extremely complex, involving a large number of targets and biological events such that a single-target drug design might not be effective to develop epi-drugs. It has also been discussed that polypharmacology is changing the paradigm in drug design from single-target approaches to multitarget drug design. In epigenetics, Franci et al. (2010) have proposed two general approaches to address polypharmacology: mono-epi and multiple-epi polypharmacology. In the former, the development of single chemical entities act as multiple ligands of hybrid molecules that would be able to modulate multiple targets simultaneously. Hybrid molecules could interact on the same target or on two different targets. In their reach review, Franci et al. (2010) point out that most of the mono-epipolypharmacology strategies developed thus far are multiple ligand molecules that combine the molecular scaffold of an HDAC inhibitor with another cancer drug. In multiple-epi polypharmacology, a single chemical entity is able to modulate more than one epigenetic target. For instance, there are compounds (simplified eosin analogs) that are able to inhibit protein arginine N-methyltransferases, histone acetyl transferases, and class III deacetylase (sirtuins) (Mai et al., 2008). The hybrid molecules have a better biological activity profile than the single-target inhibitors eosin curcumin and sirtinol. The marine-derived bromotyrosine disulfides, psammaplins, have been reported to have dual anti-DNMT and -HDAC activities (Pina et al., 2003). A derivative of psammaplin A has been reported as a triple inhibitor of epigenetic-modifying enzymes (Franci et al., 2010).

In order to guide campaigns of rational design of epigenetic multiple ligands one can use the structural information available for different enzyme classes, compare binding pockets, and propose combined or hybrid pharmacophore models that can be used as a basis for the structure-based design. At the same time, the combined pharmacophore model can be used as a query to conduct virtual screening of chemical databases. Alternatively, and following a ligand-based approach, one can use the already known epigenetic multiple ligands to conduct similarity searching in diverse or focused chemical databases or guide the synthesis of structural analogs. All these efforts would lead to an expansion of the local SAR, and can be the starting point to develop predictive models using techniques such as QSAR, support vector machines, and neural networks. Even more general, it is feasible to implement the principles and techniques of computational chemogenomics (Rognan, 2013).

It has been discussed in the literature that chemogenomics has led to the increased complexity of analyzing drug–target interactions. As a consequence, ligand-based and structure-based virtual screening strategies have been modified in order to improve their predictive limitations (Yabuuchi et al., 2011).

Chemogenomic virtual screening has emerged as a new in silico technique that evaluates drug–target interactions by exploring both protein and drug chemical spaces (Yu et al., 2012). Moreover, chemogenomic virtual screening identifies relationships between drug substructures and protein functional sites (chemogenomic features), establishing structural associations and allowing prediction of phenotypic effects (Tabei et al., 2012). Therefore, it is expected that approaches used in computational chemogenomics will be applied to epigenetic drug discovery and systematically mine the epigenetic relevant chemical space (ERCS).

Figure 2 shows a visual representation of the distribution in the chemical space of four sets of modulators of representative epigenetic targets: SIRT1 (64 compounds in green), HDAC1 (466 compounds in yellow), DNMT1 (25 compounds in red), and BRD4 (123 compounds in blue). For this figure, compounds were selected from ChEMBL, version 20 (e.g., prepared on January 14, 2015) (Gaulton et al., 2012). Molecules were selected based on the IC_{50} values obtained from biochemical assays and confidence score of nine. Compounds with unique chemical structures and IC_{50} values lower than 10 μM were

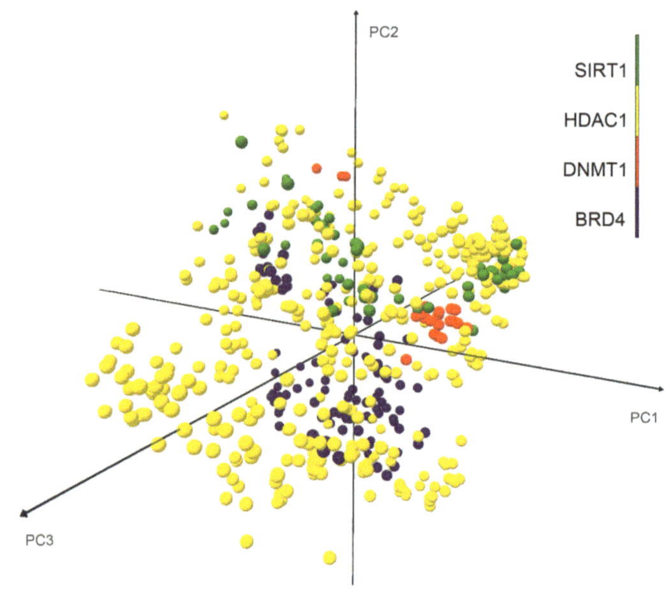

FIGURE 2

Three-dimensional representation of the chemical space of 678 compounds active with four representative epigenetic targets: SIRT1 (64 compounds in green), HDAC1 (466 compounds in yellow), DNMT1 (25 compounds in red), and BRD4 (123 compounds in blue). The visual representation of the chemical space was obtained by principal component analysis of the similarity matrix computed using MACCS keys and Tanimoto similarity. The first three principal components capture 68% of the variance. Each compound is shown as a sphere.

selected for SIRT1, DNMT1, and BRD4, and with IC_{50} values lower than 0.1 μM for HDAC1. The chemical space was generated by conducting a principal component analysis of the similarity matrix of all the 678 molecules computed with MACCs keys (166-bits) and the Tanimoto coefficient (Medina-Franco and Maggiora, 2014). A similar approach has been implemented to generate visual representations of the chemical space for compound databases of different sources including approved drugs (Kraft et al., 2013; Medina-Franco et al., 2007). The figure clearly shows the large structural diversity of the 466 active ($IC_{50} < 0.1$ μM) compounds with HDAC1 as compared to the active compounds for other data sets. This is not surprising because of the larger number of active molecules for HDAC1 available in ChEMBL (HDAC is one of the most explored epigenetic targets). Figure 2 also shows the overall different coverage of chemical space of compounds active for SIRT1 and BRD4. Finally, the molecules active versus DNMT1 roughly occupy two main regions in chemical space that partially overlap with the chemical space of HDAC1 compounds.

4. CONCLUSIONS

Computer-aided epi-drug design is rapidly growing. Thus far, computational approaches have made clear contributions to the identification, design, and development of epi-drugs and probes. Emerging concepts that impact single- or multitarget drug discovery campaigns for a number of therapeutic indications will continue to have a pronounced effect on epigenetic drug discovery. Examples of representative computational methods that form part of multidisciplinary efforts that pave the road ahead for epi-informatics are listed below. In several instances, proof-of-concept studies have been published, demonstrating the feasibility to obtain promising results.

- Epigenetic polypharmacology or multiepigenetic drug design
- Computational chemogenomics: while virtual screening using novel chemical libraries is a suitable approach to explore novel regions of the chemical space, target fishing is an attractive avenue to systematically screen the epigenetic target-relevant biological space
- Epigenetic target fishing or inverse screening
- Computer-aided drug repurposing
- Virtual screening of databases of natural products, natural product analogs, nonnatural peptides or peptides, and other collections that explore novel regions of the chemical space
- Chemoinformatic-guided design of focused chemical libraries
- Rational design of epigenetic protein–protein modulators
- Computational analysis of SAR and SmARs using novel algorithms and concepts such as proteochemometric modeling, ALM, and support vector machines

- Computer-guided design and/or compound selection of chemical libraries with complex chemical structures to achieve selectivity; this is particularly attractive for the identification of epi-probes
- Flexible docking
- Quantum mechanics calculations of metal containing epigenetic targets
- Chemoinformatic characterization of the epigenetic-relevant chemical space
- Merging the Epi- and Foodinformatics (Martinez-Mayorga and Medina-Franco, 2014) fields, e.g., computer-guided analysis of the epigenetic processes of food chemicals in the diet
- Combination of computational approaches and workflows
- New and improved combinations of computational and medicinal chemistry with high-throughput screening and other novel compound-screening technologies

ACKNOWLEDGMENTS

We thank Mariana Giralda Herrera Núñez for collecting data presented in Table 2. Authors also thank the National Autonomous University of Mexico (UNAM), grant PAIP 5000-9163 to JLM-F and the Department of Pharmacy, School of Chemistry, UNAM, for support.

REFERENCES

B-Rao, C., Subramanian, J., Sharma, S.D., 2009. Managing protein flexibility in docking and its applications. Drug Discov. Today 14, 394–400.

Bajorath, J., 2012. Modeling of activity landscapes for drug discovery. Expert Opin. Drug Discov. 7, 463–473.

Bellera, C.L., Balcazar, D.E., Vanrell, M.C., Casassa, A.F., Palestro, P.H., Gavernet, L., Labriola, C.A., Galvez, J., Bruno-Blanch, L.E., Romano, P.S., Carrillo, C., Talevi, A., 2015. Computer-guided drug repurposing: identification of trypanocidal activity of clofazimine, benidipine and saquinavir. Eur. J. Med. Chem. 93, 338–348.

Berman, H.M., Westbrook, J., Feng, Z., Gilliland, G., Bhat, T.N., Weissig, H., Shindyalov, I.N., Bourne, P.E., 2000. The protein data bank. Nucl. Acids Res. 28, 235–242.

Brewerton, S.C., 2008. The use of protein-ligand interaction fingerprints in docking. Curr. Opin. Drug Discov. Dev. 11, 356–364.

8. Computational tools and methods. Nature, 2015. http://dx.doi.org/10.1038/nature14316.

Cereto-Massagué, A., Ojeda, M.J., Valls, C., Mulero, M., Pujadas, G., Garcia-Vallve, S., 2015. Tools for in silico target fishing. Methods 71, 98–103.

Chen, S.J., Wang, Y.L., Zhou, W., Li, S.S., Peng, J.L., Shi, Z., Hu, J.C., Liu, Y.C., Ding, H., Lin, Y.Y., Li, L.J., Cheng, S.F., Liu, J.Q., Lu, T., Jiang, H.L., Liu, B., Zheng, M.Y., Luo, C., 2014. Identifying novel selective non-nucleoside DNA methyltransferase 1 inhibitors through docking-based virtual screening. J. Med. Chem. 57, 9028–9041.

Cheng, T., Li, Q., Wang, Y., Bryant, S.H., 2011. Identifying compound-target associations by combining bioactivity profile similarity search and public databases mining. J. Chem. Inf. Model. 51, 2440–2448.

Chuang, Y.C., Chang, C.H., Lin, J.T., Yang, C.N., 2015. Molecular modelling studies of sirtuin 2 inhibitors using three-dimensional structure-activity relationship analysis and molecular dynamics simulations. Mol. Biosyst. 11, 723–733.

Cruz-Monteagudo, M., Medina-Franco, J.L., Pérez-Castillo, Y., Nicolotti, O., Cordeiro, M.N.D.S., Borges, F., 2014. Activity cliffs in drug discovery: Dr. Jekyll or Mr Hyde? Drug Discov. Today 19, 1069–1080.

Da, C., Kireev, D., 2014. Structural protein–ligand interaction fingerprints (SPLIF) for structure-based virtual screening: method and benchmark study. J. Chem. Inf. Model. 54, 2555–2561.

De la Cruz-Hernandez, E., Medina-Franco, J.L., Trujillo, J., Chavez-Blanco, A., Dominguez-Gomez, G., Perez-Cardenas, E., Gonzalez-Fierro, A., Taja-Chayeb, L., Dueñas-Gonzalez, A., 2015. Ribavirin as a tri-targeted antitumor repositioned drug. Oncol. Rep. 33, 2384–2392.

Desaphy, J., Raimbaud, E., Ducrot, P., Rognan, D., 2013. Encoding protein–ligand interaction patterns in fingerprints and graphs. J. Chem. Inf. Model. 53, 623–637.

Duenas-Gonzalez, A., Garcia-Lopez, P., Herrera, L.A., Medina-Franco, J.L., Gonzalez-Fierro, A., Candelaria, M., 2008. The prince and the pauper. A tale of anticancer targeted agents. Mol. Cancer 7, 82.

Do, Q.T., Medina-Franco, J.L., Scior, T., Bernard, P., 2015. How to valorize biodiversity? Let's go hashing, extracting, filtering, mining, fishing. Planta Med. 81, 436–449.

Ebrahimi, S., Azimi, G., Akhlaghi, Y., Kompany-Zareh, M., 2013. Genetic algorithm for CoMFA setting optimization: 3D-QSAR study on alpha-aminosuberic acid derivatives as anti-cancer compounds. J. Chemom. 27, 323–329.

Feher, M., 2006. Consensus scoring for protein-ligand interactions. Drug Discov. Today 11, 421–428.

Franci, G., Miceli, M., Altucci, L., 2010. Targeting epigenetic networks with polypharmacology: a new avenue to tackle cancer. Epigenomics 2, 731–742.

Gaulton, A., Bellis, L.J., Bento, A.P., Chambers, J., Davies, M., Hersey, A., Light, Y., McGlinchey, S., Michalovich, D., Al-Lazikani, B., Overington, J.P., 2012. ChEMBL: a large-scale bioactivity database for drug discovery. Nucleic Acids Res. 40, D1100–D1107.

Gros, C., Fahy, J., Halby, L., Dufau, I., Erdmann, A., Gregoire, J.-M., Ausseil, F., Vispé, S., Arimondo, P.B., 2012. DNA methylation inhibitors in cancer: recent and future approaches. Biochimie 94, 2280–2296.

Guha, R., Van Drie, J.H., 2008. Assessing how well a modeling protocol captures a structure-activity landscape. J. Chem. Inf. Model. 48, 1716–1728.

Haupt, V.J., Schroeder, M., 2011. Old friends in new guise: repositioning of known drugs with structural bioinformatics. Brief. Bioinform. 12, 312–326.

Iyer, P., Stumpfe, D., Vogt, M., Bajorath, J., Maggiora, G.M., 2013. Activity landscapes, information theory, and structure - activity relationships. Mol. Inf. 32, 421–430.

Kaitin, K.I., 2010. Deconstructing the drug development process: the new face of innovation. Clin. Pharmacol. Ther. 87, 356–361.

Kraft, R., Kahn, A., Medina-Franco, J.L., Orlowski, M.L., Baynes, C., Lopez-Vallejo, F., Barnard, K., Maggiora, G.M., Restifo, L.L., 2013. A cell-based fascin bioassay identifies compounds with potential anti-metastasis or cognition-enhancing functions. Dis. Model. Mech. 6, 217–235.

Kuck, D., Singh, N., Lyko, F., Medina-Franco, J.L., 2010. Novel and selective DNA methyltransferase inhibitors: docking-based virtual screening and experimental evaluation. Bioorg. Med. Chem. 18, 822–829.

Li, G.-B., Yang, L.-L., Yuan, Y., Zou, J., Cao, Y., Yang, S.-Y., Xiang, R., Xiang, M., 2015. Virtual screening in small molecule discovery for epigenetic targets. Methods 71, 158–166.

Li, X.G., Zhang, Y.J., Zhang, L., Xu, Y.Y., Xu, W.F., 2014. 3D QSAR and docking studies of a series of histone deacetylase inhibitors. Med. Chem. Res. 23, 2229–2241.

López-Vallejo, F., Caulfield, T., Martínez-Mayorga, K., Giulianotti, M.A., Nefzi, A., Houghten, R.A., Medina-Franco, J.L., 2011. Integrating virtual screening and combinatorial chemistry for accelerated drug discovery. Comb. Chem. High. Throughput Screen. 14, 475–487.

López-Vallejo, F., Giulianotti, M.A., Houghten, R.A., Medina-Franco, J.L., 2012. Expanding the medicinally relevant chemical space with compound libraries. Drug Discov. Today 17, 718–726.

Lovering, F., Bikker, J., Humblet, C., 2009. Escape from flatland: increasing saturation as an approach to improving clinical success. J. Med. Chem. 52, 6752–6756.

Maggiora, G.M., 2006. On outliers and activity cliffs-why QSAR often disappoints. J. Chem. Inf. Model. 46, 1535.

Mai, A., Cheng, D., Bedford, M.T., Valente, S., Nebbioso, A., Perrone, A., Brosch, G., Sbardella, G., De Bellis, F., Miceli, M., Altucci, L., 2008. Epigenetic multiple ligands: mixed histone/protein methyltransferase, acetyltransferase, and class III deacetylase (Sirtuin) inhibitors. J. Med. Chem. 51, 2279–2290.

Martinez-Mayorga, K., Medina-Franco, J.L. (Eds.), 2014. Foodinformatics. Springer.

Medina-Franco, J., Navarrete-Vázquez, G., Méndez-Lucio, O., 2015a. Property landscape modeling is at the interface of chemoinformatics and experimental sciences. Fut. Med. Chem. 7, 1197–1211.

Medina-Franco, J.L., Yoo, J., Dueñas-Gonzalez, A., 2015b. DNA methyltransferase inhibitors for cancer therapy. In: Zheng, Y.G. (Ed.), Epigenetic Technological Applications. San Diego, Academic Press, pp. 265–290.

Medina-Franco, J.L., 2013a. Activity cliffs: facts or artifacts? Chem. Biol. Drug Des. 81, 553–556.

Medina-Franco, J.L., 2013b. Advances in computational approaches for drug discovery based on natural products. Rev. Latinoam. Quim. 41, 95–110.

Medina-Franco, J.L., Giulianotti, M.A., Welmaker, G.S., Houghten, R.A., 2013. Shifting from the single to the multitarget paradigm in drug discovery. Drug Discov. Today 18, 495–501.

Medina-Franco, J.L., Maggiora, G.M., 2014. Molecular similarity analysis. In: Bajorath, J. (Ed.), Chemoinformatics for Drug Discovery. John Wiley & Sons, Inc., pp. 343–399.

Medina-Franco, J.L., Maggiora, G.M., Giulianotti, M.A., Pinilla, C., Houghten, R.A., 2007. A similarity-based data-fusion approach to the visual characterization and comparison of compound databases. Chem. Biol. Drug Des. 70, 393–412.

Medina-Franco, J.L., Martínez-Mayorga, K., Bender, A., Marín, R.M., Giulianotti, M.A., Pinilla, C., Houghten, R.A., 2009. Characterization of activity landscapes using 2D and 3D similarity methods: consensus activity cliffs. J. Chem. Inf. Model. 49, 477–491.

Medina-Franco, J.L., Martinez-Mayorga, K., Meurice, N., 2014a. Balancing novelty with confined chemical space in modern drug discovery. Expert Opin. Drug Discov. 9, 151–165.

Medina-Franco, J.L., Méndez-Lucio, O., Martinez-Mayorga, K., 2014b. Chapter one - the interplay between molecular modeling and chemoinformatics to characterize protein–ligand and protein–protein interactions landscapes for drug discovery. In: Tatyana, K.-C. (Ed.), Advances in Protein Chemistry and Structural Biology, vol. 96. Academic Press, pp. 1–37.

Medina-Franco, J.L., Méndez-Lucio, O., Yoo, J., Dueñas, A., 2015c. Discovery and development of DNA methyltransferase inhibitors using in silico approaches. Drug Discov. Today 20, 569–577.

Medina-Franco, J.L., Yongye, A.B., López-Vallejo, F., 2012. Consensus models of activity landscapes. In: Matthias, D., Kurt, V., Danail, B. (Eds.), Statistical Modeling of Molecular Descriptors in QSAR/QSPR. Wiley-VCH, pp. 307–326.

Méndez-Lucio, O., Tran, J., Medina-Franco, J.L., Meurice, N., Muller, M., 2014. Towards drug repurposing in epigenetics: olsalazine as a novel hypomethylating compound active in a cellular context. ChemMedChem 9, 560–565.

Mullard, A., 2015. The roadmap epigenomics project opens new drug development avenues. Nat. Rev. Drug Discov. 14, 223–225.

Murray, C.W., Rees, D.C., 2009. The rise of fragment-based drug discovery. Nat. Chem. 1, 187–192.

Oprea, T.I., Nielsen, S.K., Ursu, O., Yang, J.J., Taboureau, O., Mathias, S.L., Kouskoumvekaki, I., Sklar, L.A., Bologa, C.G., 2011. Associating drugs, targets and clinical outcomes into an integrated network affords a new platform for computer-aided drug repurposing. Mol. Inf. 30, 100–111.

Pérez-Villanueva, J., Santos, R., Hernández-Campos, A., Giulianotti, M.A., Castillo, R., Medina-Franco, J.L., 2010. Towards a systematic characterization of the antiprotozoal activity landscape of benzimidazole derivatives. Bioorg. Med. Chem. 18, 7380–7391.

Pina, I.C., Gautschi, J.T., Wang, G.Y.S., Sanders, M.L., Schmitz, F.J., France, D., Cornell-Kennon, S., Sambucetti, L.C., Remiszewski, S.W., Perez, L.B., Bair, K.W., Crews, P., 2003. Psammaplins from the sponge *Pseudoceratina purpurea*: Inhibition of both histone deacetylase and DNA methyltransferase. J. Org. Chem. 68, 3866–3873.

Ragno, R., Simeoni, S., Rotili, D., Caroli, A., Botta, G., Brosch, G., Massa, S., Mai, A., 2008. Class II-selective histone deacetylase inhibitors. part 2: alignment-independent GRIND 3-D QSAR, homology and docking studies. Eur. J. Med. Chem. 43, 621–632.

Rognan, D., 2010. Structure-based approaches to target fishing and ligand profiling. Mol. Inf. 29, 176–187.

Rognan, D., 2013. Towards the next generation of computational chemogenomics tools. Mol. Inf. 32, 1029–1034.

Salentin, S., Haupt, V.J., Daminelli, S., Schroeder, M., 2014. Polypharmacology rescored: protein-ligand interaction profiles for remote binding site similarity assessment. Prog. Biophys. Mol. Biol. 116, 174–186.

Salo, H.S., Laitinen, T., Poso, A., Jarho, E., Lahtela-Kakkonen, M., 2013. Identification of novel SIRT3 inhibitor scaffolds by virtual screening. Bioorg. Med. Chem. Lett. 23, 2990–2995.

Scior, T., Bender, A., Tresadern, G., Medina-Franco, J.L., Martínez-Mayorga, K., Langer, T., Cuanalo-Contreras, K., Agrafiotis, D.K., 2012. Recognizing pitfalls in virtual screening: a critical review. J. Chem. Inf. Model. 52, 867–881.

Siedlecki, P., Boy, R.G., Musch, T., Brueckner, B., Suhai, S., Lyko, F., Zielenkiewicz, P., 2006. Discovery of two novel, small-molecule inhibitors of DNA methylation. J. Med. Chem. 49, 678–683.

Tabei, Y., Pauwels, E., Stoven, V., Takemoto, K., Yamanishi, Y., 2012. Identification of chemogenomic features from drug–target interaction networks using interpretable classifiers. Bioinformatics 28, i487–i494.

Vellore, N.A., Baron, R., 2014. Epigenetic molecular recognition: a biomolecular modeling perspective. ChemMedChem 9, 484–494.

Wei, K., Sutherland, H., Camilleri, E., Haupt, L.M., Griffiths, L.R., Gan, S.H., 2014. Computational epigenetic profiling of CpG islets in MTHFR. Mol. Biol. Rep. 41, 8285–8292.

Westermaier, Y., Barril, X., Scapozza, L., 2015. Virtual screening: an in silico tool for interlacing the chemical universe with the proteome. Methods 71, 44–57.

Willett, P., 2013. Combination of similarity rankings using data fusion. J. Chem. Inf. Mod. 53, 1–10.

Yabuuchi, H., Niijima, S., Takematsu, H., Ida, T., Hirokawa, T., Hara, T., Ogawa, T., Minowa, Y., Tsujimoto, G., Okuno, Y., 2011. Analysis of multiple compound-protein interactions reveals novel bioactive molecules. Mol. Syst. Biol. 7, 472.

Yoo, J., Choi, S., Medina-Franco, J.L., 2013. Molecular modeling studies of the novel inhibitors of DNA methyltransferases SGI-1027 and CBC12: implications for the mechanism of inhibition of DNMTs. PLoS One 8, e62152.

Yoo, J., Medina-Franco, J.L., 2011. Homology modeling, docking, and structure-based pharmacophore of inhibitors of DNA methyltransferase. J. Comput. Aided Mol. Des. 25, 555–567.

Yu, H., Chen, J., Xu, X., Li, Y., Zhao, H., Fang, Y., Li, X., Zhou, W., Wang, W., Wang, Y., 2012. A systematic prediction of multiple drug-target interactions from chemical, genomic, and pharmacological data. PLoS One 7, e37608.

Zhao, L.L., Xiang, Y.H., Song, J.L., Zhang, Z.Y., 2013. A novel two-step QSAR modeling work flow to predict selectivity and activity of HDAC inhibitors. Bioorg. Med. Chem. Lett. 23, 929–933.

Zheng, Y.G., 2015. Epigenetic Technological Applications. San Diego, Academic Press.

Index

'*Note*: Page numbers followed by "f" indicate figures and "t" indicate tables.'

E